Application of Liquid Chromatography-Mass Spectrometry and Related Techniques

Topic Editors

Chao Kang

Ronald Beckett

Basel • Beijing • Wuhan • Barcelona • Belgrade • Novi Sad • Cluj • Manchester

Topic Editors
Chao Kang
School of Chemistry and
Chemical Engineering
Guizhou University
Guiyang
China

Ronald Beckett
School of Chemistry
Monash University
Melbourne
Australia

Editorial Office
MDPI AG
Grosspeteranlage 5
4052 Basel, Switzerland

This is a reprint of the Topic, published open access by the journals *Foods* (ISSN 2304-8158), *Molecules* (ISSN 1420-3049) and *Separations* (ISSN 2297-8739), freely accessible at: https://www.mdpi.com/topics/TZL0PQ5E20.

For citation purposes, cite each article independently as indicated on the article page online and as indicated below:

Lastname, A.A.; Lastname, B.B. Article Title. *Journal Name* **Year**, *Volume Number*, Page Range.

ISBN 978-3-7258-3511-9 (Hbk)
ISBN 978-3-7258-3512-6 (PDF)
https://doi.org/10.3390/books978-3-7258-3512-6

Contents

About the Editors

Chao Kang

Chao Kang received his bachelor's degree and his doctorate in September 2006 to June 2015, respectively, from the College of Chemistry and Chemical Engineering of Hunan University and the State Key Laboratory of Chemo/Biosensing and Chemometrics; he also won the National Scholarship and other awards. From September 2015 to the present, he has taught chemistry and performed scientific research at the School of Chemistry and Chemical Engineering of Guizhou University. His research interests include chemometrics and multivariate analysis, as well as multiway analysis theory and its applications in chromatography and spectroscopy, including liquid biopsy, biological analysis, drug analysis, and food analysis. He has presided over scientific research projects such as the National Natural Science Foundation of China, the Guizhou Provincial Basic Research (Natural Science) Key Project, the Guizhou Provincial Basic Research (Natural Science) Project, the Guizhou Provincial Department of Education Yunian Science and Technology Talent Growth Project, and the Guizhou University Talent Introduction Science and Technology Fund. He has published more than 50 papers in academic journals such as *Theranostics, Biosensors and Bioelectronics, Industrial Crops and Products, Talanta, Analytica Chimica Acta,* and *Journal of Separation Science.*

Ronald Beckett

Dr Ronald Beckett worked as an academic staff member in the Department of Chemistry at Caulfield Institute of Technology, Chisholm Institute of Technology, and Monash University in Melbourne, Australia, from 1970 to 2012. His main research interest was the characterization and role of suspended particles and natural organic matter in natural waters, particularly their role in the speciation and transport of pollutants. He pioneered the use of field-flow fractionation (FFF), a separation and sizing technique, to study these materials while working with FFF's inventor, Prof Calvin Giddings from the University of Utah, USA.

separations

MDPI

Article

Qualitative and Quantitative Analysis Method of Recombinant Collagen in Complex Matrix Based on HPLC-MS/MS

Jianping Gao [1,2], Liping Jia [3], Zhenhu Guo [1,2], Liming Xu [4], Daidi Fan [3,*] and Guifeng Zhang [1,2,*]

1 State Key Laboratory of Biochemical Engineering, Institute of Process Engineering, Chinese Academy of Sciences, Beijing 100190, China; jpgao@ipe.ac.cn (J.G.); 13051838711@126.com (Z.G.)
2 Key Laboratory of Biopharmaceutical Preparation and Delivery, Institute of Process Engineering, Chinese Academy of Sciences, Beijing 100190, China
3 Xi'an Giant Biogene Technology Co., Ltd., Xi'an 710075, China; liping.jia@xajuzi.com
4 National Institutes for Food and Drug Control, Beijing 102629, China; xuliming@nifdc.org.cn
* Correspondence: fandaidi@nwu.edu.cn (D.F.); gfzhang@ipe.ac.cn (G.Z.)

Abstract: The purpose of this study is to achieve the quantitative detection of recombinant type III collagen (rh-COL-III) in dressings with complex matrix. First of all, the marker peptide (GEAGIPGVP-GAK) of rhCOL-III was identified with HPLC-MS/MS. Then, a qualitative and quantitative method based on marker peptides was established and validated. In order to obtain higher sensitivity, a pretreatment method of liquid, gel, and ointment dressings was optimized. The reference material for quantification was combined using rhCOL-III and blank matrix of each dressing. The results indicated that the relative standard deviation (RSD) of the quantitative method was 2.77%, and the RSD of intraday and interday precision was 2.76% and 2.31%, respectively. The spiking recovery rate was between 80% and 90%. The optimal pretreatment method was Tris-HCl solvent replacement. The optimal trypsin concentration for the dressing solution was 20 μg in 500 μL. The method of preparing standard substances with a blank matrix can effectively eliminate the influence of the matrix effect on the quantitative results. The average spiking recovery rates of 50 μg/mL, 100 μg/mL, and 200 μg/mL in three different dressings ranged from 80% to 120%. The quantitative detection of limit (LOD) of rhCOL-III was 1 ng/mL, 2 ng/g, and 1 ng/g in liquid, ointment, and gel dressings.

Keywords: qualitative analysis; quantitation; recombinant collagen; dressing; complex matrix; HPLC-MS/MS

Citation: Gao, J.; Jia, L.; Guo, Z.; Xu, L.; Fan, D.; Zhang, G. Qualitative and Quantitative Analysis Method of Recombinant Collagen in Complex Matrix Based on HPLC-MS/MS. *Separations* **2024**, *11*, 339. https://doi.org/10.3390/separations11120339

Academic Editors: Ronald Beckett and Chao Kang

Received: 23 October 2024
Revised: 14 November 2024
Accepted: 17 November 2024
Published: 26 November 2024

1. Introduction

Recombinant collagen, with low immunogenicity, good water solubility, and stable quality compared to tissue-derived collagen [1–3], is increasingly used in cosmetics, tissue-engineered medical products, and other medical devices. In the medical devices field, rhCOL is widely used in medical dressings as a highlighted component [4,5], which is applied in the treatment of superficial or non-chronic wounds and the surrounding skin, such as small wounds and abrasions [6–8]. In addition to rhCOL, these medical dressings also contain a complex matrix, co-solvents, and a small amount of preservatives [9,10]. Specifically, the content of rhCOL is the most important target in assessing the effectiveness of medical dressing products. However, it is difficult to determine the accurate content of rhCOL in these products. One challenge is the interference from the complex matrix. The matrix used in the dressing is very complex, most of which exists in the form of oil in water [11]. The rhCOL is encapsulated in the matrix and is difficult to release, which creates certain difficulties in the quantitative analysis of rhCOL. On the other hand, most dressings only contain trace amounts of rhCOL, making it difficult to achieve accurate quantitative analysis using conventional methods. Currently, there are no appropriate methods for the qualitative analysis and accurate quantitative determination of rhCOL in medical dressings.

At present, hydroxyproline quantification and ELISA methods are usually applied to determine collagen contents. However, there are some disadvantages to these methods Mammalian cells like CHO and HEK293 have similar translation mechanisms to human cells, which could produce rhCOL with correct modifications [12,13]. The rhCOL from other cells did not contain hydroxyproline due to a lack of hydroxylase. Thus, the quantitative detection method involving hydroxyproline is not suitable for determining rhCOL content [14]. The ELISA method assay is based on the binding reaction between the antigen and the antibody [15,16]. Currently, some types of rhCOL on the market are not full-length collagen but rather multiple repeats of a segment of the amino acid sequence [17]. If the amino acid sequence of rhCOL does not contain antigen–antibody binding sites, it can easily lead to false-negative results when using the ELISA method [18]. There are no available specific antibodies that are designed for a particular sequence of rhCOL on the market. Additionally, the content of rhCOL in different dressings varies greatly, and traces of rhCOL in complex matrices are difficult to detect. The optimization of dressing pretreatment methods and the establishment of a highly sensitive method for the detection of Re-COL in dressings are of great significance for the quality control and market regulation of dressings. However, studies on the quantification of rhCOL in dressings have rarely been reported.

In this study, the marker peptides of recombinant type III collagen (rhCOL-III) were identified with a high-performance liquid chromatography–tandem mass spectrometer (HPLC-MS/MS). Then, a sensitive quantitative detection method for rhCOL-III was established and validated. Afterwards, liquid, gel, and ointment dressings containing rhCOL-III were selected as the research objects. Specific pretreatment methods for three different dressings, the amount of trypsin, and the standard solution preparation method were optimized. Additionally, the recovery rates of rhCOL-III in three kinds of dressings with different dosages were analyzed to verify the accuracy of the pretreatment method. The content of rhCOL-III in the actual products of three kinds of dressings was further analyzed. The quantitative determination of rhCOL-III in three different dressings was achieved using this method.

2. Materials and Methods

2.1. Materials and Instruments

Materials: Liquid dressing, ointment dressing, gel dressing, and recombinant type III collagen raw material were provided by Xian Giant Biogene Co., Ltd., Xi'an, Shanxi, China. Chromatographic pure formic acid was obtained from Merck, Kennett Square, NJ, USA. Sequence-grade trypsin was obtained from Promega Corporation, Madison, WI, USA. Marker peptides (GEAGIPGVPGAK) were synthetized by QYAOBIO (ChinaPeptides Co., Ltd.), Shanghai, China, and the HPLC purity was 99.10%. All other reagents were commercially analytically pure.

Instruments: Orbitrap mass spectrometer (Exploris 480, Thermo Fisher Scientific, Waltham, MA, USA) was used to identify the marker peptide of rhCOL-III. Triple quadrupole mass spectrometer (TSQ, Quantum ACCESS MAX, Thermo Fisher Scientific, Waltham, MA, USA) was used for rhCOL-III quantification.

2.2. Design of Marker Peptide Specific for rhCOL-III

The selection principles of the marker peptide are as follows: Sequence blast alignment was performed on human type III collagen. The peptide, which appears only in rhCOL-III sequences and has no effect on its abundance in terms of chemical modification or cross-linking, can be used as the reference peptide. The reference peptide should have higher resolution in the process of chromatographic separation. The reference peptide of rhCOL-III used in this study was designed as "GEAGIPGVPGAK". It was used as reference peptide and detection target of test sample for qualification and quantitation of rhCOL III in dressings.

2.3. Marker Peptides Identification by HPLC-MS/MS

2.3.1. Sample Preparation of rhCOL-III Enzymatic Digestion

The rhCOL-III with different concentrations was prepared with 0.1 mol/L Tris-HCl (pH 8.0) solution. Then they were centrifuged at 12,000× g for 10 min. The supernatant was collected for analysis.

Sample digestion: The rhCOL-III (5 mg) was dissolved in 10 mL 0.1 mol/L Tris-HCl (pH 8.0) solution. The rhCOL-III solution (0.5 mL) was mixed with 100 µL trypsin solution (0.2 mg/mL). The mixture was incubated at 37 °C for 24 h. And 10% formic acid solution (60 µL) was added to the rhCOL-III digestion to terminate the enzymatic reaction. The enzymatic digestion was centrifuged at 12,000× g for 10 min. The supernatant was collected for marker peptide identification.

2.3.2. Marker Peptides Identification

The marker peptides in rhCOL-III digestion were identified by HPLC/MS/MS. The on-line chromatographic separation was performed by reversed-phased chromatography on a Peptide BEH C18 column (2.1 × 150 mm, 1.7 µm) (Waters, USA) by using the UHPLC (Vanquish, Thermo Fisher, USA). The mobile phase consisted of water with 0.1% formic acid (A) and acetonitrile–water solution (*v/v*, 6:4) with 0.1% formic acid (B). The gradient elution procedure was performed as follows: 0–60 min, 5–40% B; 60–85 min, 40–90% B; 85–98 min, 90% B; 98—100 min, 90—5% B; 100—110 min, 5% B. The flow rate was 0.2 mL/min, the injection volume was 10 µL, and the column temperature was held at 60 °C. The outlet of the column was introduced into an orbitrap mass spectrometer (Exploris 480, Thermo Fisher Scientific, Waltham, MA, USA). Electron spray ionization (ESI) in positive mode was used to perform the orbitrap mass spectrometry. The spray voltage was set to 3.5 kV. The capillary temperature and vaporizer temperature were 320 °C and 300 °C, respectively. The sheath gas was 19.8 mL/min. The aux gas was 5 psi. The MS scan range was set from m/z 300 to 2000 and the resolution was set to 60,000. The RF lens was 45%. The normalized AGC Target was 300%. The maximum ion injection time was 100 ms. The scan event 2 was data-dependent MS/MS, and the resolution was set to 15,000. The isolation window was m/z 1.6. The maximum ion injection time was 200 ms, and the normalized AGC Target was 100%. The MS/MS collision energy was 30%. The rhCOL-III marker peptides were identified through the SEQUEST algorithm in Protein Discoverer 2.4 software (Thermo Fisher, USA) [19].

2.4. Investigation of Quantitative Method

2.4.1. Detection Procedure of Marker Peptide in rhCOL-III

The reference peptide and trypsin-digested rhCOL-III samples were separated with a Zorbax C18 column (2.1 × 150 mm, 5 µm) (Agilent Technologies, Santa Clara, CA, USA) using an HPLC (U3000, Thermo Fisher Scientific, Waltham, MA, USA). The mobile phase consisted of water with 0.1% formic acid (A) and acetonitrile with 0.1% formic acid (B). The gradient elution procedure was performed as follows: 0–0.5 min, 5–45% B; 0.5–2 min, 45–55% B; 2–8 min, 55–65% B; 8–8.1 min, 65–5% B; 8.1–10 min, 5% B. The flow rate was 0.2 mL/min, the injection volume was 5 µL, and the column temperature was held at 30 °C. The outlet of the column was introduced into TSQ mass spectrometer (Quantumn Access MAX, Thermo Fisher Scientific, Waltham, MA, USA). The ESI positive mode was used to perform the TSQ mass spectrometry. The spray voltage was set to 3.5 kV. The capillary temperature and vaporizer temperature were 320 °C and 300 °C, respectively. The sheath gas was 19.8 mL/min. The aux gas was 5 psi. The selected reaction monitoring (SRM) was used. The monitored target ions of rhCOL-III were m/z 526.71–371.97, 625.22 (GEAGIPGVPGAK).

2.4.2. Quantitative Methodological Validation

The rhCOL-III raw material was used for quantitative methodological validation. The marker peptide with different concentrations (0.1 µg/mL–500 µg/mL) were prepared with

0.1 mol/L Tris-HCl (pH 8.0). The marker peptide reference solution was analyzed 6 times to verify repeatability. The rhCOL-III raw material was prepared at the concentration of 0.5 mg/mL for three parallel samples. Each sample was digested respectively and analyzed 6 times to determine intraday precision. One sample was used for the interday precision test, which was tested on three different days, six replicates per day. The reference peptides (0.1, 0.25, 0.5 μg) were added to rhCOL-III digestion solution for analysis of spiking recovery rate. Each sample was analyzed three times in parallel, and the recovery rate was calculated. The limit of detection (LOD) and limit of quantitation (LOQ) for rhCOL-III raw material were determined from the calibration curve as 3 S/N and 10 S/N, respectively, where S is the signal of the target peak and N is the baseline noise.

2.5. Establish Qualitative and Quantitative Analysis Methods of rhCOL-III Dressings

2.5.1. Simulated rhCOL-III Dressing Sample Preparation

The 2.0 g ointment blank dressing matrix, 2.0 g gel blank dressing matrix, and 2 mL liquid blank dressing matrix were prepared as 12 parallel samples, respectively. Each kind of dressing was divided into 4 groups, with 3 parallels per group. The 0.01 mL, 0.05 mL, 0.1 mL, and 0.2 mL of rhCOL-III raw material solutions (prepared in Section 2.3.1) were added to the four groups of blank dressings matrices. The final concentration of rhCOL-III in each group was 10 μg/mL (group A), 50 μg/mL (group B), 100 μg/mL (group C), and 200 μg/mL (group D), respectively. Then, they were mixed well.

2.5.2. Investigation of Sample Pretreatment Methods

Three kinds of dressing samples from Group C, prepared in Section 2.5.1 were used for optimization of pre-processing methods, as described below.

Acetone precipitation method: Three dressing samples were mixed with cold acetone in the ratio of 1:5 (*v/v*). Then, they were shaken well and stored at −20 °C overnight. The samples were centrifuged (12,000× *g*, 20 min). Finally, the supernatants were removed, and the precipitates were redissolved with 0.5 mL 0.1 mol/L Tris-HCl (pH 8.0).

Ultrafiltration centrifugal method: The ointment and gel dressing samples (2 g) were first mixed with 0.7 mL 4 mol/L NaOH solution and centrifuged (12,000× *g*, 20 min). Then, the syringe was used to draw up the lower layer of clear liquid in the ointment dressing. The supernatant of the gel dressing sample was collected. The collected solutions of ointment dressing, gel dressing, and the liquid dressing samples were mixed with 1 mL ultrapure water, respectively. Subsequently, the samples were transferred to ultrafiltration centrifuge tubes with molecular weight cutoffs of 3 kDa, and centrifuged (8400× *g*, 20 min). The above operation was repeated 3 times. The supernatant was collected and lyophilized. At last, the samples were redissolved with 2 mL 0.1 mol/L Tris-HCl (pH 8.0) and magnetically stirred for 20 min.

Organic solvent extraction methods: The ointment and gel dressing samples (2 g) were first mixed with 0.5 mL 2 mol/L HCl solution and centrifuged (12,000× *g*, 20 min). Then, the lower layer of the clear liquid in the ointment dressing was drawn up using syringes. The supernatant of the gel dressing sample was collected. The collected solutions of ointment dressing, gel dressing, and liquid dressing samples were mixed with dichloromethane (*v*/*v*, 1:5), respectively. The solutions were mixed well and left for 2 h. Then, the upper liquid layer was collected and lyophilized. At last, the samples were redissolved with 2 mL 0.1 mol/L Tris-HCl (pH 8.0) and magnetically stirred for 20 min.

Solvent displacement methods: Firstly, three kinds of dressing samples were lyophilized. The lyophilized ointment and gel dressing samples were redissolved with 4 mL 0.1 mol/L Tris-HCl (pH 8.0). The lyophilized liquid dressing sample was redissolved with 2 mL 0.1 mol/L Tris-HCl (pH 8.0). All samples were magnetically stirred for 20 min.

2.5.3. Investigation of Trypsin Dosage

The samples (0.5 mL) prepared in Section 2.5.2 were mixed with 5 μg or 20 μg trypsin (100 μL), respectively. Then, they were incubated at 37 °C for 24 h. In order to terminate the enzymatic reaction, 10% formic acid solution (60 μL) was added to rhCOL-III digestion. The enzymatic hydrolysate was centrifuged at 12,000× g for 10 min, and the supernatant was collected.

2.5.4. Preparation of rhCOL-III Reference Working Solution

The rhCOL-III raw material was mixed with dressing blank matrices. The mixture was used as working solution for quantitation of rhCOL-III in different dressings. Preparation of blank matrix solutions was conducted as follows: First, 5 g of blank ointment dressing matrix and blank gel dressing matrix and 5 mL of liquid dressing blank matrix were lyophilized. Then, the blank ointment dressing matrix and blank gel dressing matrix were redissolved by 10 mL 0.1 mol/L Tris-HCl (pH 8.0). The liquid dressing blank matrix was redissolved by 5 mL 0.1 mol/L Tris-HCl (pH 8.0). All the blank matrices were magnetically stirred for 20 min and centrifuged at 12,000× g for 10 min. The supernatants of the blank gel and liquid dressing matrix solutions was collected, and the lower layer of the blank ointment dressing matrix solution was drawn up by syringes.

Preparation of working solution of rhCOL-III reference: The rhCOL-III enzymatic digestion solution prepared in Section 2.3.1 was diluted to different concentrations by three blank matrix solutions, respectively. Then, they were used for quantitation of rhCOL-III in different dressings, ensuring that the marker peptide to be detected in the reference sample and test samples have the same matrix background.

2.5.5. Methodological Validation of Pretreatment Method

Sample preparation: Four groups of the samples prepared in Section 2.5.1 and three actual dressing products were used to validate the pretreatment method. First of all, they were treated using the solvent displacement methods listed in Section 2.5.2. Then, they (0.5 mL) were mixed with 20 μg trypsin (100 μL) and incubated at 37 °C for 24 h, respectively. At last, 10% formic acid solution (60 μL) was added to the digestion. The enzymatic digestion was centrifuged at 12,000× g for 10 min, and the supernatant was collected.

LOD of the method in different dressings: The LOD values of rhCOL-III in different dressings subjected to pretreatment methods were determined.

3. Results

3.1. Marker Peptide Identification of rhCOL-III

Firstly, the marker peptides of rhCOL-III were identified using HPLC-MS/MS. Figure 1 shows the total ion chromatogram of digested rhCOL-III. Subsequently, the mass spectrometric data were analyzed using BioFinder software. In order to retrieve more peptides and exclude peptides undergoing modifications in marker peptides screening, the common modifications, such as oxidation and dethiomethylation of methionine, deamidation of asparagine and glutamine, pyroglutamylation of glutamine and glutamic acid, ammonia-loss, and acetylation of the nitrogen terminal were set up in BioFinder software. The peptides identified in rhCOL-III with no modification sites and which do not exist in other animals can be used as the target peptides. Figure 2 shows the MS spectrum of m/z 526.79071 with two charges. Therefore, the molecular weight of m/z 526.79071 was 1051.566 Da, which was consistent with the molecular weight of the peptide (GEAGIPGVPGAK). Figure 3 shows the MS/MS spectrum of m/z 526.79071. The MS/MS ions in Figure 3 are match well with the theoretical product ions of the peptide (GEAGIPGVPGAK). The peptide (GEAGIPGVPGAK) had no modification sites and only existed in rhCOL-III. Therefore, the peptide (GEAGIPGVPGAK) could be used as the marker peptide of rhCOL-III.

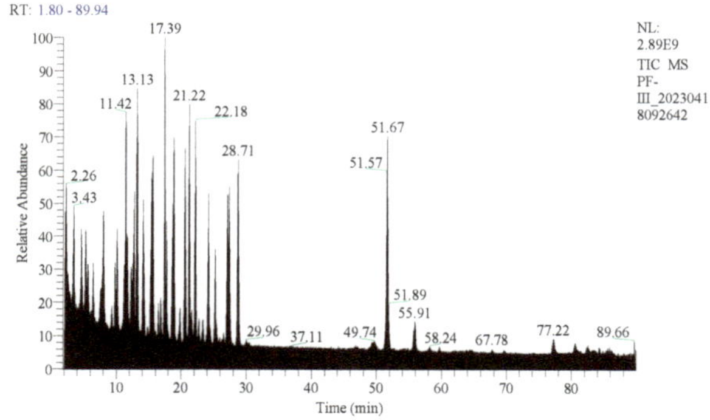

Figure 1. Total ion chromatogram of digested rhCOL-III. (The green line is used to label the retention time of the peaks).

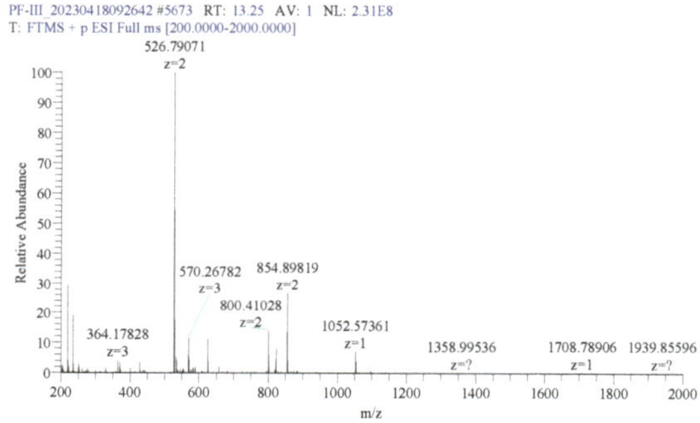

Figure 2. MS spectrum of m/z 526.79071 detected in digested rhCOL-III. (The green line is used to label the m/z of the ions).

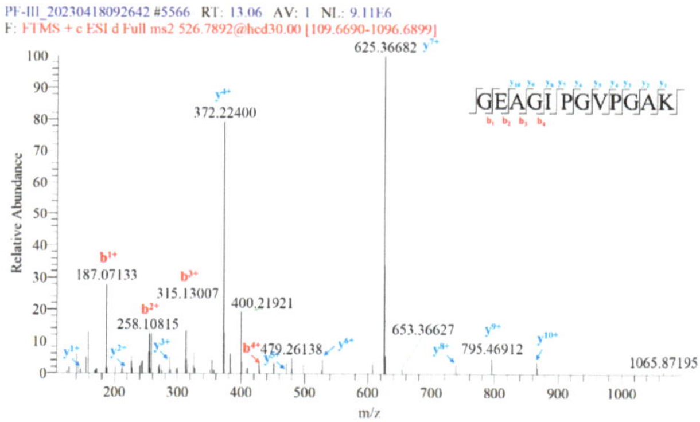

Figure 3. MS/MS spectrum of m/z 526.79071 detected in digested rhCOL-III.

3.2. *Establishment and Validation of Quantitative Method for Marker Peptide in rhCOL-III*

Marker peptides with concentrations from 0.1 µg/mL to 500 µg/mL were used as the reference solutions. Figure 4 shows that the retention time of the marker peptide (GEAGIPGVPGAK) was 3.99 min. The abscissa was the concentration of the marker peptide working solutions. The peak area of the marker peptide in working solutions was used as the ordinate. The regression equation and correlation coefficient were y = 995631x + 543070 and R^2 = 0.9999 (GEAGIPGVPGAK). In order to verify the repeatability of the quantitative method, the maker peptide reference solution was repeatedly analyzed six times. The RSD of the peak area was 2.77%, indicating the higher repeatability of this method. The intraday precision was analyzed with three rhCOL-III samples, and each sample was detected six times (Table 1). The results showed that the RSD of intraday precision was 2.76%. Then, one sample was used for the validation of interday precision and analyzed on different days (Table 2). The results showed that the RSD of the interday precision was 2.31%. Thus, the quantification method has good intraday and interday precision. Table 3 shows that the spiking recovery rate at different spiking levels was between 80% and 90%. The average spiking recovery rate was 83.15%. The RSD between the different spiking levels was 3.10%. This meets the requirements for spiking recoveries in methodological validation. The LOQ and LOD are very important in the analysis of trace substances. In this study, the marker peptide was gradually diluted and detected. The results showed that the LOQ and LOD of the marker peptide were 0.5×10^{-3} μg/mL and 0.2×10^{-3} μg/mL, respectively.

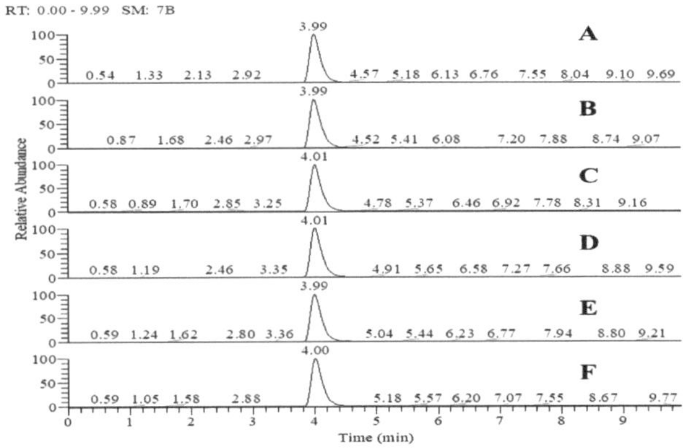

Figure 4. The MS spectrum of the marker peptide in working solutions with different concentrations (the concentrations of A, B, C, D, E, and F are 0.05 μg/mL, 0.1 μg/mL, 0.5 μg/mL, 1 μg/mL, 2.5 μg/mL, and 5 μg/mL, respectively).

Table 1. The intraday precision of quantification method for marker peptide of rhCOL-III.

Sample	Concentration (μg/mL)	Average Concentration (μg/mL)	SD	RSD (%)
S1	10.97	10.75	0.21	1.98
	10.74			
	10.42			
	10.81			
	10.62			
	10.97			
S2	10.12	10.30	0.13	1.26
	10.36			
	10.33			
	10.32			
	10.17			
	10.47			
S3	10.30	10.41	0.29	2.83
	10.40			
	10.12			
	10.98			
	10.31			
	10.34			
Average concentration (μg/mL)			10.49	
SD			0.29	
RSD			2.76	

Table 2. The interday precision of quantification method for marker peptide of rhCOL-III.

Sample	Concentration (μg/mL)	Average Concentration (μg/mL)	SD	RSD (%)
D1	10.97 10.74 10.42 10.81 10.62 10.97	10.75	0.21	1.98
D2	10.65 10.61 10.54 10.04 10.50 10.93	10.55	0.29	2.75
D3	10.31 10.77 10.35 10.63 10.85 10.59	10.59	0.22	2.04
Average concentration (μg/mL)			10.63	
SD			0.25	
RSD			2.31	

Table 3. Spiking recovery rate of quantification method for marker peptide of rhCOL-III.

NO.	Concentration (μg/mL)	Recovery Quality (μg)	Additive Quality (μg)	Spiking Recovery Rate (%)	Average Spiking Recovery Rate (%)
T-1-1	2.315	0.084	0.1	83.67	
T-1-2	2.308	0.082	0.1	82.18	82.14
T-1-3	2.300	0.081	0.1	80.56	
T-2-1	2.919	0.204	0.25	81.72	
T-2-2	2.921	0.205	0.25	81.94	84.39
T-2-3	3.016	0.224	0.25	89.50	
T-3-1	3.977	0.416	0.5	83.21	
T-3-2	3.947	0.410	0.5	81.99	82.93
T-3-3	3.987	0.418	0.5	83.58	
Average spiking recovery rate (%)					83.15
SD					2.58
RSD (%)					3.10

3.3. *Establishment of Quantitative Method of rhCOL-III in Simulated rhCOL-III Dressing*

3.3.1. Effect of Pretreatment Method on Recovery Rate of rhCOL-III

The dressing matrix was very complex and varied in different dressings. In this study, group C (prepared in Section 2.5.1) was used as the simulated dressing samples. In order to exclude the influence of matrix on the quantitative results, four different pretreatment methods, including acetone precipitation, centrifugal ultrafiltration, dichloromethane extraction, and solvent displacement, were compared. Figure 5 shows the effects of different pretreatment methods on the recoveries in simulated rhCOL-III dressings. The results showed that the recoveries of the centrifugal ultrafiltration method, acetone precipitation method, and dichloromethane extraction method were lower. In these three simulated rhCOL-III dressings, the recovery of the solvent displacement method was higher than the other three methods.

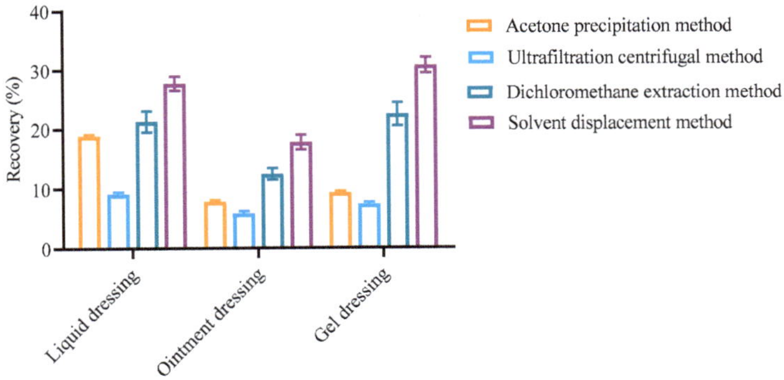

Figure 5. Effect of different pretreatment methods on the recovery of rhCOL-III ($n = 3$).

3.3.2. Effect of Trypsin Dosage on Recovery Rate of rhCOL-III

The carbomer in the three dressings can inhibit the activity of trypsin. Therefore, trypsin dosage was an important factor which could affect recovery rates. To ensure that rhCol-III in the dressing sample can be fully enzymatically hydrolyzed, excessive trypsin was added to the samples. In this study, 0.5 mL of the three simulated rhCOL-III dressing samples was mixed with 5 μg or 20 μg trypsin, respectively. The effect of enzyme dosage on the recovery of rhCOL-III was analyzed. Figure 6 shows the effect of different trypsin dosages on the recovery of rhCOL-III in simulated rhCOL-III dressings. The results showed that the recoveries of rhCOL-III using 20 μg trypsin digestion in three rhCOL-III dressings were all significantly higher than those of the 5 μg group. The results indicate that the increase in trypsin dosage could effectively increase the recovery rate of rhCOL-III in simulated rhCOL-III dressings.

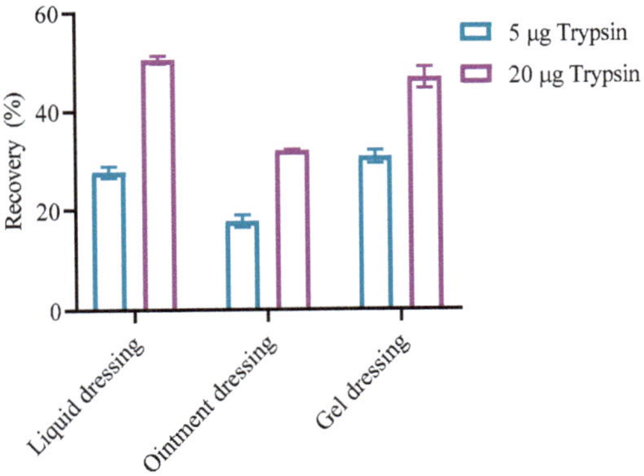

Figure 6. Effect of different trypsin dosages on the recovery of rhCOL-III in simulated rhCOL-III dressing ($n = 3$).

3.3.3. Effect of Different Reference Working Solutions on the Recovery of rhCOL-III

The matrix effect is an important factor in mass spectrometry quantification. The contents and properties of the matrix will affect the ionization efficiency of the target compounds, resulting in the enhancement, weakening, or deviation of the mass spectrum signal. Thus, an rhCOL-III reference working solution was prepared with a blank matrix, and the recovery was compared with the results of the Tris-HCl method. Figure 7 shows the recovery of rhCOL-III in three simulated dressings obtained by the working solution curve method of the blank matrix and Tris-HCl prepared as a reference. The recoveries of rhCOL-III obtained by the method of blank matrix prepared as a reference working solution curve were all between 80% and 120%, which met the requirement of the recovery rate in the methodology.

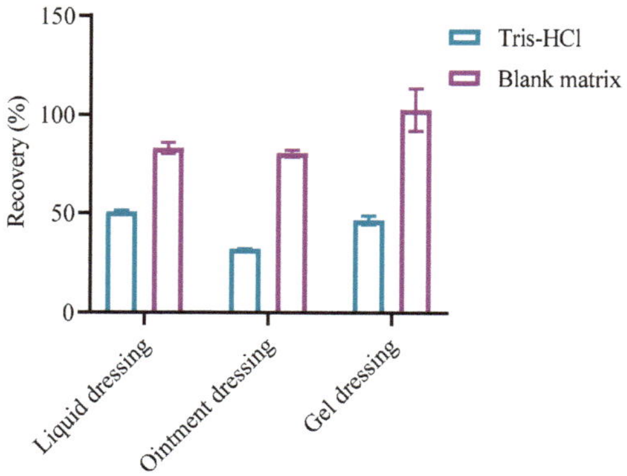

Figure 7. Effect of different reference solutions on the recovery of rhCOL-III (n = 3).

3.3.4. Methodology Validation of Pretreatment Method in Simulated Dressing Samples

To further verify the accuracy of the method, rhCOL-III was added to three kinds of dressing matrices to final concentrations of 10 μg/mL (group A), 50 μg/mL (group B), 100 μg/mL (group C), and 200 μg/mL (group D). The optimized method was used to analyze the content of rhCOL-III in the simulated dressing samples. Table 4 shows the recovery of rhCOL-III in dressing samples with different spiking levels. The results showed that the average recoveries of rhCOL-III in liquid dressings and gel dressings with spiked amounts of 10 μg/mL, 50 μg/mL, 100 μg/mL, and 200 μg/mL were all between 80% and 120%. The average recoveries of 50 μg/mL, 100 μg/mL, and 200 μg/mL in the ointment dressings were all between 80% and 120%. The average recovery of the ointment dressing with 10 μg/mL was 70.46%. The main reason for these results may be due to the semi-solid state of the ointment dressing and the uneven mixing of trace amounts of rhCOL-III in the ointment dressing.

Table 4. Recoveries of rhCOL-III in simulated dressing samples with different additive levels of rhCOL-III.

NO.	Liquid Dressing		Ointment Dressing		Gel Dressing	
	Recovery (%)	Average Recovery (%)	Recovery (%)	Average Recovery (%)	Recovery (%)	Average Recovery (%)
A-1	80.21		73.06		83.29	
A-2	80.63	87.85	70.91	70.46	80.71	80.02
A-3	79.95		67.40		76.06	
B-1	85.87		87.22		88.42	
B-2	89.32	80.26	81.14	81.87	95.59	92.01
B-3	88.34		77.24		92.03	
C-1	85.68		81.04		111.98	
C-2	79.94	83.07	78.47	80.46	90.83	102.59
C-3	83.59		81.87		104.95	
D-1	82.89		84.23		89.94	
D-2	81.73	82.78	85.49	82.69	99.34	93.66
D-3	83.73		78.37		91.72	

3.3.5. Quantification and Quantitation of rhCOL-III in Actual rhCOL-III Dressing Products

The rhCOL-III dressing products were further analyzed using the optimized method. Figure 8 shows that the MS spectra of marker peptide in dressing products were the same as in the rhCOL-III reference samples. This finding indicates that the three different dressing products contained rhCOL-III.

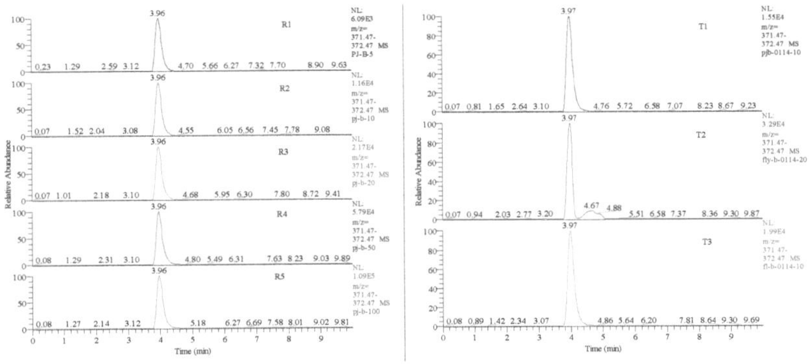

Figure 8. The MS spectra of the marker peptide in different medical dressing products (T1—liquid dressing; T2—ointment dressing; T3—gel dressing) and rhCOL-III reference samples (R1—5 µg/mL; R2—10 µg/mL; R3—20 µg/mL; R4—50 µg/mL; R5—100 µg/mL).

The theoretical rhCOL-III content in three dressing products was 100 µg/mL. Table 5 showed that the contents of rhCOL-III detected in three dressing products ranged from 80 to 112 µg/mL. The recoveries were all between 80% to 120% compared to theoretical contents.

Table 5. The content of rhCOL-III in different dressing products.

Product Name	The Content of rhCOL-III (µg/mL)	Average Content (µg/mL)
rhCOL-III liquid dressing	104.29 104.44 111.11	106.61
rhCOL-III ointment dressing	82.49 80.77 85.17	82.81
rhCOL-III gel dressing	93.53 90.32 92.00	91.95

The LOQ of the pretreatment method is an important indicator that influences the application of quantitative methods. Currently, the content of rhCOL in dressing products on the market varies widely. Some dressings only contain traces of rhCOL. In this research, we added a series of trace amounts of rhCOL-III to the blank matrices of the three dressings. Then, they were analyzed with the optimized method. The LOQ of rhCOL-III in liquid, ointment, and gel dressings with the optimized method was 1 ng/g, 2 ng/g, and 1 ng/g, respectively.

4. Discussion

In this study, a quantitative method for rhCoL based on marker peptide detection with HPLC/MS/MS was first established. Then, the accuracy of the quantitative method was verified. The RSD of repeatability was 2.77%. The RSD of intraday and interday precision was 2.76% and 2.31%, respectively. The spiking recovery rate was between 80% and 90%. Afterwards, this method was applied to the quantitative analysis of rhCOL-III contained in different dressings. Dressing matrices were complex and varied widely from one dressing matrix to another. Collagen in the dressing was mostly coated by matrices in form of oil in water, which caused difficulties in quantitative detection [20]. Therefore, pretreatment is essential for the detection of rhCOL in dressing products.

Four different pretreatment methods were compared. The recoveries of the samples prepared by centrifugal ultrafiltration method were the lowest. The main reason for this was that ointment dressings and gel dressings needed to be centrifuged to remove insoluble particulate matter to avoid clogging the ultrafiltration tube before ultrafiltration. In these two dressings, rhCOL-III was encapsulated in a matrix material in the form of oil in water. Centrifugation had difficulty breaking down this form, which resulted in part of the collagen being removed before ultrafiltration. Additionally, a part of rhCOL-III may be adsorbed on the ultrafiltration membrane during the ultrafiltration process, resulting in lower recovery. The recovery of ointment and gel dressings processed using acetone precipitation did not differ significantly from other ultrafiltration methods. This may be due to the fact that rhCOL-III in ointment and gel dressings was encapsulated by the matrices in the form of oil in water, which could not be emulsified using the acetone precipitation method. The recovery result of liquid dressings obtained by acetone precipitation was significantly higher than that of ultrafiltration. This was because the liquid dressing was in an aqueous state, and the acetone precipitation method could precipitate part of the rhCOL-III. However, the recovery rates remained relatively low. This may be due to the fact that the content of rhCOL-III in the dressings was only 100 μg/mL in general, and it was difficult for the acetone precipitation method to adequately precipitate trace amounts of this protein. The recoveries of rhCOL-III in these three dressings obtained by organic solvent extraction were significantly higher than via acetone precipitation and ultrafiltration, but the recoveries were still lower. On the one hand, the molecular weight of rhCOL-III is lower. It is soluble in organic solvents and a portion of the protein was carried away by the organic phase. On the other hand, the water-soluble matrix cannot be removed by organic solvents. Some water-soluble matrix may affect enzyme activity. Thus, rhCOL-III cannot be completely digested, resulting in a lower recovery rate. The sample with the highest recovery rate was obtained using the Tris-HCl solvent replacement method. In this method, the oil-in-water structure was destroyed through the lyophilization process to release rhCOL-III. Then, the freeze-dried sample was resolved by Tris-HCl. This was a simple step to reduce the loss of rhCOL-III caused by the pretreatment process.

In order to improve the recovery rate, the amount of trypsin was optimized. For the 0.5 mL sample solution, both 5 μg and 20 μg of trypsin were excessive. The 20 μg dosage of trypsin significantly improved the recovery. The matrices of three dressings could potentially inhibit the activity of trypsin. The content of matrix in the sample was mainly related to the sample volume. Therefore, there was no significant correlation between the amount of trypsin and the protein content in the sample; rather, it mainly related to the sample volume.

In addition to the pretreatment process, matrices are an important factor influencing quantitative detection via mass spectrometry [21,22]. During mass spectrometry ionization, matrix components in dressings may produce higher ionization energy, which inhibit the ionization of the rhCOL-III marker peptide, resulting in low detection recoveries. In addition, the matrix and rhCOL-III marker peptide may co-evaporate during ionization, forming a mixed gas stream that interferes with the signal of the target. The internal standard method is an effective method to exclude matrix effects [23]. However, the target compound was a peptide in this study. The chemically synthesized, isotopically labeled peptide was not stable in the aqueous solution and therefore degraded in a very short period of time. The method of reference working solution preparation using blank matrix can both exclude the interference of matrices and ensure the stability and accuracy of the results. The reason for this may be that the test sample and reference working solution have the same matrix background.

5. Conclusions

In this study, we first established a quantitative method of rhCOL-III based on HPLC/MS/MS. Afterwards, the quantitative method was used to analyze the rhCOL-III content in liquid, ointment, and gel dressings. We found that the pretreatment method was an important factor in the quantitative analysis of rhCOL-III dressings. Furthermore, we optimized and validated the pretreatment method of different dressings. The recoveries of different dressings treated by this method meet the quantitative methodological requirements and enable the analysis of trace collagen in different dressings. This study established an accurate quantitative method that could be used for the detection of recombinant collagen in dressings with complex matrices. We believe that this method is of great significance for the quality control and market supervision of dressing products.

Author Contributions: Conceptualization, G.Z. and D.F.; Methodology, J.G. and L.J.; Software, J.G.; Validation, L.J.; Data Curation, J.G.; Writing—Original Draft Preparation, J.G.; Writing—Review and Editing, Z.G., L.X. and D.F.; Supervision, L.J.; Project Administration, L.X.; Funding Acquisition, J.G. and G.Z. All authors have read and agreed to the published version of the manuscript.

Funding: This research was funded by the Beijing Natural Science Foundation (L234071), the Beijing Natural Science Foundation (L234021), and the Beijing Natural Science Foundation (L234070).

Data Availability Statement: The data presented in this study are available on request from the corresponding author. The data are not publicly available due to privacy reasons.

Conflicts of Interest: Authors Liping Jia and Daidi Fan were employed by the company Xi'an Giant Biogene Technology Co., Ltd. The remaining authors declare that the research was conducted in the absence of any commercial or financial relationships that could be construed as a potential conflict of interest.

References

1. Seo, W.Y.; Kim, J.H.; Baek, D.S.; Kim, S.J.; Kang, S.; Yang, W.S.; Song, J.A.; Lee, M.S.; Kim, S.; Kim, Y.S. Production of recombinant human procollagen type I C-terminal propeptide and establishment of a sandwich ELISA for quantification. *Sci. Rep.* **2017**, *7*, 15946. [CrossRef] [PubMed]
2. Fan, D.; Luo, Y.; Ma, X. A Ptsg Gene Knockout Recombinant Strain That Efficiently Expresses Human-like Collagen and Its Construction Method and Protein Expression. Patent CN103224901B, 1 January 2015.
3. Wei, C.; Liu, C.; Liu, X. Research progress on the production of recombinant collagen by genetic engineering. *Bull. Ferment. Sci. Technol.* **2021**, *50*, 1–4.
4. Guo, X.L.; Ma, Y.; Wang, H.; Yin, H.P.; Shi, X.L.; Chen, Y.Q.; Gao, G.B.; Sun, L.; Wang, J.D.; Wang, Y.B.; et al. Status and developmental trends in recombinant collagen preparation technology. *Regen. Biomater.* **2024**, *11*, rbad106. [CrossRef]
5. You, S.; Zhu, Y.; Li, H.; He, F.; Liu, S.B.; Yang, X.; Wang, L.; Zeng, H.; Dai, J.C.; Hu, L.N. Recombinant humanized collagen remodels endometrial immune microenvironment of chronic endometritis through macrophage immunomodulation. *Regen. Biomater.* **2023**, *10*, rbad033. [CrossRef]
6. Xu, L.; Liu, Y.; Tang, L.; Xiao, H.; Yang, Z.; Wang, S. Preparation of recombinant human collagen III protein hydrogels with sustained release of extracellular vesicles for skin wound healing. *Int. J. Mol. Sci.* **2022**, *23*, 6289. [CrossRef] [PubMed]
7. Wang, Y.; Zhang, Y.; Yang, Y.P.; Jin, M.Y.; Huang, S.; Zhuang, Z.M.; Zhang, T.; Cao, L.L.; Lin, X.Y.; Chen, J.; et al. Versatile dopamine-functionalized hyaluronic acid-recombinant human collagen hydrogel promoting diabetic wound healing via inflammation control and vascularization tissue regeneration. *Bioact. Mater.* **2024**, *35*, 330–345. [CrossRef]
8. Huang, J.H.; Lai, X.L.; Huang, Z.W.; Rong, Z.B.; Li, H.H.; Xie, Y.X.; Duan, L.; Xiong, J.Y.; Wang, D.P.; Zhu, S.H.; et al. Bioprinted Gelatin-Recombinant Type III Collagen Hydrogel Promotes Wound Healing. *Int. J. Bioprinting* **2022**, *8*, 517. [CrossRef]
9. Jia, Y.; Han, Y.X.; Zhang, Y.; Li, L.; Zhang, B.L.; Yan, X. Multifunctional type III recombinant human collagen incorporated sodium alginate hydrogel with sustained release of extra cellular vehicles for wound healing multimodal therapy in diabetic mice. *Regen. Ther.* **2024**, *27*, 329–341. [CrossRef]
10. Bohn, G.; Liden, B.; Schultz, G.; Yang, Q.P.; Gibson, D.J. Ovine-Based Collagen Matrix Dressing: Next-Generation Collagen Dressing for Wound Care. *Adv. Wound Care* **2016**, *5*, 1–10. [CrossRef]
11. Choi, S.M.; Ryu, H.A.; Lee, K.-M.; Kim, H.J.; Park, I.K.; Shin, H.-C.; Choi, W.J.; Lee, J.W. Development of Stabilized Growth Factor-Loaded Hyaluronate—Collagen Dressing (HCD) matrix for impaired wound healing. *Biomater. Res.* **2016**, *20*, 9. [CrossRef]
12. Gauza, W.M.; Kubisz, L.; Mielcarek, S.; Włodarczyk, D. Comparison of thermal properties of fish collagen and bovine collagen in the temperature range 298–670 K. *Mater. Sci. Eng. C Mater. Biol. Appl.* **2017**, *80*, 468–471. [CrossRef] [PubMed]
13. Lim, Y.S.; Ye, J.O.; Hwang, S.Y.; Kwak, J.Y.; Yoon, S. Marine collagen as a promising biomaterial for biomedical applications. *Mar. Drugs* **2019**, *17*, 467. [CrossRef] [PubMed]
14. Liu, W.B.; Lin, H.; Zhao, P.; Xing, L.N.; Li, J.; Wang, Z.H.; Ju, S.; Shi, X.L.; Liu, Y.H.; Deng, G.; et al. A regulatory perspective on recombinant collagen-based medical devices. *Bioact. Mater.* **2022**, *12*, 198–202. [CrossRef] [PubMed]
15. Liu, G.Y.; Kawaguchi, H.; Ogasawara, T.; Asawa, Y.; Kishimoto, J.; Takahashi, T.; Chung, U.; Yamaka, H.; Asato, H.; Nakamura, K.; et al. Optimal combination of soluble factors for tissue engineering of permanent cartilage from cultured human chondrocytes. *J. Biol. Chem.* **2007**, *282*, 20407–20415. [CrossRef] [PubMed]

16. Stoilov, I.; Starcher, B.C.; Mecham, R.P.; Broekelmann, T.J. Measurement of elastin, collagen, and total protein levels in tissues *Methods Cell Biol.* **2018**, *143*, 133–146.
17. Sai, K.P.; Babu, M. Collagen based dressings—A revies. *Burns* **2000**, *26*, 54–62. [CrossRef]
18. Song, H.D.; Zhang, S.Q.; Zhang, L.; Li, B. Effect of orally administered collagen peptides from bovine bone on skin aging in chronologically aged mice. *Nutrients* **2017**, *9*, 1209. [CrossRef]
19. Gao, J.P.; Ma, Y.; Guo, Z.H.; Zhang, Y.; Xing, F.Y.; Zhang, T.Y.; Kong, Y.J.; Luo, X.; Xu, L.M.; Zhang, G.F. Evaluating the Degradation Process of Collagen Sponge and Acellular Matrix Implants In Vivo Using the Standardized HPLC-MS/MS Method. *Separations* **2023**, *10*, 47. [CrossRef]
20. Yesiltas, B.; Moreno, P.; Gregersen, S.; Olsen, T.; Jones, N.; Hoffmann, S.; Marcatili, P.; Overgaard, M.; Hansen, E.; Jacobsen, C. Antioxidant peptides derived from potato, seaweed, microbial and spinach proteins: Oxidative stability of 5% fish oil-in-water emulsions. *Food Chem.* **2022**, *15*, 123699. [CrossRef]
21. Panuwet, P.; Hunter, R.E.; Souza, P.E.; Chen, X.Y.; Radford, S.A.; Cohen, J.R.; Marder, M.E.; Kartavenka, K.; Ryan, P.B.; Barr, D.B. Biological Matrix Effects in Quantitative Tandem Mass Spectrometry-Based Analytical Methods: Advancing Biomonitoring. *Crit. Rev. Anal. Chem.* **2016**, *46*, 93–105. [CrossRef]
22. Cortese, M.; Gigliobianco, M.; MagnonI, F.; Censi, R.; Martino, P. Compensate for or Minimize Matrix Effects? Strategies for Overcoming Matrix Effects in Liquid Chromatography-Mass Spectrometry Technique: A Tutorial Review. *Molecules* **2020**, *25*, 3048. [CrossRef] [PubMed]
23. Bronsema, K.J.; Bischoff, R.; Merbel, N.C. Internal standards in the quantitative determination of protein biopharmaceuticals using liquid chromatography coupled to mass spectrometry. *J. Chromatogr. B* **2006**, *78*, 1331–1336. [CrossRef] [PubMed]

foods

Article

Insights into Tissue-Specific Specialized Metabolism in Wampee (*Clausena lansium* (Lour.) Skeels) Varieties

Ran Zhang [1], Junjie Zhou [1], Xiaoxuan Zhang [1], Huanteng Hou [1], Xianqing Liu [1], Chenkun Yang [1,2], Shuangqian Shen [1,2] and Jie Luo [1,2,*]

1 School of Breeding and Multiplication (Sanya Institute of Breeding and Multiplication), Hainan University, Sanya 572025, China; zhangran@hainanu.edu.cn (R.Z.); sand_zhou@hainanu.edu.cn (J.Z.); 17868912065@139.com (X.Z.); naturehht@foxmail.com (H.H.); liuxq@hainanu.edu.cn (X.L.); yangchenkun@yzwlab.cn (C.Y.); shenshuangqian@yzwlab.cn (S.S.)

2 Yazhouwan National Laboratory, Sanya 572025, China

* Correspondence: jie.luo@hainanu.edu.cn

Abstract: Wampee (*Clausena lansium* (Lour.) Skeels) has natural bioactive components with diverse health benefits, but its detailed metabolism and tissue distribution are not fully understood. Here, widely targeted metabolomics analysis methods were employed to analyze the wampee fruit (peel, pulp, and seed) of 17 different varieties. A total of 1286 metabolites were annotated, including lipids, flavonoids, polyphenols, carbazole alkaloids, coumarins, and organic acids, among others. The quantitative analysis and matrix-assisted laser desorption/ionization–mass spectrometry imaging (MALDI-MSI) analysis indicated remarkable variations in metabolite categories and content in the peel, pulp, and seed of wampee fruit. Additionally, the difference analysis found that the metabolic components of peel contributed dominantly to the differences among varieties, and 7 potential biomarkers were identified. In this study, a comprehensive metabolome landscape of wampee fruit was established, which provided important information for the isolation and identification of functional components, food industry application, and nutritional improvement breeding.

Keywords: *Clausena lansium* (Lour.) Skeels; metabolic profiling; biomarker; spatial distribution; MALDI-MSI; nutritional breeding

Citation: Zhang, R.; Zhou, J.; Zhang, X.; Hou, H.; Liu, X.; Yang, C.; Shen, S.; Luo, J. Insights into Tissue-Specific Specialized Metabolism in Wampee (*Clausena lansium* (Lour.) Skeels) Varieties. *Foods* **2024**, *13*, 3092. https://doi.org/10.3390/foods13193092

Academic Editor: Cornelia Witthöft

Received: 5 September 2024
Revised: 22 September 2024
Accepted: 24 September 2024
Published: 27 September 2024

1. Introduction

Wampee (*Clausena lansium* (Lour.) Skeels) is a plant native to southern China and belongs to the Rutaceae family. It has been documented as a medicinal and edible fruit in the classics of traditional Chinese medicine in ancient China. In addition to its edible pulp, its leaves, fruit stones, and peels can also serve as potential medicinal materials with effects such as diuresis and detumescence, promoting fluid production and quenching thirst, clearing heat, and relieving cough. Therefore, it is known as the "treasure of fruits" [1]. The fruit of wampee is sweet and sour with a distinctive aroma, rich in active substances that have been widely used in the food processing industry. Apart from being consumed fresh or for medicinal purposes, it is also transformed into preserved fruits, fruit juice, pastries, etc. Currently, wampee is widely cultivated in tropical regions such as southern China, India, Australia, Thailand, and Vietnam.

For plants with medicinal and nutritional value, tissue differential analysis is indispensable, which not only helps to reveal the nutritional and medicinal value of different tissues but also helps to distinguish the precise distribution of important categories of bioactive compounds in different tissues. Researchers have examined the metabolites in different tissues of medicinal plants such as Panax notoginseng and Ginseng [2,3] and found that different types of ginsenosides accumulate in various tissues, most of which play roles in anti-inflammatory, anti-cancer, antioxidant, and immune regulation. In addition to these, metabolites in different tissues of olive (*Olea europaea* L.), *Eucommia ulmoides* Oliver, and blackberry have also been characterized for their differential profiles [4–6], and these studies are of significant importance for guiding the selection of edible parts and precision breeding. Metabolites in different tissues of wampee possess a variety of nutritional and medicinal values, such as antioxidant, antibacterial, anti-tumor, liver-protecting with lipid-lowering effects, and blood sugar-lowering activities [7–12]. So far, natural products such as carbazole alkaloids, coumarins, amide alkaloids, terpenes, and flavonoids have been successfully isolated and identified from different tissues of wampee [11,13]. However, the above studies mostly focused on the detection and analysis of various tissues and organs, such as leaves, stems, and fruits, and focused on the detection, separation, and purification analysis of extracts of a single component [7,9,14]. Due to the small number of metabolites detected and the lack of precise quantification, it is not possible to fully characterize the wampee fruit and effectively identify the differences between its fine tissues. By constructing the metabolome panorama of multiple tissues of various varieties of wampee fruit and analyzing the key metabolite biomarkers, it will greatly enrich the cognition and evaluation of the nutritional and health value of wampee fruit.

With the rapid development of metabolomics [15–18], detection instruments have also become increasingly diverse and sophisticated, such as liquid chromatography–mass spectrometry (LC–MS), gas chromatography-mass spectrometry (GC-MS), and matrix-assisted laser desorption/ionization mass spectrometry imaging (MALDI-MSI), providing opportunities for comprehensive detection of wampee metabolites. Due to the low sensitivity of non-targeted detection and the low throughput of targeted detection, a widely targeted metabolomics analysis method that combines the high throughput of non-targeted analysis with the high sensitivity of targeted analysis has become a powerful tool in metabolomics [19]. Due to the complexity of metabolites and the diversity of their chemical structures, it is not possible to analyze all metabolites simultaneously using a single platform. Currently, the metabolites detected in plants may account for less than 10% of their total quantity [20], but widely targeted metabolites aim to accurately and effectively detect as many metabolites as possible [19]. At present, a variety of plants, such as rice [17], Qingke [21], maize [22], tomato [23], pummelo [24], and citrus [25], among others, have been comprehensively characterized by widely targeted metabolites, and further studies have been conducted based on this combination with genome and transcriptome. Metabolomics plays an indispensable role in elucidating the dynamic changes of a series of biological activities as well as the comprehensive characterization of species [16]. In the field of multi-tissue metabolomics research in plants, this analytical strategy has also been used to establish a metabolite profiling database covering the entire growth period of rice, including eight main tissue organs (coleoptile, radicle, leaf, leaf sheath, stem, root, panicle, and seed). By analyzing the differences in metabolic signals among various tissues, it has been found that there are multiple metabolic signals that are specifically present in each tissue [17].

Yongxing Town of Haikou City, Hainan Province, is one of the world's original birthplaces and a gene bank for the germplasm resources of wampee. The wild wampee forest in Yongxing has a long history of growth. The germplasm resources of Yongxing wampee are abundant and unique, and they have cultivated some of the world's unique local varieties of wampee, such as Yongxing seedless wampee, Yongxing big chicken heart wampee, Yongxing small chicken heart wampee, and so on. In order to systematically clarify the differences between the three tissues (peel, pulp, and seed) of the wampee fruit, this study adopted targeted and non-targeted metabolomics methods. to detect and comprehensively analyze various categories of metabolites from 17 varieties collected and established a metabonomics database. Combined with quantitative data and spatial metabolomics detection, the differences of metabolites and metabolic networks specifically accumulated in three tissues were analyzed. In addition, key biomarkers were screened for variety differentiation. This study provided a theoretical basis for fine processing and metabolome-assisted breeding of wampee.

2. Materials and Methods

2.1. Plant Materials and Sample Preparation

Samples were collected from Yongxing Town in Haikou, one of the original habitats of the wampee fruit in the world. During the ripening period of the wampee fruit, plants with consistent phenotypes were selected for sampling. Each sample was a mixture of fruits from three different plants, which was then combined to serve as a sample for one biological replicate. Each material was set with three independent biological replicates. Detailed information on sample collection and photos of the fruits are displayed in Supplemental Table S1 and Figure S1. The ripe fruits were washed with sterile deionized water, and the peel was cut open with a sterilized scalpel. The peel and pulp were quickly separated and immediately placed in liquid nitrogen, followed by storage at −80 °C, and then subjected to vacuum freeze-drying. After freeze-drying, the pulp and seed were further separated.

The samples were ground using a grinding apparatus (MM 400; Retsch, Haan, Germany) with zirconia beads at 30 Hz for 1 min. One hundred milligrams of the powder was weighed and extracted with 0.8 mL of 70% methanol aqueous solution (methanol: H_2O, 7:3, v/v) containing 10 ppb lidocaine, and was incubated overnight at 4 °C. After centrifugation at 12,000× g for 10 min, the supernatant was filtered (SCAA-104, pore size 0.22 μm; ANPEL, Shanghai, China, (http://www.anpel.com.cn/ (accessed on 25 June 2023)) and then analyzed by LC–MS [19]. To investigate the inter-tissue differences in metabolites, samples from the same tissue of different varieties were mixed for untargeted metabolomics analysis. The samples were equally mixed into multiple quality control samples to check the stability of the instrument.

2.2. Reagents and Standards

HPLC-grade acetonitrile, acetic acid, and methanol were purchased from Unity™ Lab Services (Thermo Fisher Scientific, Waltham, MA, USA); water was purified using a Thermo Scientific LabTower EDI 15 purification system (Thermo Fisher Scientific, Waltham, MA, USA). Lidocaine, used as an internal standard in this study, was obtained from Tokyo Chemical Industry Co., Ltd., Tokyo, Japan (https://www.tokyochemical.lookchem.com/ (accessed on 15 October 2022)). Carboxymethyl cellulose (CMC) and N-(1-naphthyl) ethylenediamine dihydrochloride (NEDC) were purchased from Shanghai Macklin Biochemical Technology Co., Ltd., Shanghai, China (https://www.macklin.cn/ (accessed on 7 April 2024)), and 2,5-Dihydroxybenzoic acid (DHB) was purchased from Shanghai Yuanye Bio-Technology Co., Ltd., Shanghai, China (https://www.shyuanye.com/ (accessed on 7 April 2024)).

The imperatorin, myricitrin, and scopoletin used in this study were provided by ChemFaces, Wuhan, China (http://www.chemfaces.com/ (accessed on 18 September 2023)). Catechin, Gallocatechin, and LysoPC (18:1) were supplied by Shanghai Yuanye Bio-Technology Co., Ltd., Shanghai, China (https://www.shyuanye.com/ (accessed on 18 September 2023)). All standards were prepared by dissolving in methanol to make the stock solutions, which were then diluted to the appropriate concentration using a 70% methanol aqueous solution (methanol: H_2O, 7:3, v/v) for detection on the LC–MS. All standard stock solutions were stored at −80 °C in the dark.

2.3. LC–MS/MS Analysis of Metabolites

Samples were analyzed using a targeted method combined with a non-targeted metabolomics analysis method based on ultra-performance liquid chromatography-tandem mass spectrometry (UPLC–MS/MS) [17,19]. Non-targeted analysis was performed using an HPLC-ESI-Q-Exactive-MS/MS (Thermo Scientific Q-Exactive, Thermo Fisher Scientific, Waltham, MA, USA). Mass spectrometric detection used a HESI ion source for non-targeted metabolites profiling in full MS and ddMS2 mode to obtain the data, including the accurate masses, MS/MS fragments, and retention times. The recording conditions were as follows: source capillary, 3.2 kV; sheath gas flow rate, 7 psi; source temperature, 400 °C; scan range, m/z 100–1000. Targeted analysis of metabolites was conducted in multiple reaction monitoring (MRM) mode using an LC-ESI-Q TRAP-MS/MS (AB Sciex 6500, Applied Biosystems, Woburn, MA, USA). The ESI source operation parameters were as follows: temperature of 500; GSI, GSII, and CUR of 50, 60, and 35 psi, respectively; and IS of 5500 V in positive mode or −4500 V in negative mode; the collision gas was high. The total sMRM cycle time was set to 1.0 s, whereas the dwell time for each MRM transition was automatically adjusted according to the total cycle time, ensuring that at least 10 points were obtained for each peak. Data acquisition and analysis were carried out using SCIEX OS 3.1.5.3945 software (AB Sciex, Framingham, MA, USA). The quantification of metabolites was performed by calculating the peak area and comparing it to the standard curve drawn by the lidocaine standard.

2.4. MALDI-MSI Analysis of Metabolites

Wampee fruit was embedded in 2% carboxymethyl cellulose (CMC) and rapidly frozen. Subsequently, the frozen tissue was sectioned into 25-micrometer-thick slices at −20 °C using a Leica CM1950 cryostat (Leica Microsystems GmbH, Wetzlar, Germany). The sections were mounted onto indium tin oxide (ITO)-coated conductive glass slides (Bruker Daltonics, Bremen, Germany). Matrix solutions of 15 mg/mL DHB (2,5-dihydroxybenzoic acid, positive model) and 7 mg/mL NEDC (N-(1-naphthyl)ethylenediamine dihydrochloride, negative model) were applied using the HTX TM-Sprayer (HTX Technologie, CA, USA), following the standard protocol provided by the manufacturer for matrix solution preparation and application. MSI experiments were conducted using a timsTOF fleX MALDI-TOF/TOF (Bruker Daltonics, Bremen, Germany). The spatial resolution was set to 50 μm, with a mass range of m/z 20–1300. Data acquisition was performed using timsControl 3.1 and flexImaging 7.0 (Bruker Daltonics, Bremen, Germany), and data visualization was carried out with the SCiLS Lab 2023b 11.01.14623 software.

2.5. Classification and Identification of Metabolites

For non-targeted data, the Thermo Compound Discoverer 3.3 software was used for data extraction. By extracting molecular ions and fragment ions, an MS2 spectral tag (MS2T) library was created, which was then annotated based on the accurate m/z value, retention time (RT), and fragmentation pattern. Following the approach pioneered in plants by Morreel et al. [26], these features were used to screen the data in the literature and database. The annotated metabolites were further identified with the help of available standards. For targeted data, SCIEX OS 3.1.5.3945 software (AB Sciex, Framingham, MA, USA) was used to align peaks by retention time (RT), and the quantification of metabolites was completed by calculating the peak area.

2.6. Statistical Data Analysis

For metabolomics data analysis, principal component analysis (PCA) was conducted using the OmicStudio tools (www.omicstudio.cn/tool (accessed on 27 June 2024)). Hierarchical cluster analysis and sample repeatability assessment were performed using the R software (http://www.r-project.org/ (accessed on 13 July 2024)). Data were log2-transformed to improve normalization, followed by sample repeatability assessment. Z score normalization was applied for data normalized, which was then followed by hierarchical cluster analysis and PCA. For the comparison of absolute metabolite content across the three tissues, two-tailed Student's t-tests were used, and $p \leq 0.05$ was considered significant. Orthogonal Projection to Latent Structures Discriminant Analysis (OPLS-DA) was conducted using the Metware Cloud, a free online platform for data analysis (https://cloud.metware.cn/#/home (accessed on 27 June 2024)). The data used for computation and analysis in this study were the average values of three biological replicates.

3. Results

3.1. Non-Targeted Metabolomics Analysis in Wampee Fruit

In order to comprehensively evaluate and characterize the differences in metabolic profiles of wampee fruit, we collected 17 wampee fruit materials from different natural villages in Yongxing Town, Haikou, and performed metabolomic analysis based on non-targeted HPLC-TOF-MS.

To gain a comprehensive understanding of the changes in metabolites between different tissues and varieties of wampee fruit, we tested and analyzed tissue mixtures of different varieties, detecting nearly 30,000 mass spectrometry signals in three mixed tissues. The metabolic profile signals in the total ion current chart showed significant differences among the three tissues: peel, pulp, and seed (Figure 1A), especially between 7 and 10 min, which may be due to some lipophilic metabolites. Additionally, principal component analysis (PCA) was conducted on the three tissues based on the metabolic signals obtained from the non-targeted analysis. The PCA showed that component 1 and component 2 accounted for 36.25% and 28.94% of the variation, respectively (Figure 1B). Components 1 and 2 successfully separated the three tissues, indicating significant differences in metabolite levels among them. Furthermore, the non-targeted metabolomics analysis of different tissues helps to discover new metabolites that may not exist in traditional single tissues, suggesting that this strategy can greatly facilitate the establishment of a comprehensive metabolite database.

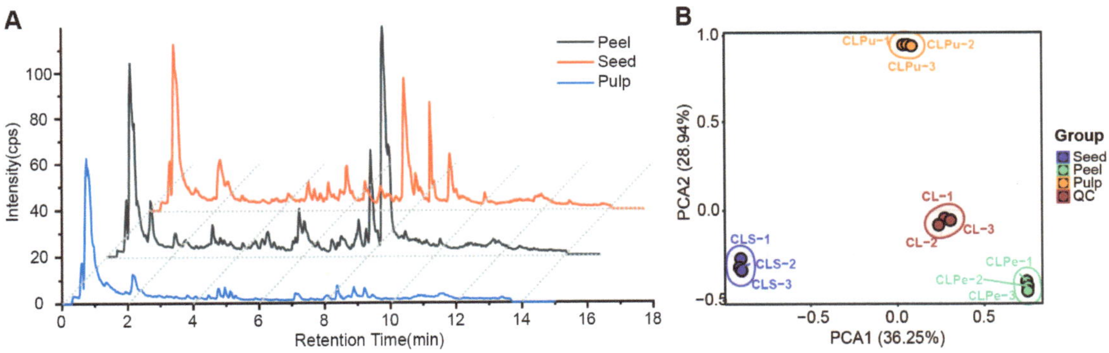

Figure 1. Analysis of metabolic variation in three tissues using ESI − Q − Exactive LC − MS/MS. (**A**) Total ion chromatography of metabolites in three tissues of the wampee fruit. (**B**) Principal component analysis (PCA) of total ion chromatography results for three tissues of the wampee fruit.

3.2. Metabolite Identification/Annotation Based on Non-Targeted Metabolic Analysis

In this study, to gain a deeper understanding of metabolites in different tissues, we used Thermo Compound Discoverer 3.3 software for data extraction from non-targeted metabolomic profiling data. After filtering for signal-to-noise ratio and deduplication, there were 12,000 metabolic signals. An MS2 spectral tag (MS2T) library was established using the obtained metabolic signals, and the MS2T spectral tags were annotated based on accurate *m*/*z* values, retention time (RT), and fragmentation patterns. These characteristics were used to screen in literature and databases and further confirmed using commercially available standards. Additionally, by applying the rules of compound fragmentation, similar compounds were inferred and identified.

Lysophosphatidylcholine (LPC) and phosphatidylcholine (PC) are important subclasses of glycerophospholipids. LPC and PC produce characteristic phosphate choline product ions at *m/z* 184.1 and then lose $N(CH_3)_3^+$ and H_2O to form fragments at *m/z* 125.0 and 86.1, respectively. For instance, the compound CLP0940 was detected at a retention time (RT) of 9.86 min (Figure 2A). CLP0940 generated a precursor ion $[M+H]^+$ at *m/z* 522.3612, and the mass spectrum showed a characteristic fragment Y_0^+ ion at *m/z*

184.0801 for CLP0940, along with fragments of *m/z* 124.9998 and 86.0960, which indicated the presence of a phosphate choline skeleton. The Z_0^+ ion at *m/z* 339.2894 $[M+H-183]^+$ corresponded to the losses of a phosphate choline moiety. These observations allowed us to characterize CLP0940 as LysoPC (18:1) (Figure 2B). By comparing the secondary mass spectrum of CLP0940 with that of the LysoPC (18:1) standard, our annotation was ultimately confirmed to be correct, with the structure and main fragmentation pathways of LysoPC (18:1) shown in Figure 2C. Based on similar fragmentation patterns, we annotated 10 LPCs and found that LPCs are more abundantly accumulated in the pulp.

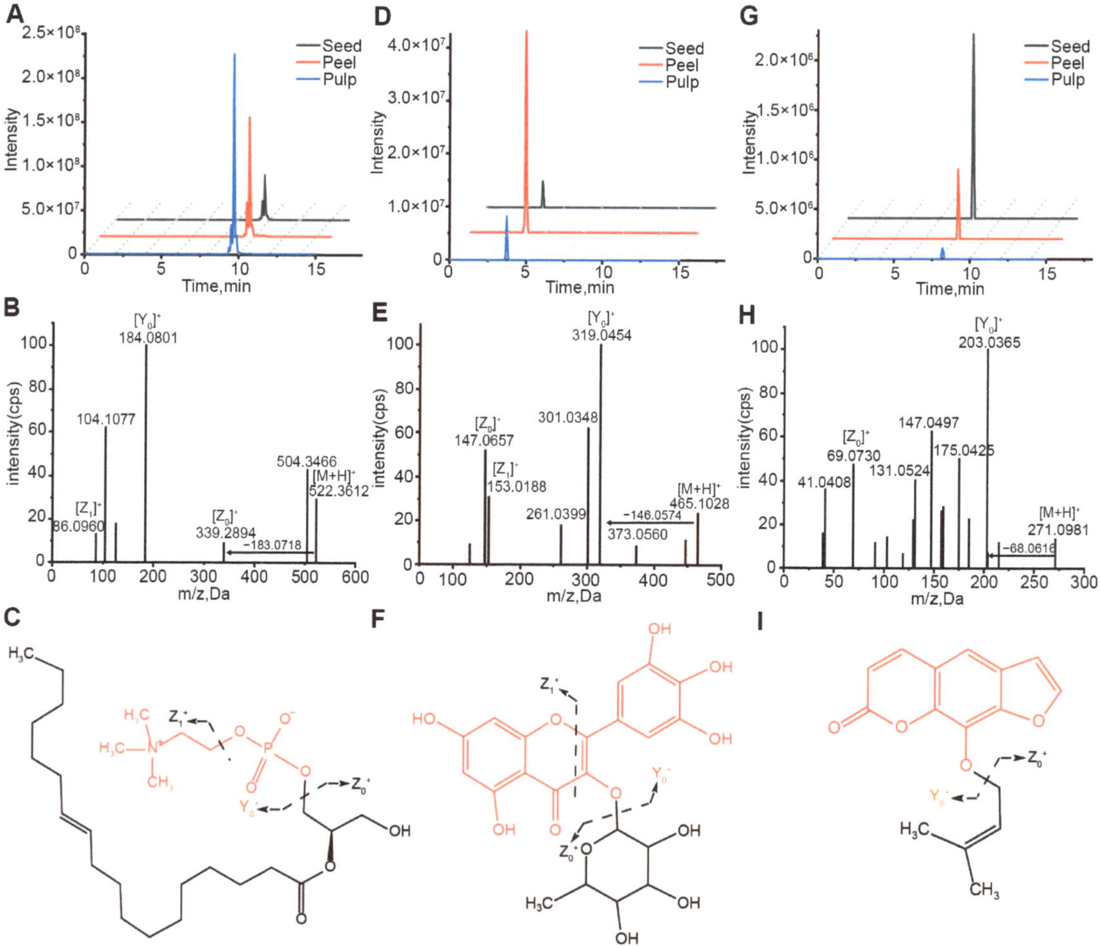

Figure 2. Detection and identification of specific LysoPC (18:1), myricitrin, and imperatorin metabolite signals by Q − Exactive LC − MS/MS. (**A**) EIC (extracted ion chromatogram) of CLP0940 at 9.86 min. (**B**) MS/MS spectra of CLP0940 at *m/z* 522.3612, and the metabolite was identified as LysoPC (18:1). (**C**) The molecular structure of the LysoPC (18:1) and its general fragmentation rules. (**D**) EIC (extracted ion chromatogram) of CLP0905 at 3.66 min. (**E**) MS/MS spectra of CLP0905 at *m/z* 465.1028, and the metabolite was identified as myricitrin. (**F**) The molecular structure of the myricitrin and its general fragmentation rules. (**G**) EIC (extracted ion chromatogram) of CLP0658 at 8.21 min. (**H**) MS/MS spectra of CLP0658 at *m/z* 271.0981, and the metabolite was identified as imperatorin. (**I**) The molecular structure of the imperatorin and its general fragmentation rules.

Flavonoid compounds are a class of secondary metabolites widely distributed in plants, including several subclasses such as anthocyanins, flavonols, flavones, flavanols, flavanones, catechins, etc. Glycosylation is a common modification of flavonoids that produces characteristic fragments of flavonoid aglycone ions, characteristic product ions of flavonoid aglycones, and modified ion fragments. For example, a compound CLP0905 (Figure 2D) was detected at a retention time (RT) of 3.66 min, where CLP0905 generated a precursor ion $[M+H]^+$ at *m/z* 465.1028. The mass spectrum showed that CLP0905 has a characteristic fragment Y_0^+ ion at *m/z* 319.0454 and characteristic fragments at *m/z* 153.0188 and 261.0399, indicating that its flavonoid aglycone is myricetin. The Z_0^+ ion at *m/z* 147.0657 $[M+H-318]^+$ corresponds to the loss of the rhamnoside modification. These observations allowed us to characterize CLP0905 as myricetin 3-O-rhamnoside, which is myricitrin (Figure 2E). By comparing the MS/MS spectrum of CLP0905 with that of the myricitrin standard, our annotation was ultimately confirmed to be correct, and the structure and main fragmentation pathways of myricitrin are shown in Figure 2F. Based on similar fragmentation patterns, we annotated 14 flavonoid glycosides and found that they accumulate more in the fruit peel. Catechins are an important subclass of flavonoids. Catechin subclasses produce a fragment Y_0^+ ion at *m/z* 139.04. Based on this, by comparing with the standard, we identified catechin and (+)-gallocatechin (GC), and their extracted ion current maps, MS/MS spectra, and fragmentation patterns are shown in Supplemental Figure S2A–F, and it was found that they also accumulate more in the fruit peel.

Coumarin compounds are a class of metabolites with broad pharmacological activities in the Clausena genus, which are categorized into different types based on their structure, such as simple coumarins, furanocoumarins, and pyranocoumarins. Furanocoumarins produce a characteristic product ion at *m/z* 203.04. For instance, the compound CLP0658 was detected at a retention time (RT) of 8.21 min (Figure 2G). CLP0658 generated a precursor ion $[M+H]^+$ at *m/z* 271.0981, and the mass spectrum showed a characteristic fragment Y_0^+ ion for CLP0658 at *m/z* 203.0365, along with a Z_0^+ ion at *m/z* 69.0730 $[M+H-203]^+$ corresponding to the loss of the furanocoumarin substituent. These observations allowed us to characterize CLP0658 as imperatorin (Figure 2H). By comparing the secondary mass spectrum of CLP0658 with that of the imperatorin standard, the annotation was ultimately confirmed to be correct, with the structure and main fragmentation pathways of Imperatorin shown in Figure 2I. Based on similar fragmentation patterns, we annotated six furanocoumarins and found that they are more abundantly accumulated in the seeds and peels of some varieties. Simple coumarins produce a characteristic product ion at *m/z* 163.04. For example, the compound CLP0720 was detected at an RT of 3.81 min (Supplemental Figure S2G). CLP0720 generated a precursor ion $[M+H]^+$ at *m/z* 193.0528, and the mass spectrum showed a characteristic fragment Y_0^+ ion for CLP0720 at *m/z* 178.0266, along with the characteristic coumarin product ion Z_0^+ at m/z 163.04. These observations led us to infer that it might be a coumarin metabolite but not a typical monoterpene substitution or isoprenyl substitution; instead, it is a simpler form of substitution. By comparing with the standard, we confirmed this metabolite as scopoletin (Supplemental Figure S2H), with its structure and main fragmentation pathways shown in Supplemental Figure S2I.

By summarizing and annotating other types of metabolites in this manner, we have identified a total of 1286 specific metabolites, mainly consisting of flavonoids, lipids, benzene and its derivatives, and organic acids, with their classification and proportion as shown in Figure 3A, that is, 14.2% flavonoids, 15.4% lipids, 15.3% benzene and its derivatives, 8.8% organic acids, etc. This study is the first to quantify a large number of lipids, phenolic amides, coumarins, and terpenoids in multiple tissues of wampee fruit, such as some glycerophospholipids, dCMP, myricitrin, sudachiin C, lansamide 4, and other compounds that have been identified for the first time in wampee fruit.

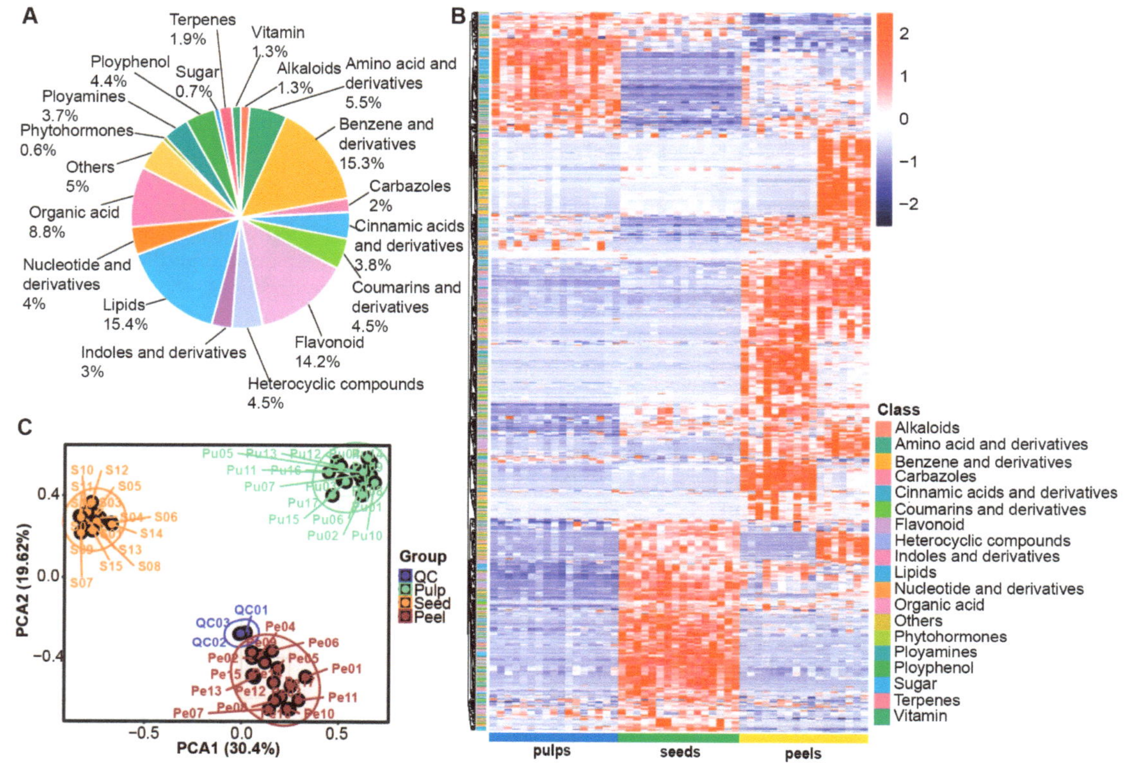

Figure 3. Summary of metabolic profiling and tissue variability analysis. (**A**) Categorical pie chart of the 1286 annotated metabolites. (**B**) Hierarchically clustered heatmap of the 1286 annotated metabolites from 50 wampee samples. (**C**) PCA results for the metabolome data from 50 wampee samples.

3.3. *Targeted Metabolomics Analysis in Wampee Fruit*

To analyze the accumulation patterns of metabolites in different varieties and tissues of wampee fruit, we conducted targeted quantitative detection of 1286 metabolites in three tissues of 17 varieties of wampee fruits using LC–MS. First, a correlation analysis was performed on the results, which revealed good repeatability among the three replicates, a high correlation within the tissue, and reliable metabolomics detection results, indicating that the data are suitable for further analysis (Supplemental Figure S3A). Subsequently, we performed PCA and found that samples from the same tissue clustered together, indicating that tissue specificity is greater than variety specificity (Figure 3C). Next, we conducted hierarchical cluster analysis on the content of 1286 metabolites in 50 different samples. Hierarchical cluster analysis groups data by constructing a multi-level nested classification system, which allows different compounds to be clustered together due to their similar accumulation patterns in different varieties and tissues. A heatmap can more clearly and intuitively show the differential accumulation of substances. The results showed that there is a significant tissue difference in metabolite accumulation, with higher lipid accumulation in the pulp, higher accumulation of flavonoids in the peel, and a rich accumulation of various metabolites in the seed, such as flavonoids, coumarins, and carbazole alkaloids. In addition, there are considerable differences in metabolites between different varieties in the peel, such as lipids, coumarins, and terpenes (Figure 3B).

To compare the accumulation patterns of metabolites in each tissue, we generated a Venn diagram by screening differential metabolites, which shows the intersection and differential accumulation of metabolites between different tissues (Supplemental Figure S3B).

Consistent with the hierarchical clustering results, the number of metabolites in the peel is the highest. Furthermore, compared to the pulp, the peel is richer in flavonoids, carbazole alkaloids, quinic acid and its derivatives, and indole compounds (Supplemental Figure S4A). Compared to the seed, the peel is richer in glycerophospholipids, quinic acid and its derivatives, and indole compounds (Supplemental Figure S4B). When comparing the pulp to the seed, the pulp is richer in glycerophospholipids, while the seed is richer in anthocyanins, flavonoids, and terpenes (Supplemental Figure S4C). Some information on differential metabolites can be found in Supplemental Table S2.

3.4. Tissue and Variety Analysis of Wampee Fruit Metabolites Based on LC–MS

3.4.1. Accumulation Patterns of Metabolites in the Pulps of Wampee

Based on the hierarchical clustering analysis results of the 1286 metabolites, we found that the pulp has the richest accumulation of lipids. In this study, lipids were categorized into subclasses such as fatty acids, fatty acyls, glycerolipids, lysophosphatidylcholine (LPC), lysophosphatidylethanolamine (LPE), phosphatidylcholine (PC), phosphatidylethanolamine (PE), sphingolipids, and steroids. Therefore, we conducted hierarchical cluster analysis on the 198 detected lipid subclass metabolites and found that the accumulation of glycerophospholipids in the pulp is relatively abundant (Figure 4A). At the same time, we found that some lipids have differential accumulation in the peel among different varieties, such as penidienone, 2,4,6-octatriyn-1-ol, pyreudione B, and diatretyne 3. We further conducted hierarchical cluster analysis on the glycerophospholipids and found that LPC, LPE, and PE mainly accumulate in the pulp, while PC mainly accumulates in the peel (Supplemental Figure S5A). Further research on the absolute content of the four subclasses of glycerophospholipids showed that LPC, LPE, and PE have significantly higher accumulation levels in the pulp, followed by the peel (Figures 4B,C and S5C). In contrast, the accumulation of PC in the pulp is lower than in the peel (Supplemental Figure S5B). LPC and LPE have emulsifying effects, which can promote the digestion and absorption of lipids in the human body and improve the efficiency of nutritional utilization [27], which is consistent with the rich nutritional value of the pulp. In addition to this, the pulp is also rich in organic acids, such as citric acid, and amino acids, such as 5-oxoproline and gamma-glutamyl glutamine (Figure 4D).

3.4.2. Accumulation Patterns of Metabolites in the Peels of Wampee

Based on the hierarchical clustering analysis results of the 1286 metabolites, we found that the accumulation of flavonoids in the peel is the most significant. Therefore, we conducted hierarchical cluster analysis on the 182 detected flavonoid metabolites and found that flavonoids are most abundantly accumulated in the peel, followed by the seed (Figure 5A). To explore the specific accumulation of each subclass within flavonoids, we performed hierarchical cluster analysis on the main subclasses of flavonoids and found that the number of metabolites such as anthocyanins, flavones, and flavonols that are specifically accumulated in the peel and seed is considerable, but the accumulation patterns differ (Supplemental Figure S6A–C). At the same time, most of the catechin, flavonoid glycosides, and flavanone metabolites accumulate in the peel, with a smaller portion accumulating in the seed (Supplemental Figure S6D–F). Additionally, based on the absolute content, we screened metabolites that are specifically accumulated in the peel with a content greater than 3 micrograms per milliliter and have a difference of more than 10 times compared to the seed (Figure 5B), such as catechin, L-epicatechin, (+)-gallocatechin (GC), epigallocatechin (EGC), and other catechin metabolites, as well as myricetin, sudachiin C, myricetin 3-neohesperidoside, and other flavonoid glycosides, which play an important role in health maintenance and promotion [28], including antioxidant [29], regulation of blood lipid metabolism, anti-inflammatory [30], and prevention of cardiovascular and cerebrovascular diseases [31]. At the same time, we screened metabolites with a specific accumulation in the seed based on their absolute content, with a content greater than 1 microgram per milliliter and a difference of more than 2 times compared to the peel (Supplemental

Figure S8), such as cynaroside, kaempferol, nicotiflorin, luteolin O-malonylhexoside, and other flavonoids and their derivatives, which have important roles in antioxidant and anti-inflammatory activities. Furthermore, metabolites that specifically accumulate in the peel include quinic acid and its derivatives, polyphenols, and terpenes (Figure 5C). For instance, parthenolide is a sesquiterpene lactone [32] that possesses functions such as anti-cancer [33], anti-inflammatory, and antioxidant activities [34]; clitocybin A has the functions of scavenging reactive oxygen species and anti-wrinkle [35]; picrotoxinin has neurotoxic effects and can be used as an insecticide, which may be related to the peel's antioxidant and pest-resistant properties. Interestingly, some polyphenols and terpenes exhibit differential accumulation among different varieties of the peel (Supplemental Figure S7A–C), such as picrotoxin, cnidioside A, lansamide 4, aspergentisyl A, and quizalofop-methyl.

3.4.3. Accumulation Patterns of Metabolites in the Seeds of Wampee

Based on the hierarchical clustering analysis results of the 1286 metabolites, we conducted cluster analysis on the detected coumarins, carbazole alkaloids, and the widely reported clausenamide, which is known for its therapeutic effects on Alzheimer's disease (Figure 6A–C). We found that these categories have certain specific accumulations in the seed and that there are significant differences in the accumulation of these metabolites among different varieties of the peel, which will provide an important reference for finding wampee fruit varieties with high medicinal value. To explore the accumulation of these metabolites in the seed, we conducted statistical analysis by calculating their absolute content (Figure 6D–F). We found that the most specifically accumulated coumarin in the seed is scopoletin, which has been found in a variety of medicinal and edible plants and plays an important role in the treatment of various diseases [36]. Coumarins such as phellopletrin, marmesin, imperatorin, and heraclenol, which are accumulated in larger amounts, have significant effects in anti-inflammatory and anti-tumor aspects [37–39]. Carbazole alkaloids with medicinal values such as anti-tumor, anti-inflammatory, and antioxidant effects accumulate most abundantly in the seed. For instance, heptaphylline, which can reach up to 6 μg/g in the seed, has been specifically reported to have effects in anti-malarial treatments [40]. Other carbazole alkaloids like mukonine, murrayanine, murrayamine A, and bicyclomahanimbine, although not explicitly reported, show potential for medicinal value. Among the clausenamides, (−)clausenamide has been demonstrated to improve cognitive deficits [41] and holds promise as a candidate drug for the treatment of Alzheimer's disease and other neurodegenerative diseases [42]. The study of the accumulation pattern of important metabolites in the seed is of significant importance for the comprehensive characterization of medicinally valuable metabolites in wampee fruit.

3.5. Spatial Distribution Visualization of Metabolites in Wampee Fruit Based on MALDI-MSI

To explore the actual spatial distribution of differential metabolites in the three tissues of wampee fruit, we conducted precise mass measurement and mass fragment analysis of the compounds through MALDI-MSI to identify some differential metabolites in the three tissues of the wampee fruit, such as organic acids, alkaloids, coumarins, and lipids. The information on identified and some unidentified metabolites can be found in Supplemental Table S3.

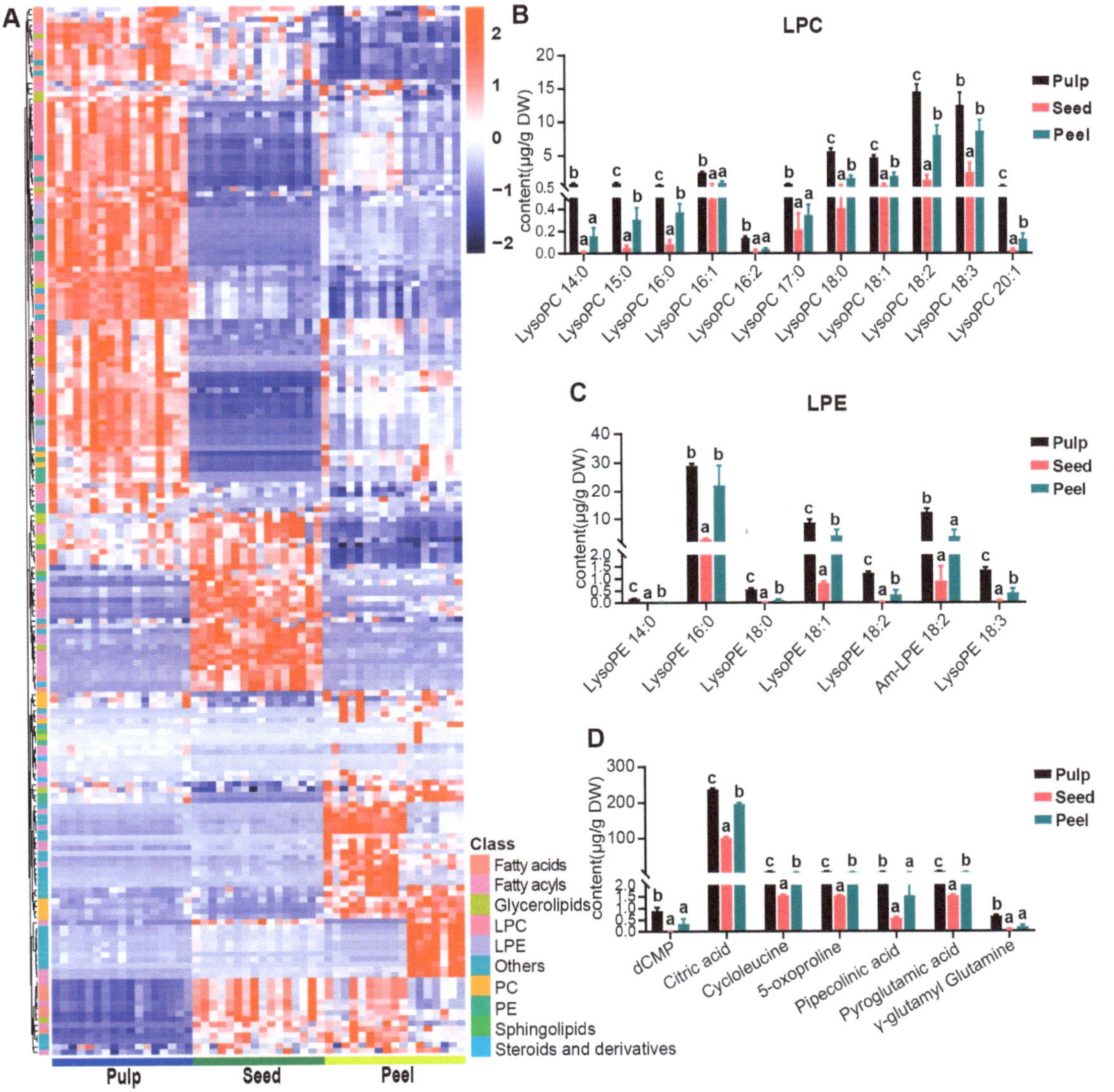

Figure 4. Distribution of lipids in different wampee fruit tissues. (**A**) Hierarchically clustered heatmap of the 198 lipid metabolites from 50 wampee samples. (**B**,**C**) Bar plots showing the contents of LPC and LPE content of various wampee fruit tissues, respectively. (**D**) Bar plots showing the content of substances that accumulate more in the pulps of wampee compared to other tissues, such as organic acids and amino acids. When using the abc notation to indicate significant differences, the same letter indicates no significant difference between groups ($p > 0.05$), and different letters indicate significant differences ($p < 0.05$).

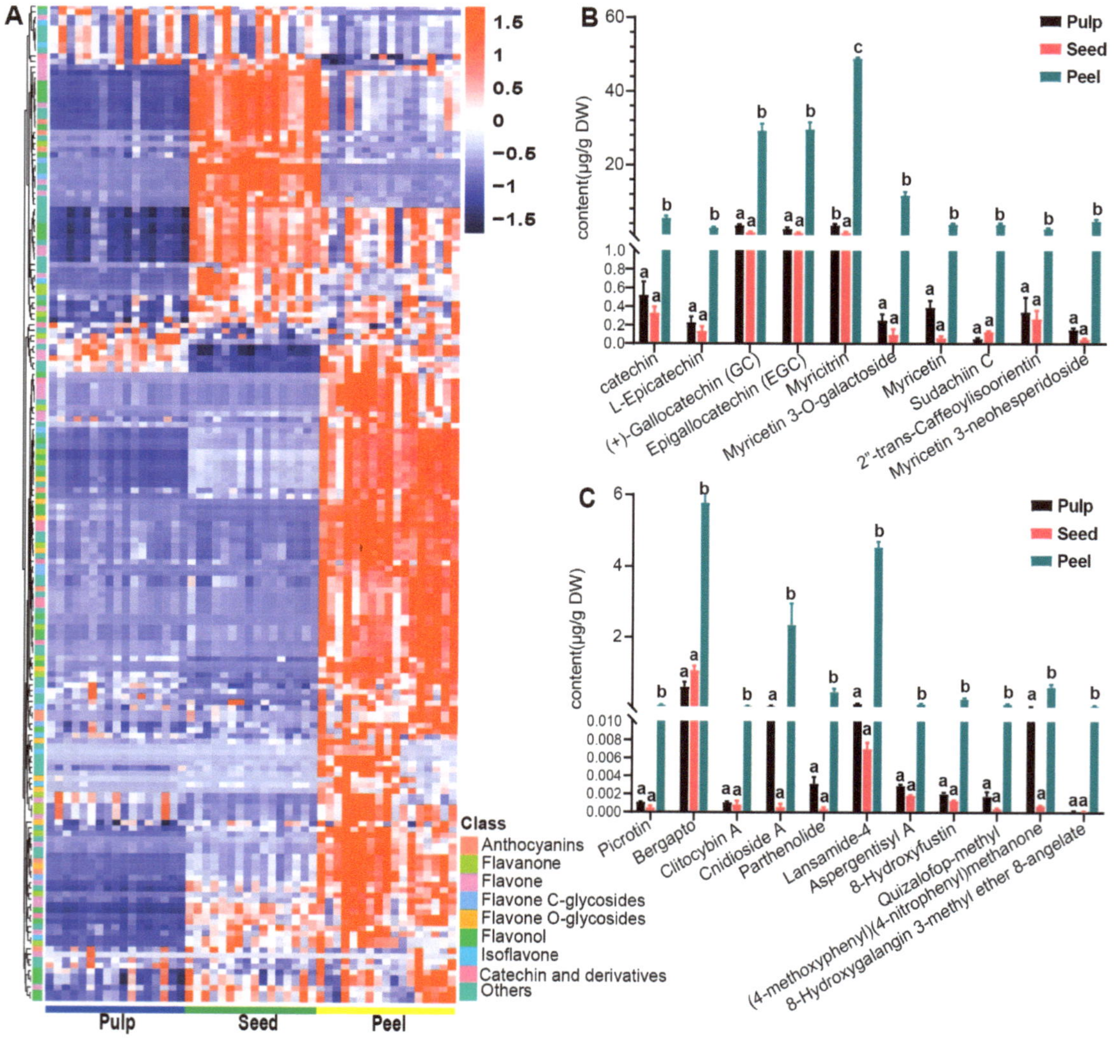

Figure 5. Distribution of flavonoid in different wampee fruit tissues. (**A**) Hierarchically clustered heatmap of the 182 flavonoid metabolites from 50 wampee samples. (**B**) Bar plots showing the flavonoids that are more abundant in the peel of wampee. (**C**) Bar plots showing the content of substances that accumulate more in the peels of wampee compared to other tissues, such as terpenes and polyphenol. When using the abc notation to indicate significant differences, the same letter indicates no significant difference between groups ($p > 0.05$), and different letters indicate significant differences ($p < 0.05$).

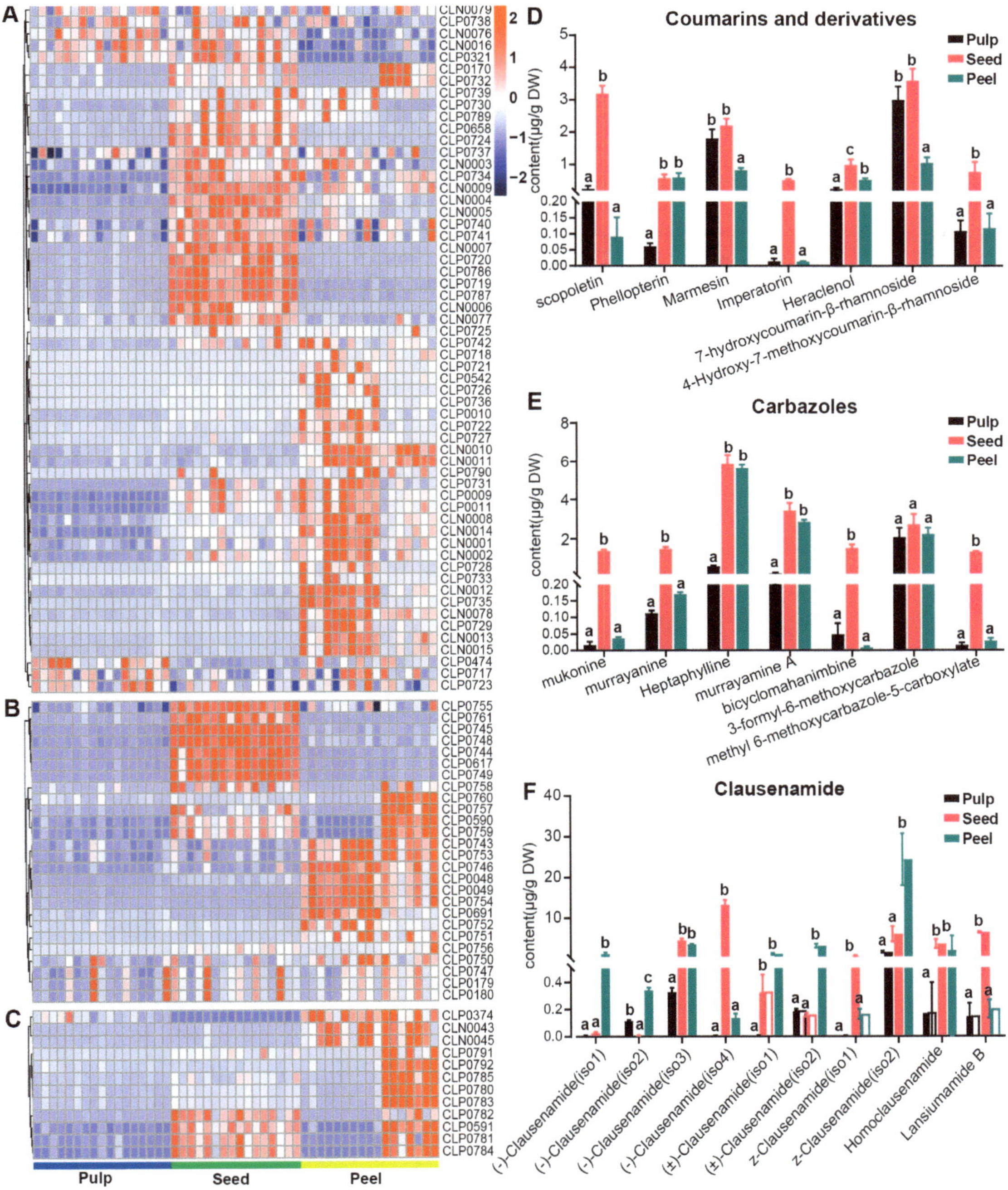

Figure 6. Distribution of coumarins, carbazoles, and clausenamide in different wampee fruit tissues. (**A–C**) Hierarchically clustered heatmap of the coumarins, carbazoles, and clausenamide metabolites from 50 wampee samples. (**D–F**) Bar plots showing the coumarins, carbazoles, and clausenamide content of various wampee fruit tissues, respectively. When using the abc notation to indicate significant differences, the same letter indicates no significant difference between groups ($p > 0.05$), and different letters indicate significant differences ($p < 0.05$).

In the pulp, we identified citric acid and LysoPC 18:3, which, from a spatial distribution perspective, were found to accumulate mainly in the pulp (Figure 7), consistent with our LC–MS results (Figure 4B,D), and closely related to the nutritional value and taste of the pulp. In the peel, we identified bergaptol, which, from a spatial distribution perspective, accumulates primarily in the peel, seed coat, and placental tissue, with a higher accumulation in the peel (Figure 7), which is basically consistent with our LC–MS results (Figure 5C), but its distribution pattern is more refined and intuitive in the MALDI-MSI results. Bergaptol, as a coumarin, plays an important role in anti-inflammatory, antioxidant, and anti-tumor activities [43]. In the seed, we identified coumarin and two alkaloids; marmesin mainly accumulates in the seed coat and placental tissue (Figure 7), which further enriches our LC–MS results (Figure 6D), and marmesin has an important role in regulating angiogenesis and anti-tumor activities [44,45]. Lansiumamide B and homoclausenamide are alkaloids isolated from wampee; lansiumamide B mainly accumulates in the seed, while homoclausenamide accumulates in the seed, peel, and pulp but is more abundant in the seed (Figure 7), consistent with our LC–MS results (Figure 6F), and lansiumamide B has antifungal and anti-obesity effects [46,47]. There are also some unidentified substances whose fragment information and spatial distribution are shown in Supplemental Table S3 and Supplemental Figure S9, awaiting further identification.

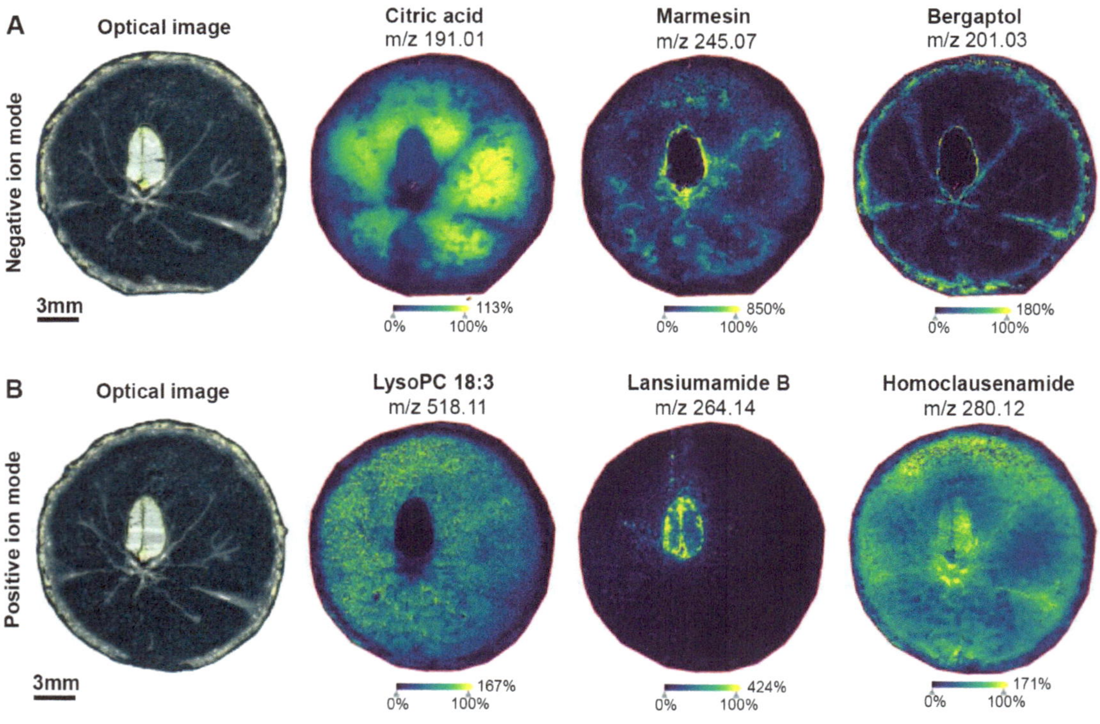

Figure 7. The optical image of wampee fruit slices and the spatial distribution heatmaps of various metabolites in MALDI-MSI. (**A**) The spatial distribution heatmaps of citric acid, marmesin, and bergaptol in negative ion mode. (**B**) The spatial distribution heatmaps of LysoPC 18:3, lansiumamide B, and homoclausenamide in positive ion mode. The distributions are displayed as heat maps, with the color code ranging from blue (low) to yellow (high). Images were exported from the SCiLS Lab software.

3.6. Identify Different Varieties of Biomarker Probes

Through cluster analysis, it was found that there are significant differences in the accumulation of some metabolites in different varieties of the peel, including lipids, polyphenols, terpenes, coumarins, and carbazole alkaloids (Supplemental Figure S10). To further explore the biomarkers of variety differences in wampee fruit, we conducted orthogonal projections to latent structure discriminant analysis (OPLS-DA) [48]. The OPLS-DA score plot of the scatter diagram shows significant differences between the two groups (Figure 8A). The permutation plot of OPLS-DA shows that the R2Y and Q2 values are similar and close to 1, indicating that the model has strong explanatory and predictive power for the data (Figure 8B). Therefore, the OPLS-DA model used in this study is robust and repeatable.

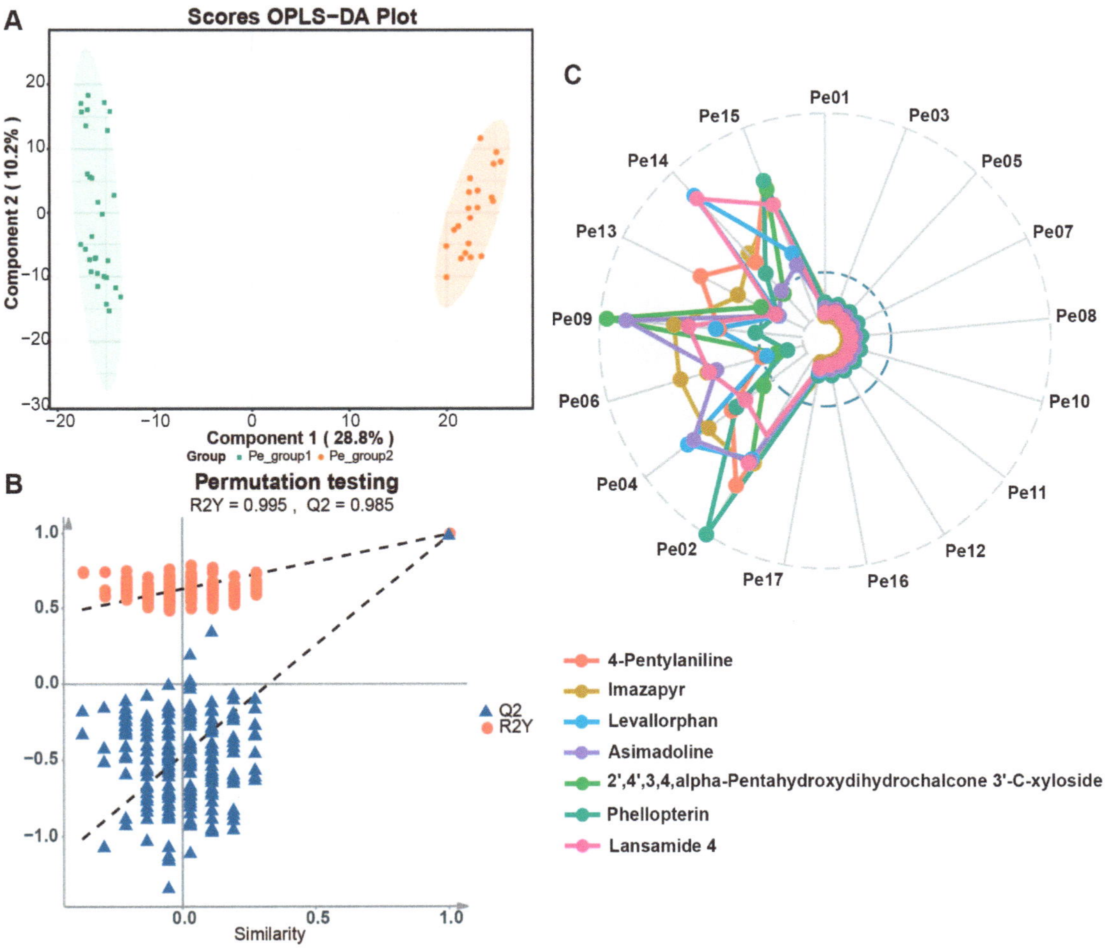

Figure 8. Identify different varieties of biomarker probes. (**A**) Orthogonal projections to latent structures discriminant analysis (OPLS-DA) of different wampee fruit tissues. (**B**) Permutations plot of the OPLS-DA model for the binary grouping. (**C**) Radar chart of six selected biomarkers from different categories.

The results of multivariate statistical analysis indicate that the peel contributes significantly to distinguishing different varieties of wampee fruit. Through variance analysis, we screened potential biomarkers with significant changes, with p values ≤ 0.05,

VIP values ≥ 1, and FC values > 50 or FC values < 0.02. A total of 59 potential biomarkers for different varieties of wampee fruit were identified (Supplemental Table S4). The selected differential metabolites include 27 benzene and its derivatives, 1 cinnamic acid and its derivative, 4 heterocyclic compounds, 1 indole and its derivative, 2 alkaloids, 3 coumarins, 4 lipids, 1 nucleotide and its derivative, 6 organic acids, 3 polyamines, 3 polyphenols, and 4 other metabolites. At the same time, by applying further stringent selection criteria (fold change (FC) value > 200 or FC value < 0.005), we identified 7 biomarkers (4-pentylaniline, imazapyr, levallorphan, asimadoline, 2′,4′,3,4,alpha-pentahydroxydihydrochalcone 3′-C-xyloside, phellopterin, lansamide 4) and created a radar chart (Figure 8C), which clearly divided the wampee varieties into two groups. Pe01, Pe03, Pe05, Pe07, Pe08, Pe10, Pe11, Pe12, Pe16, and Pe17 formed one group, while Pe02, Pe04, Pe06, Pe09, Pe13, Pe14, and Pe15 formed another group. The metabolites in the peels of these two groups of varieties exhibit different accumulation patterns in categories of compounds with potential for medicinal value development, such as polyphenols, terpenes, coumarins, carbazole alkaloids, and benzene and its derivatives. This is of significant reference value for exploring the medicinal value of wampee metabolites and for variety breeding.

4. Discussion

Wampee is native to the southern regions of China and has a long history of consumption and medicinal use in the country. The fruit of the wampee is rich in sugars, organic acids, lipids, polyphenols, flavonoids, and other compounds, providing a wealth of nutrients for the human body and possessing health benefits and medicinal value [9,49,50]. The nutritional and medicinal components and their content differences in the three tissues of the wampee fruit—peel, pulp, and seed—directly affect the commercial value and development potential of wampee, especially in areas rich in wampee germplasm resources like Yongxing. However, the current detection and comprehensive characterization of metabolites in the three tissues of the wampee fruit are insufficient, which limits the development and utilization of the wampee fruit. To gain a comprehensive understanding of the metabolic differences among the three tissues in wampee fruit, we employed non-targeted LC–MS to detect and perform PCA analysis on the peel, pulp, and seed of the wampee. The results revealed significant differences in the metabolic profiles between the three tissues (Figure 1A). To better quantify the key metabolites in wampee, we utilized the previously established widely targeted metabolomics method to construct a wampee metabolite database that includes 1286 metabolites, which is the most comprehensive wampee metabolomics database known to date, and conducted quantitative analysis on the metabolites from different varieties and different tissues collected from Yongxing Town.

Research on the metabolites of wampee has always been a major direction in wampee studies, especially in exploring the physiological functions of its metabolites. It has been shown that the ethyl acetate extract of the pericarp of wampee peel has very strong antioxidant activity and anti-cancer activity, which is higher than that of the synthetic antioxidant butylated hydroxy toluene (BHT) and the conventional anti-cancer drug cisplatin [12]. In addition, some researchers have studied the chemical composition of essential oils (EOs) from the leaves and peel of wampee and found that EOs have a wide range of antifungal activities and have the potential to be an anticandida drug [7]. A researcher studied the branch and leaves of wampee and isolated nine carbazole alkaloids, five of which showed varying degrees of resistance to tumor cells [11]. Some researchers fed polyphenol extracts of wampee leaves to rats and found that they significantly improved lipid disorders, protected the liver, and also had the effect of lowering fasting blood glucose [9]. A number of recent studies have shown that there are a large number of neuroprotective metabolites in the fruits and leaves of wampee [51], particularly dominated by clausenamide, which is expected to be a promising candidate for the treatment of Alzheimer's disease and other neurodegenerative disorders [42], while a more detailed study has found that (−)clausenamide, but not (+)clausenamide, has an ameliorating effect on cognitive deficits [41]. However, current research on wampee is primarily focused on the overall functionality of extracts, such as wampee polyphenol extracts [9], volatile wampee essential oils [7], and wampee methanol extracts [14]. As well as the isolation and purification of metabolites from the extracts to identify new functional compounds, such as the lipid-lowering and liver-protecting metabolite zetaclausenamid [10] and neuroprotective metabolites containing a variety of new carbazole alkaloids [52,53]. In addition, there have been preliminary attempts to study the metabolites of different tissues of wampee. Ruiyi Fan and colleagues used liquid chromatography–mass spectrometry (LC–MS) for non-targeted metabolomics analysis of the leaves, barks, flowers, peels, pulps, and seeds of the wampee and identified 62 potential biomarkers to distinguish between different tissues of the wampee [54]. However, due to the small number of metabolites detected and the inability to accurately quantify them, there was no effective way to illustrate the differences between the wampee tissues, and a comprehensive characterization of this important medicinal and edible fruit could not be achieved.

In comparison, the wampee metabolite database we have established covers various categories of compounds and uses targeted metabolomics methods for precise quantification, effectively providing a comprehensive characterization of the three tissues of the wampee fruit. Additionally, our analysis found that the peel contributes the most to variety differences. By constructing an OPLS-DA model with the differences in the peel, we identified seven potential biomarkers that distinguish different varieties of wampee, providing ideas for the breeding of wampee varieties and the exploration of more substances with medicinal value. At the same time, this study also provides an important reference for the development, utilization, and protection of germplasm resources of Yongxing wampee.

In recent years, with the development of mass spectrometry imaging, an increasing number of metabolites have been characterized spatially in different species, such as rice [55], strawberries [56], and coffee beans [57]. In the case of wampee, researchers have conducted spatial mass spectrometry imaging studies on the fruit, stems, and leaves of wampee, comparing the spatial distribution of some alkaloids, coumarins, sugars, and organic acids in different tissues [58]. In comparison, our study combines MALDI-MSI with LC–MS for a more in-depth spatial imaging analysis of three tissues (peel, pulp, and seed) of the wampee fruit, which can more effectively characterize the differences in metabolites among the three tissues.

Wampee, as a fruit with both medicinal and edible properties, has great potential for development in terms of nutrition and medicinal uses. Wampee possesses effects such as antioxidant, anti-tumor, lipid lowering, liver protection, and blood sugar lowering. However, it is not very clear whether these functions are exerted in the edible parts or the medicinal parts of the fruit. This study provides a comprehensive characterization of the wampee fruit and finds that the flesh is rich in lipids, especially LPC and LPE, which can promote the digestion and absorption of lipids by the human body. Therefore, the flesh may play a role in lipid lowering and liver protection. Additionally, the flesh is rich in organic acids and amino acids, indicating that it has very high nutritional value. Moreover, the study finds that the skin is rich in flavonoids and terpenes, which typically have roles in antioxidant, anti-inflammatory, and lipid metabolism regulation. Furthermore, the core is found to be rich in carbazole alkaloids and coumarins, which play important roles in anti-tumor and anti-cancer effects and are also important categories for exploring medicinal value metabolites. Thus, this study characterizes the functions of different parts of the wampee fruit, providing a basis for the refined processing of wampee fruit. At the same time, this study characterizes the nutritional and medicinal values of wampee fruit, offering a reference for the exploration of metabolites with health care functions and medicinal value. It is of great significance for the development and utilization of wampee fruit.

In summary, we have comprehensively characterized the metabolites in the three tissues of multiple varieties of wampee fruit using the broad-targeted metabolomics approach with LC–MS and combined this with MALDI-MSI for an intuitive display of the spatial distribution of wampee fruit metabolites. These efforts provide significant reference value for the development and utilization of wampee.

5. Conclusions

In this study, by combining LC–MS and MALDI-MSI, we conducted a systematic detection and comprehensive characterization of metabolites in the three tissues of fruits from 17 wampee varieties in Yongxing Town. We found that the pulp is rich in lipids, amino acids, and organic acids, which may be related to its nutritional value; the peel is rich in flavonoids, quinic acid, polyphenols, and terpenes, which may be related to its antioxidant and anti-inflammatory properties; and the seed is rich in carbazole alkaloids, coumarins, and clausenamides, which may be related to its medicinal value. Additionally, we found that the peel contributes the most to the differences between varieties, especially some compounds with medicinal potential. All of this provides reference value for the development and utilization of Yongxing wampee, as well as for variety breeding and the protection of germplasm resources.

Supplementary Materials: The following supporting information can be downloaded at: https://www.mdpi.com/article/10.3390/foods13193092/s1, Figure S1. Photos of Wampee Fruit Varieties Involved in This Study. Figure S2. Detection and identification of specific catechin, gallocatechin, and scopoletin metabolite signals by ESI-Q-Exactive LC-MS/MS. Figure S3. Assessment of sample repeatability and Venn diagram of the three tissues. Figure S4. Volcano plots of metabolites between pairwise comparisons of the three types of tissues. Figure S5. Distribution of glycerophospholipids in different wampee tissues. Figure S6. Distribution of flavonoid subclasses in different wampee tissues. Figure S7. Distribution of polyphenols, terpenes, quinate and its derivatives in different wampee tissues. Figure S8. Bar plots showing the flavonoids that are more abundant in the seeds of wampee. Figure S9. The optical image of wampee fruit slices and the spatial distribution heatmaps of various metabolites in MALDI-MSI. Figure S10. Hierarchically clustered heatmap of the 345 annotated metabolites from 17 wampee peel samples. Table S1. The information of wampee variety used in this study. Table S2. Selected partial differential metabolites exist between and among the three tissues of the wampee fruit. Table S3. Identification of metabolites in wampee by MALDI-MS analysis. Table S4. Identification of potential biomarkers for different varieties of wampee.

Author Contributions: R.Z. resources, data curation, investigation, methodology, software, visualization, writing—original draft; J.Z. resources, methodology, software, validation, visualization; X.Z. methodology, software, validation; H.H. methodology, software, validation; X.L. resources, methodology; C.Y. supervision, methodology, software, validation, writing—review and editing, funding acquisition; S.S. supervision, methodology, validation, visualization, writing—review and editing, project administration, funding acquisition; J.L. supervision, methodology, writing—review and editing, project administration, funding acquisition. All authors have read and agreed to the published version of the manuscript.

Funding: This work was supported by the Project of Sanya Yazhou Bay Science and Technology City (SCKJ-JYRC-2022-06), the Hainan Provincial Natural Science Foundation of China (323MS019, 324QN196 and 324QN209), the Postdoctoral Fellowship Program of CPSF (GZB20230187), the China Postdoctoral Science Foundation (2022M710991, 2023M730896), Innovative Research Projects for Postgraduates in Hainan Province in 2023 (Qhys2023-240), the Hainan Provincial Academician Innovation Platform Project (HD-YSZX-202003), and the Hainan University Startup Fund (KYQD(ZR) 1866), the "111" Project (No. D20024), the Hainan Province Science and Technology Talent Innovation Project (KJRC2023D10).

Institutional Review Board Statement: Not applicable.

Informed Consent Statement: Not applicable.

Data Availability Statement: The original contributions presented in the study are included in the article/Supplementary Materials, further inquiries can be directed to the corresponding author.

Conflicts of Interest: The authors declare that they have no known competing financial interests or personal relationships that could have appeared to influence the work reported in this paper.

References

1. Phachonpai, W.; Tongun, T. Cognition enhancing effects of *Clausena lansium* (Lour.) peel extract attenuate chronic restraint stress-induced memory deficit in rats. *Heliyon* **2021**, *7*, e07003. [CrossRef] [PubMed]
2. Chen, W.; Balan, P.; Popovich, D.G. Comparison of Ginsenoside Components of Various Tissues of New Zealand Forest-Grown Asian Ginseng (*Panax Ginseng*) and American Ginseng (*Panax Quinquefolium* L.). *Biomolecules* **2020**, *10*, 372. [CrossRef] [PubMed]
3. Shi, R.; Xiong, B.; He, S.; Liu, C.; Ben-Asher, J.; Horowitz, A.R.; Wang, S.; He, X. Comparative metabolic profiling of root, leaf, fruit, and stem tissues of *Panax notoginseng*. *Int. J. Food Prop.* **2022**, *25*, 1132–1145. [CrossRef]
4. Chen, J.; Wang, W.; Kong, J.; Yue, Y.; Dong, Y.; Zhang, J.; Liu, L. Application of UHPLC-Q-TOF MS based untargeted metabolomics reveals variation and correlation amongst different tissues of Eucommia ulmoides Oliver. *Microchem. J.* **2022**, *172*, 106919. [CrossRef]
5. Rao, G.; Liu, X.; Zha, W.; Wu, W.; Zhang, J. Metabolomics reveals variation and correlation among different tissues of olive (*Olea europaea* L.). *Biol. Open* **2017**, *6*, 1317–1323. [CrossRef]
6. Wu, Y.; Huang, X.; Yang, H.; Zhang, S.; Lyu, L.; Li, W.; Wu, W. Analysis of flavonoid-related metabolites in different tissues and fruit developmental stages of blackberry based on metabolome analysis. *Food Res. Int.* **2023**, *163*, 112313. [CrossRef]
7. He, X.; Zhang, L.; Chen, J.; Sui, J.; Yi, G.; Wu, J.; Ma, Y. Correlation between Chemical Composition and Antifungal Activity of *Clausena lansium* Essential Oil against *Candida* spp. *Molecules* **2019**, *24*, 1394. [CrossRef]
8. Jiang, H.; Wang, C.; Fan, L.; Yang, K.; Feng, J.; Geng, Z.; Xu, J.; Deng, Z.; Du, S.; Yin, H. Cytotoxic Constituents from the Stems of *Clausena lansium* (Lour.) Skeels. *Molecules* **2013**, *18*, 10768–10775. [CrossRef]
9. Kong, F.; Su, Z.; Guo, X.; Zeng, F.; Bi, Y. Antidiabetic and Lipid-Lowering Effects of the Polyphenol Extracts from the Leaves of *Clausena lansium* (Lour.) Skeels on Streptozotocin-Induced Type 2 Diabetic Rats. *J. Food Sci.* **2018**, *83*, 212–220. [CrossRef]
10. Ma, N.; Wu, K.; Huang, L. An elegant synthesis of Zetaclausenamide. *Eur. J. Med. Chem.* **2008**, *43*, 893–896. [CrossRef]
11. Peng, W.-W.; Zheng, L.-X.; Ji, C.-J.; Shi, X.-G.; Xiong, Z.-H.; Shangguan, X.-C. Carbazole alkaloids isolated from the branch and leaf extracts of Clausena lansium. *Chin. J. Nat. Med.* **2018**, *16*, 509–512. [CrossRef] [PubMed]
12. Prasad, K.N.; Hao, J.; Yi, C.; Zhang, D.; Qiu, S.; Jiang, Y.; Zhang, M.; Chen, F. Antioxidant and Anticancer Activities of Wampee (*Clausena lansium*(Lour.) Skeels) Peel. *J. Biomed. Biotechnol.* **2009**, *2009*, 612805. [CrossRef]
13. Ao, H.; Jia, X.; Dong, L.; Zhang, R.; Liu, L.; Huang, F.; Wang, L.; Zhang, M. A new benzofuran glycoside from the fruit of *Clausena lansium*. *Nat. Prod. Res.* **2022**, *36*, 501–507. [CrossRef] [PubMed]
14. Adebajo, A.C.; Iwalewa, E.O.; Obuotor, E.M.; Ibikunle, G.F.; Omisore, N.O.; Adewunmi, C.O.; Obaparusi, O.O.; Klaes, M.; Adetogun, G.E.; Schmidt, T.J.; et al. Pharmacological properties of the extract and some isolated compounds of *Clausena lansium* stem bark: Anti-trichomonal, antidiabetic, anti-inflammatory, hepatoprotective and antioxidant effects. *J. Ethnopharmacol.* **2009**, *122*, 10–19. [CrossRef] [PubMed]
15. Huttenhower, C.; Finn, R.D.; McHardy, A.C. Challenges and opportunities in sharing microbiome data and analyses. *Nat. Microbiol.* **2023**, *8*, 1960–1970. [CrossRef]
16. Shen, S.; Zhan, C.; Yang, C.; Fernie, A.R.; Luo, J. Metabolomics-centered mining of plant metabolic diversity and function: Past decade and future perspectives. *Mol. Plant* **2023**, *16*, 43–63. [CrossRef]
17. Yang, C.; Shen, S.; Zhou, S.; Li, Y.; Mao, Y.; Zhou, J.; Shi, Y.; An, L.; Zhou, Q.; Peng, W.; et al. Rice metabolic regulatory network spanning the entire life cycle. *Mol. Plant* **2022**, *15*, 258–275. [CrossRef]
18. Yao, J.; Chu, Q.; Guo, X.; Shao, W.; Shang, N.; Luo, K.; Li, X.; Chen, H.; Cheng, Q.; Mo, F.; et al. Spatiotemporal transcriptomic landscape of rice embryonic cells during seed germination. *Dev. Cell* **2024**, *59*, 2320–2332.e5. [CrossRef]
19. Chen, W.; Gong, L.; Guo, Z.; Wang, W.; Zhang, H.; Liu, X.; Yu, S.; Xiong, L.; Luo, J. A Novel Integrated Method for Large-Scale Detection, Identification, and Quantification of Widely Targeted Metabolites: Application in the Study of Rice Metabolomics. *Mol. Plant* **2013**, *6*, 1769–1780. [CrossRef] [PubMed]
20. Trethewey, R.N. Metabolite profiling as an aid to metabolic engineering in plants. *Curr. Opin. Plant Biol.* **2004**, *7*, 196–201. [CrossRef] [PubMed]
21. Yu, K.; Wei, L.; Yuan, H.; Zhang, W.; Zeng, X.; Wang, B.; Wang, Y. Genetic architecture of inducible and constitutive metabolic profile related to drought resistance in qingke (Tibetan hulless barley). *Front. Plant Sci.* **2022**, *13*, 1076000. [CrossRef] [PubMed]
22. Wen, W.; Li, D.; Li, X.; Gao, Y.; Li, W.; Li, H.; Liu, J.; Liu, H.; Chen, W.; Luo, J.; et al. Metabolome-based genome-wide association study of maize kernel leads to novel biochemical insights. *Nat. Commun.* **2014**, *5*, 3438. [CrossRef] [PubMed]
23. Zhu, G.; Wang, S.; Huang, Z.; Zhang, S.; Liao, Q.; Zhang, C.; Lin, T.; Qin, M.; Peng, M.; Yang, C.; et al. Rewiring of the Fruit Metabolome in Tomato Breeding. *Cell* **2018**, *172*, 249–261.e12. [CrossRef] [PubMed]
24. Shen, S.; Wang, S.; Yang, C.; Wang, C.; Zhou, Q.; Zhou, S.; Zhang, R.; Li, Y.; Wang, Z.; Dai, L.; et al. Elucidation of the melitidin biosynthesis pathway in pummelo. *J. Integr. Plant Biol.* **2023**, *65*, 2505–2518. [CrossRef]
25. Wang, F.; Chen, L.; Chen, S.; Chen, H.; Liu, Y. Characterization of two closely related citrus cultivars using UPLC-ESI-MS/MS-based widely targeted metabolomics. *PLoS ONE* **2021**, *16*, e0254759. [CrossRef] [PubMed]
26. Morreel, K.; Saeys, Y.; Dima, O.; Lu, F.; Van de Peer, Y.; Vanholme, R.; Ralph, J.; Vanholme, B.; Boerjan, W. Systematic Structural Characterization of Metabolites in Arabidopsis via Candidate Substrate-Product Pair Networks. *The Plant Cell* **2014**, *26*, 929–945. [CrossRef] [PubMed]
27. Bauer, E.; Jakob, S.; Mosenthin, R. Principles of Physiology of Lipid Digestion. *Asian-Australasian J. Anim. Sci.* **2005**, *18*, 282–295. [CrossRef]

28. Shen, N.; Wang, T.; Gan, Q.; Liu, S.; Wang, L.; Jin, B. Plant flavonoids: Classification, distribution, biosynthesis, and antioxidant activity. *Food Chem.* **2022**, *383*, 132531. [CrossRef]
29. Grzesik, M.; Naparło, K.; Bartosz, G.; Sadowska-Bartosz, I. Antioxidant properties of catechins: Comparison with other antioxidants. *Food Chem.* **2018**, *241*, 480–492. [CrossRef]
30. Kim, J.M.; Heo, H.J. The roles of catechins in regulation of systemic inflammation. *Food Sci. Biotechnol.* **2022**, *31*, 957–970. [CrossRef] [PubMed]
31. Shaterzadeh-Yazdi, H.; Farkhondeh, T.; Samarghandian, S. An Overview on Cardiovascular Protective Effects of Catechins. *Cardiovasc. Hematol. Disord. Drug Targets* **2018**, *17*, 154–160. [CrossRef] [PubMed]
32. Wu, J.; Wang, X.; Hou, Y.; Gong, P. Bioactivities and the structural modification of Parthenolide: A review. *Med. Chem. Res.* **2024**, *33*, 221–238. [CrossRef]
33. Zhu, S.; Sun, P.; Bennett, S.; Charlesworth, O.; Tan, R.; Peng, X.; Gu, Q.; Kujan, O.; Xu, J. The therapeutic effect and mechanism of parthenolide in skeletal disease, cancers, and cytokine storm. *Front. Pharmacol.* **2023**, *14*, 1111218. [CrossRef] [PubMed]
34. Freund, R.R.A.; Gobrecht, P.; Fischer, D.; Arndt, H.-D. Advances in chemistry and bioactivity of parthenolide. *Nat. Prod. Rep.* **2020**, *37*, 541–565. [CrossRef]
35. Yoo, K.-D.; Park, E.-S.; Lim, Y.; Kang, S.-I.; Yoo, S.-H.; Won, H.-H.; Kim, Y.-H.; Yoo, I.-D.; Yoo, H.-S.; Hong, J.T.; et al. Clitocybin A, a Novel Isoindolinone, From the Mushroom Clitocybe aurantiaca, Inhibits Cell Proliferation Through G1 Phase Arrest by Regulating the PI3K/Akt Cascade in Vascular Smooth Muscle Cells. *J. Pharmacol. Sci.* **2012**, *118*, 171–177. [CrossRef] [PubMed]
36. Gao, X.; Li, X.; Zhang, C.; Bai, C. Scopoletin: A review of its pharmacology, pharmacokinetics, and toxicity. *Front. Pharmacol.* **2024**, *15*, 1268464. [CrossRef]
37. Guo, A.; Lin, J.; Zhong, P.; Chen, J.; Wang, L.; Lin, X.; Feng, M. Phellopterin attenuates ovarian cancer proliferation and chemoresistance by inhibiting the PU.1/CLEC5A/PI3K-AKT feedback loop. *Toxicol. Appl. Pharmacol.* **2023**, *477*, 116691. [CrossRef] [PubMed]
38. Wang, C.; Huang, J.; Zhou, Z.; Xu, P.; Shi, J.; Yang, Y.; Tong, S.; Hu, H. Coumarins from Jinhua Finger Citron: Separation by Liquid–Liquid Chromatography and Potential Antitumor Activity. *Molecules* **2023**, *28*, 6917. [CrossRef]
39. Xu, X.P.; Su, Y.; Pan, Y.; Shen, M.M.; Liu, D.C.; Liu, Z.H.; Chen, D.; Wu, J. The therapeutic effect of phellopterin on colitis-associated cancer and its effects on TLR4/NF-κB pathway and macrophage M2 polarization. *Cell. Mol. Biol.* **2023**, *69*, 51–57. [CrossRef]
40. Kumatia, E.K.; Zoiku, F.K.; Asase, A.; Tung, N.H. Anti-malarial activity of the alkaloid, heptaphylline, and the furanocoumarin, imperatorin, from Clausena anisata against human Plasmodium falciparum malaria parasites: Ex vivo trophozoitocidal, schizonticidal and gametocytocidal approach. *Malar. J.* **2023**, *22*, 264. [CrossRef]
41. Ning, N.; Hu, J.-F.; Sun, J.-D.; Han, N.; Zhang, J.-T.; Chen, N.-H. (−)Clausenamide facilitates synaptic transmission at hippocampal Schaffer collateral-CA1 synapses. *Eur. J. Pharmacol.* **2012**, *682*, 50–55. [CrossRef] [PubMed]
42. Chu, S.; Zhang, J. Recent advances in the study of (−) clausenamide: Chemistry, biological activities and mechanism of action. *Acta Pharm. Sin. B* **2014**, *4*, 417–423. [CrossRef] [PubMed]
43. Wu, J.; Zhang, J.; Xie, Q.; He, X.; Guo, Z.; Zheng, B.; Wang, S.; Yang, Q.; Du, C. Bergaptol Alleviates LPS-Induced Neuroinflammation, Neurological Damage and Cognitive Impairment via Regulating the JAK2/STAT3/p65 Pathway. *J. Inflamm. Res.* **2022**, *15*, 6199–6211. [CrossRef]
44. Kim, J.H.; Kim, J.-K.; Ahn, E.-K.; Ko, H.-J.; Cho, Y.-R.; Lee, C.H.; Kim, Y.K.; Bae, G.-U.; Oh, J.S.; Seo, D.-W. Marmesin is a novel angiogenesis inhibitor: Regulatory effect and molecular mechanism on endothelial cell fate and angiogenesis. *Cancer Lett.* **2015**, *369*, 323–330. [CrossRef] [PubMed]
45. Wang, Q.; Zhong, S.; Wu, H.; Wu, Q. In vitro anti-cancer effect of marmesin by suppression of PI3K/Akt pathway in esophagus cancer cells. *Esophagus* **2022**, *19*, 163–174. [CrossRef]
46. Huang, L.; Li, D.; Xu, Y.-S.; Feng, Z.; Meng, F.; Zhang, Q.-W.; Gan, L.-S.; Lin, L.-G. Clausoxamine, an alkaloid possessing a 1,3-oxazine-4-one ring from the seeds of *Clausena lansium* and the anti-obesity effect of lansiumamide B. *RSC Adv.* **2017**, *7*, 46900–46905. [CrossRef]
47. Xu, H.; Chen, T.; Huang, L.; Shen, Q.; Lian, Z.; Shi, Y.; Ouyang, M.-A.; Song, L. Synthesis and Fungicidal Activity of Lansiumamide A and B and Their Derivatives. *Molecules* **2018**, *23*, 1499. [CrossRef] [PubMed]
48. Thévenot, E.A.; Roux, A.; Xu, Y.; Ezan, E.; Junot, C. Analysis of the Human Adult Urinary Metabolome Variations with Age, Body Mass Index, and Gender by Implementing a Comprehensive Workflow for Univariate and OPLS Statistical Analyses. *J. Proteome Res.* **2015**, *14*, 3322–3335. [CrossRef] [PubMed]
49. Chang, X.; Lu, Y.; Lin, Z.; Qiu, J.; Guo, X.; Pan, J.; Abbasi, A.M. Impact of Leaf Development Stages on Polyphenolics Profile and Antioxidant Activity in *Clausena lansium* (Lour.) Skeels. *BioMed Res. Int.* **2018**, *2018*, 7093691. [CrossRef]
50. Wu, H.; Min, T.; Li, X.; Li, L.; Lai, F.; Tang, Y.; Yang, X. Physicochemical properties and antioxidant activities of acidic polysaccharides from wampee seeds. *Int. J. Biol. Macromol.* **2013**, *59*, 90–95. [CrossRef]
51. Liu, Y.; Guo, J.-M.; Liu, Y.-Y.; Hu, S.; Yan, G.; Qiang, L.; Fu, Y.-H. Carbazole Alkaloids with Potential Neuroprotective Activities from the Fruits of *Clausena lansium*. *J. Agric. Food Chem.* **2019**, *67*, 5764–5771. [CrossRef] [PubMed]
52. Liu, H.; Li, C.; Yang, J.; Ning, N.; Si, Y.-K.; Li, L.; Chen, N.; Zhao, Q.; Zhang, D.-M. Carbazole Alkaloids from the Stems of *Clausena lansium*. *J. Nat. Prod.* **2012**, *75*, 677–682. [CrossRef] [PubMed]

53. Liu, Y.-P.; Guo, J.-M.; Wang, X.-P.; Liu, Y.-Y.; Zhang, W.; Wang, T.; Qiang, L.; Fu, Y.-H. Geranylated carbazole alkaloids with potential neuroprotective activities from the stems and leaves of *Clausena lansium*. *Bioorganic Chem.* **2019**, *92*, 103278. [CrossRef] [PubMed]
54. Fan, R.; Peng, C.; Zhang, X.; Qiu, D.; Mao, G.; Lu, Y.; Zeng, J. A comparative UPLC-Q-Orbitrap-MS untargeted metabolomics investigation of different parts of *Clausena lansium* (Lour.) Skeels. *Food Sci. Nutr.* **2020**, *8*, 5811–5822. [CrossRef] [PubMed]
55. Kamjijam, B.; Suwannaporn, P.; Bednarz, H.; Na Jom, K.; Niehaus, K. Elevation of gamma-aminobutyric acid (GABA) and essential amino acids in vacuum impregnation mediated germinated rice traced by MALDI imaging. *Food Chem.* **2021**, *365*, 130399. [CrossRef] [PubMed]
56. Enomoto, H. Mass Spectrometry Imaging of Flavonols and Ellagic Acid Glycosides in Ripe Strawberry Fruit. *Molecules* **2020**, *25*, 4600. [CrossRef]
57. Li, N.; Dong, J.; Dong, C.; Han, Y.; Liu, H.; Du, F.; Nie, H. Spatial Distribution of Endogenous Molecules in Coffee Beans by Atmospheric Pressure Matrix-Assisted Laser Desorption/Ionization Mass Spectrometry Imaging. *J. Am. Soc. Mass Spectrom.* **2020**, *31*, 2503–2510. [CrossRef]
58. Tang, X.; Zhao, M.; Chen, Z.; Huang, J.; Chen, Y.; Wang, F.; Wan, K. Visualizing the spatial distribution of metabolites in *Clausena lansium* (Lour.) skeels using matrix-assisted laser desorption/ionization mass spectrometry imaging. *Phytochemistry* **2021**, *192*, 112930. [CrossRef]

separations

MDPI

Article

Development and Validation of a Novel Method Using QuEChERS and UHPLC-MS-MS for the Determination of Multiple Emerging Fungicides in Surface Waters

Pulasthi Serasinghe [1,*], Deni Taleski [2], Hao T. K. Nguyen [2], Dayanthi Nugegoda [1] and Vincent Pettigrove [1]

[1] Aquatic Environmental Stress Research Group, RMIT University, Melbourne, VIC 3000, Australia; dayanthi.nugegoda@rmit.edu.au (D.N.); vincent.pettigrove@rmit.edu.au (V.P.)
[2] National Measurement Institute, Port Melbourne, VIC 3207, Australia; deni.taleski@measurement.gov.au (D.T.); hao.nguyen@measurement.gov.au (H.T.K.N.)
* Correspondence: pulasthi.serasinghe@rmit.edu.au

Abstract: The increasing global reliance on pesticides for agricultural pest control has raised significant environmental concerns, particularly due to inadequate monitoring of emerging chemicals in surface waters. This study addresses the potential contamination of aquatic ecosystems by developing and validating a method for detecting trace amounts of four recently registered fungicides: three succinate dehydrogenase inhibitors (fluopyram, penthiopyrad, pydiflumetofen) and fluopicolide, a structurally related fungicide. Employing QuEChERS-based sample extraction combined with ultra-high-performance liquid chromatography (UHPLC-MS-MS), this method achieves detection limits of 0.1 to 0.2 µg/L, with recovery rates between 90% and 110%, and intra-day relative standard deviation values well within the acceptable range of less than 20%. Applied to surface grab water samples from the greater Melbourne area, Australia, the method successfully identified all four fungicides at trace levels, including a notable high concentration of fluopyram (7.3 µg/L) during autumn, with the others intermittently detected at lower concentrations. This study represents the first documented instance of quantifiable detections of these four fungicides in Australian surface water systems. Given their high toxicity to several organisms and the limited global data on these substances, our findings underscore the critical need for continuous monitoring to inform strategies to safeguard aquatic ecosystems from these chemicals.

Keywords: SDHI fungicides; QuEChERS; fluopyram; surface water; emerging pesticides

Citation: Serasinghe, P.; Taleski, D.; Nguyen, H.T.K.; Nugegoda, D.; Pettigrove, V. Development and Validation of a Novel Method Using QuEChERS and UHPLC-MS-MS for the Determination of Multiple Emerging Fungicides in Surface Waters. *Separations* **2024**, *11*, 279. https://doi.org/10.3390/separations11100279

Academic Editor: Grzegorz Boczkaj

Received: 30 August 2024
Revised: 20 September 2024
Accepted: 21 September 2024
Published: 24 September 2024

1. Introduction

In the last decade, global pesticide usage has significantly increased. Pesticides are commonly used for controlling pests like rodents, weeds, fungi, parasites, and disease vectors, thus safeguarding global agriculture [1]. Pesticides can help maintain and increase commercial agricultural production [2]. However, increased usage, coupled with the broad range of pesticide chemicals and classes administered, has contributed to adverse effects on non-target biota and ecosystems [3].

Unlike other environmental pollutants, pesticides are developed with inherent toxicities associated with the host/pests and specified for use within a target environment. Pesticides are generally associated with long half-lives and environmental persistence and mobility within the environment leading to the possibility of off-target exposure, heightened toxicity, and bioaccumulation. In addition, pesticide contamination of surface water systems can occur in processes such as leaching, spray drift, surface runoff, and drainage through agricultural and industrial application including point source contamination [4].

The widespread and repeated usage of pesticides with similar modes of action may contribute to emergence of resistance in pesticide hosts/pests. As such, incorporation of multiple pesticides with novel modes of action are required as part of control programs to prevent and offset the chance of pesticide resistance [5,6]. Succinate dehydrogenase inhibitors (SDHIs) are one of the most widely utilized classes of fungicides globally [7]. As such, over the years the SDHI class has seen continuous additions, including the following: fluopyram, penthiopyrad, and pydiflumetofen, which were developed within the last two decades [8,9]. The succinate dehydrogenase (SDH) complex, the molecular target of these fungicides, plays a vital role in the energy metabolism within target molds and fungi [10]. However, this complex and its significance extend to nearly all existing eukaryotes, including non-target organisms. The mode of action of these fungicides is not species-specific, and this has raised concerns regarding their potential toxicity to non-target organisms and their broader environmental impact [10,11]. Based on their physiochemical properties, SDHI fungicides are slightly polar with low-to-moderate soil and water mobilities (Figure 1). However, previous studies have demonstrated their propensity to contaminate groundwaters [12,13]. Furthermore, several reports have demonstrated the occurrence and prevalence of SDHI fungicides in surface water on a global scale across various water systems, [14,15]. Fluopicolide is another fungicide that shares similar properties to members of the SDHI category; however, its precise mode of action is not well understood (Figure 1). Similarly, residues of this fungicide and its metabolites have also been reported in several water systems across the European regions [12,16]. To our knowledge, only one published study is available confirming the detection of SDHI-class and similar fungicides, such as fluopicolide, in surface water systems in Australia [17].

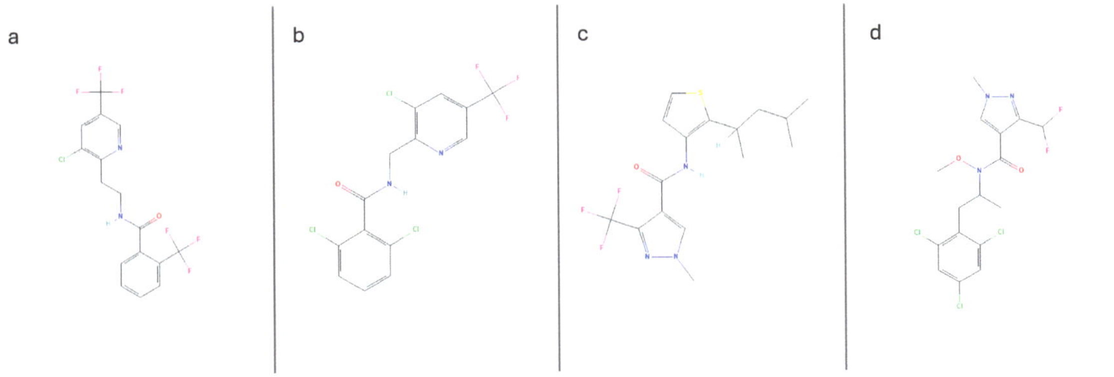

Figure 1. Chemical structures of the 4 analytes under investigation, fluopyram (**a**), fluopicolide (**b**), penthiopyrad (**c**), pydiflumetofen (**d**), structure data adapted from the PubChem database [18].

Fungicides classified as SDHI can effectively inhibit SDH activity in various non-target species, including earthworms and humans [10]. Importantly, non-target aquatic and marine organisms may display a heightened sensitivity to these fungicides, thereby posing a potential risk due to their documented high toxicity [10,19].

Given these findings and the widespread use of SDHI fungicides, with the potential to contaminate aquatic environments, there is a distinct probability for adverse effects on non-target organisms. Adverse effects have been observed in cases of acute exposure, including developmental malformations, oxidative stress, and endocrine disruption within fish [20,21]. Additionally, chronic and sublethal exposure to multiple SDHI fungicides has been linked to issues such as mitochondrial dysfunction, metabolic disorders, visual impairments, and motor impairments within non-target aquatic organisms, such as fish and aquatic invertebrates [22,23].

Over the past decade, numerous fungicides have gained registration and approval for use in Victoria, Australia. Among them are several belonging to the novel SDHI class, along with others that share similar properties. These fungicides have collectively been approved for diverse applications across a broad spectrum of crops, including brassica and bulb vegetables, citrus fruits, berries, and nuts [24]. Additionally, these have been approved for use in commercial and industrial settings, such as on turfs and lawns. Given the limited availability of detection and toxicity studies, it is important to understand the presence of these fungicides in surface systems and potential impacts on non-target aquatic organisms.

Chromatographic separation used in tandem with mass spectrometric detection has enabled the analysis of trace level organic or chemical contaminants in several complex matrices including water, soils, and biota. The current study explores the use of liquid chromatography coupled to targeted multiple reaction monitoring (MRM) mass spectroscopy in the detection of polar molecules where the monitoring of select transitions can increase specificity, remove matrix interferences, and afford improved sensitivity and detection limits [25].

Prior to chromatographic analysis, sample extraction and pre-treatment are necessary steps which may directly impact the overall sensitivity and selectivity of the analytical method for compound analysis within a sample matrix. Some of the most established and commonly used sample extraction techniques include solid-phase extraction (SPE), liquid–liquid extraction (LLE) and solid-phase microextraction (SPME). Solid-phase extraction remains a widely used and accepted technique for residue extraction from across multiple matrices including water [26]. Offline SPE techniques are often linked to lengthy and laborious procedures, while online methods, although more accurate and efficient, tend to be considerably expensive and restricted to small sample batches per instrumental run [27,28]. The QuEChERS (quick, easy, cheap, effective, rugged, and safe) method offers a simple, rapid, and efficient technique for aiding liquid–liquid extraction in a variety of difficult matrices such as food in suspect and non-target residue screening. With various modifications to its methodology over the years, its applicability in pesticide residue analysis has expanded to different matrices, including environmental soil [29], sediment [30], and agricultural samples [31]. However, its application in the extraction of pesticides from surface water systems is limited [26], even more so for novel and emerging pesticides such as those from the SDHI class. Only a limited number of studies have developed and applied QuEChERS-based methodologies to quantify SDHI fungicides within aquatic matrices such as agricultural water [31,32].

This study has a dual purpose: firstly, to create and validate an optimized analytical method for detecting four recently registered fungicides in surface water systems within Australia; and secondly, to utilize the validated method for analyzing surface water samples from waterways in the Greater Melbourne area (GMA). The objective is to quantify the presence of the four SDHI fungicides, thereby enhancing our understanding of their occurrence in local waterways.

2. Materials and Methods

2.1. Reagents and Chemicals

Four high-purity fungicide standards were obtained for this experiment: fluopyram (LGC Standards, Wesel, Germany), fluopicolide, penthiopyrad (AccuStandard, New Haven, CT, USA), and pydiflumetofen (HPC Standards GmbH, Cunnersdorf, Germany). All standards were supplied through Novachem (Novachem Pty Ltd., Heidelberg West, Australia). Each standard was purchased as 100 μg/mL in acetonitrile derived from the neat product at a purity of 99.90% or greater. High-performance liquid chromatography (HPLC)-grade reagents included acetonitrile, methanol, formic acid, and ammonium formate (Merck, Mountain Highway, Australia).

2.2. Preparation of Standard Solution

A working standard mixture of 10 μg/mL consisting of all four fungicides from their respective stock solutions was prepared via serial dilution in HPLC-grade acetonitrile and stored at −20 °C until analysis within amber glass ampoules. This solution was used for sample spiking and preparation of calibration curves.

2.3. Blank Matrices

For the method development work, blank surface water samples were collected from a catchment located within the GMA. Water samples were collected from a clean site where residues of fungicides to be tested were unlikely to be present. The site water samples were collected in 1 L amber glass bottles with screw caps. The water samples were stored at −18 to −22 °C upon arrival at the laboratory. Sample collection and storage conditions were based on US EPA Method 1699 [33].

2.4. Sample Preparation and Extraction

A 10 mL sample was added into a 50 mL centrifuge tube followed by 10 mL of the extraction solvent, 1% formic acid (*v*/*v*) acetonitrile. Sample extraction was performed using a QuEChERS extraction pouch—EN method (Part No. 5982-0650) containing the following: 4 g $MgSO_4$, 1 g NaCl, 1 g sodium citrate dihydrate, and 0.5 g sodium hydrogen citrate sesquihydrate. Then, the sample tubes were vortexed (Vortex-Genie 2, Scientific Industries incorporated, New York, NY, USA) and placed on a horizontal shaker (IKA-KS260 Basic, IKA, San Diego, CA, USA) for 2 min. The samples were then centrifuged (Thermo Scientific, Heraeus, Hanau, Germany, Megafuge 40) at 5100 rpm at 2 °C for 5 min. Then, 1 mL of the supernatant was transferred into a 4 mL microcentrifuge tube (prepacked with 150 mg $MgSO_4$, 50 mg primary secondary amine (PSA), 50 mg octadecyl (CEC18)). The tubes were vortexed for 1 min and centrifuged at 5100 rpm at 2 °C for 5 min. A 650 µL aliquot was then filtered through 0.2 µm filtration tube, and a 450 µL aliquot of this filtrate was combined with the internal standard—triphenyl phosphate (TPP) (1 µg/L in acetonitrile). Finally, 300 µL of sample was transferred into 2 mL amber LC-MS vials fitted with glass inserts for subsequent analysis.

2.5. Quality Control

Quality control was established and maintained throughout all experiments by incorporating internal standard-spiked blanks for each batch of samples to be analyzed. To further ensure accuracy, a quality control process involved running a low-level spike (20 µg/L) and a reagent blank (acetonitrile) was incorporated after every 5 samples within every instrumental run. These measures were implemented in every batch to monitor for carry-over and instrument performance during the analysis. In addition, blank surface water samples without internal standards were introduced at the beginning of each run to confirm the absence of any residual target analytes in the sample matrix and prevent potential carry-over from highly concentrated spiked samples.

2.6. Instrumentation and Software

Chromatographic analysis of the pesticides was performed using a Waters ACQUITY UPLC H-class system (Waters, Australia), consisting of a Waters Xevo TQ-S mass spectrometer utilizing multiple reaction monitoring. An ACQUITY UPLC BEH C18 column (2.1 × 100 mm: 1.7 µm) (Waters Corporation) was utilized for chromatographic separation. Several combinations of mobile phases A and B were tested for this analysis. Peaks of each target analyte were observed under varied mobile phase combinations. The following composition was chosen as the optimal mobile phase composition yielding the highest peak intensities and relative abundance. It consisted of 10 mM ammonium formate in Milli-Q water containing 0.1% formic acid (*v*/*v*), while mobile phase B comprised pure methanol. The injection was performed at a flow rate 0.4 mL/min, with an injection volume of 2.0 µL. The instrumental run time for each sample was 5.50 min. The mobile phase gradient profile was as follows: 0–0.10 min 95% A, 5% B; 0.10–0.30 min 30% A, 70% B; 0.30–3.0 min 2% A, 98% B; 3.0–5.0 min 95% A, 5% B. The instrument was run using a positive electrospray ionization (ESI) mode for all compounds with scheduled multiple reaction monitoring (MRM) used for all analyses with the following conditions: capillary voltage (3 kV), cone voltage (30 V), source temperature (150 °C), desolvation temperature (400 °C), cone gas flow (150 L/h), desolvation gas flow (1000 L/h). The UHPLC-MS/MS conditions for the four target fungicides are shown in Table 1. The conditions were provided by the Quanpedia database (Waters Corporation, Milford, MA, USA) and used without further optimization. Data acquisition and processing were conducted using MassLynx 4.2 and Target Lynx XS 4.2 software (Waters Corporation, Milford, MA, USA).

Table 1. Ultra-high-performance liquid chromatography parameters for the detection of four fungicides in positive ESI mode.

Pesticides	R.T (min)	Precursor Ion (*m*/*z*)	Quantification Transition (*m*/*z*)	D.P (V)	C.E (V)	Qualifier Transition (*m*/*z*)	D.P (V)	C.E (V)
Fluopicolide	2.62	383.0	172.9	40	20	365	40	15
Fluopyram	2.67	397.1	144.9	25	60	207.9	25	24
Penthiopyrad	2.91	360.0	176.9	10	43	256	14	29
Pydiflumetofen	3.19	425.9	166.4	110	30	192.9	110	40
Triphenylphosphate	2.99	326.7	76.9	2	36	151.9	2	22

Retention time (R.T) expressed in minutes, delustering potential (D.P) expressed in volts, collision energy (C.E) expressed in volts.

2.7. Method Validation and Matrix Effects

The method validation criteria in the European Commission SANTE/12682/2019 guidelines [34] for method linearity, matrix effects, limit of detection (LOD) and quantification (LOQ), accuracy (percentage recovery), and precision (relative standard deviation %) were followed.

Method linearity was determined using matrix-matched calibration curves produced using spiked blank samples at 0, 0.5, 5, 10, 20, 50 (for fluopyram and fluopicolide), and up to 100 μg/L (penthiopyrad and pydiflumetofen).

Method recovery was determined using seven replicate extractions and analysis at five (fluopyram, fluopicolide) and six (penthiopyrad, pydiflumetofen) spiked concentrations, respectively. Calculations were performed using peak areas according to Equation (1), where C_1 is the concentration of analyte in the spiked sample, C_2 is the concentration of analyte within the blank sample, and C_3 is the known concentration added to the sample [35–37]. The acceptable recovery range is between 70 and 120% with relative standard deviation (% RSD) less than 20% [34]. Intra-day accuracy and precision were evaluated using seven replicates at $n = 5$ (fluopyram, fluopicolide) and 6 (penthiopyrad, pydiflumetofen) spiked concentrations, respectively.

$$Recovery(\%) = \frac{(C_1 - C_2)}{C_3} \times 100 \tag{1}$$

The lowest spike level meeting the method's performance and acceptability requirements throughout the validation process is defined as the limit of quantification (LOQ) [34]. As such, the LOQ for method sensitivity was established using the lowest calibration level spike for each analyte. The LOD and LOQ were calculated according to equations as specified by Magnusson [38]. In Equations (2) and (3), S is the standard deviation of the average ($n = 7$) replicates of a spiked low-level blank which undergoes extraction and analysis.

$$\mathrm{LOD} = 3 \times S \tag{2}$$

$$\mathrm{LOQ} = 10 \times S \tag{3}$$

The LOD and LOQ were calculated as the 3- and 10-times corresponding standard deviation (SD) of seven replicate analyses of a low-level spike which meet acceptable recoveries (70–120%) and precision (RSD $\leq$ 20%). The determination of the measurement uncertainty (MU) is a requirement under the ISO/IEC 17025 [34] guidelines stated within the SANTE/12682/2019 [39] guidelines. Utilizing validation/QC data available through intra-lab experiments, an estimation of MU (u') and expanded uncertainty (U) were calculated for each pesticide analyzed in this study.

$$Relative\ bias = \frac{measured\ concentration - spiked\ concentration}{spiked\ concentration} \times 100\% \quad (4)$$

$$u'(bias) = \sqrt{mean^2{}_{bias} + SD.P^2_{bias}} \quad (5)$$

$$u'(precision) = RSD_{wR} \quad (6)$$

$$u' = \sqrt{u'(bias)^2 + u'(precision)^2} \quad (7)$$

$$u' = \sqrt{mean^2_{bias} + SD.P^2_{bias} + RSD^2_{rW}} \quad (8)$$

The following definitions [39] were used in the above equations for the derivation of MU and its expanded uncertainty for each compound. The variable "$mean_{bias}$" is the mean of the relative bias, $SD.P^2_{bias}$ is the population standard deviation of relative bias, RSD_{wR} is defined as the within-laboratory reproducibility, and $u'(bias)$ is the uncertainty associated with method and laboratory bias estimated through proficiency testing data [39].

The uncertainty associated with spiked concentrations, $u'(C_{ref})$ is assumed negligible when certified analytical standards and calibrated/verified volumetric balances are used in the preparation of spiked samples [39]. As such, only the $mean_{bias}$ and $SD.P^2_{bias}$ were used for calculating $u'(bias)$ within this study.

Finally, the expanded uncertainty of the method, expressed as (U), was obtained by multiplying combined uncertainty (u') with a coverage factor k = 2 at a 95% confidence level.

$$U = k \times u' \quad (9)$$

2.8. Applications to Surface Water Samples

After the optimization and validation, this method was applied to surface water samples from one site within the GMA (Figure 2) collected during the periods of May, June, and September 2023. This site was previously linked to the detection of all four fungicides in a recent semi-quantitative passive sampling study [17]. The catchment sampled was in a location with intensive agricultural and residential land uses (Figure 2). Grab water samples of 1.0 L were taken in triplicate during each sampling month. These samples were stored at 2 °C overnight prior to analysis.

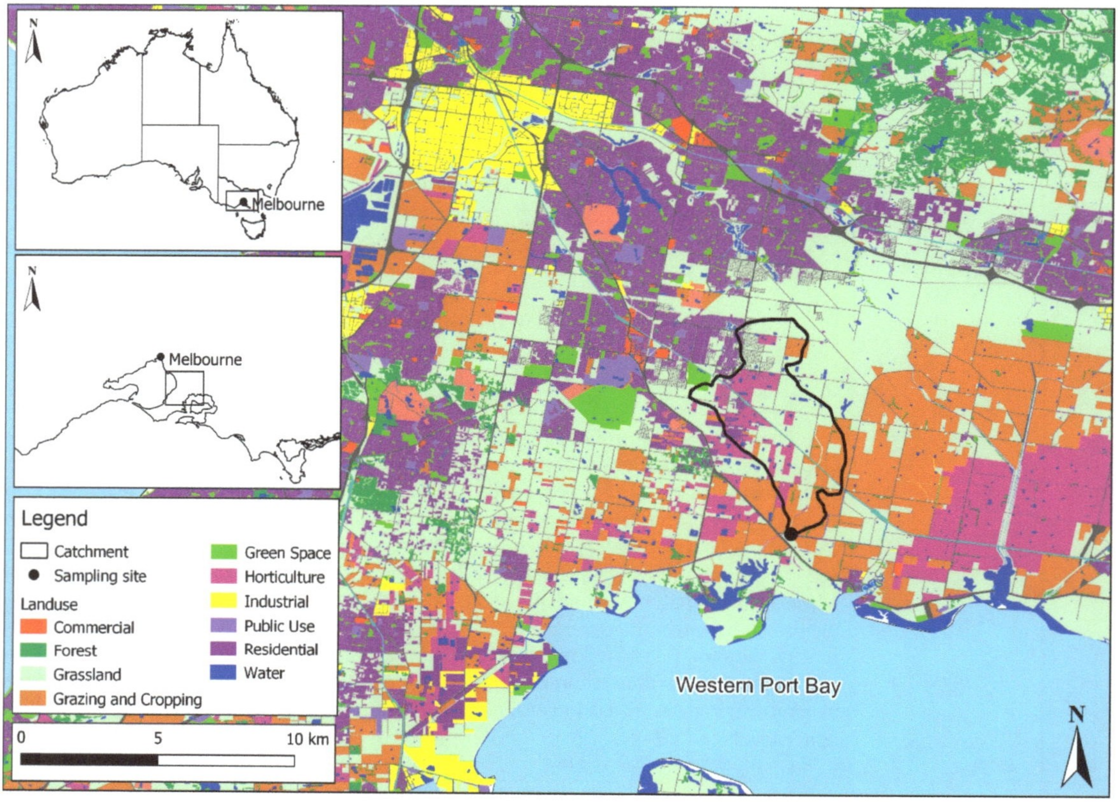

Figure 2. The location of surface water sampling site (n = 1) for method application within the Greater Melbourne area, Victoria, Australia. The land use layers showcased in this image were provided by Melbourne Water and the Victorian Department of Environment, Land, Water, and Planning (DELWP) through Spatial Economics [40].

3. Results and Discussion

3.1. Instrumental Method Optimization

Several adjustments were made to optimize the instrumental method to yield satisfactory results for key parameters such as chromatographic peak shapes, separations, and recoveries for all four target compounds.

The utilization of formic acid and ammonium formate as additives in the mobile phase has been widely explored in various studies, with reported enhancements in peak intensity, separation of target compounds, and sensitivity of LC-ESI-MS detection [36,38,41]. Consequently, a mobile phase composition of 0.1% formic acid (v/v) and 10 mM ammonium formate in Milli-Q water (mobile phase A) was chosen. During the initial optimization studies, it was observed that methanol as the second mobile phase yielded higher relative abundance and peak intensities compared to solvents such as pure acetonitrile. Similar findings were reported in a multi-residue analysis study of pesticides employing QuEChERS for water samples [36]. Therefore, pure methanol (mobile phase B) was selected as the final optimized configuration for mobile phase B.

UHPLC instruments exhibit superior peak separations, lower flow rates, and faster run times compared to their more conventionally used predecessor, high-performance liquid chromatography (HPLC) instruments [42–45]. Method optimization and validation experiments for this study were conducted using a UHPLC system, with the flow rate adjusted to 0.4 mL/min, resulting in optimal separation of all target pesticides through gradient elution during a 5.5-min run (Figure 3). Additional MRM chromatograms, displaying the qualifier transitions for each compound at 50 μg/L and the quantification transitions at 5.0 μg/L, are available in Supplementary Figures S1 and S2, respectively. In this study, a variety of injection volumes were tested on the UHPLC instrument, which has the advantage of superior resolution and sensitivity relative to HPLC instrumentation [46,47]. The injection volumes of 0.5, 1, 2, and 5 μL were trialed and although larger injection volumes are associated with increased sensitivity and lower limits of detection [48,49], 2 μL was selected as providing adequate sensitivity without impairing analyte resolution when used as the injection solvent in acetonitrile.

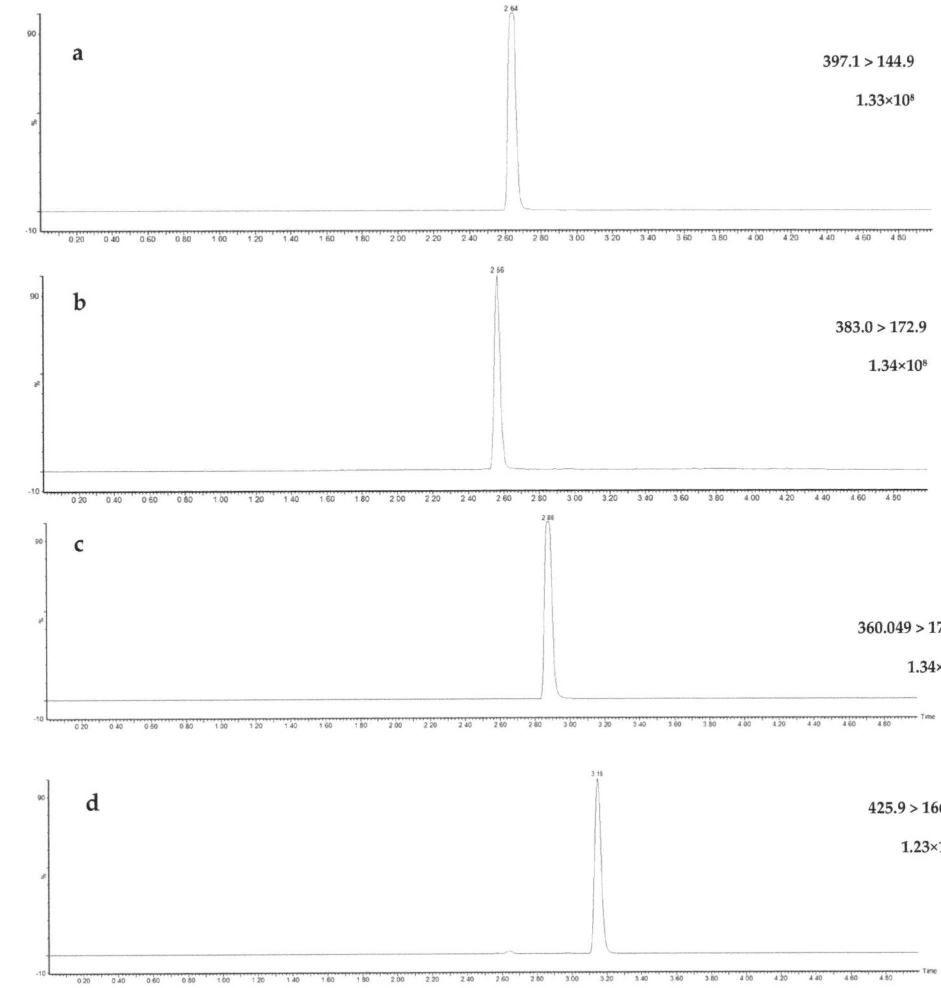

Figure 3. Ultra-high-performance liquid chromatography multiple reaction monitoring chromatogram (quantification) for the four fungicides—fluopyram (**a**), fluopicolide (**b**), penthiopyrad (**c**), pydiflumetofen (**d**)—in spiked surface water (50 μg/L), their respective quantification ion transition including raw count intensities.

3.2. Sample Preparation and Extraction

The initial method development experiments were performed using ultrapure Milli-Q water. Optimization studies and subsequent analyses, including method validation, used blank surface water samples with non-detectable residues of the compounds of interest. This assurance was attained through the analysis of matrix blanks using the optimized conditions, which exhibited low background noise across all analytes. These blank surface water samples are representative of a typical matrix within shallow surface water with low flow, influenced by agricultural land use applications in proximity.

The QuEChERS sample preparation and extraction procedure can be modified and optimized based on the compounds of interest, matrix, and instrumental conditions. Various configurations of buffering salts and sorbents for QuEChERS clean-up are available; however, most of the literature on QuEChERS methods is focused on the analysis of pesticide residues with food, sediment, and soil matrices [50,51]. Only a limited number of examples of QuEChERS extractions from water matrices, particularly in the context of micropollutant analysis, are available [26,52].

As such, experiments were carried out to trial this clean-up technique for application within the surface water matrices for pesticide residue analysis. In this study, 1 g of NaCl was added along with 4 g $MgSO_4$ during sample extraction phase for the blank surface water. This combination is widely employed in QuEChERS extraction for various contaminants, including fungicides, and is expected to enhance the selectivity of the analyzed compounds [32,53,54].

This study utilized sorbent conditions similar to those optimized by a recent study which focused on determining pydiflumetofen residues in multiple matrices, including paddy field water [32]. The researchers found the addition of PSA (150 mg), $MgSO_4$, and C18 (50 mg) exhibited the highest efficiency in purification. Notably, the addition of 150 mg of PSA resulted in high recoveries in their sorbent comparison study. A similar result in the use of PSA was observed in this study, where 50 mg PSA achieved satisfactory recoveries (>90%). It is recommended that future optimization studies include an addition of 150 mg of PSA under the same experimental conditions to assess its impact on overall fungicide recoveries.

Other refinements reported in the literature, such as utilizing graphite carbon black (GCB) instead of the C18 QuEChERS additive, have yielded modest improvements in recoveries across a diverse range of pesticide classes [31]. However, this approach was not trialed in the current study. Acetonitrile with 0.1% formic acid (v/v) was used as the extraction solvent, as it was also found to be suitable for pydiflumetofen in a study analyzing this compound in paddy field water [32].

3.3. Method Validation

3.3.1. Linearity and Limit of Quantification

As advised within the EU guidelines [39], validation experiments were carried out using matrix-matched calibration curves for more accurate results, excluding influences of matrix effects. Method performances for the four fungicides in blank surface water are summarized in Tables 2 and 3. Internal standard, triphenyl phosphate, was used to normalize data obtained during method development and validation experiments. Satisfactory linearities were observed for each pesticide with coefficients of determination: $R^2 > 0.99$. During the development process using matrix-matched calibration curves, we found that the high sensitivity of the instrument affected the linearity of results. For fluopicolide and fluopyram, the linearity was maintained only up to 50 μg/L. In contrast, penthiopyrad and pydiflumetofen showed linear results up to 100 μg/L, which was the maximum concentration of the calibration curve. Method LOQ values for each of the fungicides are within the range of 0.2–0.6 μg/L (Table 2).

Table 2. Limit of detection, limit of quantification, and method uncertainty parameters of the four target fungicides within the blank surface water samples.

Pesticides	LOD (µg/L)	LOQ (µg/L)	Mean Bias (%)	SD. P Bias (%)	RMS Bias (%)	u' (%)	U (%)
Fluopicolide	0.1	0.3	4	8	9	11	22
Fluopyram	0.1	0.3	4	7	8	10	20
Penthiopyrad	0.1	0.4	−4	7	8	11	21
Pydiflumetofen	0.2	0.6	−2	12	12	18	35

Limit of detection (LOD) expressed in µg/L, limit of quantification (LOQ) expressed in µg/L, population standard deviation (SD. P) bias expressed as a percentage (%), root mean square (RMS) bias expressed as a percentage (%), combined uncertainty (u') expressed as a percentage, expanded uncertainty (U) at 95% confidence.

Table 3. Linearity, precision, mean recovery, and intra-day relative standard deviation values of the four target fungicides within the blank surface water samples.

		Spiked Level (µg/L)													
		Mean Recovery (%)							Intra-Day RSD (%)						
Pesticide	**Linearity (R^2)**	**0.5**	**1**	**2**	**5**	**25**	**50**	**100**	**0.5**	**1**	**2**	**5**	**25**	**50**	**100**
Fluopicolide	0.999	95.3	106.7	108.8	106.3	105.8	103.2	-	6.1	5	7.4	6.3	9.2	7.2	-
Fluopyram	0.999	96.6	105.9	107.8	105.4	104.9	102.8	-	4.5	6.3	6.5	5.7	8.1	5.1	-
		0.5	1	2	5	25	50	100	0.5	1	2	5	25	50	100
Penthiopyrad	0.999	90.4	95.6	100	98.1	96.1	96.4	101.2	7.9	6.3	2.8	8.5	9.2	6.9	6.5
Pydiflumetofen	0.998	94.3	90	100.3	105.5	100.2	100.2	98.3	12.3	6.8	7.4	13.1	18.7	12.2	12.9

All mean recoveries and intra-day relative standard deviation values were calculated using seven replicates for each spiked concentration level. Values for 100 µg/L for fluopicolide and fluopyram were not included due to acceptable linearity not observed at this level.

3.3.2. Precision and Accuracy

The determined LOD values for individual pesticides were within the range of 0.1–0.2 (Table 2). All compounds had excellent recoveries at tested spike levels within blank surface water samples (Table 3) and were within the acceptable range of 70–120%. Method repeatability was determined through seven parallel analyses and results were expressed as intra-day RSD % values, with all values at ≤20%, which is within the acceptable range for environmental analysis [36,55].

The results obtained during the validation experiments showed the developed method can successfully recover all target analytes within spiked blank environmental surface water samples at each tested range. An MRM chromatogram from the analysis of a standard mixture of the four compounds, presented in Figure 3, showcases the transitions used for quantification of each analyte. An example chromatogram for a blank surface water sample is provided (Figure 4), showing no detectable residues of any target analytes within its matrix.

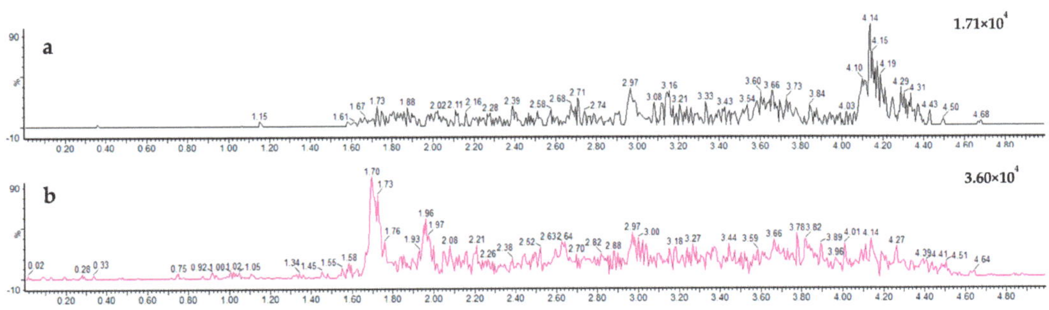

Figure 4. *Cont.*

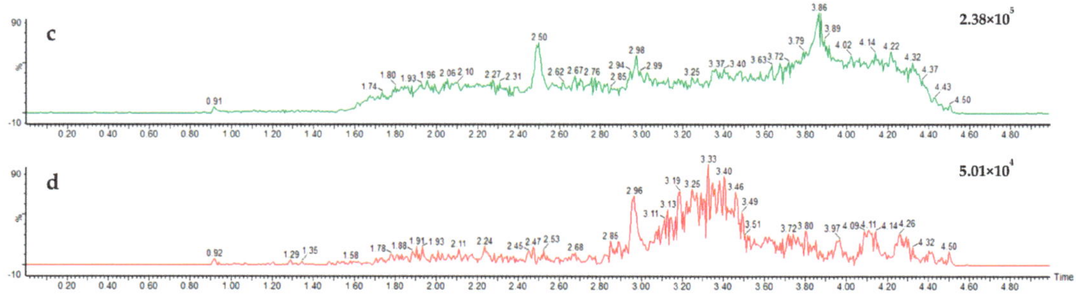

Figure 4. Ultra-high-performance liquid chromatography multiple reaction monitoring chromatogram of a blank surface water sample showing non-detectable residues for all four target fungicides—pydiflumetofen (**a**), fluopyram (**b**), fluopicolide (**c**), penthiopyrad (**d**)—within its matrix.

3.3.3. Estimation of Measurement Uncertainty

Measurement uncertainty serves as a quantitative indicator for analytical data, prescribed by ISO/IEC 17025 [34]. In this study, we employed an empirical method to calculate uncertainty, incorporating reproducible results, including linearity, recovery, and precision, to determine a total uncertainty value. The values for combined and expanded uncertainty are presented in Table 2. Expanded uncertainty values at a 95% confidence level ranged from 20 to 35%. Pydiflumetofen exhibited the highest U value, potentially attributed to poorer ionization and sensitivity relative to other analytes, particularly at lower spiking levels (Table 2).

Despite this, all four compounds demonstrated a mean bias of less than 20% across the calibration range, with each expanded uncertainty value equal to or less than 35%. These results align with the requirements outlined in SANTE 12682/2019 guidelines [39], which stipulate that a default value of 50% should not be exceeded.

The notable variation in uncertainty associated with pydiflumetofen, in comparison to the other three fungicides, may be linked to differences in its chromatographic profile, interactions with the matrix, and stability. A comparable study utilized a QuEChERS-based LC-MS-MS method for a water matrix; however, information on method bias and uncertainty was not provided [32].

Ideally, the trials for assessing accuracy should differ from those for estimating bias. Preferably, the bias assessments should rely on an external, independent source such as certified reference materials (CRMs) and proficiency testing (PT) reference values [39]. Since these were not available for the fungicides analyzed, ongoing precision results from validation experiments were used for estimating measurement uncertainty within this study.

Therefore, for a comprehensive future study, an experiment utilizing isotopically labelled standards and certified reference material for each compound could be employed to provide a more accurate estimate of uncertainty or to lower bias in the extraction and associated recoveries to achieve lower expanded uncertainty values in our method.

3.4. Method Application to Real-World Samples

Our method for analyzing samples revealed the presence of all four pesticides in surface water samples collected over a three-month period. The four fungicides investigated in this study are currently registered and approved for use in Victoria, Australia, a status that has been maintained for at least the past two years according to the Australian Pesticides and Veterinary Medicines Authority [24]. These fungicides are currently registered for a number of pests for a range of hosts and land uses. This includes a broad range of vegetables including brassica and bulb varieties as well as fruits such as different citruses and berries [24].

The mean concentrations of each pesticide found in the water samples are presented in Table 4. The highest concentration among all detected pesticides occurred in May 2023, with fluopyram reaching a mean concentration of 7.3 µg/L; however, no detectable levels were observed in July and September. Fluopicolide was only detected at a trace level of 0.15 µg/L in September, while pydiflumetofen was found at a trace level of 0.19 µg/L, exclusively in June. Penthiopyrad consistently showed trace levels across all three sampling periods. An example MRM chromatogram from a sample collected in May 2023, highlighting analyte detections, is available in Supplementary Figure S3.

Table 4. The mean pesticide concentrations within surface water samples from the Greater Melbourne area.

	Mean Pesticide Concentration (µg/L)			
Month	**Fluopyram**	**Fluopicolide**	**Penthiopyrad**	**Pydiflumetofen**
May	7.3	ND	0.11 *	ND
June	ND	ND	0.14 *	0.19 *
Sept	ND	0.15 *	0.16 *	ND

Mean pesticide concentrations were calculated using triplicate sample replicates (n = 3) for each pesticide. * Trace level detection with value falling between LOD and LOQ. ND refers to non-detection due to values that are less than LOD.

Fluopyram is authorized in Victoria for application on various fruits, including apples and strawberries. The increased concentrations observed in May at our sampling site, situated near numerous strawberry fields typically cultivated during early autumn and late winter, may be linked to this result. A similar hypothesis was explored in another study where consistent occurrences of fluopyram detected within winter periods coincided with greenhouse cultivation, as well as apple and strawberry harvesting [14]. This study did not conduct sampling during the previous summer and spring periods; however, monitoring the presence of fluopyram throughout the entire year is recommended—especially with its potential to occur at higher concentrations in spring and summer periods coinciding with increases agricultural use [56].

Fluopyram has been detected in both surface and groundwater systems globally, with maximum reported concentrations reaching 0.3 µg/L [15], 1.5 µg/L [14], and 6.0 µg/L [57] within surface water systems and >0.1 µg/L in groundwater [58]. The concentration detected in this study is comparatively higher than other reported values (Table 4).

Fluopyram is moderately mobile in soil and potentially contaminates surface waters through runoff or groundwater leaching [59,60]. It is expected to be more persistent in water–sediment environments and be stable within aquatic systems under anaerobic conditions [59]. As discussed above, its heightened concentrations in May could be directly linked to its use during this time or during previous summer periods, on produce such as strawberries. Non-detectable levels in the following months may be due to its binding to sediment, fast breakdown, and low persistency associated with surface water. Fluopyram can potentially decompose into toxic fluorine-containing compounds that may contaminate the environment [59,61]. Additionally, as highlighted by Hu, et al. [62] the photolytic degradation products of fluopyram can potentially induce higher toxicity than the parent compound [63]. The study by Li, et al. [64] showed that fluopyram and three of its breakdown products were toxic to non-target aquatic organisms including fish and Daphnia. Further toxicity studies on the effects of fluopyram on non-target aquatic species is not currently available, especially at chronic and sublethal concentrations [10].

The detection of fluopyram at 7 µg/L suggests low persistence in flowing water systems. However, its relatively high concentration could still impact local aquatic organisms, especially considering that chronic exposure levels as low as 30 µg/L have previously been reported to affect fish [64]. As such, in addition to long-term monitoring studies, a detailed ecotoxicological study involving both acute and low-level concentrations including fluopyram and its degradation products is recommended to better understand risks to local waterways.

Fluopicolide residues were detected at trace concentrations of up to 0.15 µg/L in September. Similarly, several global studies have reported fluopicolide residues, like fluopyram, in surface water systems, with levels as low as 0.02 µg/L [12,57,65]. Our study also found low-level detections (Table 4). Similar to fungicides from the SDHI class, the benzamide class, fluopicolide, has also been associated with moderate toxicities to non-target aquatic species including zebrafish and Daphnia and chronic effects to tadpoles, including developmental deformities [66–68].

Furthermore, there have been multiple reports on the presence of 2,6-dichlorobenzamide (BAM), a potential transformative product of fluopicolide, occurring with water systems [16,69,70]. Accordingly, consideration of the presence of fluopicolide and its degradation products, such as BAM, is warranted as part of further monitoring studies. In particular, consideration of pesticide breakdown products may reveal unknown or higher toxicological properties compared to the parent compound [71].

Penthiopyrad was detected at trace levels in the range of 0.11–0.16 µg/L during the three sampling periods. Pydiflumetofen was only detected in June at a trace concentration of 0.19 µg/L. Environmental detections of penthiopyrad and pydiflumetofen, especially within water systems, are very limited. As such, no comparison within surface water systems could be conducted. This may be attributed to their relatively recent entry into the pesticide market and limited analytical capabilities within monitoring programs [20,72]. Penthiopyrad has been associated with acute toxicity to fish. Several studies have highlighted the toxic effects induced by penthiopyrad through its mode of action, potentially exhibiting mitochondrial, metabolic disruption and developmental abnormalities within tested species such as zebrafish [73]. Although occurring at trace levels, its consistent presence could be associated with its long hydrolysis half-life (46–68 d) within aquatic systems [74]. While data on toxic effects of pydiflumetofen are limited, similar to other fungicides discussed within this study, it is also highly toxic to fish [75].

Both pydiflumetofen and penthiopyrad are chiral pesticides, existing as enantiomers. However, limited studies have investigated their enantioselective activities, including toxicities of specific isomers [76,77]. Available toxicological data on aquatic organisms recommend exploring their enantioselective effects for risk assessments [77]. Given the presence of pydiflumetofen in June and the continuous presence of penthiopyrad throughout all sampling periods in this study, further monitoring of these pesticides in surface water systems is justified. The proposed methodology may be used in conjunction with detailed future analysis, including novel enantioselective separations, as demonstrated by another recent study [72].

4. Conclusions

This study developed an optimized QuEChERS-UHPLC-MS/MS method to detect four recently registered fungicides in surface waters in Victoria, Australia. These fungicides were selected due to their limited global and Australian data and their high toxicity to aquatic organisms. The method was validated using blank surface water samples with matrix-matched calibrations, showing excellent linearity. Detection limits ranged from 0.1 to 0.2 µg/L, with recoveries between 90% and 108.8%, and precision from 2.8% to 18.7%. Expanded uncertainties for each compound were $\leq$35%. When applied to surface water samples from the Victorian GMA, all four fungicides were detected and accurately quantified for the first time, highlighting their persistence in aqueous environments. These findings underscore the need for ultra-trace fungicide monitoring, especially for novel compounds with limited detection and ecotoxicological data. The method will aid in assessing ecotoxicological risks and may inform regulatory monitoring for these and similar classes of fungicides in aquatic environments.

Supplementary Materials: The following supporting information can be downloaded at: https://www.mdpi.com/article/10.3390/separations11100279/s1, Figure S1: Ultra-high performance liquid chromatography multiple reaction monitoring chromatogram of four fungicides -fluopyram (a), fluopicolide (b), penthiopyrad (c), pydiflumetofen (d) in spiked surface water sample (50 μg/L), showcasing their qualifier ion transition including their respective raw count intensities; Figure S2: Ultra-high performance liquid chromatography multiple reaction monitoring chromatogram of four fungicides -fluopyram (a), fluopicolide (b), penthiopyrad (c), pydiflumetofen (d) and internal standard -triphenylphosphate (e)in spiked surface water sample (5 μg/L), showcasing their quantification transition including their respective raw count intensities; Figure S3: Ultra-high performance liquid chromatography multiple reaction monitoring chromatogram of four fungicides -fluopyram (a), fluopicolide (b), penthiopyrad (c), pydiflumetofen (d) in contaminated surface water grab sample collected in May 2023, showcasing their quantification transition including their respective raw count intensities.

Author Contributions: P.S.: Conceptualization, Methodology, Formal analysis, Investigation, Data curation, Writing—original draft, Visualization. D.T.: Investigation, Writing—review and editing, Data curation. H.T.K.N.: Investigation, Validation, Resources, Project administration. D.N.: Supervision, Methodology, Writing—review and editing. V.P.: Conceptualization, Methodology, Formal analysis, Investigation, Writing—original draft, Visualization, Supervision, Project administration, Funding acquisition. All authors have read and agreed to the published version of the manuscript.

Funding: This research was funded by a stipend funded by Melbourne Water through the Aquatic Pollution Prevention (A3P) partnership with RMIT University.

Data Availability Statement: The data presented in this study are available within the article and its Supplementary Materials.

Acknowledgments: We would like to thank Rhys Coleman (Melbourne Water) for providing this research opportunity and his comments on the draft manuscript, Saman Buddhadasa for coordinating and supporting the collaboration between NMI and RMIT University, and a special thanks to Monica Tewman for her invaluable support in gathering and managing key documents for the copyright and submission requirements. We would also like to thank the Aquatic Environmental Stress Research Group (AQUEST), RMIT for their assistance in field and laboratory work conducted during this research.

Conflicts of Interest: The authors declare no conflicts of interest.

References

1. Vryzas, Z.; Ramwell, C.; Sans, C. Pesticide prioritization approaches and limitations in environmental monitoring studies: From Europe to Latin America and the Caribbean. *Environ. Int.* **2020**, *143*, 105917. [CrossRef]
2. Sharma, A.; Kumar, V.; Shahzad, B.; Tanveer, M.; Sidhu, G.P.S.; Handa, N.; Kohli, S.K.; Yadav, P.; Bali, A.S.; Parihar, R.D.; et al. Worldwide pesticide usage and its impacts on ecosystem. *SN Appl. Sci.* **2019**, *1*, 1446. [CrossRef]
3. Kadiru, S.; Patil, S.; D'Souza, R. Effect of pesticide toxicity in aquatic environments: A recent review. *Int. J. Fish Aquat. Stud* **2022**, *10*, 113–118. [CrossRef]
4. Cuevas, N.; Martins, M.; Costa, P.M. Risk assessment of pesticides in estuaries: A review addressing the persistence of an old problem in complex environments. *Ecotoxicology* **2018**, *27*, 1008–1018. [CrossRef] [PubMed]
5. Mei, L.C.; Chen, H.M.; Dong, A.Y.; Huang, G.Y.; Liu, Y.W.; Zhang, X.; Wang, W.; Hao, G.F.; Yang, G.F. Pesticide Informatics Platform (PIP): An International Platform for Pesticide Discovery, Residue, and Risk Evaluation. *J Agric Food Chem* **2022**, *70*, 6617–6623. [CrossRef]
6. Das, S.K. Mode of action of pesticides and the novel trends—A critical review. *Int. Res. J. Agric. Sci. Soil Sci.* **2013**, *3*, 393–401.
7. Duarte Hospital, C.; Tête, A.; Debizet, K.; Imler, J.; Tomkiewicz-Raulet, C.; Blanc, E.B.; Barouki, R.; Coumoul, X.; Bortoli, S. SDHi fungicides: An example of mitotoxic pesticides targeting the succinate dehydrogenase complex. *Environ. Int.* **2023**, *180*, 108219. [CrossRef]
8. Umetsu, N.; Shirai, Y. Development of novel pesticides in the 21st century. *J. Pestic. Sci.* **2020**, *45*, 54–74. [CrossRef]
9. US EPA. *US EPA—Pesticides—Fact Sheet for Fluopicolide*; US EPA: Washington, DC, USA, 2007.
10. Yanicostas, C.; Soussi-Yanicostas, N. SDHI Fungicide Toxicity and Associated Adverse Outcome Pathways: What Can Zebrafish Tell Us? *Int. J. Mol. Sci.* **2021**, *22*, 12362. [CrossRef]
11. Marchand, P.A. EU Chemical Plant Protection Products in 2023: Current State and Perspectives. *Agrochemicals* **2023**, *2*, 106–117. [CrossRef]
12. Gulkowska, A.; Buerge, I.; Poiger, T. Online solid phase extraction LC–MS/MS method for the analysis of succinate dehydrogenase inhibitor fungicides and its applicability to surface water samples. *Anal. Bioanal. Chem.* **2014**, *406*, 6419–6427. [CrossRef] [PubMed]

13. Reilly, T.J.; Smalling, K.L.; Orlando, J.L.; Kuivila, K.M. Occurrence of boscalid and other selected fungicides in surface water and groundwater in three targeted use areas in the United States. *Chemosphere* **2012**, *89*, 228–234. [CrossRef] [PubMed]
14. la Cecilia, D.; Dax, A.; Ehmann, H.; Koster, M.; Singer, H.; Stamm, C. Continuous high-frequency pesticide monitoring in a small tile-drained agricultural stream to reveal diel concentration fluctuations in dry periods. *Front. Water* **2023**, *4*, 1062198. [CrossRef]
15. Nanusha, M.Y.; Frøkjær, E.E.; Liigand, J.; Christensen, M.R.; Hansen, H.R.; Hansen, M. Unravelling the occurrence of trace contaminants in surface waters using semi-quantitative suspected non-target screening analyses. *Environ. Pollut.* **2022**, *315*, 120346. [CrossRef]
16. Lapworth, D.J.; Baran, N.; Stuart, M.E.; Manamsa, K.; Talbot, J. Persistent and emerging micro-organic contaminants in Chalk groundwater of England and France. *Environ. Pollut.* **2015**, *203*, 214–225. [CrossRef]
17. Serasinghe, P.; Nguyen, H.T.K.; Hepburn, C.; Nugegoda, D.; Pettigrove, V. Use of passive sampling and high-resolution mass spectrometry for screening emerging pesticides of concern within surface waters. *J. Hazard. Mater. Adv.* **2024**, *13*, 100408. [CrossRef]
18. Kim, S.; Chen, J.; Cheng, T.; Gindulyte, A.; He, J.; He, S.; Li, Q.; Shoemaker, B.A.; Thiessen, P.A.; Yu, B.; et al. PubChem 2019 update: Improved access to chemical data. *Nucleic Acids Res.* **2018**, *47*, D1102–D1109. [CrossRef]
19. Xiao, P.; Liu, X.; Zhang, H.; Li, W. Chronic toxic effects of isoflucypram on reproduction and intestinal energy metabolism in zebrafish (*Danio rerio*). *Environ. Pollut.* **2022**, *315*, 120479. [CrossRef]
20. Qian, L.; Qi, S.; Cao, F.; Zhang, J.; Li, C.; Song, M.; Wang, C. Effects of penthiopyrad on the development and behaviour of zebrafish in early-life stages. *Chemosphere* **2019**, *214*, 184–194. [CrossRef]
21. Teng, M.; Zhu, W.; Wang, D.; Yan, J.; Qi, S.; Song, M.; Wang, C. Acute exposure of zebrafish embryo (*Danio rerio*) to flutolanil reveals its developmental mechanism of toxicity via disrupting the thyroid system and metabolism. *Environ. Pollut.* **2018**, *242*, 1157–1165. [CrossRef]
22. Qian, L.; Qi, S.; Wang, Z.; Magnuson, J.T.; Volz, D.C.; Schlenk, D.; Jiang, J.; Wang, C. Environmentally relevant concentrations of boscalid exposure affects the neurobehavioral response of zebrafish by disrupting visual and nervous systems. *J. Hazard. Mater.* **2021**, *404*, 124083. [CrossRef] [PubMed]
23. Yang, G.; Lv, L.; Di, S.; Li, X.; Weng, H.; Wang, X.; Wang, Y. Combined toxic impacts of thiamethoxam and four pesticides on the rare minnow (Gobiocypris rarus). *Environ. Sci. Pollut. Res.* **2021**, *28*, 5407–5416. [CrossRef] [PubMed]
24. APVMA. Public Chemical Registration Information System—PUBCRIS. Available online: https://portal.apvma.gov.au/pubcris (accessed on 18 June 2020).
25. Beretov, J.; Wasinger, V.C.; Graham, P.H.; Millar, E.K.; Kearsley, J.H.; Li, Y. Proteomics for breast cancer urine biomarkers. *Adv. Clin. Chem.* **2014**, *63*, 123–167. [CrossRef]
26. Hajeb, P.; Zhu, L.; Bossi, R.; Vorkamp, K. Sample preparation techniques for suspect and non-target screening of emerging contaminants. *Chemosphere* **2022**, *287*, 132306. [CrossRef] [PubMed]
27. Suseela, M.N.L.; Viswanadh, M.K.; Mehata, A.K.; Priya, V.; Vikas; Setia, A.; Malik, A.K.; Gokul, P.; Selvin, J.; Muthu, M.S. Advances in solid-phase extraction techniques: Role of nanosorbents for the enrichment of antibiotics for analytical quantification. *J. Chromatogr. A* **2023**, *1695*, 463937. [CrossRef]
28. Samsidar, A.; Siddiquee, S.; Shaarani, S.M. A review of extraction, analytical and advanced methods for determination of pesticides in environment and foodstuffs. *Trends Food Sci. Technol.* **2018**, *71*, 188–201. [CrossRef]
29. Lee, Y.J.; Choi, J.H.; Abd El-Aty, A.; Chung, H.S.; Lee, H.S.; Kim, S.W.; Rahman, M.M.; Park, B.J.; Kim, J.E.; Shin, H.C. Development of a single-run analytical method for the detection of ten multiclass emerging contaminants in agricultural soil using an acetate-buffered QuEChERS method coupled with LC–MS/MS. *J. Sep. Sci.* **2017**, *40*, 415–423. [CrossRef]
30. Li, L.; Yin, Y.; Zheng, G.; Liu, S.; Zhao, C.; Xie, W.; Ma, L.; Shan, Q.; Dai, X.; Wei, L. Determination of multiclass herbicides in sediments and aquatic products using QuECHERS combined with ultra-high performance liquid chromatography-tandem mass spectrometry (UHPLC-MS/MS) and its application to risk assessment of rice-fish co-culture system in China. *Microchem. J.* **2021**, *170*, 106628. [CrossRef]
31. Song, N.-E.; Jung, Y.S.; Choi, J.Y.; Koo, M.; Choi, H.-K.; Seo, D.-H.; Lim, T.-G.; Nam, T.G. Development and application of a multi-residue method to determine pesticides in agricultural water using QuEChERS extraction and LC-MS/MS analysis. *Separations* **2020**, *7*, 52. [CrossRef]
32. Bian, C.; Gao, M.; Liu, L.; Zhou, W.; Li, Y.; Wan, C.; Li, B. Determination of Pydiflumetofen Residues in Rice and its Environment by an Optimized QuEChERS Method Coupled with HPLC-MS. *Bull. Environ. Contam. Toxicol.* **2021**, *107*, 239–247. [CrossRef]
33. Englert, B. *Method 1699: Pesticides in Water, Soil, Sediment, Biosolids, and Tissue by HRGC/HRMS*; US Environmental Protection Agency (EPA): Washington, DC, USA, 2007; pp. 1–96.
34. *ISO/IEC 17025*; General Requirements for the Competence of Testing and Calibration Laboratories. International Organisation for Standardisation, Croatian Standard Institute: Zagreb, Croatia, 2007.
35. Koçyiğit, H.; Sinanoğlu, F. Method validation for the analysis of pesticide residue in aqueous environment. *Environ. Monit. Assess* **2020**, *192*, 567. [CrossRef] [PubMed]
36. Zaidon, S.Z.; Ho, Y.B.; Hamsan, H.; Hashim, Z.; Saari, N.; Praveena, S.M. Improved QuEChERS and solid phase extraction for multi-residue analysis of pesticides in paddy soil and water using ultra-high performance liquid chromatography tandem mass spectrometry. *Microchem. J.* **2019**, *145*, 614–621. [CrossRef]

37. Magnusson, B. *The Fitness for Purpose of Analytical Methods: A Laboratory Guide to Method Validation and Related Topics (2014)*; EurachemI: Lisboa, Portugal, 2014; p. 2.
38. European Commission. *Guidance Document on Analytical Quality Control and Method Validation Procedures for Pesticide Residues and Analysis in Food and Feed*; SANTE/12682/2019; European Commission: Brussels, Belgium, 2020.
39. Spatial Economics. Land Use Projections Greater Metropolitan Melbourne 2016 to 2051. 2016. Available online: https://www.portphillip.vic.gov.au/media/iu5h5tdo/11-2-att-1-draft-yarra-catchment-integrated-water-management-plan.pdf (accessed on 20 September 2024).
40. Wang, S.; Wang, X.; He, Q.; Lin, H.; Chang, H.; Sun, H.; Liu, Y. Simultaneous Determination of Seven Pesticides and Metabolite Residues in Litchi and Longan through High-Performance Liquid Chromatography-Tandem Mass Spectrometry with Modified QuEChERS. *Molecules* **2022**, *27*, 5737. [CrossRef] [PubMed]
41. Rizzetti, T.M.; Kemmerich, M.; Martins, M.L.; Prestes, O.D.; Adaime, M.B.; Zanella, R. Optimization of a QuEChERS based method by means of central composite design for pesticide multiresidue determination in orange juice by UHPLC–MS/MS. *Food Chem.* **2016**, *196*, 25–33. [CrossRef]
42. Su, Y.; Lu, J.; Liu, J.; Li, F.; Wang, N.; Lei, H.; Shen, X. Optimization of a QuEChERS-LC-MS/MS method for 51 pesticide residues followed by determination of the residue levels and dietary intake risk assessment in foodstuffs. *Food Chem* **2024**, *434*, 137467. [CrossRef]
43. Guillarme, D.; Dong, M. UHPLC, Part II: Benefits. *LCGC N. Am.* **2017**, *35*, 486–495.
44. Behnoush, B.; Sheikhazadi, A.; Bazmi, E.; Fattahi, A.; Sheikhazadi, E.; Saberi Anary, S.H. Comparison of UHPLC and HPLC in benzodiazepines analysis of postmortem samples: A case-control study. *Medicine* **2015**, *94*, e640. [CrossRef]
45. Frenich, A.G.; Romero-González, R.; del Mar Aguilera-Luiz, M. Comprehensive analysis of toxics (pesticides, veterinary drugs and mycotoxins) in food by UHPLC-MS. *TrAC Trends Anal. Chem.* **2014**, *63*, 158–169. [CrossRef]
46. Swetha, S.R.; Bhavya, S.K.; Mounika, C. A Review on Comparative study of HPLC and UPLC. *Res. J. Pharm. Technol.* **2020**, *13*, 1570–1574. [CrossRef]
47. Chawla, G.; Ranjan, C. Principle, instrumentation, and applications of UPLC: A novel technique of liquid chromatography. *Open Chem. J.* **2016**, *3*. [CrossRef]
48. Ren, D.-B.; Yang, Z.-H.; Liang, Y.-Z.; Fan, W.; Ding, Q. Effects of injection volume on chromatographic features and resolution in the process of counter-current chromatography. *J. Chromatogr. A* **2013**, *1277*, 7–14. [CrossRef] [PubMed]
49. Hogenboom, A.C.; Hofman, M.P.; Kok, S.J.; Niessen, W.M.; Brinkman, U.A. Determination of pesticides in vegetables using large-volume injection column liquid chromatography-electrospray tandem mass spectrometry. *J. Chromatogr. A* **2000**, *892*, 379–390. [CrossRef] [PubMed]
50. Acosta-Dacal, A.; Rial-Berriel, C.; Díaz-Díaz, R.; Bernal-Suárez, M.d.M.; Luzardo, O.P. Optimization and validation of a QuEChERS-based method for the simultaneous environmental monitoring of 218 pesticide residues in clay loam soil. *Sci. Total Environ.* **2021**, *753*, 142015. [CrossRef] [PubMed]
51. Golge, O.; Koluman, A.; Kabak, B. Validation of a Modified QuEChERS Method for the Determination of 167 Pesticides in Milk and Milk Products by LC-MS/MS. *Food Anal. Methods* **2018**, *11*, 1122–1148. [CrossRef]
52. Berlioz-Barbier, A.; Vauchez, A.; Wiest, L.; Baudot, R.; Vulliet, E.; Cren-Olivé, C. Multi-residue analysis of emerging pollutants in sediment using QuEChERS-based extraction followed by LC-MS/MS analysis. *Anal. Bioanal. Chem.* **2014**, *406*, 1259–1266. [CrossRef]
53. Jing, W.; Nakano, K.; Shen, Z.; Okuda, T. Optimization of the QuEChERS extraction method to determine Polycyclic Aromatic Hydrocarbons (PAHs) in powder aerosol particles collected by cyclone. *Environ. Technol. Innov.* **2023**, *31*, 103141. [CrossRef]
54. Perestrelo, R.; Silva, P.; Porto-Figueira, P.; Pereira, J.A.M.; Silva, C.; Medina, S.; Câmara, J.S. QuEChERS - Fundamentals, relevant improvements, applications and future trends. *Anal. Chim. Acta* **2019**, *1070*, 1–28. [CrossRef]
55. Ho, Y.B.; Zakaria, M.P.; Latif, P.A.; Saari, N. Simultaneous determination of veterinary antibiotics and hormone in broiler manure, soil and manure compost by liquid chromatography–tandem mass spectrometry. *J. Chromatogr. A* **2012**, *1262*, 160–168. [CrossRef]
56. Wightwick, A.M.; Bui, A.D.; Zhang, P.; Rose, G.; Allinson, M.; Myers, J.H.; Reichman, S.M.; Menzies, N.W.; Pettigrove, V.; Allinson, G. Environmental fate of fungicides in surface waters of a horticultural-production catchment in southeastern Australia. *Arch. Environ. Contam. Toxicol.* **2012**, *62*, 380–390. [CrossRef]
57. Spycher, S.; Mangold, S.; Doppler, T.; Junghans, M.; Wittmer, I.; Stamm, C.; Singer, H. Pesticide Risks in Small Streams—How to Get as Close as Possible to the Stress Imposed on Aquatic Organisms. *Environ. Sci. Technol.* **2018**, *52*, 4526–4535. [CrossRef]
58. Sjerps, R.M.A.; Kooij, P.J.F.; van Loon, A.; Van Wezel, A.P. Occurrence of pesticides in Dutch drinking water sources. *Chemosphere* **2019**, *235*, 510–518. [CrossRef] [PubMed]
59. Rathod, P.; Shah, P.; Parmar, K.; Kalasariya, R. The Fate of Fluopyram in the Soil–Water–Plant Ecosystem: A Review. *Rev. Environ. Contam. Toxicol.* **2022**, *260*, 1. [CrossRef]
60. APVMA. *Submission to the Productivity Commission inquiry into Regulation of Australian Agriculture*; APVMA: Armidale, Australia, 2016.
61. Dong, B.; Hu, J. Photodegradation of the novel fungicide fluopyram in aqueous solution: Kinetics, transformation products, and toxicity evolvement. *Environ. Sci. Pollut. Res.* **2016**, *23*, 19096–19106. [CrossRef] [PubMed]
62. Hu, J.; Pang, K.; Dong, B.Z. Mechanism and identify photolysis products of fluopyram under TiO2: Experiments, DFT and ab initio Molecular dynamics study. *SDRP J. Earth Sci. Environ. Stud.* **2019**, *4*, 681–690. [CrossRef]

63. Zhan, J.; Liang, Y.; Liu, D.; Liu, C.; Liu, H.; Wang, P.; Zhou, Z. Organochlorine pesticide acetofenate and its hydrolytic metabolite in rabbits: Enantioselective metabolism and cytotoxicity. *Pestic. Biochem. Physiol.* **2018**, *145*, 76–83. [CrossRef]
64. Li, C.; Yuan, S.; Jiang, F.; Xie, Y.; Guo, Y.; Yu, H.; Cheng, Y.; Qian, H.; Yao, W. Degradation of fluopyram in water under ozone enhanced microbubbles: Kinetics, degradation products, reaction mechanism, and toxicity evaluation. *Chemosphere* **2020**, *258*, 127216. [CrossRef]
65. Pazikowska-Sapota, G.; Galer-Tatarowicz, K.; Dembska, G.; Wojtkiewicz, M.; Duljas, E.; Pietrzak, S.; Dzierzbicka-Glowacka, L.A. The impact of pesticides used at the agricultural land of the Puck commune on the environment of the Puck Bay. *PeerJ* **2020**, *8*, e8789. [CrossRef]
66. Wen, S.; Wang, Y.; Wang, X.; Liu, C.; Xue, Y.; Liu, C.; Wang, J.; Xia, X. Fluopicolide-Induced Oxidative Stress and DNA Damage in the Earthworm Eisenia foetida. *Toxics* **2023**, *11*, 808. [CrossRef]
67. Jin, L.; Yun, G.; Wei, M.; Kaiyun, W.; Hui, X.; Jie, L. Acute toxicity of fluopicolide to 9 kinds of environmental organisms and its bioaccumulation in zebrafish. *Asian J. Ecotoxicol.* **2016**, *6*, 296–305.
68. Lin, J.; Wang, H.; Wang, K.; Fan, W.; Xu, H.; Liu, J. Toxic effects of fluopicolide on zebrafish. *China Environ. Sci.* **2014**, *34*, 3230–3236.
69. Lopez, B.; Ollivier, P.; Togola, A.; Baran, N.; Ghestem, J.-P. Screening of French groundwater for regulated and emerging contaminants. *Sci. Total Environ.* **2015**, *518–519*, 562–573. [CrossRef] [PubMed]
70. Postigo, C.; Barceló, D. Synthetic organic compounds and their transformation products in groundwater: Occurrence, fate and mitigation. *Sci. Total Environ.* **2015**, *503–504*, 32–47. [CrossRef] [PubMed]
71. Adasakaveg, E.J.; Michailides, T.; Eskalen, A. *Fungicides, Bactericides, Biocontrols, and Natural Products for Deciduous Tree Fruit and Nut, Citrus, Strawberry, and Vine Crops in California*; University of California Agriculture and Natural Resources: Davis, CA, USA, 2022.
72. Wang, Z.; Liu, S.; Zhao, X.; Tian, B.; Sun, X.; Zhang, J.; Gao, Y.; Shi, H.; Wang, M. Enantioseparation and stereoselective dissipation of the novel chiral fungicide pydiflumetofen by ultra-high-performance liquid chromatography tandem mass spectrometry. *Ecotoxicol. Envioron. Saf.* **2021**, *207*, 111221. [CrossRef] [PubMed]
73. Ren, B.; Liang, H.; Li, L.; Li, Y.; Liang, H.; Zhao, T.; Chen, H.; Zhao, Y. Enantioselective toxic effects of the novel chiral antifungal agrochemical penthiopyrad in the early life stage of zebrafish (*Danio rerio*). *Chem.-Biol. Interact.* **2023**, *369*, 110252. [CrossRef]
74. Yang, G.; Liu, Z.; Lan, T.; Dou, L.; Zhang, K. Potential enantioselectivity of the hydrolysation and photolysation of the chiral agrochemical penthiopyrad in aquatic environments. *Environ. Sci. Water Res. Technol.* **2021**, *7*, 1220–1229. [CrossRef]
75. Wang, Z.; Tan, Y.; Li, Y.; Duan, J.; Wu, Q.; Li, R.; Shi, H.; Wang, M. Comprehensive study of pydiflumetofen in *Danio rerio*: Enantioselective insight into the toxic mechanism and fate. *Environ. Int.* **2022**, *167*, 107406. [CrossRef]
76. Di, S.; Cang, T.; Liu, Z.; Xie, Y.; Zhao, H.; Qi, P.; Wang, Z.; Xu, H.; Wang, X. Comprehensive evaluation of chiral pydiflumetofen from the perspective of reducing environmental risks. *Sci. Total Environ.* **2022**, *826*, 154033. [CrossRef]
77. Ren, B.; Zhao, T.; Li, Y.; Liang, H.; Zhao, Y.; Chen, H.; Li, L.; Liang, H. Enantioselective bioaccumulation and toxicity of the novel chiral antifungal agrochemical penthiopyrad in zebrafish (*Danio rerio*). *Ecotoxicol. Environ. Saf.* **2021**, *228*, 113010. [CrossRef]

molecules

MDPI

Article

Dissipation and Dietary Risk Assessment of the Fungicide Pyraclostrobin in Apples Using Ultra-High Performance Liquid Chromatography–Mass Spectrometry

Bin Wang [1], Lei Shi [2], Pengcheng Ren [1], Shu Qin [1], Jindong Li [1] and Junli Cao [1,*]

[1] Shanxi Center for Testing of Functional Agro-Products, Shanxi Agricultural University, Taiyuan 030031, China
[2] Shanxi Institute for Functional Food, Shanxi Agricultural University, Taiyuan 030031, China
* Correspondence: caojunli@sxau.edu.cn; Tel.: +86-0351-7965708; Fax: +86-0351-7639301

Abstract: The fungicide pyraclostrobin is the main measure used to control apple alternaria blotch in production. To evaluate the potential dietary risks for consumers, the dissipation and terminal residues of pyraclostrobin were investigated using ultra-high performance liquid chromatography–mass spectrometry (UHPLC–MS/MS). Pyraclostrobin in apples was extracted by acetonitrile with 2% ammonia and then purified using primary secondary amine (PSA) and graphitized carbon black (GCB). The method showed good linearity within the concentration range of 0.005–0.1 mg L^{-1}, with a coefficient of determination (R^2) $\geq$ 0.9958. The recoveries ranged from 96.0% to 103.8%, with relative standard deviations (RSDs) between 0.8% and 2.3%. The limit of quantification (LOQ) was 0.01 mg kg^{-1}. Pyraclostrobin dispersible oil suspension was applied in 12 apple fields across China according to good agricultural practices (GAPs). In Beijing and Shandong, the dissipation of pyraclostrobin followed first-order kinetic equations, with a half-life of 11 days. The terminal residues ranged from <0.01 to 0.09 mg kg^{-1}. The national estimated daily intake (NEDI) of pyraclostrobin was compared with the acceptable daily intake (ADI), resulting in risk quotient (RQ_c) of 80.8%. These results suggest that pyraclostrobin poses a low health risk to consumers under GAP conditions and according to recommended dosages.

Keywords: pyraclostrobin; UHPLC–MS/MS; dissipation; terminal residues; risk assessment

Citation: Wang, B.; Shi, L.; Ren, P.; Qin, S.; Li, J.; Cao, J. Dissipation and Dietary Risk Assessment of the Fungicide Pyraclostrobin in Apples Using Ultra-High Performance Liquid Chromatography–Mass Spectrometry. *Molecules* **2024**, *29*, 4434. https://doi.org/10.3390/molecules29184434

Academic Editor: José Bernal del Nozal

Received: 15 August 2024
Revised: 4 September 2024
Accepted: 14 September 2024
Published: 18 September 2024

1. Introduction

Apples are the most popular fruits among consumers, known for being rich in antioxidants, trace elements, vitamins, and other nutrients [1,2]. China is the world's largest apple producer and consumer, with a planting area of 2,129,134 hectares in 2022, accounting for about half of the world's apple production (https://www.fao.org/faostat/en/#compare (accessed on 29 July 2024)). However, apples are attacked by pests and diseases during their growth cycle, and pesticides need to be applied multiple times to protect farmers' income. Pyraclostrobin (Figure 1), *N*-(2-phenyl)-*N*-methoxycarbamate, belongs to the broad-spectrum methacrylate fungicide used to control apple alternaria mali Roberts. It has antibacterial activity against almost all fungal plant pathogens by blocking the electron transfer between pathogen cytochromes, making it impossible for mitochondria to provide energy (ATP) to cells, ultimately leading to cell death [3].

There are 791 registered pyraclostrobin products in China, including 297 single-dose formulations (http://www.chinapesticide.org.cn/zwb/dataCenter (accessed on 29 July 2024)). However, excessive use of pesticides leads to pesticide residues, which seriously affect the quality of apple products and may increase potential risks to human health. Pyraclostrobin's toxicity is widely recognized, as it initiates oxidative stress and apoptosis in zebrafish embryos [4,5], disrupts the antioxidant system in earthworms [6,7] (leading to irreversible toxicity), induces glutathione transferase activity and affects immobility in

Daphnia magna [8], and has the potential to induce oxidative DNA damage and mitochondrial dysfunction in human hepatocytes [9]. The maximum residue limits (MRLs) of pyraclostrobin in apples set by the European Union, China, the United States, South Korea, and Japan are 0.5, 0.5, 1.5, 0.2, and 1 mg kg^{-1}, respectively.

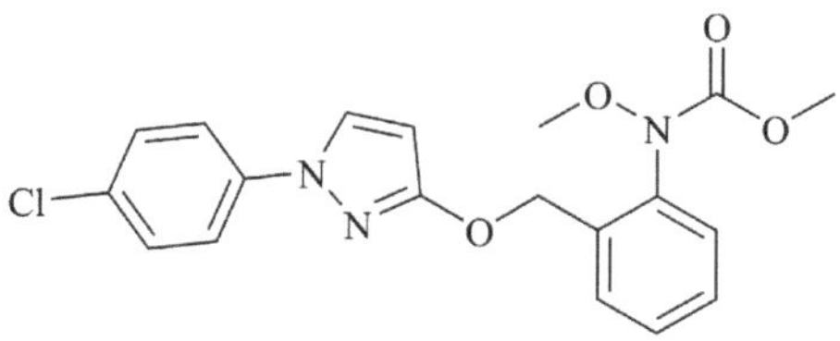

Figure 1. Structure of pyraclostrobin.

To avoid the potential harm of pyraclostrobin to public health, it is necessary to comprehensively study the residues of pyraclostrobin in apples and conduct a dietary risk assessment. As far as we know, there have been some reports of pyraclostrobin residues in fruits such as grapes [10,11], apricots [12], lemons [13], watermelons [14], bananas [15,16], strawberries [17,18], citrus [19], and apples [20]; information which is crucial for ensuring food safety. In these studies, LC–MS/MS and HPLC techniques were typically used to detect pyraclostrobin according to the different characteristics of the fruits. Fan et al. [21] studied pyraclostrobin's dissipation and terminal residues in apple samples through field trials in the Beijing, Shandong, and Anhui provinces in China based on HPLC–MS/MS. With its increasing use, dietary risk assessments of pyraclostrobin residues in fruits are necessary to determine whether excessive use poses a hazard to consumers. However, the dietary risk assessment of pyrazole in apples under Good Agricultural Practice conditions has not been reported.

In this study, the residues of pyraclostrobin in 12 representative apples in China were determined, aiming to establish an accurate, simple, and sensitive UHPLC–MS/MS method for the determination of pyraclostrobin residues in apples; to assess the dietary intake risk of pyraclostrobin based on the final residues; to provide guidance for the dietary risk assessment of pyraclostrobin residues in apples; and to provide an effective analytical method for the determination of pyraclostrobin in apples.

2. Results and Discussion

2.1. Method Validation

QuEChERS pretreatment has become the preferred method for determining pesticide residues in food. In this study, a modified QuEChERS method was used to extract pyraclostrobin from apples, using 2% ammonia and acetonitrile as the extraction solvent, followed by sodium chloride (NaCl) salting out, PSA and GCB dispersive solid phase extraction cleanup, and finally ultra-high performance liquid chromatography–mass spectrometry (UHPLC–MS/MS) detection. As in our study, Li et al. used a modified QuEChERS method to extract pyraclostrobin, methyl thiophanate, carbendazim, and tebuconazole from apples [22]. As shown in Figure 2, pyraclostrobin was well separated, with a retention time of 4.54 min. The linearity, sensitivity, precision, accuracy, and matrix effect (ME) of the established method were verified according to the guidelines on pesticide residue trials (NY/T 788-2018) [23] and the European Union SANTE guidelines (SANTE/11312/2021) [24].

The accuracy and precision of the method were verified by five repeated recovery experiments at three spiked levels. As shown in Table 1, the recoveries of pyraclostrobin in apples were 96.0–103.8%, and the RSD was 0.8–2.3%, which conformed to the European Union SANTE guidelines (SANTE/11312/2021). The limit of quantification (LOQ) was defined as the minimum spiked level (0.01 mg kg^{-1}). The recoveries of the method established by Fan et al. for detecting pyraclostrobin in apples ranged from 88.1% to

105.2%, with RSDs below 5.1% [21]. In contrast, the method we established was more accurate and stable, with recoveries ranging from 96.0 to 103.8% and RSDs below 2.3%.

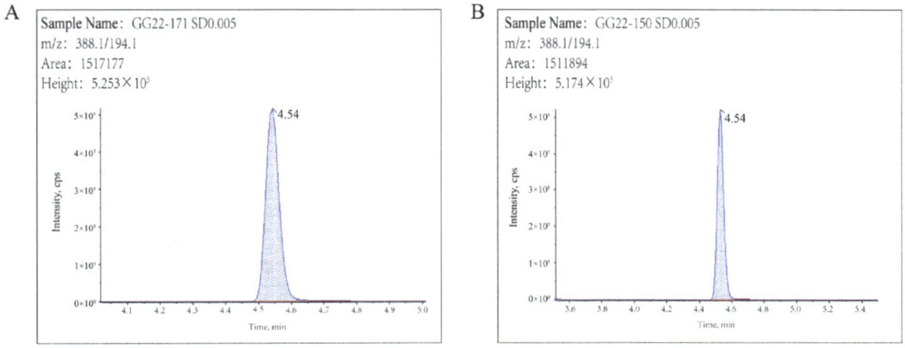

Figure 2. Chromatograms of 0.005 mg kg^{-1} pyraclostrobin in solvent (**A**) and blank apple (**B**).

Table 1. Recoveries of pyraclostrobin in apples.

Spiked Level (mg kg^{-1})	Recoveries (%)						RSD (%)
	1	2	3	4	5	Average	
0.01	100.1	99.7	101.3	103.8	97.0	100.4	2.3
0.1	96.9	99.2	100.5	99.8	100.2	99.3	1.4
0.5	97.8	98.4	96.6	97.6	96.0	97.3	0.8

ME refers to the interference of sample impurities with target compounds during ionization, thus affecting quantification accuracy [25]. ME may negatively impact the analysis of pyraclostrobin and must be evaluated during method validation. ME was usually evaluated by comparing the slopes of the matrix-matched standard curve to the slopes of the solvent standard curve. The pyraclostrobin standard solution was diluted with blank apple matrix (apples without pyraclostrobin treatment) and solvent to prepare 0.005, 0.01, 0.02, 0.05, and 0.1 mg L^{-1} standard solutions, which were measured by high-performance chromatography–tandem mass spectrometry. Blank apple samples and solvent were extracted and analyzed according to the established method, and no background interference was found at the retention time of pyraclostrobin. The matrix-matching calibration curve was $y = 2.45253 \times 10^8x + 3.62017 \times 10^5$, $R^2 = 0.9972$ and the solvent calibration curve was $y = 2.46766 \times 10^8x + 4.21324 \times 10^5$, $R^2 = 0.9958$, indicating an excellent linear relationship. The ME calculation result was −0.6%, indicating that the apple sample matrix had a weak inhibitory effect on pyraclostrobin. The ME was determined by the type of compound, extraction method and matrix type. There was considerable variability in matrix effects between different fruits. The ME of pyraclostrobin was 40% in strawberries [26] and 12.63% in pomegranates [27], using QuEChERS sample preparation combined with ultra-performance liquid chromatography–tandem mass spectrometry (UPLC–MS/MS). This study used the matrix-matching standard calibration method to eliminate the matrix effect.

In conclusion, the method established in this study was reliable, sensitive, and can be used to quantify pyraclostrobin in real apple samples.

2.2. Dissipation

The validated method was applied to detect pyraclostrobin residues in field samples. 59 g/L pyraclostrobin dispersible oil suspension 800 times was sprayed three times, according to the maximum recommended dosage of 73.75 mg kg^{-1}, with an interval of 7 days between applications. The initial depositions of pyraclostrobin in Shanxi, Beijing,

Shandong, and Henan were 0.014, 0.069, 0.082, and 0.090 mg kg^{-1}, respectively. The original deposition of pyraclostrobin on apples in Shanxi was significantly lower than in the other provinces. A variety of factors can affect the residual behavior of pesticides on crops, including the nature of the pesticide, fruit variety, climate, etc. The experimental apples in Shanxi were planted in 2013 and the trees are older than those in the other three places. The apples in Beijing and Shandong were planted in 2015, and those in Henan were planted in 2016. During the experiment, the average temperature (22.3 °C) and rainfall (202.4 mm) in Shanxi were higher than those in Beijing, where the average temperature was 21.2 °C and the average rainfall was 4.4 mm. Variances in the initial deposition of pyraclostrobin may be caused by soil type, apple variety, and temperature [28,29]. During the experiment, the average temperature in the four locations ranged from 21.2 °C to 27.9 °C. The apples in Shanxi, Beijing, Shandong, and Henan were mainly local varieties, namely Red Fuji, Red Fuji, Gala, and Red Star, respectively. There were significant differences in soil types among the Shanxi, Beijing, Shandong, and Henan provinces, with pH values of 8.48, 7.9, 6.9, and 7.9, cation exchange capacities of 13.92 cmol/kg, 13.8 cmol/kg, 9.3 cmol/kg, and 7.68 cmol/kg, and organic matter contents of 1.83%, 1.17%, 0.98%, and 1.29%, respectively. In previous studies, the half-life of pyraclostrobin was 12.8 days in banana pulp [15], 6.6–11.8 days in waxberries [30], 1.79–2.48 days in citrus [19], 17.8–25.9 days in grapes [10], and 3.27 days in watermelons [14]. Pyraclostrobin degraded rapidly in Shanxi and Anhui, and the residues were lower than LOQ at 28 days. Figure 3 shows the dissipation curves of pyraclostrobin in apples in Beijing and Shandong. The residues of pyraclostrobin in apples in Beijing and Shandong gradually decreased with time, consistent with the first-order kinetic equation. The dissipation equations were $C_t = 0.069 \times 10^{-0.061t}$ ($R^2 = 0.886$) and $C_t = 0.082 \times 10^{-0.061t}$ ($R^2 = 0.7842$); the half-life of pyraclostrobin was 11 days. Shandong and Beijing belong to a temperate semi-humid continental monsoon climate, while Shanxi and Henan belong to a temperate continental monsoon climate. Pyraclostrobin degrades rapidly in apples in Shanxi and Henan, and the half-life of pyraclostrobin in apples in Shandong and Beijing is 11 days, which may be related to climate. Consistent with our results, Magdalena et al. reported that the half-life of pyraclostrobin in apples was 11.5 days [31]. In another report, the original deposition of pyraclostrobin in apples was 0.07 to 0.53 mg kg^{-1}, and the half-life was 4.3 to 8.3 days [32]. The existing research results showed that the half-life of pyraclostrobin in fruits was less than 30 days, which means that it is a degradable pesticide.

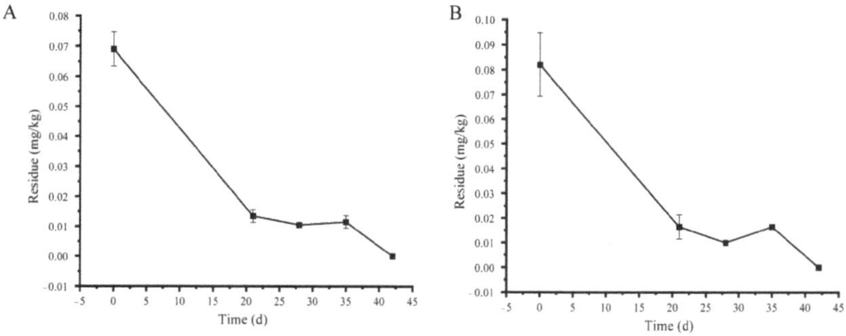

Figure 3. Dissipation curves of pyraclostrobin in apples in Beijing (**A**) and Shandong (**B**).

2.3. Final Residue

The samples were tested in batches according to their arrival time, and two quality control samples were added to each analysis. As shown in Table 2, the quality control results showed that the recoveries of pyraclostrobin in apples were 90.8–106.1%, indicating that the established detection method for pyraclostrobin was stable and accurate and the analysis results of the field trial samples were reliable.

Table 2. Quality control (QC) of pyraclostrobin in real sample detection.

Spiked Level mg kg^{-1}	QC Sample Analysis Date	Recoveries (%)		
		1	2	Average
0.10	31 August 2022	95.1	96.4	95.7
	21 September 2022	95.8	97.0	96.4
	13 October 2022	103.6	106.1	104.8
	24 October 2022	90.8	95.1	92.9
	3 November 2022	96.0	98.9	97.4

Harvest intervals of 28 and 35 days were designed to monitor terminal residues of pyraclostrobin and recommend pre-harvest intervals (PHIs) for commercial formulations applied in apples. As shown in Table 3, at PHIs of 28 and 35 days, the terminal residues of pyraclostrobin in apples were <0.010–0.070 mg kg^{-1} and <0.010–0.048 mg kg^{-1}, which were lower than the MRL (0.5 mg kg^{-1}) recommended by China [33]. Therefore, according to the maximum recommended dosage of 73.75 mg kg^{-1} for spraying, 59 g L^{-1} pyraclostrobin dispersed oil suspension should be applied three times with an interval of 7 days between applications, and the recommended PHI is 28 days. These data can provide a reference for the rational use of pyraclostrobin in apples.

Table 3. The final residue of pyraclostrobin in apple samples.

Location	Terminal Residue (mg kg^{-1})				Supervised Trials Median Residue (STMR, mg kg^{-1})
	PHI = 28 days		PHI = 35 days		
	1	2	1	2	
Shenyang, Liaoning province	0.013	<0.010	<0.010	<0.010	0.012
Jinzhong, Shanxi Province	<0.010	<0.010	0.02	0.01	
Yuncheng, Shanxi Province	0.048	0.028	0.016	0.012	
Dingxi, Gansu province	0.013	0.012	<0.010	<0.010	
Yinchuan, Ningxia Province	0.031	0.042	0.022	0.016	
Changping, Beijing	0.011	<0.010	<0.010	0.013	
Tai'an, Shandong Province	0.01	<0.010	0.017	0.016	
Qingdao, Shandong Province	0.06	0.07	0.048	0.047	
Xinxiang, Henan Province	<0.010	<0.010	<0.010	<0.010	
Zhumadian, Henan province	<0.010	<0.010	<0.010	<0.010	
Suzhou, Anhui Province	0.019	0.014	0.013	<0.010	
Kunming, Yunnan Province	0.041	0.035	0.033	0.026	

2.4. Dietary Risk Assessment

As a method, chronic risk quotient (RQ_c) was the most commonly used parameter in dietary risk assessments of pesticide residues [34]. RQ_c was calculated by comparing the NEDI (national estimated daily intake) with the product of average body weight (bw) and ADI (acceptable daily intake). The weight of an Chinese adult was considered to be 63 kg [35]. An RQ_c greater than 100% indicates an unacceptable level of risk, whereas an RQc below 100% indicates an acceptable level of risk, and the larger the RQ_c, the higher the risk [36]. The NEDI was calculated based on the typical Chinese dietary structure and the supervised trials median residue (STMR) or MRL of pyraclostrobin in registered crops. The ADI value of pyraclostrobin in GB2763-2021 was 0.03 mg kg^{-1} bw [33]. The MRL set by China was given priority. The MRL set by CAC and the United States was referenced if missing. The STMR of pyraclostrobin for 28 days was 0.012 mg kg^{-1}. As shown in Table 4, the RQ_c of pyraclostrobin (80.8%) was less than 100%, indicating that pyraclostrobin will not pose a long-term risk to ordinary Chinese consumers when used in apples at the recommended dose and consumed according to the typical Chinese dietary structure.

Table 4. Chronic dietary risk assessment of pyraclostrobin.

Food Classification	Daily Consumption of a Particular Food (F_i, kg)	Reference Residue Limits (mg kg^{-1})	National Estimated Daily Intake (NEDI, mg)	Acceptable Daily Intake × Average Body Weight (mg)	Chronic Risk Quotient (RQ_c, %)
Rice and its products	0.2399	1 (China)	0.2399		
Flour and its products	0.1385	0.2 (China)	0.0277		
Other cereals	0.0233	0.05 (China)	0.001165		
Tubers	0.0495	0.05 (China)	0.002475		
Dried beans (products)	0.016	0.2 (China)	0.0032		
Dark vegetables	0.0915	2 (China)	0.183		
light vegetables	0.1837	5 (China)	0.9185		
Pickles	0.0103				
Fruits	0.0457	0.012 (STMR, China)	0.0005484	ADI × 63	
Nuts	0.0039				
Livestock and poultry	0.0795				
Milk and dairy products	0.0263				
Egg and its products	0.0236				
Egg and its products	0.0301				
Vegetable oil	0.0327	0.1 (China)	0.00327		
Animal oil	0.0087				
Sugar, starch	0.0044				
Salt	0.012	10 (China)	0.12		
Salt	0.009	3 (China)	0.027		
Total	1.0286		1.5268	1.8900	80.8%

3. Materials and Methods

3.1. Chemicals and Reagents

The certified reference standards of pyraclostrobin (purity 99.57%) were purchased from Dr. Ehrenstorfer (Augsburg, Germany). HPLC-grade acetonitrile was purchased from the Tedia Company, Inc. (Fairfield, CT, USA). LC–MS grade acetonitrile was provided by Thermo Fisher Scientific (Shanghai, China). Analytical grade ammonia was purchased from the Tianjin Damao Chemical Reagent Factory (Tianjin, China). Analytical grade anhydrous magnesium sulfate ($MgSO_4$), sodium chloride (NaCl), disodium hydrogen citrate ($C_6H_6Na_2O_7$), sodium citrate ($C_6H_5Na_3O_7$), primary secondary amine (PSA, 40–60 μm), and graphitized carbon black (GCB, 40–60 μm) were purchased from Shimadzu Laboratory Equipment Co., Ltd. (Shanghai, China).

Weigh 10 ± 0.1 mg pyraclostrobin standard in a 10 mL volumetric flask and dissolve in mass spectrometry grade acetonitrile to prepare a standard stock solution (1000 mg L^{-1}). Dilute the above standard stock solution with acetonitrile to prepare a 100 mg L^{-1} standard solution. Store at −18 °C until use. Dilute the above standard solution with acetonitrile and blank apple samples to prepare a series of standard solutions and matrix-matched standard solutions of 5, 10, 20, 50, and 100 μg L^{-1}.

3.2. Chromatography Conditions

Pyraclostrobin was determined by ultra-high performance liquid chromatography–tandem triple quadrupole mass spectrometry (SCIEXAB TripleQuad 4500) (Foster City, CA, USA) in multiple reaction monitoring (MRM) mode and mass spectrometry analysis was performed in positive ion mode (ESI+). Pyraclostrobin was separated by ACQUITY UPLC® HSST3 column (100 mm × 2.1 mm, 1.8 μm) (Waters, Milford, MA, USA), and the column temperature was 40 °C. The mobile phase was 4 mmol/L ammonium acetate aqueous solution (A) and methanol (B) containing 0.1% formic acid. The flow rate was 0.3 mL min^{-1} and the injection volume was 2 μL. The ion source temperature was 150 °C, and the cone voltage was 50 °C. Two daughter ions were selected as quantitative and qualitative ions: m/z 388.10 → 194.1 (18 eV) and 388.10 → 163.1 (36 eV).

3.3. Extraction and Purification Procedures

The QuEChERS method was developed by the United States Department of Agriculture in 2003 [37] and has been widely used in detecting pesticide residues in apples [38,39]. In this study, a modified QuEChERS method was used to extract pyraclostrobin residues in apples. 10.00 g apple samples were weighed into a 50 mL centrifuge tube. 20 mL acetonitrile containing 2% ammonia water was added to the centrifuge tube and ultrasonicated for 5 min. Then 4 g $MgSO_4$, 1 g NaCl, 0.5 g disodium hydrogen citrate, and 1 g sodium citrate were added to each sample, oscillated for 1 min, and centrifuged at 3000 r min^{-1} for 3 min. 1.5 mL of supernatant was placed in a 2 mL centrifuge tube containing 142.5 mg $MgSO_4$, 20 mg PSA, and 7.5 mg GCB, vortexed at 2500 r min^{-1} for 1 min, and centrifuged at 5000 r min^{-1} for 2 min. The supernatant was removed by syringe, passed through a 0.22 μm organic filter membrane, and stored in a sample injection vial for detection.

3.4. Method Validation

The method's linearity, limit of quantification (LOQ), matrix effect, accuracy, and precision were validated according to the EU guidance document.

The pyraclostrobin standard solution was diluted to five concentrations (5, 10, 20, 50, 100 μg L^{-1}) using blank apples and LC–MS grade acetonitrile to obtain matrix-matched and solvent-based calibration curves. Impurities in the sample matrix can interfere with the ionization of the target compound, thus affecting the accuracy of quantification. The matrix effect (ME%) was calculated as follows:

$$\mathrm{ME}(\%) = \frac{\mathrm{Smatrix} - \mathrm{Ssolvent}}{\mathrm{Ssolvent}} \times 100\% \quad (1)$$

where S_{matrix} and $S_{solvent}$ represent the slopes of the matrix-matched standard curve and the solvent standard curve, respectively.

The accuracy and precision of the method were evaluated by spike recovery experiment. The pyraclostrobin standard solution was spiked into blank apple samples at three concentrations (0.01, 0.1, and 0.5 mg kg^{-1}) and five replicates were performed at each concentration level and allowed to stand for 1 h. The spiked samples were extracted and cleaned up according to the sample extraction and cleanup procedures are described in Section 3.3. The limit of quantification (LOQ) was defined as the lowest spike concentration.

3.5. Field Trials

Following the guidelines on pesticide residue trials (NY/T 788-2018) issued by the Ministry of Agriculture and Rural Affairs of China, the open field experiment was conducted in 12 different locations in China, including Shenyang, Liaoning (123.7 E, 42.7 N); Jinzhong, Shanxi (112.68 E, 37.55 N); Yuncheng, Shanxi (110.54 E, 35.06 N); Dingxi, Gansu (103.77 E, 35.64 N); Yinchuan, Ningxia (106.05 E, 38.48 N); Beijing (116.28 E, 40.22 N); Tai'an, Shandong (116.91 E, 36.00 N); Qingdao, Shandong (120.57 E, 36.41 N); Xinxiang, Henan (113.76 E, 34.98 N); Zhumadian, Henan (114.09 E, 33.07 N); Suzhou, Anhui (116.72 E, 344.30 N); and Kunming, Yunnan (102.37 E, 24.63 N). At least 4 apple trees should be in each experimental plot, with isolation zones between plots and a blank control set up. The average temperature and total precipitation information in 12 locations are shown in Table S1.

Terminal residue experiments were carried out in 12 locations. 59 g L^{-1} pyraclostrobin dispersible oil suspension 800 times was sprayed 3 times according to the recommended dosage of 73.75 mg kg^{-1}, with intervals of 7 days. The control group was sprayed with water without pesticides. Mature apple samples were randomly collected from the experimental plots on the 28th and 35th days after the last spraying, with 2 independent samples collected each time. Detailed sample collection information is shown in Table S2.

Dissipation experiments were conducted in Jinzhong, Shanxi Province; Beijing and Tai'an, Shandong Province; and Xinxiang, Henan Province. 59 g L^{-1} pyraclostrobin dispersible oil suspension 800 times was sprayed three times, according to the maximum recommended dosage of 73.75 mg kg^{-1}, with intervals of 7 days between applications. Apple samples were randomly collected from each plot at 0 (2 h), 21, 28, 35, and 42 days after the last spraying to evaluate the dissipation kinetics of pyraclostrobin. All samples were chopped, mixed, and divided into quarters, and two samples of no less than 200 g were taken, then labeled and stored in a −18°C freezer for further analysis. Detailed sample collection information is shown in Table S3.

3.6. Dissipation of Pyraclostrobin

The dissipation curve of pyraclostrobin was fitted by the first-order rate equation

$$C_t = C_0 \times \exp(-kt) \quad (2)$$

$$DT_{50} = \ln 2/K \quad (3)$$

where

C_t and C_0 represent the residual concentration (mg kg^{-1}) at time t (d) and the original deposition amount of pyraclostrobin;
k is the dissipation rate constant (d^{-1}); and
DT_{50} indicates the degradation half-life of pyraclostrobin (d).

3.7. Chronic Dietary Risk Assessment

The chronic risk quotient (RQ_c) was used to study the chronic dietary risk of pyraclostrobin, and the calculation formula was as follows:

$$NEDI = \sum(STMR \times Fi) \quad (4)$$

$$RQ_c = \frac{NEDI}{ADI \times bw} \quad (5)$$

where

NEDI (mg kg^{-1}) is the national estimated daily intake;
STMR (mg kg^{-1}) is the supervised trials median residue of pyraclostrobin in apples;
Fi (kg) is the reference dietary intake;
ADI (mg kg^{-1} bw) is the acceptable daily intake; and
bw (63 kg) is the average body weight.

4. Conclusions

In this study, a sensitive and effective QuEChERS-UHPLC–MS/MS method was established to detect pyraclostrobin residues in apples. The established method showed satisfactory linearity, selectivity, trueness and precision validation parameters. The recoveries were 96.0–103.8% with RSD values below 2.3%. Field trials were conducted according to Good Agricultural Practice. Dissipation tests at four sites showed that the original deposition of pyraclostrobin in apples was between <0.010 and 0.093 mg kg^{-1}, and the DT_{50} was 11 days in Beijing and Shandong. The results of field trials at 12 sites showed that the residues of pyraclostrobin in all samples were between <0.010 and 0.070 mg kg^{-1}, which was lower than the MRL of 0.5 mg kg^{-1} recommended by China. Considering all registered crops for pyraclostrobin for dietary risk assessment, the calculated results showed that the threat to consumers at the recommended dose was negligible, with an RQ_c of 80.8%. Combined with the relevant MRL regulations and dietary intake risks, the PHI of 59 g L^{-1} pyraclostrobin dispersible oil suspension should be 28 days. This study can provide comprehensive risk assessment guidance for the rational use of pyraclostrobin in apple ecosystems.

Supplementary Materials: The following supporting information can be downloaded at https://www.mdpi.com/article/10.3390/molecules29184434/s1, Table S1: Average temperature and total precipitation information; Table S2: Apple sample collection for terminal residue experiment; Table S3: Apple sample collection for dissipation experiment.

Author Contributions: Conceptualization, S.Q.; methodology, L.S. and J.L.; validation, B.W.; formal analysis, B.W.; resources, J.L.; data curation, P.R.; writing—original draft, B.W.; writing—review and editing, J.C.; visualization, P.R. and J.L.; project administration, S.Q.; and funding acquisition, L.S. and S.Q. All authors have read and agreed to the published version of the manuscript.

Funding: This research was funded by 'Transformation of technical achievements in processing characteristic coarse cereals into staple food' (YDZJSX2022C012), the Shanxi Key R&D Project (202202130501011), and the Shanxi Applied Basic Research Program Science–Youth Technology Research Fund (202103021223149).

Institutional Review Board Statement: Not applicable.

Informed Consent Statement: Not applicable.

Data Availability Statement: The data presented in this study are available from the authors upon request.

Acknowledgments: We would like to thank Zhang Junwen, Su Hailing, and Zhu Li'ao for their support in this field trial and our colleague Wang Xia and Jin jing for technical support.

Conflicts of Interest: The authors declare no conflict of interest.

References

1. Koutsos, A.; Tuohy, K.M.; Lovegrove, J.A. Apples and cardiovascular health—Is the gut microbiota a core consideration? *Nutrients* **2015**, *7*, 3959–3998. [CrossRef] [PubMed]
2. Hyson, D.A. A comprehensive review of apples and apple components and their relationship to human health. *Adv. Nutr.* **2011**, *2*, 408–420. [CrossRef] [PubMed]
3. Balba, H. Review of strobilurin fungicide chemicals. *J. Environ. Sci. Health Part B* **2007**, *42*, 441–451. [CrossRef]
4. Huang, X.; Yang, S.; Li, B.; Wang, A.; Li, H.; Li, X.; Luo, J.; Liu, F.; Mu, W. Comparative toxicity of multiple exposure routes of pyraclostrobin in adult zebrafish (*Danio rerio*). *Sci. Total Environ.* **2021**, *777*, 145957. [CrossRef] [PubMed]
5. Zhang, C.; Wang, J.; Zhang, S.; Zhu, L.; Du, Z.; Wang, J. Acute and subchronic toxicity of pyraclostrobin in zebrafish (*Danio rerio*). *Chemosphere* **2017**, *188*, 510–516. [CrossRef]
6. Hou, K.; Shi, B.; Liu, Y.; Lu, C.; Li, D.; Du, Z.; Li, B.; Zhu, L. Toxicity evaluation of pyraclostrobin exposure in farmland soils and co-exposure with nZnO to *Eisenia fetida*. *J. Hazard. Mater.* **2022**, *433*, 128794. [CrossRef]
7. Ma, J.; Cheng, C.; Du, Z.; Li, B.; Wang, J.; Wang, J.; Wang, Z.; Zhu, L. Toxicological effects of pyraclostrobin on the antioxidant defense system and DNA damage in earthworms (*Eisenia fetida*). *Ecol. Indic.* **2019**, *101*, 111–116. [CrossRef]
8. Cui, F.; Chai, T.; Liu, X.; Wang, C. Toxicity of three strobilurins (kresoxim-methyl, pyraclostrobin, and trifloxystrobin) on *Daphnia magna*. *Environ. Toxicol. Chem.* **2017**, *36*, 182–189. [CrossRef] [PubMed]
9. Wu, M.; Bian, J.; Han, S.; Zhang, C.; Xu, W.; Tao, L.; Li, Z.; Zhang, Y. Characterization of hepatotoxic effects induced by pyraclostrobin in human HepG2 cells and zebrafish larvae. *Chemosphere* **2023**, *340*, 139732. [CrossRef]
10. Chen, X.; He, S.; Gao, Y.; Ma, Y.; Hu, J.; Liu, X. Dissipation behavior, residue distribution and dietary risk assessment of field-incurred boscalid and pyraclostrobin in grape and grape field soil via MWCNTs-based QuEChERS using an RRLC-QqQ-MS/MS technique. *Food Chem.* **2019**, *274*, 291–297. [CrossRef]
11. Zhao, P.; Liu, R.; Yuan, L. Dissipation, Residue and Human Dietary Risk Assessment of Pyraclostrobin and Cyazofamid in Grapes Using an HPLC-UV Detector. *Foods* **2024**, *13*, 314. [CrossRef] [PubMed]
12. Dost, K.; Oksuz, M.; Cittan, M.; Mutlu, B.; Tural, B. Determination of boscalid, pyraclostrobin and trifloxystrobin in dried grape and apricot by HPLC/UV method. *J. Food Compos. Anal.* **2023**, *115*, 104926. [CrossRef]
13. Dominguez, A.N.; Jimenez, L.E.; Alvarez, R.M.S. Rapid detection of pyraclostrobin fungicide residues in lemon with surface-enhanced Raman spectroscopy. *J. Food Meas. Charact.* **2023**, *17*, 6350–6362. [CrossRef]
14. Lv, L.; Su, Y.; Dong, B.; Lu, W.; Hu, J.; Liu, X. Dissipation Residue Behaviors and Dietary Risk Assessment of Boscalid and Pyraclostrobin in Watermelon by HPLC-MS/MS. *Molecules* **2022**, *27*, 4410. [CrossRef] [PubMed]
15. Lin, S.; Yang, L.; Zheng, Q.; Wang, Y.; Cheng, D.; Zhang, Z. Dissipation and distribution of pyraclostrobin in bananas at different temperature and a risk assessment of dietary intake. *Int. J. Environ. Anal. Chem.* **2022**, *102*, 5798–5810. [CrossRef]

16. Yang, M.; Zhang, J.; Zhang, J.; Rashid, M.; Zhong, G.; Liu, J. The control effect of fungicide pyraclostrobin against freckle disease of banana and its residue dynamics under field conditions. *J. Environ. Sci. Health Part B* **2018**, *53*, 615–621. [CrossRef]
17. Malhat, F.; Saber, E.-S.; Shokr, S.A.E.; Ahmed, M.T.; Amin, A.E.-S. Consumer safety evaluation of pyraclostrobin residues in strawberry using liquid chromatography tandem mass spectrometry (LC-MS/MS): An Egyptian profile. *Regul. Toxicol. Pharmacol.* **2019**, *108*, 104450. [CrossRef]
18. Valera-Tarifa, N.; Santiago-Valverde, R.; Hernández-Torres, E.; Martínez-Vidal, J.; Garrido-Frenich, A. Development and full validation of a multiresidue method for the analysis of a wide range of pesticides in processed fruit by UHPLC-MS/MS. *Food Chem.* **2020**, *315*, 126304. [CrossRef]
19. Liu, X.; Yang, Y.; Chen, Y.; Zhang, Q.; Lu, P.; Hu, D. Dissipation, residues and risk assessment of oxine-copper and pyraclostrobin in citrus. *Food Addit. Contam. Part A Chem. Anal. Control Expo. Risk Assess.* **2019**, *36*, 1538–1550. [CrossRef]
20. Zarębska, M.; Hordyjewicz-Baran, Z.; Wasilewski, T.; Zajszły-Turko, E.; Stanek, N. A new LC-MS method for evaluating the efficacy of pesticide residue removal from fruit surfaces by washing agents. *Processes* **2022**, *10*, 793. [CrossRef]
21. Fan, X.; Zhao, S.; Chen, X.; Hu, J. Simultaneous Determination of Pyraclostrobin, Prochloraz, and its Metabolite in Apple and Soil Via RRLC-MS/MS. *Food Anal. Methods* **2018**, *11*, 1312–1320. [CrossRef]
22. Li, P.; Sun, P.; Dong, X.; Li, B. Residue analysis and kinetics modeling of thiophanate-methyl, carbendazim, tebuconazole and pyraclostrobin in apple tree bark using QuEChERS/HPLC–VWD. *Biomed. Chromatogr.* **2020**, *34*, e4851. [CrossRef] [PubMed]
23. Ministry of Agriculture and Rural Affairs of the People's Republic of China. *Guideline for the Testing of Pesticide Residues in Crops*; China Agriculture Press: Beijing, China, 2018.
24. European Commission. *Guidance Document on Analytical Quality Control and Method Validation Procedures for Pesticides Residues and Analysis in Food and Feed*; European Commission: Brussels, Belgium, 2021.
25. Matuszewski, B.K.; Constanzer, M.; Chavez-Eng, C. Strategies for the assessment of matrix effect in quantitative bioanalytical methods based on HPLC−MS/MS. *Anal. Chem.* **2003**, *75*, 3019–3030. [CrossRef]
26. Wang, H.; Ping, H.; Liu, Q.; Han, P.; Guo, X. Determination of Pesticide Residues in Strawberries by Ultra-performance Liquid Chromatography-Tandem Mass Spectrometry. *Food Anal. Methods* **2022**, *15*, 85–95. [CrossRef]
27. Naik, R.H.; Pallavi, M.S.; Nandini; Shwetha, A.; Bheemanna, M.; Nidoni, R.U. Simultaneous determination of pesticide residues in pomegranate whole fruit and arils using LC-MS/MS. *Food Chem.* **2022**, *387*, 132865. [CrossRef]
28. Fan, X.; Zhao, S.; Hu, J. Dissipation behavior and dietary risk assessment of lambda-cyhalothrin, thiamethoxam and its metabolite clothianidin in apple after open field application. *Regul. Toxicol. Pharmacol.* **2019**, *101*, 135–141. [CrossRef]
29. Tian, F.; Lu, J.; Qiao, C.; Wang, C.; Pang, T.; Guo, L.; Li, J.; Pang, R.; Xie, H. Dissipation behavior and risk assessment of imidacloprid and its metabolites in apple from field to products. *Chemosphere* **2024**, *359*, 142309. [CrossRef]
30. Jianzhong, Y.; Liezhong, C.; Jiayin, H.; Ruixian, Y.; Xiuqing, H.; Xueping, Z. Residue and dissipation dynamics of pyraclostrobin in waxberry (*Myrica rubra*) and soil. *Chin. J. Pestic. Sci.* **2020**, *22*, 857–863.
31. Podbielska, M.; Szpyrka, E.; Piechowicz, B.; Sadło, S.; Sudoł, M. Assessment of boscalid and pyraclostrobin disappearance and behavior following application of effective microorganisms on apples. *J. Environ. Sci. Health Part B* **2018**, *53*, 652–660. [CrossRef]
32. Yan, Q.; Pengfei, D.; Yue, Z.; Shanshan, W.; Maojun, J.; Jing, W. Determination and dissipation dynamics of cuppric nonyl phenolsulfonate and pyraclostrobin in apples and soil. *Chin. J. Pestic. Sci.* **2017**, *19*, 76–83.
33. *GB 2763-2021*; National Food Safety Standard-Maximum Residue Limits for Pesticides in Food. Ministry of Agriculture and Rural Affairs of the People's Republic of China: Beijing, China, 2021.
34. Dong, M.; Ma, L.; Zhan, X.; Chen, J.; Huang, L.; Wang, W.; Zhao, L. Dissipation rates and residue levels of diflubenzuron and difenoconazole on peaches and dietary risk assessment. *Regul. Toxicol. Pharmacol.* **2019**, *108*, 104447. [CrossRef] [PubMed]
35. Li, R.; Men, X.; Li, R.; Liu, T.; Liang, H.; Fang, F.; Sun-Waterhouse, D.; Wang, Y. Residue behaviors and dietary risk of cyazofamid in turnip, onion and romaine lettuce assessed by a QuEChERS-LC-MS/MS method. *Food Sci. Hum. Wellness* **2023**, *12*, 1538–1544. [CrossRef]
36. Dong, X.; Tong, Z.; Chu, Y.; Sun, M.; Wang, M.; Gao, T.; Duan, J. Dissipation of prothioconazole and its metabolite prothioconazole-desthio in rice fields and risk assessment of its dietary intake. *J. Agric. Food Chem.* **2019**, *67*, 6458–6465. [CrossRef] [PubMed]
37. Anastassiades, M.; Lehotay, S.J.; Štajnbaher, D.; Schenck, F.J. Fast and easy multiresidue method employing acetonitrile extraction/partitioning and "dispersive solid-phase extraction" for the determination of pesticide residues in produce. *J. AOAC Int.* **2003**, *86*, 412–431. [CrossRef]
38. Lehotay, S.J.; Son, K.A.; Kwon, H.; Koesukwiwat, U.; Fu, W.; Mastovska, K.; Hoh, E.; Leepipatpiboon, N. Comparison of QuEChERS sample preparation methods for the analysis of pesticide residues in fruits and vegetables. *J. Chromatogr. A* **2010**, *1217*, 2548–2560. [CrossRef]
39. Machado, I.; Gérez, N.; Pistón, M.; Heinzen, H.; Cesio, M.V. Determination of pesticide residues in globe artichoke leaves and fruits by GC–MS and LC–MS/MS using the same QuEChERS procedure. *Food Chem.* **2017**, *227*, 227–236. [CrossRef]

foods MDPI

Article

Metabolomics Reveals Glycerophospholipids, Peptides, and Flavonoids Contributing to Breast Meat Flavor and Benefit Properties of Beijing-You Chicken

Jian Zhang, Xia Chen, Jing Cao, Ailian Geng, Qin Chu, Zhixun Yan, Yao Zhang and Huagui Liu *

Institute of Animal Husbandry and Veterinary Medicine, Beijing Academy of Agriculture and Forestry Sciences, Beijing 100097, China; zjcau@126.com (J.Z.); chenxia_91@163.com (X.C.); caojing2046555@163.com (J.C.); ailiangengcau@126.com (A.G.); chuqinsd@163.com (Q.C.); yanzhixun2008@sina.com (Z.Y.); duguyimeng1@126.com (Y.Z.)

* Correspondence: liuhuagui@baafs.net.cn

Abstract: Unique metabolites contribute to the performance of meat flavor and potential function. In this study, UHPLC-Q Exactive HF-X-based metabolomics and multivariate analysis were applied to explore the characteristic metabolites in the breast meat of Beijing-You chicken (BYC) aged 150, 300, and 450 days (D150, D300, and D450). Based on the criteria of variable importance in the projection (VIP) > 1 and $p < 0.05$, a total of 154 and 97 differential metabolites (DMs) were screened out compared with D450 (D450 vs. D150, D450 vs. D300), respectively. In general, the relative content of carnosine, L-L-homoglutathione, demethyloleuropein, neohesperidin dihydrochalcone, 7-chloro-2-(3,4-dimethoxyphenyl)-3,5-dihydroxy-6,8-dimethoxy-4H-chromen-4-one, glycerophospholipids, exhibited the highest abundance at D450, while balenine, anserine, L-beta-aspartyl-L-leucine, glutathione, oxidized glutathione, stearoylcarnitine, ganoderic acid alpha, oleuroside, Lysoglycerophospholipid species (LGP) presented a downward trend with age. These 210 DMs were involved in 10 significantly enriched pathways related to the synthesis and metabolism of amino acids, peptides, and glycerophospholipid, such as glutathione metabolism, histidine metabolism, glycerophospholipid metabolism, arginine biosynthesis, tyrosine metabolism, and lysine degradation. In conclusion, this work could not only facilitate a better understanding of the differences of chicken flavor and benefit properties with age, but also provide potential valuable bioactive compounds for further research.

Keywords: Beijing-You chicken; metabolomics; glycerophospholipids; peptides; flavonoids; breast meat

Citation: Zhang, J.; Chen, X.; Cao, J.; Geng, A.; Chu, Q.; Yan, Z.; Zhang, Y.; Liu, H. Metabolomics Reveals Glycerophospholipids, Peptides, and Flavonoids Contributing to Breast Meat Flavor and Benefit Properties of Beijing-You Chicken. *Foods* **2024**, *13*, 2549. https://doi.org/10.3390/foods13162549

Academic Editor: Yonathan Asikin

Received: 19 July 2024
Revised: 11 August 2024
Accepted: 13 August 2024
Published: 15 August 2024

1. Introduction

Foods that contain special quality nutrients/bioactive compounds continues to attract significant attention in recent scientific studies, and are referred to as functional foods due to their health and wellbeing characteristics [1]. In the face of increasing demand for functional products, there is a great need to identify and evaluate bioactive compounds in food.

Most of the available evidence suggests that poultry meat represents a safe, low-fat source of high-quality dietary protein, which could provide the nutrients humans need [2]. Chicken, especially older indigenous native breeds, is one of the most important sources of functional food containing a wide range of bioactive compounds including carnosine (Car), anserine (Ans), balenine (Bal), glutathione, and a series of lipid molecular species [3]. For example, Silkie chicken, a famous Chinese indigenous chicken breed with prosperous medicinal and nutritional value, was referred to as a functional food product in China [4,5].

In recent years, with the development of metabolite separation, bioinformatics platforms, and mass-spectrometry resolution, the metabolomics approach has been comprehensively applied not only to characterize bioactive compounds, flavor, and nutrients in meat, particularly in chicken [6,7], but also to comprehensively understand the dynamic biochemical changes in metabolites caused by specific factors, such as animal age [8], animal genetic background [9], postmortem aging [10], and meat treatment processing [11]. A series of metabolites, such as estradiol, lipid molecules, and fatty acids, have been identified as characteristic bioactive compounds responsible for the health-promoting effects of Silkie chicken [5]. Using HPLC–QTRAP–MS-based metabolomics, the metabolites in breast meat of Beijing-You chicken are affected by age, and arginine biosynthesis, purine metabolism, alanine, aspartic acid, and glutamic acid metabolism are further suggested as important metabolic pathways [12]. Moreover, chickens with older age are highly preferred over younger ones by consumers due to their special bioactive compounds, and significant health benefits, which are widely accepted in Chinese culture [13,14]. Revealing these functional metabolites of chickens of different ages is thus fascinating for the potential health benefits of chickens with older age to be further realized, whereas studies focusing on the metabolic profiling of chickens associated with age have not yet been thoroughly studied, particularly for indigenous native breeds with age prolonged more than 450 days.

Therefore, this study selected Beijing-You chicken (BYC), a famous indigenous chicken breed with excellent meat quality, as a model system to explore the dynamic differences in metabolic profiling in breast meat throughout a wide age spectrum for a duration of 450 days through LC–MS/MS analysis strategies. Characterization and discrimination of metabolic profiles of BYC breast meat between different ages were conducted. Moreover, based on chicken at 450 days, the potential pathways and bioactive compounds of lipids, peptides, and flavonoids were further provided. This study provides an important foundation to dissect the mechanisms involved in dynamic alternations of bioactive compounds and pathways in BYC breast meat due to age.

2. Materials and Methods

2.1. Ethics Statement

The experimental procedures and animal welfare practices were conducted in accordance with the guidelines for experimental animals established by the Ministry of Science and Technology (Beijing, China). The experiment was approved by the Institute of Animal Husbandry and Veterinary Medicine, Beijing Academy of Agriculture and Forestry Sciences (Beijing, China) (No. BAAFS-IAHVM20190115).

2.2. Animals and Samples Collection

The detailed information on animals and sample preparation was consistent with that of our previous studies [15,16]. In brief, a total of 90 1-day-old female BYC birds with the same genetic backgrounds were provided by the Institute of Animal Husbandry and Veterinary Medicine, Beijing Academy of Agriculture and Forestry Sciences. These birds were evenly divided into three groups and raised under identical environmental and nutritional conditions with free access to feed and water throughout the whole rearing period. At each sampling age (D150, D300, and D450), 15 birds were randomly selected from each group, electrically stunned and killed by exsanguination, respectively. The samples from the left fillet (pectoralis major) were collected, snap-frozen in liquid nitrogen, and stored at −80 °C until UHPLC-Q Exactive HF-X metabolomics analysis was carried out. On the other hand, the right fillet samples were stored at 4 °C for measuring of meat quality characteristics.

2.3. Meat Quality Characteristics

The meat dry matter content was determined by the relative weight difference before and after drying using a freeze-dryer (Millrock Technology, Kingston, NY, USA). Then, about 5 g dried meat powder per sample was used to calculate IMF using Soxhlet extraction with petroleum ether according to the method of Zerehdaran et al. [17]. In addition, 1 g degreased meat lyophilized powder was used to obtain the composition of free amino acids and some peptides, as described by Chen et al. [18], using an amino acid analyzer (L-8900, Hitachi Ltd., Tokyo, Japan).

2.4. Sample Preparation of Metabolomics Study

The extraction method of metabolites was adopted according to the report by Li et al. [19]. In brief, 400 μL of methanol/water = 4:1 (*v/v*) mixture, containing 0.02 mg/mL of internal standard (L-2-chlorophenylalanine), was used to extract metabolites from the meat samples (50 mg) in a 2 mL centrifuge tube with a 6 mm diameter grinding ball. A Wonbio-96c frozen tissue grinder (Shanghai Wando Biotechnology co., LTD, Shanghai, China) was applied for 6 min (−10 °C at 50 Hz) to grind the samples, and then they were low-temperature ultrasonic extracted for 30 min (5 °C at 40 kHz). Subsequently, the samples were placed at −20 °C for 30 min, and centrifuged at 13,000× *g* for 15 min at 4 °C. Finally, the supernatant was transferred into an injection vial for LC–MS/MS analysis. Moreover, to ensure the reliability and stability of the analysis, a quality control (QC) sample was inserted into every 5–15 analysis samples. QC samples were prepared by mixing 20 μL extracts from each sample. Both the injection and detection methods of the QC samples are consistent with those of the normal samples.

2.5. UHPLC–MS/MS Analysis

A Thermos UHPLC-Q Exactive HF-X system equipped with an ACQUITY HSS T3 column (100 mm × 2.1 mm i.d. × 1.8 μm; Waters, Milford, CT, USA) was applied for the metabolomics profiling analysis in positive mode and negative mode. The sample extracts were separated with solvent A (0.1% formic acid in water/acetonitrile = 95:5 (*v/v*)) and solvent B (0.1% formic acid in acetonitrile/isopropanol/water = 47.5:47.5:5 (*v/v*)). The injection volume was 2 μL, and the column temperature was set at 40 °C. The data-dependent acquisition mode (DDA) was applied for data acquisition. Top 10 mother ions were collected by DDA. Before sample tested, based on the operational manual of Thermofisher Orbitrap, the instrument was calibrated with a blank sample (pure water) to identify and correct the deviation in the experimental operation.

The experimental parameters were as follows: the temperatures of ion source and capillary were set to 425 °C and 325 °C, respectively. The normalized collision energy was set to 20–40–60 V rolling for MS/MS. The spray voltage was set to (+) 3500 and (−) 3500 V. The resolutions of MS spectra and MS/MS were 60,000 and 7500 over a mass range of 70–1050 m/z, respectively. In addition, the gas flow rate of sheath and Aux were set at 50 arb and 13 arb, respectively.

2.6. Data Statistics and Bioinformatics Analysis

The software of Progenesis QI 2.3 (Waters Corporation, Milford, CT, USA) was applied to pretreat the LC/MS raw data. After the peaks of internal standard, noise, column bleed, and derivative reagent were removed from the three-dimensional data matrix, the metabolites were identified by searching the main database of HMDB (http://www.hmdb.ca/), Metlin (https://metlin.scripps.edu/) and the Majorbio Database. In brief, the mass tolerance between the measured m/z values and the exact mass of the components of interest was ±10 ppm. Metabolites with an MS/MS fragments score above 30 were considered as confidently identified. After searching the database, the data matrix was uploaded to Majorbio Cloud Platform (www.majorbio.com) for further data analysis through the "ropls" (Version 1.6.2) R package. Subsequently, the final normalized data matrix was obtained after filtering and normalization. Partial least squares discriminant analysis (PLS-DA) was applied to obtain an overview of the metabolic data, general clustering, trends, and visualized outliers. In addition, orthogonal partial least squares discriminant analysis (OPLS-DA) was adopted to detect the global metabolic changes between different comparable groups and provide the variable importance in the projection (VIP) for the model. In addition, 7-cycle interactive validation and the Student's *t*-test were performed. The model parameters of R2 and Q2 were employed to evaluate the model validity avoiding the risk of over-fitting. To explore the changes in metabolite composition compared with D450, two different comparisons of metabolite composition for D450 vs. D150 and D450 vs. D300 were performed, respectively. The metabolites with the variable importance in the projection (VIP) score > 1, obtained by the OPLS-DA model, and p-value of t-test < 0.05 were identified as differential metabolites (DMs). The receiver operating characteristic (ROC) curves were determined and the under the ROC curve (AUC) was calculated to detect the efficiency of these DMs in distinguishing meat from different developmental stages. Kyoto Encyclopedia of Genes and Genomes (KEGG) metabolic pathway analysis and metabolic pathway analysis (MetPA) were adopted to further evaluate these DMs. The –log(p) values and pathway impact values of all matched pathways were visualized by plotting on the Y-axis and X-axis, respectively. In addition, data on meat quality characteristics were analyzed by the general linear model procedure of SAS (version 9.2, SAS Institute Inc., Cary, NC, USA), with the factor of age being set as the main effect. Tukey's method was applied to identify significant differences between LSmeans ($p < 0.05$).

3. Results

3.1. Characteristics of Meat Quality during the Developmental Process

The birds' live body weight and breast meat quality characteristics of intramuscular fat (IMF), dry matter, peptides, and free amino acids (FAAs) of BYC at three different stages are shown in Table 1. The live body weight and dry matter increased significantly with age ($p < 0.05$). As expected, IMF of D150 was lower ($p < 0.05$) than either D300 or D450 samples, which did not differ from each other ($p > 0.05$). Carnosine presented the highest at D450, followed by D300 and then D150 ($p < 0.05$). However, anserine displayed the opposite trend. Regarding the FAAs, 17 out of 19 FAAs presented significant differences ($p < 0.05$) at various stages, except for serine and proline ($p > 0.05$). Furthermore, total free amino acids at D150 were lower than those at either D300 or D450, which did not differ from each other ($p > 0.05$).

Table 1. Characteristics of chicken breast meat quality at three different ages (mean ± SD, n = 15) [1].

Traits [2]	D150	D300	D450
live body weight (g)	1344 ± 63 ^{c}	1951 ± 266 ^{b}	2228 ± 294 ^{a}
dry matter (%)	24.91 ± 0.99 ^{c}	28.45 ± 0.97 ^{b}	29.19 ± 1.02 ^{a}
IMF (%)	1.36 ± 0.53 ^{b}	3.10 ± 0.90 ^{a}	3.25 ± 1.55 ^{a}
Peptides			
Ans (mg/g)	9.54 ± 0.58 ^{a}	7.58 ± 0.97 ^{b}	6.77 ± 1.11 ^{c}
Car (mg/g)	1.36 ± 0.74 ^{c}	4.63 ± 1.21 ^{b}	5.70 ± 1.12 ^{a}
Free Amino Acids			
P-Ser (mg/g)	0.019 ± 0.004 ^{a}	0.01 ± 0.003 ^{b}	0.018 ± 0.005 ^{a}
Tau (mg/g)	0.083 ± 0.021 ^{a}	0.048 ± 0.007 ^{b}	0.048 ± 0.007 ^{b}
Asp (mg/g)	0.01 ± 0.013 ^{b}	0.013 ± 0.005 ^{b}	0.036 ± 0.012 ^{a}
Thr (mg/g)	0.026 ± 0.01 ^{b}	0.068 ± 0.052 ^{a}	0.042 ± 0.015 ^{b}
Ser (mg/g)	0.049 ± 0.012	0.056 ± 0.012	0.049 ± 0.006
Glu (mg/g)	0.061 ± 0.02 ^{b}	0.085 ± 0.029 ^{a}	0.07 ± 0.025 ^{ab}
Gly (mg/g)	0.031 ± 0.009 ^{c}	0.045 ± 0.013 ^{a}	0.038 ± 0.007 ^{b}
Ala (mg/g)	0.052 ± 0.016 ^{c}	0.081 ± 0.02 ^{a}	0.068 ± 0.013 ^{b}
Val (mg/g)	0.018 ± 0.006 ^{b}	0.037 ± 0.009 ^{a}	0.03 ± 0.007 ^{c}
Met (mg/g)	0.003 ± 0.001 ^{b}	0.003 ± 0.002 ^{b}	0.009 ± 0.002 ^{a}
Ile (mg/g)	0.003 ± 0.001 ^{b}	0.003 ± 0.002 ^{b}	0.009 ± 0.002 ^{a}
Leu (mg/g)	0.017 ± 0.003 ^{b}	0.042 ± 0.01 ^{a}	0.038 ± 0.008 ^{a}
Tyr (mg/g)	0.016 ± 0.002 ^{b}	0.021 ± 0.005 ^{a}	0.022 ± 0.003 ^{a}
Phe (mg/g)	0.011 ± 0.003 ^{b}	0.019 ± 0.004 ^{a}	0.02 ± 0.005 ^{a}
b-Ala (mg/g)	0.052 ± 0.014 ^{a}	0.024 ± 0.011 ^{b}	0.023 ± 0.013 ^{b}
Lys (mg/g)	0.032 ± 0.015 ^{b}	0.047 ± 0.014 ^{a}	0.033 ± 0.009 ^{b}
His (mg/g)	0.007 ± 0.001 ^{b}	0.019 ± 0.005 ^{a}	0.017 ± 0.003 ^{a}
Arg (mg/g)	0.024 ± 0.01 ^{c}	0.052 ± 0.017 ^{b}	0.065 ± 0.014 ^{a}

Table 1. *Cont.*

Traits [2]	D150	D300	D450
Pro (mg/g)	0.021 ± 0.03	0.029 ± 0.005	0.022 ± 0.004
Total Amino Acids (mg/g)	0.535 ± 0.072 [b]	0.701 ± 0.129 [a]	0.657 ± 0.084 [a]

[1] Values within a row followed by different superscript letters (a–c) differ significantly ($p \leq 0.05$). [2] IMF: intramuscular fat; Ans: anserine; Car: carnosine; P-Ser: O-Phosphoserine; Tau: taurine; Asp: aspartic acid; Thr: threonine; Ser: serine; Glu: glutamic acid; Gly: glycine; Ala: alanine; Val: valine; Met: methionine; Ile: isoleucine; Leu: leucine; Tyr: tyrosine; Phe: phenylalanine; b-Ala: β-alanine; Lys: lysine; His: histidine; Arg: arginine; Pro: proline. D150: day of 150; D300: day of 300; D450: day of 450.

3.2. Metabolic Profiling and PLS-DA Results

UHPLC-Q Exactive HF-X metabolomics was applied to address the changes in the breast muscle metabolic profiles of BYC at 150, 300, and 450 days (D150, D300, and D450), during the developmental process. All metabolites identified by an ion mode of positive and negative were integrated for the following analysis. A total of 536 metabolites were detected (Table S1), including 73 metabolite classes according to the human metabolome database (HMDB), such as carboxylic acids and derivatives, glycerophospholipids, fatty acyls, organooxygen compounds, prenol lipids, steroids and steroid derivatives (Table S2). Thereinto, carboxylic acids and derivatives, glycerophospholipids, and fatty acyls were the top three classes in the volume of metabolites.

These metabolites of BYC during the developmental process were further analyzed using PLS-DA, a multivariate statistical analysis method, with supervised pattern recognition. A clear separation between the D150, D300, and D450 stages was observed (Figure S1A), indicating that the metabolic profiles of breast muscle varied with age. Furthermore, the parameters of R2Y (0.937) and Q2 (0.799) were all more than 0.50, demonstrating that the model exhibited both good cumulative interpretation ability and strong predictive ability. Moreover, the negative intercept of Q2 regression line and the red line (R2) was always higher than the blue line (Q2), further validating the reliability of these models (Figure S1B).

3.3. Identification of Characteristic Metabolites

To elucidate the changes in metabolites relative to D450 of BYC breast muscle, differential metabolite analyses were explored in two comparisons between D450 and D150 (D450 vs. D150), as well as between D450 and D300 (D450 vs. D300). Based on the criteria of VIP > 1 and $p < 0.05$, a total of 154 and 97 metabolites were screened out, including 44 and 47 up-regulated and 110 and 50 down-regulated in the comparisons of D450 vs. D150 and D450 vs. D300, respectively (Table S3). These differential metabolites (DMs) were further visualized by volcano plots (Figure 1). Additionally, 41 common different metabolites were detected between these two comparisons by a Venn plot analysis (Figure S2A). Ultimately, 210 unique metabolites were further identified and classified according to the HMDB database (Figure S2B). These 210 differentially accumulated metabolites were predominantly categorized as carboxylic acids and derivatives (48, 25.13%), glycerophospholipids (42, 21.99%), fatty acyls (18, 9.42%), organooxygen compounds (11, 5.76%), prenol lipids (8, 4.19%), steroid and steroid derivatives (5, 2.62%), peptidomimetics (5, 2.62%) and flavonoids (4, 2.09%) (Figure S2B). According to the weighed coefficients of the OPLS-DA model, the top 20 DMs were further screened out for the comparisons of D450 vs. D150 and D450 vs. D300, respectively (Figure 2). These DMs, with VIP scores more than 2, were considered as biomarkers associated with meat quality for further study.

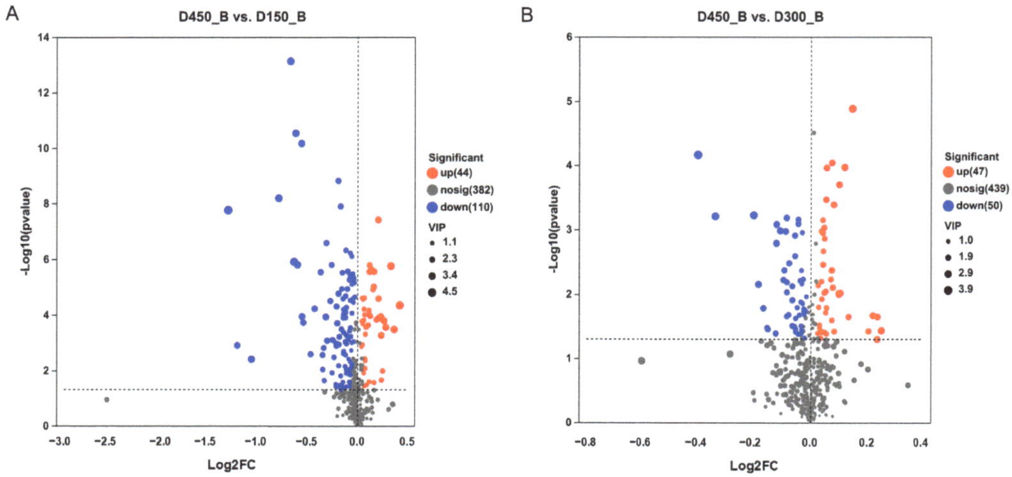

Figure 1. Volcano plot of p values between different ages according to D450 (D450 vs. D150, D450 vs. D300). The red and blue dots represent the significantly up-regulated and down-regulated metabolites, respectively, while gray dots represent no significantly differential metabolites. The dot size represents the value of variable importance in projection (VIP). (**A**) D450_B vs. D150_B; (**B**) D450_B vs. D300_B. D150_B: breast muscle at 150 days, D300_B: breast muscle at 300 days, D450_B: breast muscle at 450 days.

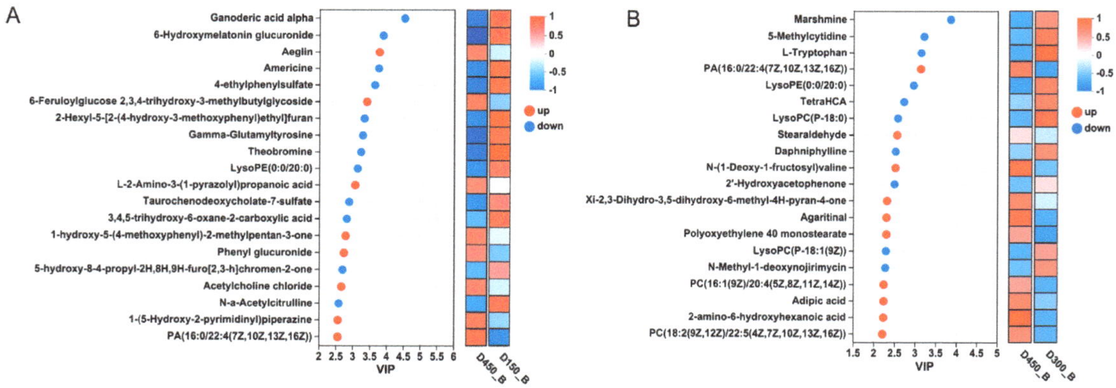

Figure 2. Variable importance in projection (VIP) scores based on the weighted coefficients of OPLS-DA model were applied to rank the top 20 DMs contributed to the metabolites discrimination between different ages according to D450. (**A**) D450_B vs. D150_B; (**B**) D450_B vs. D300_B. D150_B: breast muscle at 150 days, D300_B: breast muscle at 300 days, D450_B: breast muscle at 450 days.

In addition, to further study the classification and changing trend in DMs during developmental progress, the expression pattern of metabolites classified as lipids and lipid-like molecules, organic acids and derivatives, organooxygen compounds, and flavonoids, were further visualized through hierarchical clustering and heatmap analyses (Figures 3–5).

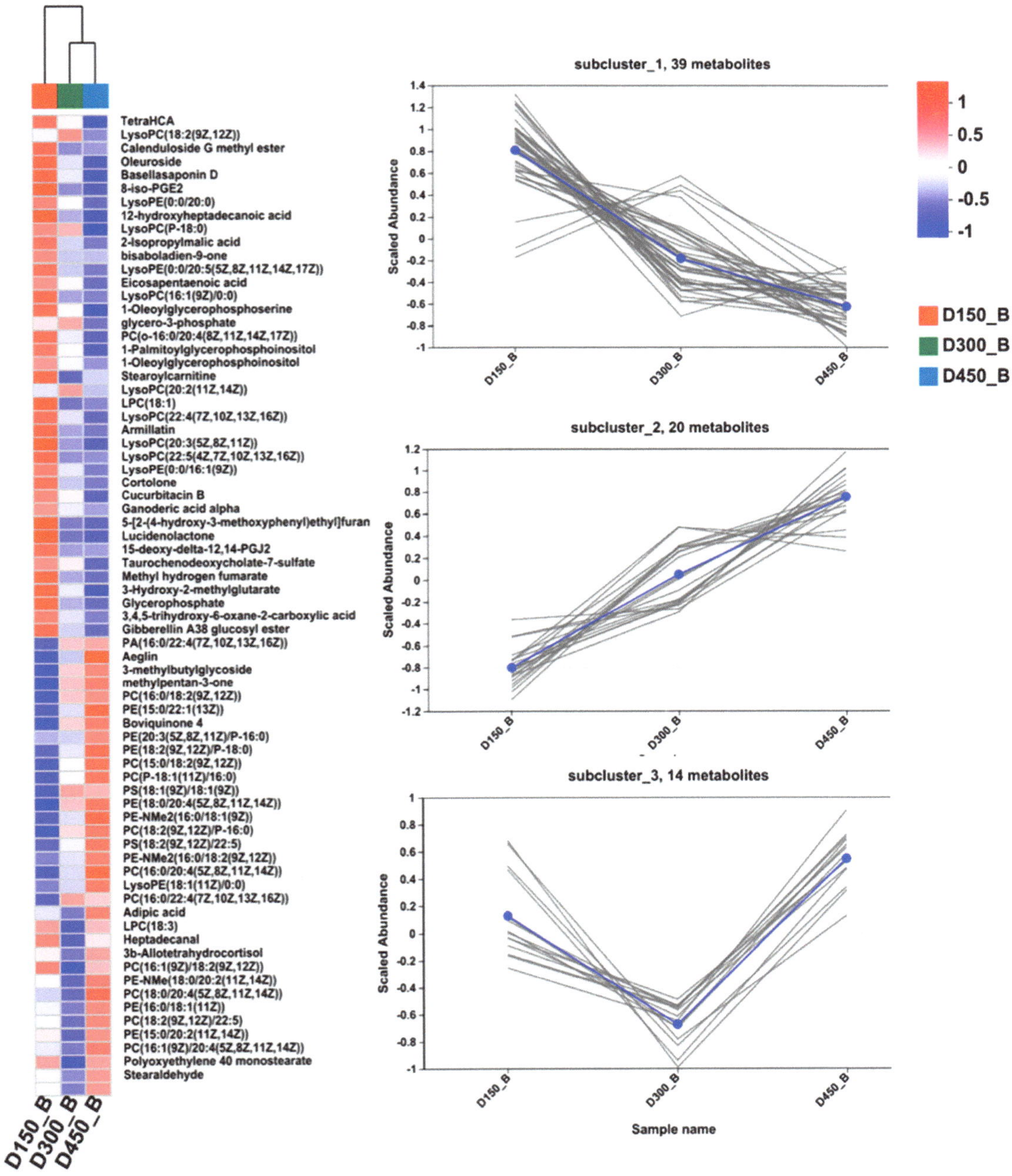

Figure 3. Heatmap (**left**) and subclustering (**right**) analysis of the expression patterns of the DMs of lipids and lipid-like molecules according to human metabolome database (HMDB) in breast muscle of Beijing-You chicken at three different age stages. D150_B: breast muscle at 150 days, D300_B: breast muscle at 300 days, D450_B: breast muscle at 450 days. Each gray line in the subclustering graph represents a differential metabolite, and the blue line represents the average expression level of all differential metabolites in the individual subcluster.

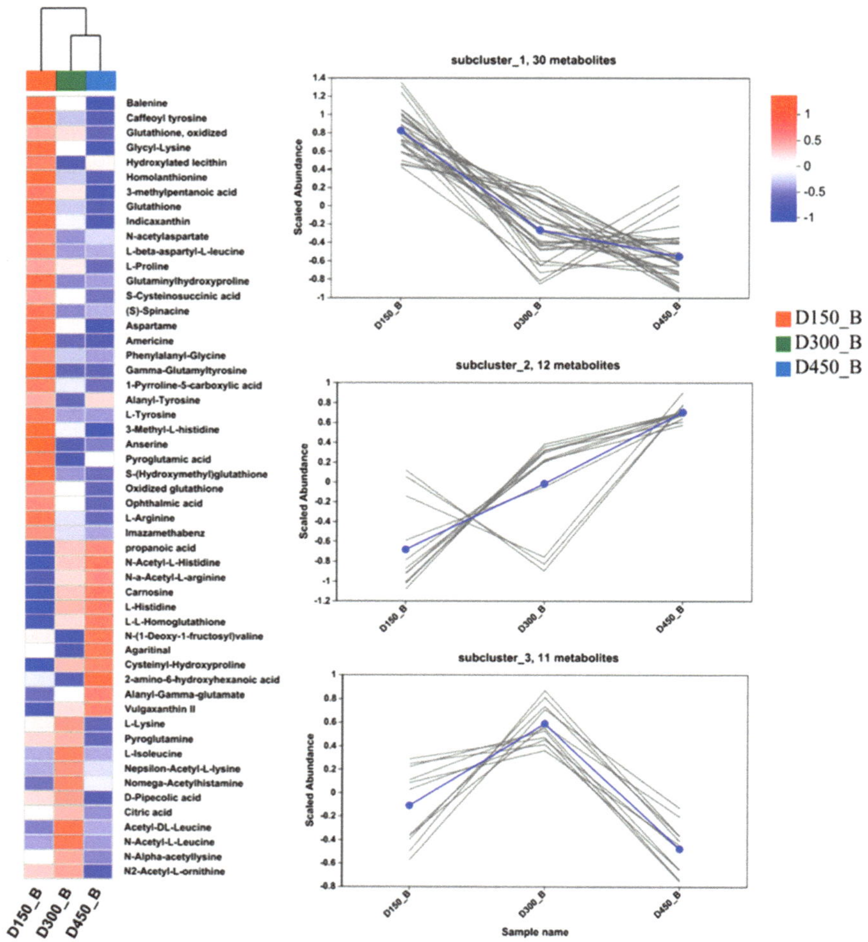

Figure 4. Heatmap (**left**) and subclustering (**right**) analysis of the expression patterns of the DMs of organic acids and derivatives based on human metabolome database (HMDB) in breast muscle of Beijing-You chicken at three different age stages. D150_B: breast muscle at 150 days, D300_B: breast muscle at 300 days, D450_B: breast muscle at 450 days. Each gray line in the subclustering graph represents a differential metabolite, and the blue line represents the average expression level of all differential metabolites in the individual subcluster.

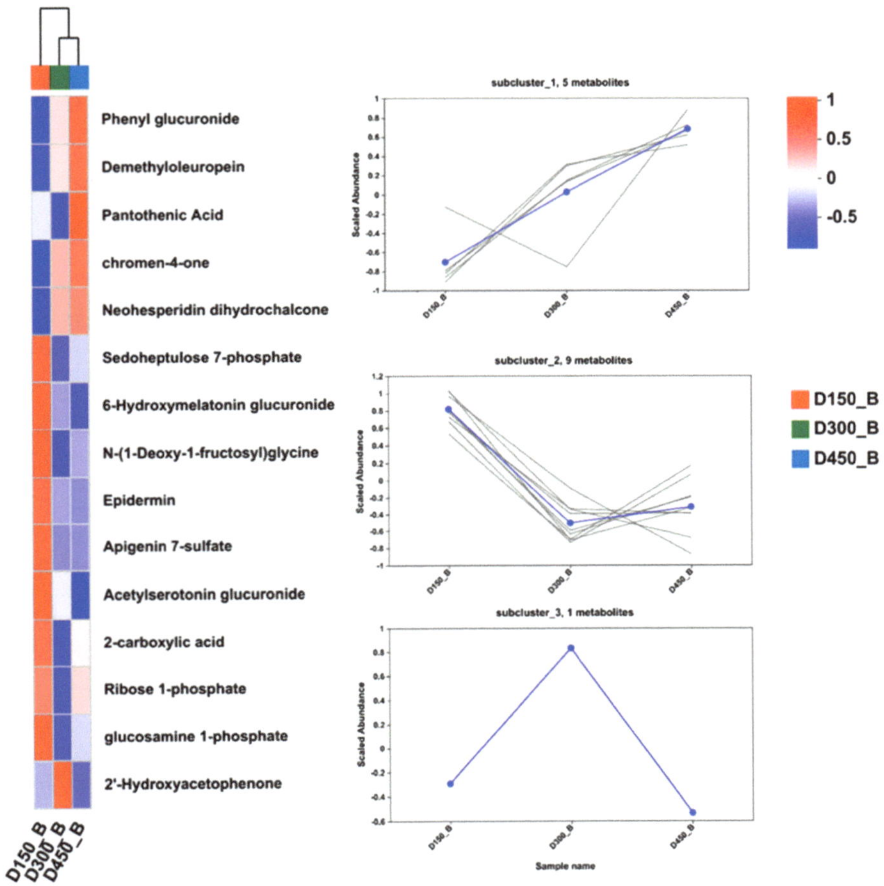

Figure 5. Heatmap (**left**) and subclustering (**right**) analysis of the expression patterns of the DMs of organooxygen compounds and flavonoids based on human metabolome database (HMDB) in breast muscle of Beijing-You chicken at three different age stages. D150_B: breast muscle at 150 days, D300_B: breast muscle at 300 days, D450_B: breast muscle at 450 days. Each gray line in the subclustering graph represents a differential metabolite, and the blue line represents the average expression level of all differential metabolites in the individual subcluster.

3.3.1. Lipids and Lipid-like Molecules

In this study, lipids and lipid-like molecules were the most abundant metabolites. A total of 73 lipids and lipid-like molecules were identified, comprising 42 glycerophospholipids, 18 fatty acyls, 8 prenol lipids, as well as 5 steroids and steroid derivatives (Table S4). The most abundant lipid and lipid-like molecules in BYC breast meat are glycerophospholipids, containing five subclasses including phosphocholines (PC, 20 species), phosphoethanolamines (PE, 13 species), phosphates (4 species), phosphoserines (PS, 3 species), and phosphoinositols (PI, 2 species). Additionally, the expression patterns of those 73 lipids and lipid-like molecules from D150 to D450 are shown in Figure 3. On the whole, the expression patterns of lipid-related DMs at D150 were more distinguished from those at D300 and D450, which agreed with the IMF results of this study (Table 1). In addition, these DMs could be further assigned into three distinct subclusters based on the K-means clustering result during the developmental process. Subcluster 1 (the largest subcluster, 39 DMs), including LPC (18:1), Lysoglycerophospholipid species (LGP), glycerophosphate, eicosapentaenoic acid, 1-oleoylglycerophosphoserine, stearoylcarnitine, ganoderic acid alpha, oleuroside and methyl hydrogen fumarate, presented a consistent down-regulation trend from D150 to D450. However, the opposite pattern was observed in subcluster 2 (20 DMs), such as in glycerophospholipids (PC, PE, PS, and PA), aeglin, and boviquinone 4. In addition, the relative levels of 14 DMs in subcluster 3, including stearaldehyde, and some glycerophospholipids (PC and PE), decreased to the minimums at D300 and then increased. The above results suggest that the factor of age could impose extreme effects on lipids and lipid-like molecules.

3.3.2. Organic Acids and Derivatives

Following lipids and lipid-like molecules, a total of 53 organic acids and derivatives were detected, consisting of 46 amino acids, peptide and analogues, 5 hybrid peptides, 1 carboxylic acid derivative, and 1 tricarboxylic acid (Table S5), suggesting that amino acids, peptide and analogues are the main compounds of organic acids and derivatives. Figure 4 illustrates the expression patterns of these 53 DMs from D150 to D450. In general, the expression patterns of these 53 DMs of organic acids were also more similar between D300 and D450, which is consistent with the results for free amino acids in this study (Table 1). These DMs could also be further classed into three distinct subclusters according to the expression pattern. Subcluster one (30 DMs), including balenine, anserine, L-beta-aspartyl-L-leucine, glutathione, oxidized glutathione, glycyl-lysine, alanyl-tyrosine, s-cysteinosuccinic acid, L-proline, L-tyrosine, L-arginine, and indicaxanthin, showed a lasting downregulation from D150 to D450. In addition, the contents of 12 DMs, clustered in subcluster two, such as carnosine, L-L-homoglutathione, vulgaxanthin II, alanyl-gamma-glutamate, cysteinyl-hydroxyproline, N-a-acetyl-L-arginine, N-acetyl-L-histidine, N-(1-deoxy-1-fructosyl)valine, 2-amino-6-hydroxyhexanoic acid, and L-histidine, were increased with age. In contrast, the contents of 11 DMs, clustered in subcluster three, including L-lysine, L-isoleucine, pyroglutamine, N-alpha-acetyllysine, N2-acetyl-L-ornithine, acetyl-DL-leucine, and N-acetyl-L-leucine, achieved the maximum at D300. This result indicates that amino acids, peptide and analogues, along with hybrid peptides, are greatly affected by age.

3.3.3. Organooxygen Compounds and Flavonoids

In this study, 11 organooxygen compounds and 4 flavonoids were detected, respectively (Table S6). The expression patterns of these 15 DMs from D150 to D450 are illustrated in Figure 5. These DMs could be further classed into three distinct subclusters according to the expression pattern. Subcluster one (5 DMs), including neohesperidin dihydrochalcone, demethyloleuropein, 7-chloro-2-(3,4-dimethoxyphenyl)-3,5-dihydroxy-6,8-dimethoxy-4H-chromen-4-one, and phenyl glucuronide, showed a steady upregulation from D150 to D450. In addition, the contents of 6-hydroxymelatonin glucuronide, epidermin, apigenin 7-sulfate, L-histidine, and sedoheptulose 7-phosphate, coming from subcluster two, showed down-regulated significantly from D150 to D300, then no significant change was detected from D300 to D450, while only one differential metabolite (2′-Hydroxyacetophenone) was detected and assigned to subcluster three, characterized by achieving maximum relative content at D300.

3.3.4. Receiver Operating Characteristic Curve Analysis

The ROC curve was employed to evaluate the predictive performance of the aforementioned DMs based on D450 of BYC breast muscle. Some metabolites with AUC ≥ 0.95 are shown in Figure 6.

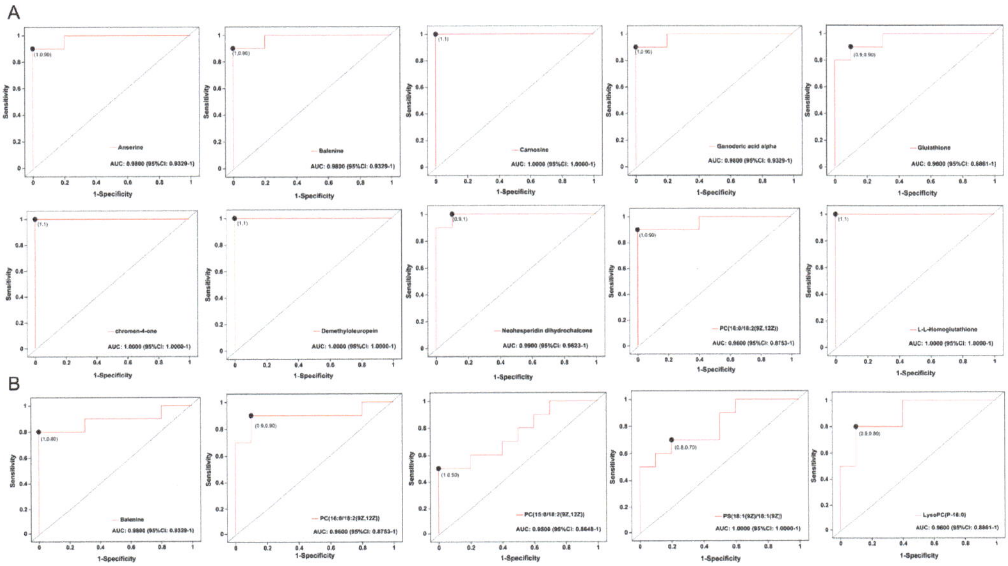

Figure 6. Receiver operating characteristic (ROC) curves for the metabolites between different ages comparison according to D450 with AUC $\geq$ 0.95. (**A**) D450 vs. D150; (**B**) D450 vs. D300. D150: day of 150; D300: day of 300; D450: day of 450.

3.4. KEGG Pathway Enrichment and Topology Analysis

KEGG pathway enrichment and pathway topology analysis were employed to explore the biological pathways involved in DMs due to different ages. Based on the criteria of impact value of pathway impact more than 0.1, there were eight most impacted metabolic pathways between the comparison of D450 vs. D150, including glycerophospholipid metabolism, taurine and hypotaurine metabolism, glutathione metabolism, histidine metabolism, arginine and proline metabolism, tryptophan metabolism, arginine biosynthesis, tyrosine metabolism, whereas three metabolic pathways, including pantothenate and CoA biosynthesis, tryptophan metabolism, and lysine degradation were detected between D300 and D450 (Figure 7). These enriched pathways were mainly involved in the synthesis and metabolism of amino acid, peptide, and glycerophospholipid.

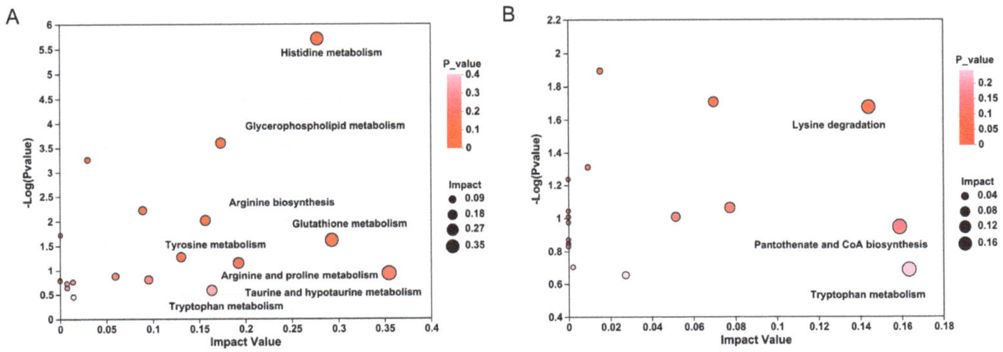

Figure 7. KEGG topology analysis of differential metabolites between different ages according to D450. (**A**) D450 vs. D150; (**B**) D450 vs. D300. The *X*-axis and *Y*-axis represent the pathway impact and pathway enrichment, respectively. The size and color of bubble stand for impact value and *p*-value, respectively. D150: day of 150; D300: day of 300; D450: day of 450.

4. Discussion

4.1. Lipid Compounds

Prior studies have demonstrated a prominent number of phospholipids in the muscles and these play a central role in the formation of characteristic volatile flavors of meat products [20–23]. In addition, phospholipids are the key constituents of membrane involving in a broad range of cellular functions, including signal transduction and regulation of the transport process [24]. According to the results of this study, the main DM components classified as lipids and lipid-like molecules in BYC breast meat were PC (20 species) and PE (13 species). It was noteworthy that polyunsaturated fatty acids (PUFAs), such as docosapentaenoic acid (DPA), eicosapentaenoic acid (EPA), arachidonic acid (AA), and linoleic acid (LA), were not free fatty acids. They predominantly deposited in the glycerophospholipids instead. These results are consistent with previous studies focusing on Chinese indigenous chicken breeds, such as Taihe black-boned silky fowl, Guangyuan grey chicken, Tibetan chicken, and Jiuyuan black chicken [4,25]. Furthermore, the BYC breast meat at D450 had a significantly higher content of PUFA-enriched lipids (Table S4), such as PC(18:2(9Z,12Z)/22:5(4Z,7Z,10Z,13Z,16Z)), PC(16:0/22:4(7Z,10Z,13Z,16Z)), PE(15:0/22:1(13Z)), PC(16:0/20:4(5Z,8Z,11Z,14Z)), PC(16:0/18:2(9Z,12Z)), PC(P-18:1(11Z)/16:0), which was consistent with our previous study, which reported that the breast meat of BYC at D450 presents a feature with a significantly higher concentration of PUFA as compared with that at D150 and D300 [15]. Notably, Lysoglycerophospholipid species (LGP) exhibited a lasting down-regulation trend from D150 to D450, except for LysoPE(18:1(11Z)/0:0) and LPC(18:3). However, PC- and PE-related glycerophospholipids exhibited the highest abundance at D450, except for PC(o-16:0/20:4(8Z,11Z,14Z,17Z)). This finding might be related to much higher levels of hydrolase and enzyme activation during the later growth process. Ge et al. [12] found that LysoPC(18:1), LysoPC(18:2), and LysoPC(16:0) in BYC breast meat increased from 56 days to 120 days, different from our results, which might be due to age differences. Altogether, these results suggested that having significantly higher PUFA-enriched glycerophospholipids of BYC at D450 was the root cause of the better properties of meat flavor and of health benefits.

Besides glycerophospholipids, some compounds of prenol lipids, such as ganoderic acid alpha and oleuroside, were also identified. Ganoderic acid alpha (GAA) is a distinguished bioactive compound belonging to the lanostane triterpenes, which presents a broad range of pharmacological attributes, such as anticancer, antioxidant, and anti-inflammatory [26,27]. GAA has been documented as possessing a hepatoprotective effect on liver injury and hepatic toxicity [28,29]. In this study, GAA was further identified with the highest VIP value (4.53) based on the OPLS-DA analysis (Figure 2), which could be considered as a prominent bioactive compound accounting for the nutritional and medicinal properties of BYC.

4.2. Peptide Compounds

Numerous experimental studies have demonstrated that peptides are widely regarded as components with special functional effects on flavor development, unique aromas and tastes during the thermal processing of foods [30,31]. Peptides containing glutamate or aspartate always present an umami taste in food [32]. In our study, besides the common amino acids, DMs classed as peptides were primarily identified, such as carnosine, anserine, balenine, L-beta-aspartyl-L-leucine, L-L-homoglutathione, glutathione (γ-L-glutamyl-L-cysteinyl-glycine; GSH), oxidized glutathione (GSSG) and cysteinyl-hydroxyproline. Of these, metabolites of L-beta-aspartyl-L-leucine, L-L-homoglutathione, GSH, and GSSG containing the residues of glutamyl or aspartyl, might be associated with the umami taste of BYC. In addition to an umami function, glutathione takes part in a series of physiological processes, including antioxidant function, regulation of cell cycle and cellular redox balance, a storage form, and a transport form of cysteine avoiding autoxidation [33]. In addition, GSH is involved in the formation of a sulfurous odor after heating [34].

Carnosine (β-alanyl-L-histidine) and its methylated analogs anserine (β-alanyl-1-methyl-L-histidine) and balenine (β-alanyl-3-methyl-L-histidine), known as histidine-containing imidazole dipeptides, have been demonstrated as accounting for the biological benefits of antioxidant, antifatigue, antiglycation and the buffering effect in muscle tissues [35–37]. It has been confirmed that histidine-containing imidazole dipeptides are present in chicken meat at high concentrations [38]. Kojima et al. [39] reported that Silky Fowl (*Gallus gallus dommesticus*) had 1.6- and 1.9-fold higher carnosine contents in thigh and breast meat compared with broilers. Moreover, dietary supplementation with meat with a high content of carnosine could prevent age-related diseases [40]. In this study, carnosine presented the highest at D450, followed by D300 and then D150, while anserine and balenine decreased with age, which was consistent with our results on meat quality characteristics (Table 1) and previous reports in Wuding chicken, Daheng broilers, and BYC [12,13,25]. All these data consistently confirm a decreasing ratio of anserine to carnosine in chicken with increasing age, which might be used as an excellent discriminator between chickens with different ages, and the relationship between this ratio and bird age requires further study. Taken together, in the case of the wide benefits of peptides as mentioned above, products of BYC could work as natural peptide sources to play a pivotal role in providing health benefits to consumers.

4.3. Flavonoid Compounds

It has been demonstrated that flavonoid compounds, with diverse biological presence, possess great health benefits for the human body, including antioxidant, antibacterial, anticancer, anti-inflammatory, anti-aging, and immunomodulatory activities [41,42]. During olive maturation, demethyloleuropein is reported to be a precursor for oleacein, which possesses a powerful antioxidant activity [43,44]. Neohesperidin dihydrochalcone, belonging to the citrus flavonoids, is characterized by its anti-inflammatory, antioxidant, and sweetness potential, and has been used against osteoporosis and osteoarthritis in bone health [45] and as a food additive (sweetener) [46]. Derivatives of chroman-4-one, classed as flavanones, have been applied to regulate cellular metabolism and scavenge free radicals [47,48]. Apigenin, belonging to the flavone subclass, is reported to halt cellular proliferation in human breast and liver cancer cells [49,50]. In this study, four flavonoids compounds were identified as DMs according to D450, including demethyloleuropein, neohesperidin dihydrochalcone, 7-chloro-2-(3,4-dimethoxyphenyl)-3,5-dihydroxy-6,8-dimethoxy-4H-chromen-4-one (a chromen-4-one derivative), and apigenin 7-sulfate (Table S6). The concentration of demethyloleuropein, neohesperidin dihydrochalcone, and the chromen-4-one derivative showed an increasing trend throughout the developmental process of BYC, while the abundance of apigenin 7-sulfate declined from D150 to D300, and afterwards remained in a stable state from D300 to D450 (Figure 5). Overall, these results indicate that the abundance of flavonoid compounds exists at maximum at D450 in the breast meat of chicken, which accounts for the high nutritional value of BYC, particularly for chicken with older age.

5. Conclusions

In this study, the dynamic alterations of metabolite composition of BYC breast meat during the developmental process were evaluated by UHPLC-Q Exactive HF-X-based metabolomics and multivariate analysis. A total of 210 DMs were determined as the main discriminatory components throughout a wide age spectrum over 450 days. Carnosine, L-L-homoglutathione, chromen-4-one, neohesperidin dihydrochalcone, demethyloleuropein, and glycerophospholipids exhibited the highest abundance at D450. However, balenine, anserine, L-beta-aspartyl-L-leucine, glutathione, oxidized glutathione, ganoderic acid alpha, and Lysoglycerophospholipid species (LGP) exhibited a lasting down-regulation trend from D150 to D450. Additionally, AUC values of these most discriminant metabolites were further approved by AUC $\geq$ 0.95. A total of 10 pathways, including glycerophospholipid metabolism, glutathione metabolism, histidine metabolism, pantothenate and CoA biosynthesis, affecting the properties of meat flavor and health benefit through synthesis and metabolism compounds of glycerophospholipids, peptides, and flavonoids were further identified. Our study elucidates the alternations of metabolic profiling due to age and the potential bioactive compounds of BYC.

Supplementary Materials: The following supporting information can be downloaded at: https://www.mdpi.com/article/10.3390/foods13162549/s1, Figure S1: The partial least squares discriminant analysis (PLS-DA) between different age stages (A). PLS-DA models validated by permutation tests (B). R2Y(cum) = 0.937, Q2(cum) = 0.799. D150_B: breast muscle at 150 days, D300_B: breast muscle at 300 days, D450_B: breast muscle at 450 days; Figure S2: Venn diagram of differential metabolites (A). Categorization of differential metabolites based on human metabolome database (HMDB) (B). D450 vs. D150: the comparison stages between D450 and D150, D450 vs. D300: the comparison stages between D450 and D300; Table S1: The information of total identified metabolites; Table S2: Summary of identified metabolites in breast muscle of Beijing-You chicken during the developmental process; Table S3: The lists of differential metabolites for the comparisons of D450 vs. D150, and D450 vs. D300; Table S4: Differential metabolites classed to lipids and lipid-like molecules at D150, D300 and D450; Table S5: Differential metabolites classed to organic acids and derivatives at D150, D300 and D450; Table S6: Differential metabolites classed to organooxygen compounds and flavonoids at D150, D300 and D450.

Author Contributions: Conceptualization, J.Z. and H.L.; methodology, J.Z. and X.C.; software, J.C.; validation, J.Z., X.C. and H.L.; formal analysis, A.G.; resources, Q.C.; data curation, Z.Y. and Y.Z.; writing—original draft preparation, J.Z.; writing—review and editing, J.Z.; visualization, X.C.; supervision, H.L.; project administration, H.L.; funding acquisition, J.Z., X.C., A.G. and H.L. All authors have read and agreed to the published version of the manuscript.

Funding: This research was funded by the Project of Reform and Development of Beijing Academy of Agriculture and Forestry Sciences (XMS202410, XMS202404), Beijing Indigenous Animal Genetic Resources Protection Project (202203310001), Youth Fund of Beijing Academy of Agriculture and Forestry Sciences (QNJJ202422), Earmarked Fund for CARS (CARS-41-Z04), and Beijing Innovation Consortium of Agriculture Research System (BAIC06-2023).

Institutional Review Board Statement: The animal study protocol was approved by the Institute of Animal Husbandry and Veterinary Medicine, Beijing Academy of Agriculture and Forestry Sciences (Beijing, China) (No. BAAFS-IAHVM20190115) (15 January 2019).

Informed Consent Statement: Not applicable.

Data Availability Statement: The original contributions presented in the study are included in the article/Supplementary Materials, further inquiries can be directed to the corresponding author.

Acknowledgments: The authors acknowledge the support of bioinformation analysis provided by Majorbio Cloud Platform (www.majorbio.com) (accessed on 5 July 2024).

Conflicts of Interest: The authors declare no conflicts of interest.

References

1. Charoensin, S.; Laopaiboon, B.; Boonkum, W.; Phetcharaburanin, J.; Villareal, M.O.; Isoda, H.; Duangjinda, M. Thai Native Chicken as a Potential Functional Meat Source Rich in Anserine, Anserine/Carnosine, and Antioxidant Substances. *Animals* **2021**, *11*, 902. [CrossRef] [PubMed]
2. Salter, A.M. The effects of meat consumption on global health. *Rev. Sci. Tech.* **2018**, *37*, 47–55. [CrossRef]
3. Ishihara, K.; Watanabe, R.; Kato, T.; Seko, T.; Matsuda, T.; Omura, Y.; Shigemura, Y.; Kawabata, Y.; Maegawa, T. Isolation of balenine from opah (*Lampris megalopsis*) muscle and comparison of antioxidant and iron-chelating activities with other major imidazole dipeptides. *Food Chem.* **2021**, *364*, 130343. [CrossRef]
4. Mi, S.; Shang, K.; Jia, W.; Zhang, C.-H.; Li, X.; Fan, Y.-Q.; Wang, H. Characterization and discrimination of Taihe black-boned silky fowl (*Gallus gallus domesticus* Brisson) muscles using LC/MS-based lipidomics. *Food Res. Int.* **2018**, *109*, 187–195. [CrossRef]
5. Yang, X.; Tang, C.; Ma, B.; Zhao, Q.; Jia, Y.; Meng, Q.; Qin, Y.; Zhang, J. Identification of Characteristic Bioactive Compounds in Silkie Chickens, Their Effects on Meat Quality, and Their Gene Regulatory Network. *Foods* **2024**, *13*, 969. [CrossRef] [PubMed]
6. Muroya, S.; Ueda, S.; Komatsu, T.; Miyakawa, T.; Ertbjerg, P. MEATabolomics: Muscle and meat metabolomics in domestic animals. *Metabolites* **2020**, *10*, 188. [CrossRef]

7. Ramanathan, R.; Kiyimba, F.; Suman, S.P.; Mafi, G.G. The potential of metabolomics in meat science: Current applications, trends, and challenges. *J. Proteomics* **2023**, *283–284*, 104926. [CrossRef] [PubMed]
8. Gu, T.; Duan, M.; Chen, L.; Tian, Y.; Xu, W.; Zeng, T.; Lu, L. Proteomic-metabolomic combination analysis reveals novel biomarkers of meat quality that differ between young and older ducks. *Poult. Sci.* **2024**, *103*, 103530. [CrossRef]
9. Chen, X.; Cao, J.; Geng, A.; Zhang, X.; Wang, H.; Chu, Q.; Yan, Z.; Zhang, Y.; Liu, H.; Zhang, J. Integration of GC-MS and LC-MS for metabolite characteristics of thigh meat between fast- and slow-growing broilers at marketable age. *Food Chem.* **2023**, *403*, 134362. [CrossRef]
10. Zhang, M.; Sun, L.; Su, R.; Corazzin, M.; Yang, Z.; Dou, L.; Hu, G.; Zhang, Y.; Liu, T.; Guo, Y.; et al. Widely targeted metabolomic analysis reveals the dynamic changes of metabolites during postmortem chilled aging in Mongolian sheep. *Food Chem.* **2024**, *431*, 137035. [CrossRef] [PubMed]
11. Wang, H.; Wu, B.; Zhang, J.; Liu, Y.; Zhang, M.; Chen, L.; Zhao, W.; Kan, H.; Cao, C. Bamboo shoots improve the nutritional and sensory quality, and change flavor composition of chicken soup. *Food Chem. X* **2024**, *21*, 101140. [CrossRef] [PubMed]
12. Ge, Y.; Gai, K.; Zheng, L.; Chen, Y.; Wang, L.; Qi, X.; Xing, K.; Wang, X.; Xiao, L.; Ni, H.; et al. HPLC-QTRAP-MS-based metabolomics approach investigates the formation mechanisms of meat quality and flavor of Beijing You chicken. *Food Chem. X* **2022**, *17*, 100550. [CrossRef] [PubMed]
13. Xiao, Z.; Ge, C.; Zhou, G.; Zhang, W.; Liao, G. 1H NMR-based metabolic characterization of Chinese Wuding chicken meat. *Food Chem.* **2019**, *274*, 574–582. [CrossRef]
14. Deng, S.; Xing, T.; Li, C.; Xu, X.; Zhou, G. The Effect of Breed and Age on the Growth Performance, Carcass Traits and Metabolic Profile in Breast Muscle of Chinese Indigenous Chickens. *Foods* **2022**, *11*, 483. [CrossRef] [PubMed]
15. Zhang, J.; Zhuang, H.; Cao, J.; Geng, A.; Wang, H.; Chu, Q.; Yan, Z.; Zhang, X.; Zhang, Y.; Liu, H. Breast Meat Fatty Acid Profiling and Proteomic Analysis of Beijing-You Chicken During the Laying Period. *Front. Vet. Sci.* **2022**, *9*, 908862. [CrossRef]
16. Zhang, J.; Chen, X.; Cao, J.; Chang, C.; Geng, A.; Wang, H.; Chu, Q.; Yan, Z.; Zhang, Y.; Liu, H. Proteomic Profiling of Thigh Meat at Different Ages of Chicken for Meat Quality and Development. *Foods* **2023**, *12*, 2901. [CrossRef]
17. Zerehdaran, S.; Vereijken, A.L.J.; van Arendonk, J.A.M.; van der Waaijt, E.H. Estimation of genetic parameters for fat deposition and carcass traits in broilers. *Poult. Sci.* **2004**, *83*, 521–525. [CrossRef]
18. Chen, X.; Cao, J.; Chang, C.; Geng, A.; Wang, H.; Chu, Q.; Yan, Z.; Zhang, X.; Zhang, Y.; Liu, H.; et al. Effects of Age on Compounds, Metabolites and Meat Quality in Beijing-You Chicken Breast Meat. *Animals* **2023**, *13*, 3419. [CrossRef] [PubMed]
19. Li, C.; Al-Dalali, S.; Zhou, H.; Xu, B. Influence of curing on the metabolite profile of water-boiled salted duck. *Food Chem.* **2022**, *397*, 133752. [CrossRef]
20. Mottram, D.S. Flavour formation in meat and meat products: A review. *Food Chem.* **1998**, *62*, 415–424. [CrossRef]
21. Wood, J.D.; Enser, M.; Fisher, A.V.; Nute, G.R.; Sheard, P.R.; Richardson, R.I.; Hughes, S.I.; Whittington, F.M. Fat deposition, fatty acid composition and meat quality: A review. *Meat Sci.* **2008**, *78*, 343–358. [CrossRef] [PubMed]
22. Petrova, I.; Aasen, I.M.; Rustad, T.; Eikevik, T.M. Manufacture of dry-cured ham: A review. Part 1. Biochemical changes during the technological process. *Eur. Food. Res. Technol.* **2015**, *241*, 587–599. [CrossRef]
23. Fan, M.; Xiao, Q.; Xie, J.; Cheng, J.; Sun, B.; Du, W.; Wang, Y.; Wang, T. Aroma compounds in chicken broths of Beijing Youji and commercial broilers. *J. Agric. Food Chem.* **2018**, *66*, 10242–10251. [CrossRef] [PubMed]
24. Carrizo, D.; Chevallier, O.P.; Woodside, J.V.; Brennan, S.F.; Cantwell, M.M.; Cuskelly, G.; Elliott, C.T. Untargeted metabolomic analysis of human serum samples associated with exposure levels of Persistent organic pollutants indicate important perturbations in Sphingolipids and Glycerophospholipids levels. *Chemosphere* **2017**, *168*, 731–738. [CrossRef] [PubMed]
25. Li, J.; Li, Z.; Ran, J.; Yang, C.; Lin, Z.; Liu, Y. LC/MS-based lipidomics to characterize breed-specific and tissue-specific lipid composition of chicken meat and abdominal fat. *LWT* **2022**, *163*, 113611. [CrossRef]
26. Hsu, C.-L.; Yen, G.-C. Ganoderic Acid and Lucidenic Acid (Triterpenoid). *Enzymes* **2014**, *36*, 33–56. [CrossRef] [PubMed]
27. Gill, B.S.; Navgeet; Kumar, S. Antioxidant potential of ganoderic acid in Notch-1 protein in neuroblastoma. *Mol. Cell Biochem.* **2019**, *456*, 1–14. [CrossRef]
28. Zheng, C.; Lv, S.; Ye, J.; Zou, L.; Zhu, K.; Li, H.; Dong, Y.; Li, L. Metabolomic Insights into the Mechanisms of Ganoderic Acid: Protection against α-Amanitin-Induced Liver Injury. *Metabolites* **2023**, *13*, 1164. [CrossRef]
29. Xu, L.; Yan, L.; Huang, S. Ganoderic acid A against cyclophosphamide-induced hepatic toxicity in mice. *J. Biochem. Mol. Toxicol.* **2019**, *33*, e22271. [CrossRef]
30. Ogasawara, M.; Katsumata, T.; Egi, M. Taste properties of Maillard-reaction products prepared from 1000 to 5000 Da peptide. *Food Chem.* **2006**, *99*, 600–604. [CrossRef]
31. Zhou, R.; Grant, J.; Goldberg, E.M.; Ryland, D.; Aliani, M. Investigation of low molecular weight peptides (<1 kDa) in chicken meat and their contribution to meat flavor formation. *J. Sci. Food Agric.* **2019**, *99*, 1728–1739. [CrossRef] [PubMed]
32. Tamura, M.; Nakatsuka, T.; Tada, M.; Kawasaki, Y.; Kikuchi, E.; Okai, H. The relationship between taste and primary structure of "delicious peptide" (Lys-Gly-Asp-Glu-GluSer-Leu-Ala) from beef soup. *Agric. Biol. Chem.* **1989**, *53*, 319–325. [CrossRef]
33. Kubát, M.; Roušarová, E.; Roušar, T.; Česla, P. Recent advances in separation methods for characterization of glutathione metabolism and dietary supplementation. *TrAC Trends Anal. Chem.* **2024**, *176*, 117751. [CrossRef]
34. Mecchi, E.P.; Pippen, E.L.; Lineweaver, H. Origin of Hydrogen Sulfide in Heated Chicken Muscle. *J. Food Sci.* **1964**, *29*, 393–399. [CrossRef]

35. Carnegie, P.R.; Ilic, M.Z.; Etheridge, M.O.; Collins, M.G. Improved high-performance liquid chromatographic method for analysis of histidine dipeptides anserine, carnosine and balenine present in fresh meat. *J. Chromatogr.* **1983**, *261*, 153–157. [CrossRef]
36. Peiretti, P.G.; Medana, C.; Visentin, S.; Giancotti, V.; Zunino, V.; Meineri, G. Determination of carnosine, anserine, homocarnosine, pentosidine and thiobarbituric acid reactive substances contents in meat from different animal species. *Food Chem.* **2011**, *126*, 1939–1947. [CrossRef] [PubMed]
37. Freund, M.A.; Chen, B.; Decker, E.A. The Inhibition of Advanced Glycation End Products by Carnosine and Other Natural Dipeptides to Reduce Diabetic and Age-Related Complications. *Compr. Rev. Food Sci. Food Saf.* **2018**, *17*, 1367–1378. [CrossRef] [PubMed]
38. Barbaresi, S.; Maertens, L.; Claeys, E.; Derave, W.; De Smet, S. Differences in muscle histidine-containing dipeptides in broilers. *J. Sci. Food Agric.* **2019**, *99*, 5680–5686. [CrossRef] [PubMed]
39. Kojima, S.; Saegusa, H.; Sakata, M. Histidine-containing dipeptide concentration and antioxidant effects of meat extracts from silky fowl: Comparison with meat-type chicken breast and thigh meats. *Food Sci. Technol. Res.* **2014**, *20*, 621–628. [CrossRef]
40. Hipkiss, A.R.; Cartwright, S.P.; Bromley, C.; Gross, S.R.; Bill, R.M. Carnosine: Can understanding its actions on energy metabolism and protein homeostasis inform its therapeutic potential? *Chem. Cent. J.* **2013**, *7*, 38. [CrossRef]
41. Yang, B.; Liu, H.; Yang, J.; Gupta, V.K.; Jiang, Y. New insights on bioactivities and biosynthesis of flavonoid glycosides. *Trends Food Sci. Technol.* **2018**, *79*, 116–124. [CrossRef]
42. Chen, L.; Cao, H.; Huang, Q.; Xiao, J.; Teng, H. Absorption, metabolism and bioavailability of flavonoids: A review. *Crit. Rev. Food Sci. Nutr.* **2022**, *62*, 7730–7742. [CrossRef] [PubMed]
43. Amiot, M.-J.; Fleuriet, A.; Macheix, J.-J. Accumulation of oleuropein derivatives during olive maturation. *Phytochemistry* **1989**, *28*, 67–69. [CrossRef]
44. Czerwińska, M.; Kiss, A.K.; Naruszewicz, M. A comparison of antioxidant activities of oleuropein and its dialdehydic derivative from olive oil, oleacein. *Food Chem.* **2012**, *131*, 940–947. [CrossRef]
45. Ortiz, A.C.; Fideles, S.O.M.; Reis, C.H.B.; Bellini, M.Z.; Pereira, E.S.B.M.; Pilon, J.P.G.; de Marchi, M.Â.; Detregiachi, C.R.P.; Flato, U.A.P.; Trazzi, B.F.M.; et al. Therapeutic Effects of Citrus Flavonoids Neohesperidin, Hesperidin and Its Aglycone, Hesperetin on Bone Health. *Biomolecules* **2022**, *12*, 626. [CrossRef]
46. Xia, X.; Fu, J.; Song, X.; Shi, Q.; Su, C.; Song, E.; Song, Y. Neohesperidin dihydrochalcone down-regulates MyD88-dependent and -independent signaling by inhibiting endotoxin-induced trafficking of TLR4 to lipid rafts. *Free Radic. Biol. Med.* **2015**, *89*, 522–532. [CrossRef] [PubMed]
47. Gordo, J.; Máximo, P.; Cabrita, E.; Lourenço, A.; Oliva, A.; Almeida, J.; Filipe, M.; Cruz, P.; Barcia, R.; Santos, M.; et al. Thymus mastichina: Chemical constituents and their anti-cancer activity. *Nat. Prod. Commun.* **2012**, *7*, 1491–1494. [CrossRef]
48. Kumar, D.; Sharma, P.; Singh, H.; Nepali, K.; Gupta, G.K.; Jain, S.K.; Ntie-Kang, F. The value of pyrans as anticancer scafolds in medicinal chemistry. *RSC Adv.* **2017**, *7*, 36977–36999. [CrossRef]
49. Nabavi, S.M.; Habtemariam, S.; Daglia, M.; Nabavi, S.F. Apigenin and Breast Cancers: From Chemistry to Medicine. *Anticancer Agents Med. Chem.* **2015**, *15*, 728–735. [CrossRef]
50. Huang, C.; Wei, Y.-X.; Shen, M.-C.; Tu, Y.-H.; Wang, C.-C.; Huang, H.-C. Chrysin, abundant in Morinda citrifolia fruit water-EtOAc extracts, combined with apigenin synergistically induced apoptosis and inhibited migration in human breast and liver cancer cells. *J. Agric. Food Chem.* **2016**, *64*, 4235–4245. [CrossRef]

molecules

MDPI

Article

Flavonoid Profiles in the Pulp of Different Lemon Cultivars and Their Antioxidant Activity Based on UPLC–Q–TOF–MS

Zhixiang Liu [1,*], Peng Wang [2], Chengcheng Liu [1] and Xin Tang [3]

1 Institute of Herbgenomics, Chengdu University of Traditional Chinese Medicine, Chengdu 611137, China; l18674128836@163.com
2 Department of Pharmacy, Chengdu University of Traditional Chinese Medicine, Chengdu 611137, China; wangdapeng19981026@163.com
3 Chongqing Key Laboratory of the Innovative Chinese Materia Medica & Health Intervention, Chongqing Academy of Chinese Materia Medica, Chongqing College of Traditional Chinese Medicine, Chongqing 400065, China; tangxin@cqctcm.edu.cn
* Correspondence: zhixiangliu88@cdutcm.edu.cn

Abstract: Previous studies have indicated that there may be differences among the varieties of lemon flavonoids, but the details have not yet been made clear, which limits the comprehensive use of different cultivated lemon varieties. In this study, ultra-performance liquid chromatography–quadrupole–time-of-flight–mass spectrometry (UPLC–Q–TOF–MS) and ultraviolet–visible spectroscopy (UV–Vis) were used to investigate the types and contents of flavonoids in the flesh of the main cultivated variety (Eureka) and five common lemon varieties, as well as their in vitro antioxidant activity. A total of 21 compounds were identified, five of which were common compounds. Among them, Verna, Lisbon, and Bearss each have characteristic components that can serve as potential criteria for variety identification. Each of the six varieties of lemon has strong antioxidant activity. The antioxidant activity of different lemon varieties is related to flavonoids. Therefore, Eureka and the other five varieties of lemon are good natural antioxidants, and the cultivation and industrial production of lemons should consider the needs and selection of suitable varieties.

Keywords: lemon; different cultivated varieties; antioxidant activity; flavonoids; UPLC–Q–TOF–MS

Citation: Liu, Z.; Wang, P.; Liu, C.; Tang, X. Flavonoid Profiles in the Pulp of Different Lemon Cultivars and Their Antioxidant Activity Based on UPLC–Q–TOF–MS. *Molecules* **2024**, *29*, 3464. https://doi.org/10.3390/molecules29153464

Academic Editor: Valeria Patricia Sülsen

Received: 23 June 2024
Revised: 20 July 2024
Accepted: 21 July 2024
Published: 24 July 2024

1. Introduction

Lemon (*Citrus limon*) is one of the main cultivated citrus plants in the world; it is grown in more than 60 countries and regions (FAO Statistics, 2021) and is popular among consumers for its unique flavour [1,2]. Previous studies have shown that lemons likely originated in China or India from the three ancestors of the citrus genus, namely C. Reticulata, pomelo, and citron, that were hybridized [2–4]. Then, various locally cultivated varieties (e.g., Eureka, Lisbon, Femminello, and Fino) were produced after continued human intervention. These lemon varieties are important for industrial production. In addition to being used as fruits, lemons and their extracts are also processed for the production of other commodities, such as juice/juice beverages, preserved fruits, lemon essential oils, and food supplements. Recent studies have shown that lemons are rich in polyphenolic compounds, including flavonoids, phenolic acids, limonoids, and terpenoids, which have antioxidant, anti-inflammatory, antitumour, and lipid-lowering properties. They are widely used in fields such as medicine, food, and cosmetics [5–10]. However, research on different lemon varieties is limited, which leads to confusion in the cultivation and industrial production of lemon varieties. Currently, Eureka is the most commonly cultivated variety, and more research and understanding are needed to fully utilize different lemon varieties.

Flavonoids are the main phenolics in citrus plants and are the main compounds that exert biological activity [11,12]. Flavonoids that have been isolated and identified from lemons include eriodictyl, hesperidin, hesperetin, naringin, apigenin, diosmin, quercetin, lipocitrin, spinacetin, and neohesperidin. These components have been shown to be associated with lemon's anti-inflammatory, antioxidant, antibacterial, antitumour, and lipid-lowering properties [5,6,8,10–12]. However, current research has focused on the chemical composition and pharmacological effects of individual lemon varieties, or on the composition of essential oils from the peels of multiple varieties of lemon. There have been no studies on the distribution of flavonoids in the flesh of different varieties of lemon fruit. Therefore, it is necessary to further study the composition and content of flavonoids in different varieties of lemon pulp.

Flavonoids play an important role in antioxidant activity and provide many of the health benefits of healthy food. The biological activity, composition, and content of flavonoids are closely related [13–15]. Previous studies have shown that lemon peel extract and lemon essential oil have good antioxidant activity [6,7,13]. The differences in the antioxidant activities of flavonoids in different varieties of lemon pulp are still unclear.

The aim of this study was to clarify the differences in flavonoid compounds and antioxidant activities in different varieties of lemon pulp. The composition and content of flavonoids in lemon pulp were measured using five representative lemon varieties and one main cultivated variety (Eureka). Then, the differences in antioxidant activity were compared, and the correlations between different flavonoids and antioxidant activity were analysed.

2. Results and Discussion

2.1. Total Phenolic Content

The detection results of TPC in different varieties of lemon pulp are shown in Table 1. The results showed that the total phenolic content in the lemon pulp ranged from 27.60 to 46.19 mg GAE/g DW, and the total phenolic content of the main cultivated variety (Eureka) was 35.59 mg GAE/g DW. The total phenolic content in the Fino lemon was the highest, followed by that in Lisbon, Femminello, and Verna lemons, all of which were significantly greater than the total phenolic content in Eureka and Bearss lemons ($p < 0.05$). This may be related to the extraction method and testing variety of phenolic compounds [14,16,17]. Compared with the distribution of phenolic compounds in different parts of lemon fruit [18], our results support that lemon pulp, especially that of Fino and Lisbon lemons, is a good source of phenolic compounds.

Table 1. The TPC and TFC in different varieties of lemon.

Sample	TPC (mg GAE/g DW)	TFC (mg RE/g DW)
FN	46.19 ± 1.08 a	5.27 ± 0.23 a
LS	41.06 ± 0.92 b	4.27 ± 0.11 c
WE	37.76 ± 1.25 c	2.43 ± 0.06 e
FM	37.91 ± 0.72 c	4.66 ± 0.03 b
EU	35.59 ± 0.77 d	3.74 ± 0.34 d
BE	27.60 ± 0.17 e	2.52 ± 0.06 e

Note: The different lowercase letters indicate significant differences ($p < 0.05$). FN: Fino; LS: Lisbon; WE: Verna; FM: Femminello; EU: Eureka; BE: Bearss.

2.2. Total Flavonoid Content

The detection results of TFC in different varieties of lemon pulp are shown in Table 1. The results showed that the total flavonoid content in the lemon pulp ranged from 2.52 to 5.27 mg RE/g DW, and the total flavonoid content in the main cultivated variety (Eureka) was 3.74 mg RE/g DW. The TFC in the Fino lemon was the highest, followed by that in Femminello, Lisbon, and Eureka lemons, all of which were significantly greater than the TFC in Verna and Bearss lemons ($p < 0.05$). This may be related to the distribution of flavonoids in citrus fruits and the tested varieties [5,8,17,19–21]. Flavonoids are most abundant in the peel of citrus fruits and have a lower content in the flesh. Our results indicate that the composition of flavonoids strongly depends on the lemon variety, and further analysis is needed to determine the differences in the types and contents of flavonoids.

2.3. *Flavonoid Compositions and Contents*

Metabolomics is widely used in fields such as botany, medicine, food, and traditional herbal medicine due to its high information content, simple operation, universal metabolites, and overall reflection of metabolites in the body [15,22–24]. Metabolite profiling research, especially high-resolution mass spectrometry-based, highly selective metabolomics, has been widely applied in citrus research [23,25–27]. In this study, we used UPLC–Q–TOF–MS to determine the composition and content of flavonoids in different varieties of lemon pulp, and the results are represented by the relative content of each component (Table 2). A total of 21 compounds were detected, with 11, 10, 13, 11, 10, and 14 flavonoid components detected from Eureka, Verna, Fino, Femminello, Lisbon, and Bearss, respectively. There are five common components, namely hesperidin, narirutin, 5,7-dihydroxy-3′-methoxyflavone-4′-*O*-β-D-glucoside, (2*S*)-5,7-dihydroxy-6-methoxyflavanone-7-*O*-β-D-glucopyranoside., and swertiajaponin. Hesperidin and narirutin are the main flavonoid components of citrus fruits (pomelo, lime, sweet orange, citrus, and lemon), are closely related to lemon's antioxidant, anti-inflammatory, and lipid-lowering properties, and are indicator components for the quality control of the traditional Chinese medicine Chenpi (Citri reticulatae pericarpium) [15,28–30]. The content of hesperidin in the different varieties of lemons ranged from 12.37% to 23.55%. Lisbon had the highest content, while Eureka had the lowest content. Similarly, the contents of narirutin in Eureka (2.68%), Verna (2.32%), and Fino (2.13%) lemons were significantly greater than those in Lisbon (0.20%), Femminello (0.20%), and Bearss (0.38%) lemons. In addition, Verna is the only cultivar with quercetin-3-gentiobioside-7-glucoside in quantifiable proportions. The unique components in the Lisbon lemon are 5,7,3′,4′-tetramethoxyflavones. There are five unique components in the Bearss lemon, including vicenin-II, luteolin-7,4′-*O*-β-D-diglucoside, violanthin, limocitrin, and nobiletin, but with relatively low contents. The differences in the composition and content of flavonoids in different varieties of lemon suggest that their potential biological activities also vary.

Principal component analysis (PCA) of samples helps us understand the differences between group samples and the degree of variation within group samples, as shown in Figure 1. In the 3D PCA graph, samples from different groups are clustered separately, indicating that the metabolic profiles of different lemon varieties are different. The clustering of samples within the same group indicates a uniform distribution of metabolites, which confirms the repeatability and reliability of this experiment. According to the 2D PCA plot, Verna was significantly different from the five other lemons, indicating the uniqueness of its flavonoid compounds. To visualize the distribution of flavonoids in different varieties of lemon, we conducted a cluster heatmap analysis, as shown in Figure 2. Verna can be clustered separately from the other five types of lemons, consistent with the PCA results, and some flavonoid compounds have relatively high contents. The clustering of samples within the same group indicates that the composition and content of flavonoids in lemon pulp may be influenced by genotype.

Table 2. Volatile organic compounds identified in dry peel of Zangju at different ripening stages, sampled by GC–IMS.

No.	Compounds	Formula	tR (min)	Ion Adduction	Parent Ion (*m*/*z*)	Error (ppm)	Fragmentation Profile (*m*/*z*)	Relative Content (%)					
								EU	WE	FN	FM	LS	BE
1	Hesperidin	$C_{28}H_{34}O_{15}$	11.50	$[M-H]^-$	609.1825	0.0	593.1551, 301.0720, 286.0489	12.37	21.76	18.46	12.66	23.55	20.55
2	5,7-Dihydroxy-3′-methoxyflavone-4′-*O*-β-D-glucoside	$C_{22}H_{22}O_{11}$	10.53	$[M-H]^-$	461.1100	2.3	371.0802, 298.0491, 195.0671	1.12	0.44	0.85	0.36	0.22	0.03
3	(2*S*)-5,7-Dihydroxy-6-methoxyflavanone-7-*O*-β-D-glucopyranoside	$C_{22}H_{24}O_{10}$	10.35	$[M+H]^+$	463.1240	1.1	427.1048, 343.0831, 303.0523	1.70	4.92	4.42	2.89	4.13	7.57
4	Swertiajaponin	$C_{22}H_{22}O_{11}$	11.49	$[M+H]^+$	449.1442	0.0	301.0716, 263.0569, 153.0192	0.98	0.46	0.85	0.36	0.52	0.65
5	Kuwanon E	$C_{25}H_{28}O_6$	9.33	$[M-H]^-$	639.1592	3.9	595.1668, 287.0562, 151.0032	5.06	0.55	2.18	6.14	6.54	-
6	(2*R*,3*R*)-Taxifolin-7-*O*-α-L-rhamnopyranosyl(1→6)-β-D-glucopyranoside	$C_{27}H_{32}O_{16}$	6.37	$[M-H]^-$	625.1783	1.4	317.0676, 289.0737, 125.0239	26.30	-	-	34.64	11.41	30.64
7	Isosakuranetin-7-rutinoside	$C_{28}H_{34}O_{14}$	17.56	$[M-H]^-$	469.1880	2.5	229.1245, 227.1443, 179.0361	5.12	0.60	0.51	-	2.53	0.26
8	Narirutin	$C_{27}H_{32}O_{14}$	13.53	$[M-H]^-$	593.1891	2.5	559.1505, 519.1178, 285.0776	2.68	2.32	2.13	0.20	0.20	0.38
9	Isoscoparin-3′-*O*-glucopyranoside	$C_{28}H_{32}O_{16}$	4.28	$[M-H]^-$	611.1625	1.3	303.0533, 285.0419, 125.0208	0.51	-	1.07	0.17	-	0.62
10	2′-*O*-Acetylrutin	$C_{29}H_{32}O_{17}$	6.62	$[M+H]^+$	625.1787	3.9	487.1239, 355.0835, 289.0741	0.04	-	0.11	-	-	0.26
11	Isorhamnetin-3-*O*-glucoside-7-*O*-rhamnoside	$C_{28}H_{32}O_{16}$	7.03	$[M-H]^-$	623.1616	−0.3	503.1213, 413.0893, 383.0782	9.34	8.73	9.18	10.74	8.01	-
12	2″-*O*-rhamnosylvitexin	$C_{27}H_{32}O_{15}$	12.31	$[M-H]^-$	651.1584	2.7	579.1382, 375.0734, 360.0497	-	8.73	9.18	10.74	-	5.68
13	Kaempferol-3-(2G-rhamnosylrutinoside)	$C_{33}H_{40}O_{19}$	9.97	$[M-H]^-$	623.1638	3.3	593.1512, 285.0405, 283.0266	-	0.01	0.01	0.02	-	-
14	Quercetin-3-gentiobioside-7-glucoside	$C_{33}H_{40}O_{22}$	9.31	$[M-H]^-$	595.1666	−0.3	459.1161, 287.0564, 151.0034	-	35.72	-	-	-	-
15	Luteolin-7,4′-*O*-β-D-Diglucoside	$C_{27}H_{30}O_{16}$	4.29	$[M-H]^-$	341.1044	3.8	285.0427, 134.0379, 125.0217	-	-	-	-	-	0.05
16	3′,4′,5,7-Tetramethoxyflavone	$C_{19}H_{18}O_6$	1.72	$[M-H]^-$	833.1964	−3.5	391.0342, 217.0168, 87.0093	-	-	-	-	11.41	-
17	Vicenin-II	$C_{27}H_{30}O_{15}$	9.32	$[M+H]^+$	435.1290	0.9	289.07160, 153.0197, 135.0453	-	-	-	-	-	0.38
18	4,5,7-Trihydroxyflavone-3-*O*-β-D-glucopyranoside	$C_{21}H_{22}O_{10}$	6.88	$[M-H]^-$	785.2182	4.6	577.1583, 431.1019, 269.0467	-	-	2.67	1.53	0.35	-
19	Violanthin	$C_{27}H_{30}O_{14}$	6.06	$[M-H]^-$	593.1519	1.2	473.1112, 383.0791, 353.0680	-	-	-	-	-	0.99
20	Limocitrin	$C_{17}H_{14}O_8$	10.28	$[M-H]^-$	609.1474	2.1	461.1112, 341.0680, 298.0489	-	-	-	-	-	0.79
21	Nobiletin	$C_{21}H_{22}O_8$	11.21	$[M+H]^+$	579.1711	0.4	433.1141, 271.0605, 85.0304	-	-	-	-	-	0.39

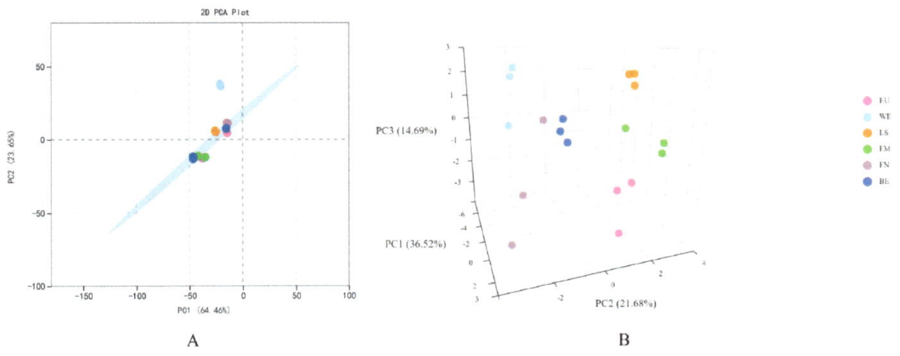

Figure 1. Principal component analysis results chart. (**A**) 2D PCA plot; (**B**) 3D PCA plot. EU: Eureka; WE: Verna; LS: Lisbon; FM: Femminello; FN: Fino; BE: Bearss.

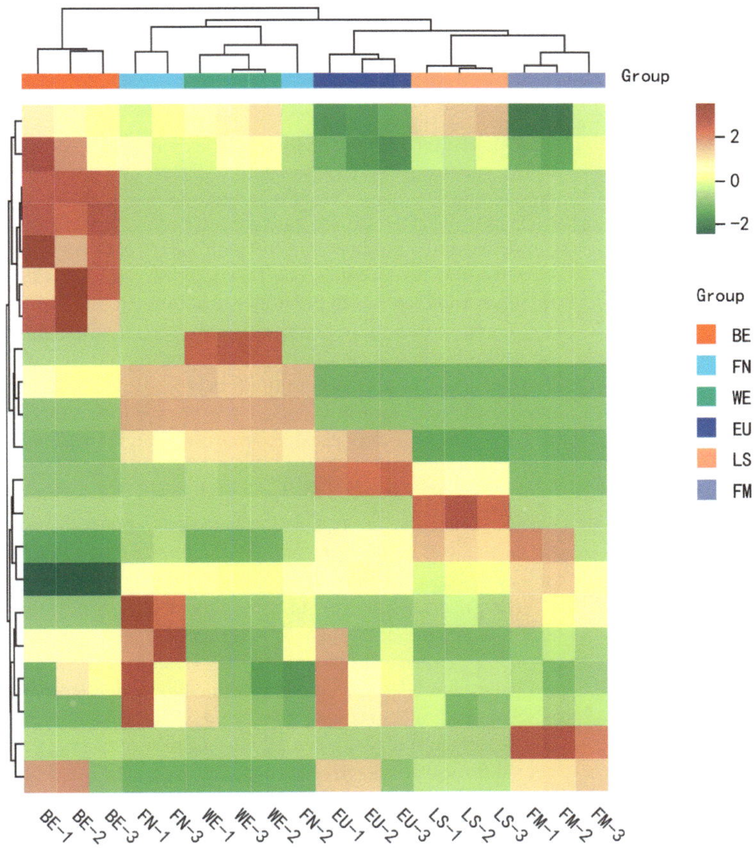

Figure 2. The cluster heatmap of flavonoids in different lemon varieties. BE: Bearss; FN: Fino; WE: Verna; EU: Eureka; LS: Lisbon; FM: Femminello.

2.4. Antioxidant Activity

The antioxidant activity of the six lemon varieties was evaluated using DPPH, ABTS, and FRAP, as shown in Figure 3. The results showed that the IC_{50} values of the DPPH and ABTS clearance rates of these lemons were 2.10–3.79 mg/mL and 3.62–6.49 mg/mL, respectively, with a total antioxidant capacity of 97.80–114.35 mg $FeSO_4$/g DW, indicating differences in antioxidant activity among the different lemon varieties. The DPPH clearance rate of the Fino lemon was the highest, followed by those of the Lisbon, Verna, Femminello, and Eureka lemons, all of which were significantly greater than that of the Bearss lemon ($p < 0.05$). The ABTS clearance rate of the Verna lemon was the highest, followed by those of the Fino, Femminello, Lisbon, and Eureka lemons, all of which were significantly greater than that of the Bearss lemon ($p < 0.05$). The total antioxidant capacities of the Lisbon and Verna lemons were comparable, as were those of the Fino and Femminello lemons. The total antioxidant capacity of the Bearss lemon was significantly lower than that of the other five lemon varieties.

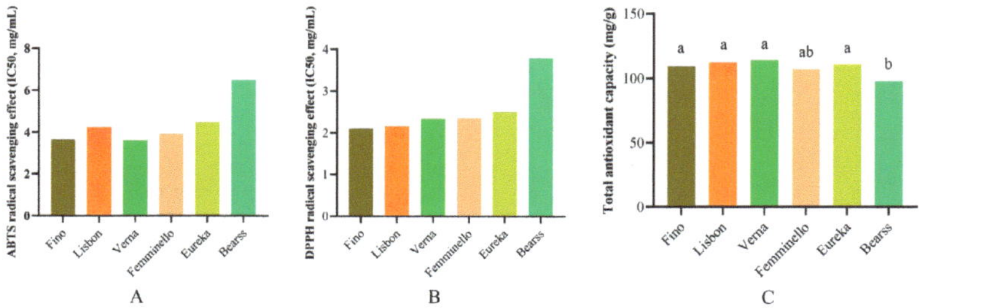

Figure 3. The antioxidant activity of the different lemon varieties. (**A**) ABTS measurement; (**B**) DPPH measurement; (**C**) The FRAP method was used to determine the total antioxidant activity of the different lemon varieties. The different lowercase letters indicate significant differences ($p < 0.05$).

Our results indicate that the antioxidant activity reflected by different methods is not entirely consistent, which may be related to phenolic substances. There are differences in the types and contents of phenolic substances in different varieties of lemon pulp, resulting in different antioxidant activity detection results for different indicators [14,22]. Overall, the differences in antioxidant activity among the different varieties of lemon pulp were consistent with the trends in the total phenolic and flavonoid contents. The antioxidant activity of the different varieties of lemons is lower than that of oranges but greater than that of fruits such as pomelo, grapefruit, pear, watermelon, and mango [14,15,23].

2.5. Correlations between TPC, TFC, Flavonoids, and Antioxidant Activity

To investigate the correlation between flavonoids and antioxidant differences in different varieties of lemon, we conducted a Pearson correlation analysis, as shown in Figure 4. (2*S*)-5,7-Dihydroxy-6-methoxyflavanone-7-*O*-β-D-glucopyranoside, (2*R*,3*R*)-taxifolin-7-*O*-α-L-rhamnopyranosyl(1→6)-β-D-glucopyranoside, vicenin II, violanthin, limocitrin, nobiletin, and luteolin-7,4′-*O*-β-D-diglucoside were significantly positively correlated with the ABTS free radical scavenging rate, while isorhamnetin-3-*O*-glucoside-7-*O*-rhamnoside and kaempferol-3-(2G-rhamnosylrutinoside) were significantly negatively correlated with the ABTS free radical scavenging rate ($|r| > 0.5$, $p < 0.05$). (2*S*)-5,7-Dihydroxy-6-methoxyflavanone-7-*O*-β-D-glucopyranoside, vicenin II, violanthin, limocitrin, and luteolin-7,4′-*O*-β-D-diglucoside were significantly positively correlated with the DPPH free radical scavenging rate, and kuwanon E and isorhamnetin-3-*O*-glucoside-7-*O*-rhamnoside were significantly negatively correlated with the DPPH free radical scavenging rate.

Violanthin, limocitrin, nobiletin, luteolin-7,4′-*O*-β-D-diglucoside, vicenin II, (2*R*,3*R*)-taxifolin-7-*O*-α-L-rhamnopyranosyl(1→6)-β-D-glucopyranoside, and (2*S*)-5,7-dihydroxy-6-methoxyflavanone-7-*O*-β-D-glucopyranoside were significantly negatively correlated with FRAP. Interestingly, the TPC and TFC of the different varieties of lemon were significantly negatively correlated with the DPPH and ABTS free radical scavenging rates and significantly positively correlated with the FRAP rate. These results indicate that the composition and content of flavonoids in different varieties of lemons are different, which affects their antioxidant activity.

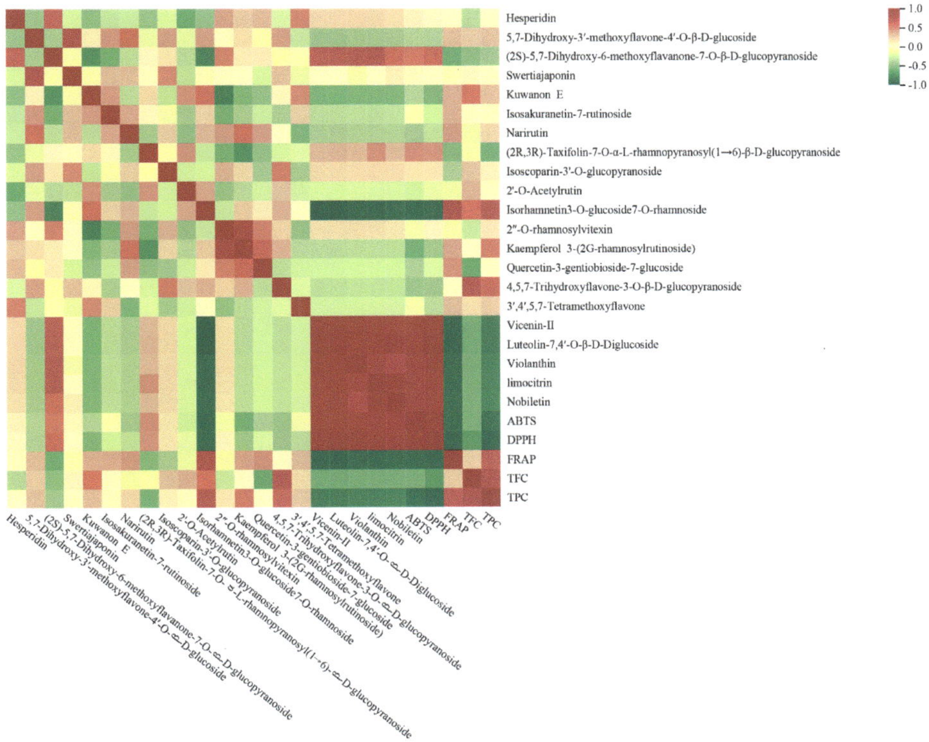

Figure 4. Pearson correlation analysis heatmap of flavonoids in different lemon varieties.

3. Materials and Methods

3.1. Plant Material

All lemon samples were collected from the planting base of the Lemon Industry Bureau in Anyue County, Sichuan Province, China, in November 2023. Four adjacent lemon trees of high yield and stable quality were randomly selected, and 6 fresh fruits of similar size that were free from pests and diseases were randomly selected from each tree. The varieties (Eureka, Lisbon, Femminello, Fino, Bearss, and Verna) were numbered and transported to the laboratory using dry ice. The samples were washed, the peel and pulp were manually separated, and the pulp was freeze-dried using the following method: The lemon pulp was evenly spread in a single layer and placed in a vacuum freeze dryer, cooled to −30 °C, and pre-frozen for 5 h before the heating plate temperature of the freeze dryer was set to 50 °C and the condenser temperature to −50 °C. After sealing, the pump was set to a pressure of 10 Pa and the sample dried for 72 h. The pulp was crushed with a grinder after drying, and all the samples were stored in a dry place at room temperature.

3.2. Chemicals

Methanol, sodium nitrite, aluminium nitrate trihydrate, sodium hydroxide, anhydrous sodium carbonate, anhydrous ethanol, potassium persulfate (analytical pure, Chengdu Cologne Chemical Co., Ltd., Chengdu, China). Rutin (chromatographically pure, Sichuan Vicki Biotechnology Co., Ltd., Chengdu, China). Gallic acid (chromatographically pure, Chengdu Ruifensidan Biotechnology Co., Ltd., Chengdu, China). Folin phenol reagent (1 mol/mL, Feijing Biotechnology Co., Ltd., Shanghai, China). 2,2-Diphenyl-1-Picrylhydrazyl (DPPH) and 2,2′-azino-bis(3-ethylbenzothiazoline-6-sulfonate) (ABTS) were both purchased from Shanghai Macklin Biochemical Technology Co., Ltd., Shanghai, China. Total antioxidant capacity (T-AOC) assay kit (ferric ion reducing antioxidant power, FRAP, Nanjing Jiancheng Biotechnology Research Institute, Nanjing, China).

3.3. *Sample Preparation and Extraction*

The sample processing followed the method of Ledesma Escobar et al. [31]. Lemon pulp powder (0.15 g) was accurately weighed, 5 mL of 80% methanol solution was added, and the mixture was mixed evenly. The sample was extracted using ultrasonication for 60 min (vortexing every 15 min). Then, the mixture was centrifuged for 15 min (4000 rpm/min), the supernatant was collected, and 0.22 μM filter membrane filtration was used. The filtrate was stored in an injection bottle for ultra-performance liquid chromatography–quadrupole–time-of-flight–mass spectrometry (UPLC–Q–TOF–MS) analysis.

3.4. *Measurement of Total Phenolic Content*

The Folin-Ciocalteu colorimetric method [32] was used to determine the total phenolic content (TPC), and the results are presented as mg gallic acid equivalents (GAE)/g dry weight (DW) of the pulp sample. Gallic acid solutions with concentrations of 0, 0.004, 0.005, 0.006, 0.008, 0.009, and 0.01 mg/mL were prepared. Then, 0.25 mL of phenol reagent, 4.75 mL of 20% sodium carbonate solution, 0.5 mL of gallic acid solution, or different concentrations of lemon extract were added, the solution was mixed well, 80% methanol was added to bring the volume to 15 mL, and the mixture was placed in a water bath at 75 °C for 30 min. The solution was cooled to room temperature, and the absorbance was measured at 760 nm. The standard curve was $y = 10.697x - 0.0324$ ($R^2 = 0.9994$).

3.5. *Measurement of Total Flavonoid Content*

The total flavonoid content (TFC) was determined using the sodium nitrite–aluminium nitrate–sodium hydroxide colorimetric method [33]. A 0.2 mg/mL rutin solution consisting of 0 mL, 0.5 mL, 1.0 mL, 2 mL, 2.5 mL, 4.0 mL, 5.0 mL, or 3.0 mL of lemon extract was prepared. Then, the following solutions were added: 0.75 mL of 5% sodium nitrite, 0.75 mL of 10% aluminium nitrate, 6 mL of 4% sodium hydroxide, and, finally, 80% methanol to adjust the volume to 15 mL. After each addition, the mixture was mixed well and allowed to stand for 6 min, and for 15 min after methanol addition. The absorbance was measured at 510 nm. The standard curve was $y = 12.55x - 0.0095$ ($R^2 = 0.9998$). The results are presented as mg rutin equivalent (RE)/g DW of the pulp sample.

3.6. *UPLC–Q–TOF–MS Conditions*

This study was analysed using the UPLC–Q–TOF–MS system (SynaptXS, Waters, Milford, MA, USA). The chromatographic conditions: (1) Column: ACQUITY UPLC CSH C_{18}, 1.7 μm, 2.1 mm ×100 mm; (2) Mobile phase: A phase was ultrapure water, B phase was acetonitrile; (3) Elution gradient: 0~1.0 min 2% (B); 1.0~3.0 min 2% → 7% (B); 3.0~8.0 min 7% → 11% (B); 8.0~13.0 min 11% → 15% (B); 13.0~16.0 min 15% → 30% (B); 16.0~20.0 min 30% → 36%; 20.0~26.0 min, 36% → 70% (B); 26.0~30.0 min 70% → 85% (B); 30.0~31.0 min 85% → 95% (B); 31.0–33.0 min 95% → 2% (B); (4) Flow rate of 0.3 mL/min; Column temperature of 35 °C; Injection volume 2 μL.

Mass spectrum condition: positive and negative ion mode of electric spray ion source was adopted, capillary voltage: 3000 V; source temperature: 120 °C; desolvent gas temperature: 450 V; air curtain flow rate: 50.0 L/h; desolvent gas flow rate: 800.0 L/h; nebulizer gas flow rate: 6.0 Bar; low-end resolution: 4.7; high-end resolution: 15.0; IMS gas flow rate: 90.00 mL/min.

The collected raw data were imported into Compound Discoverer 3.0 software, the peak areas and peak alignments were extracted, and the measured spectra of secondary fragments were matched with the mzCloud network database and the local traditional Chinese medicine ingredient database (OTCML). The set filtering parameters for the matching results were as follows: peak area threshold value: 80,000, primary and secondary quality deviation: 5 ppm, and matching degree score: above 80. Compounds were analysed and identified by comparing the filtered ions with the compound information in the database.

3.7. Antioxidant Activity

DPPH. The DPPH assay was conducted following the protocol outlined by Wang et al. [15] with adjustments. The lemon extracts of different concentrations (100 μL) were combined with DPPH solution (0.008 mg/mL) (100 μL) in a 96-well plate. The mixtures underwent a 30-min reaction at room temperature in darkness, and the measurement of absorbance value at 517 nm was conducted using a microplate reader. In the control group, substituting the DPPH solution with 80% methanol, the blank group was similarly treated with 80% methanol. The IC_{50} value represents the DPPH clearance rate.

ABTS. Refer to Wang et al.'s method for ABTS determination, which was followed with slight modifications [15]. The $K_2S_2O_8$ solution (2.6 mmol/L) was added to the ABTS solution (7.00 mmol/L) and allowed to react thoroughly in a cool place for 12–16 h. The solution obtained after dilution of the mixture seven times with 80% methanol (with an absorbance of 7.00 ± 0.02 at 734 nm) was used to prepare the ABTS working solution. Following that, 25 μL aliquots of lemon extract of different concentrations were introduced into the 96-well plate, the working solution of ABTS, a volume of 175 μL, was thoroughly mixed and allowed to react in darkness for a duration of 40 min, followed by measuring the absorbance value at 734 nm using a microplate reader; within the control group, substituting ABTS solution with 80% methanol, the blank group was also treated with 80% methanol. The IC_{50} value represents the ABTS clearance rate.

FRAP. The FRAP clearance rate was determined using a total antioxidant assay kit (Nanjing Jiancheng Bioengineering Research Institute, China). Overall, the standard determination curves of $FeSO_4$ solutions with concentrations of 0.15, 0.3, 0.6, 0.9, 1.2, and 1.5 mmol/L were prepared. Then, 180 μL of FRAP working fluid was introduced into the wells of the 96-well plate, and 5 μL each of different concentrations of $FeSO_4$ solutions and lemon extracts were added. This was followed by the addition of 5 μL of each solution in the blank group; the solution extraction was substituted with distilled water. Incubation of the mixture was carried out at 37 °C for 3–5 min, and the absorbance value was measured at 593 nm using a microplate reader. The standard curve is: $y = 0.007x - 0.0067$, $R^2 = 0.9974$. The results are presented as mg $FeSO_4$/g DW of the pulp sample. The standard curve is: $y = 0.007x - 0.0067$, $R^2 = 0.9974$.

3.8. Statistical Analysis

The statistical function prcomp in R version 3.3.1 for principal component analysis, and the ComplexHeatmap package in R version 3.3.1 for hierarchical cluster analysis (HCA) and production of a heatmap were used. GraphPad Prism 8.4.3 (GraphPad Software company, La Jolla, CA, USA) was used for statistical analysis, plotting bar charts, and ANOVA analysis ($p < 0.05$ indicated statistical significance). All results are the average of three experiments.

4. Conclusions

This study evaluated the differences in TPC, TFC, and antioxidant activity among different lemon varieties. Fino had the highest total phenolic content, followed by Lisbon, Femminello, and Eureka, all of which were significantly greater than those of Verna and Bearss. Fino lemon had the highest total flavonoid content, followed by Femminello, Lisbon, and Eureka, all of which were significantly greater than those of Verna and Bearss. Fino had the highest DPPH clearance rate, followed by Lisbon, Verna, Femminello, and Eureka, all of which were significantly greater than that of Bearss. The ABTS clearance rate of Verna was the highest, followed by those of Fino, Femminello, Lisbon, and Eureka, all of which were significantly greater than that of Bearss. The total antioxidant capacities of Lisbon and Verna were comparable, as were those of Fino and Femminello. The total antioxidant capacity of Bearss was significantly lower than that of the other five lemon varieties.

Furthermore, UPLC–Q–TOF–MS was used to determine the composition and content of flavonoids in different varieties of lemon pulp. A total of 21 compounds were detected, five of which were common components, such as hesperidin, narirutin, and swertiajaponin. Among them, Verna, Lisbon, and Bearss each had characteristic components that can serve as potential criteria for variety identification. The Pearson correlation analysis results indicate that vicenin-II, luteolin-7,4′-*O*-β-D-diglucoside, violanthin, limocitrin, and nobiletin are associated with antioxidant activity. Overall, our research revealed that Eureka and five other types of lemon are good natural antioxidants. There are differences in the composition and content of flavonoids in the different varieties of lemon, and the cultivation and industrial production of lemons should consider the needs and selection of suitable varieties.

Author Contributions: Z.L.: funding acquisition, conceptualization, investigation, formal analysis, data curation, and writing–original draft. P.W.: conceptualization, investigation, methodology, and formal analysis. C.L.: investigation, formal analysis, resources, and software. X.T.: methodology, data curation, project administration, and writing–review and editing. All authors have read and agreed to the published version of the manuscript.

Funding: This research was supported by the National Science Fund for Distinguished Young Scholars (No. 82003888) and the Chengdu University of Traditional Chinese Medicine Youth Talent Program (No. QJRC2022055).

Institutional Review Board Statement: Not applicable.

Informed Consent Statement: Not applicable.

Data Availability Statement: Data are contained within the article materials.

Acknowledgments: The authors would like to thank all those who contributed directly or indirectly to the project. In addition, we appreciate the support and materials provided by the Lemon Industry Bureau of Anyue County, Sichuan Province, China, for this study.

Conflicts of Interest: The authors declare that they have no known competing financial interests or personal relationships that could appear to have influenced the work reported in this paper.

References

1. Matsumoto, T.; Watanabe, T. Isolation and structure elucidation of constituents of *Citrus limon*, *Isodon japonicus*, and *Lansium domesticum* as the cancer prevention agents. *Genes Environ.* **2020**, *42*, 17. [CrossRef] [PubMed]
2. Bao, Y.; Zeng, Z.; Yao, W.; Chen, X.; Jiang, M.; Sehrish, A.; Wu, B.; Powell, C.A.; Chen, B.; Xu, J.; et al. A gap-free and haplotype-resolved lemon genome provides insights into flavor synthesis and huanglongbing (HLB) tolerance. *Hortic. Res.* **2023**, *10*, uhad020. [CrossRef]
3. Wu, G.A.; Terol, J.; Ibanez, V.; López-García, A.; Pérez-Román, E.; Borredá, C.; Domingo, C.; Tadeo, F.R.; Carbonell-Caballero, J.; Alonso, R.; et al. Genomics of the origin and evolution of Citrus. *Nature* **2018**, *554*, 311–316. [CrossRef] [PubMed]
4. Huang, Y.; He, J.; Xu, Y.; Zheng, W.; Wang, S.; Chen, P.; Zeng, B.; Yang, S.; Jiang, X.; Liu, Z.; et al. Pangenome analysis provides insight into the evolution of the orange subfamily and a key gene for citric acid accumulation in citrus fruits. *Nat. Genet.* **2023**, *55*, 1964–1975. [CrossRef]
5. Liu, N.; Li, X.; Zhao, P.; Zhang, X.; Qiao, O.; Huang, L.; Guo, L.; Gao, W. A review of chemical constituents and health-promoting effects of citrus peels. *Food Chem.* **2021**, *365*, 130585. [CrossRef]
6. Nasirian, V.S.; Shahidi, S.; Tahermansouri, H.; Chekin, F. Application of graphene oxide in the adsorption and extraction of bioactive compounds from lemon peel. *Food Sci. Nutr.* **2021**, *9*, 3852–3862. [CrossRef] [PubMed]
7. Imeneo, V.; Romeo, R.; De Bruno, A.; Piscopo, A. Green-sustainable extraction techniques for the recovery of antioxidant compounds from "*Citrus limon*" by-products. *J. Environ. Sci. Health B* **2022**, *57*, 220–232. [CrossRef]
8. Chaves, J.O.; Sanches, V.L.; Viganó, J.; Mesquita, L.M.d.S.; de Souza, M.C.; da Silva, L.C.; Acunha, T.; Faccioli, L.H.; Rostagno, M.A. Integration of pressurized liquid extraction and in-line solid-phase extraction to simultaneously extract and concentrate phenolic compounds from lemon peel (*Citrus limon* L.). *Food Res. Int.* **2022**, *157*, 111252. [CrossRef]
9. Ou, Q.; Zhao, J.; Sun, Y.; Zhao, Y.; Zhang, B. Utilization of Lemon Peel for the Production of Vinegar by a Combination of Alcoholic and Acetic Fermentations. *Foods* **2023**, *12*, 2488. [CrossRef]
10. Sánchez-Bravo, P.; Martínez-Tomé, J.; Hernández, F.; Sendra, E.; Noguera-Artiaga, L. Conventional vs. Organic: Evaluation of Nutritional, Functional and Sensory Quality of *Citrus limon*. *Foods* **2023**, *12*, 4304. [CrossRef] [PubMed]
11. Maleki, S.J.; Crespo, J.F.; Cabanillas, B. Anti-inflammatory effects of flavonoids. *Food Chem.* **2019**, *299*, 125124. [CrossRef] [PubMed]

12. Klimek-Szczykutowicz, M.; Szopa, A.; Ekiert, H. *Citrus limon* (Lemon) Phenomenon-A Review of the Chemistry, Pharmacological Properties, Applications in the Modern Pharmaceutical, Food, and Cosmetics Industries, and Biotechnological Studies. *Plants* **2020**, *9*, 119. [CrossRef] [PubMed]
13. Park, Y.S.; Kim, I.D.; Dhungana, S.K.; Park, E.J.; Park, J.J.; Kim, J.H.; Shin, D.H. Quality Characteristics and Antioxidant Potential of Lemon (*Citrus limon* Burm. f.) Seed Oil Extracted by Different Methods. *Front. Nutr.* **2021**, *8*, 644406. [CrossRef] [PubMed]
14. Deng, M.; Dong, L.; Jia, X.; Huang, F.; Chi, J.; Muhammad, Z.; Ma, Q.; Zhao, D.; Zhang, M.; Zhang, R. The flavonoid profiles in the pulp of different pomelo (*Citrus grandis* L. Osbeck) and grapefruit (*Citrus paradisi* Mcfad) cultivars and their in vitro bioactivity. *Food Chem. X* **2022**, *15*, 100368. [CrossRef] [PubMed]
15. Wang, P.; Wang, H.; Xiao, Y.; Zou, J.; Chen, H.; Chen, L.; Wang, F.; Hu, Y.; Liu, Y. Insights into metabolic characteristics and biological activity changes in Zangju (*Citrus reticulata* cv. Manau Gan) peel at different maturity stages through UPLC-MS/MS-based metabolomics. *Food Chem. X* **2024**, *21*, 101197. [CrossRef]
16. Zhang, R.; Zeng, Q.; Deng, Y.; Zhang, M.; Wei, Z.; Zhang, Y.; Tang, X. Phenolic profiles and antioxidant activity of litchi pulp of different cultivars cultivated in Southern China. *Food Chem.* **2013**, *136*, 1169–1176. [CrossRef]
17. Chen, Y.; Pan, H.; Hao, S.; Pan, D.; Wang, G.; Yu, W. Evaluation of phenolic composition and antioxidant properties of different varieties of Chinese citrus. *Food Chem.* **2021**, *364*, 130413. [CrossRef]
18. Xi, W.; Lu, J.; Qun, J.; Jiao, B. Characterization of phenolic profile and antioxidant capacity of different fruit part from lemon (*Citrus limon* Burm.) cultivars. *J. Food Sci. Technol.* **2017**, *54*, 1108–1118. [CrossRef]
19. Salahuddin, M.A.H.; Ismail, A.; Kassim, N.K.; Hamid, M.; Ali, M.S.M. Phenolic profiling and evaluation of in vitro antioxidant, alpha-glucosidase and alpha-amylase inhibitory activities of *Lepisanthes fruticosa* (Roxb) Leenh fruit extracts. *Food Chem.* **2020**, *331*, 127240. [CrossRef]
20. Özcan, M.M.; Ghafoor, K.; Al Juhaimi, F.; Uslu, N.; Babiker, E.E.; Ahmed, I.A.M.; Almusallam, I.A. Influence of drying techniques on bioactive properties, phenolic compounds and fatty acid compositions of dried lemon and orange peel powders. *J. Food Sci. Technol.* **2021**, *58*, 147–158. [CrossRef]
21. Deng, M.; Zhang, R.; Zhang, S.; Lu, Q.; Dong, L.; Huang, F.; Jia, X.; Ma, Q.; Chi, J.; Zhao, D.; et al. The flavonoid profiles in different tissue parts of Shatianyu (*Citrus grandis* L. Osbeck) and their in vitro bioactivity. *LWT* **2023**, *180*, 114712. [CrossRef]
22. Zhang, A.; Sun, H.; Wang, Z.; Sun, W.; Wang, P.; Wang, X. Metabolomics: Towards understanding traditional Chinese medicine. *Planta Med.* **2010**, *76*, 2026–2035. [CrossRef] [PubMed]
23. Dadwal, V.; Joshi, R.; Gupta, M. A comparative metabolomic investigation in fruit sections of *Citrus medica* L. and *Citrus maxima* L. detecting potential bioactive metabolites using UHPLC-QTOF-IMS. *Food Res. Int.* **2022**, *157*, 111486. [CrossRef]
24. Wang, F.; Hu, Y.; Chen, H.; Chen, L.; Liu, Y. Exploring the roles of microorganisms and metabolites in the 30-year aging process of the dried pericarps of *Citrus reticulata* 'Chachi' based on high-throughput sequencing and comparative metabolomics. *Food Res. Int.* **2023**, *172*, 113117. [CrossRef] [PubMed]
25. Yan, H.; Pu, Z.-J.; Zhang, Z.-Y.; Zhou, G.-S.; Zou, D.-Q.; Guo, S.; Li, C.; Zhan, Z.-L.; Duan, J.-A. Research on Biomarkers of Different Growth Periods and Different Drying Processes of *Citrus wilsonii* Tanaka Based on Plant Metabolomics. *Front. Plant Sci.* **2021**, *12*, 700367. [CrossRef] [PubMed]
26. Wang, Z.; Gmitter, F.G.; Grosser, J.W.; Wang, Y. Natural Sweeteners and Sweetness-Enhancing Compounds Identified in Citrus Using an Efficient Metabolomics-Based Screening Strategy. *J. Agric. Food Chem.* **2022**, *70*, 10593–10603. [CrossRef]
27. He, G.; Chen, X.; Hou, X.; Yu, X.; Han, M.; Qiu, S.; Li, Y.; Qin, S.; Wang, F. UPLC-Q-TOF/MS-based metabolomic analysis reveals the effects of asomate on the citrus fruit. *Curr. Res. Food Sci.* **2023**, *6*, 100523. [CrossRef] [PubMed]
28. Gómez-Mejía, E.; Rosales-Conrado, N.; León-González, M.E.; Madrid, Y. Citrus peels waste as a source of value-added compounds: Extraction and quantification of bioactive polyphenols. *Food Chem.* **2019**, *295*, 289–299. [CrossRef]
29. National Pharmacopoeia Commission. *Pharmacopoeia of People's Republic of China*; China Medical Science Press: Beijing, China, 2020.
30. Singh, B.; Singh, J.P.; Kaur, A.; Singh, N. Phenolic composition, antioxidant potential and health benefits of citrus peel. *Food Res. Int.* **2020**, *132*, 109114. [CrossRef]
31. Ledesma-Escobar, C.A.; Priego-Capote, F.; de Castro, M.D.L. Characterization of lemon (*Citrus limon*) polar extract by liquid chromatography-tandem mass spectrometry in high resolution mode. *J. Mass Spectrom.* **2015**, *50*, 1196–1205. [CrossRef]
32. Dewanto, V.; Wu, X.; Adom, K.K.; Liu, R.H. Thermal processing enhances the nutritional value of tomatoes by increasing total antioxidant activity. *J. Agric. Food Chem.* **2002**, *50*, 3010–3014. [CrossRef] [PubMed]
33. Chen, H.; Ouyang, K.; Jiang, Y.; Yang, Z.; Hu, W.; Xiong, L.; Wang, N.; Liu, X.; Wang, W. Constituent analysis of the ethanol extracts of *Chimonanthus nitens* Oliv. leaves and their inhibitory effect on alpha-glucosidase activity. *Int. J. Biol. Macromol.* **2017**, *98*, 829–836. [CrossRef] [PubMed]

foods

MDPI

Article

Characterization of Meat Metabolites and Lipids in Shanghai Local Pig Breeds Revealed by LC–MS-Based Method

Jun Gao [1,2,3,4], Lingwei Sun [1,2,3,4], Weilong Tu [1,2,4], Mengqian Cao [3,5], Shushan Zhang [1,2,3,4], Jiehuan Xu [1,3,4], Mengqian He [1,3,4], Defu Zhang [1,2,3,4], Jianjun Dai [1,2,3,4], Xiao Wu [6,*] and Caifeng Wu [1,2,3,*]

1 Institute of Animal Husbandry and Veterinary Science, Shanghai Academy of Agricultural Sciences, Shanghai 201106, China; gaojun@saas.sh.cn (J.G.); sunlingwei1987@126.com (L.S.); tuweilong@saas.sh.cn (W.T.); smalltreexj@126.com (S.Z.); jiehuanxu810@163.com (J.X.); he1037247863@163.com (M.H.); zhangdefuzdf@163.com (D.Z.); daijianjun@saas.sh.cn (J.D.)

2 Key Laboratory of Livestock and Poultry Resources (Pig) Evaluation and Utilization, Ministry of Agriculture and Rural Affairs, Shanghai 201106, China

3 Shanghai Municipal Key Laboratory of Agri-Genetics and Breeding, Shanghai 201106, China; 13887223118@163.com

4 Shanghai Engineering Research Center of Breeding Pig, Shanghai 201106, China

5 College of Fisheries and Life Science, Shanghai Ocean University, Shanghai 201306, China

6 Biotechnology Research Institute, Shanghai Academy of Agriculture Sciences, Shanghai 201106, China

* Correspondence: wuxiao@saas.sh.cn (X.W.); wucaifengwcf@163.com (C.W.)

Abstract: The meat of local livestock breeds often has unique qualities and flavors. In this study, three Shanghai native pig breeds (MSZ, SWT, and SHB) exhibited better meat quality traits than globalized commercial pig breeds (DLY). Subsequently, metabolomic and lipidomic differences in the longissimus dorsi (L) and gluteus (T) muscles of the Shanghai native pig breeds and DLY pig breed were compared using liquid chromatography–mass spectrometry (LC–MS). The results demonstrated that the metabolites mainly consisted of (28.16%) lipids and lipid-like molecules, and (25.87%) organic acids and their derivatives were the two most dominant groups. Hundreds of differential expression metabolites were identified in every compared group, respectively. One-way ANOVA was applied to test the significance between multiple groups. Among the 20 most abundant differential metabolites, L-carnitine was significantly different in the muscles of the four pig breeds (p-value = 7.322×10^{-11}). It was significantly higher in the L and T muscles of the two indigenous black pig breeds (MSZ and SWT) than in the DLY pigs (p-value < 0.001). Similarly, lipidomic analysis revealed the PA (18:0/18:2) was significantly more abundant in the muscle of these two black breeds than that in the DLY breed (p-value < 0.001). These specific metabolites and lipids might influence the meat quality and taste properties and lead to customer preferences. Therefore, this study provided insights into the characterization of meat metabolites and lipids in Shanghai native pig breeds.

Keywords: metabolomics; lipidomics; liquid chromatography–mass spectrometry; longissimus dorsi muscle; gluteus muscle

Citation: Gao, J.; Sun, L.; Tu, W.; Cao, M.; Zhang, S.; Xu, J.; He, M.; Zhang, D.; Dai, J.; Wu, X.; et al. Characterization of Meat Metabolites and Lipids in Shanghai Local Pig Breeds Revealed by LC–MS-Based Method. *Foods* **2024**, *13*, 2327. https://doi.org/10.3390/foods13152327

Academic Editor: Severino Matias De Alencara

Received: 10 June 2024
Revised: 18 July 2024
Accepted: 22 July 2024
Published: 24 July 2024

1. Introduction

Livestock and poultry products are an important source of protein in the human diet. Among them, pork, as one of the most important sources of meat, plays a pivotal role in residential meat consumption. China is a major pork producer and consumer in the world, and it is reported that China's pork production has reached 52.96 million tons in 2020, accounting for more than 59.6% of meat production [1]. The Duroc × Landrace × Yorkshire crossbreed (DLY) has the biggest market share in China and is characterized by high leanness, fast growth, and high feed conversion efficiency [2,3]. However, the meat of the DLY breed has relatively inferior quality characteristics that include PSE (pale, soft, and exudative) or DFD (dark, firm, dry), which tend to cause consumer dissatisfaction [4,5].

Genetic defects and pre-slaughter stress have been identified as important factors [6,7]. It has been shown that myocytes lose water, and the reflectance of incident light decreases, which can lead to paleness [6]. Alternatively, when exposed to high temperatures, muscle protein denaturation is accelerated, leading to paleness [8,9] and reduced water holding capacity (WHC) [9,10]. Liu et al. applied metabolomics analysis to study the difference in meat quality between the Chuanzang black (CB) pig and DLY breed [11].

Chinese native pig breeds have their own unique characteristics, such as a higher intramuscular fat content (IMF) and better meat quality. It might be formed by long-term artificial or natural selection [3,5]. The content and composition of free amino acids and the fatty acid in the muscle play an important role in determining the nutritional profile and directly affect its taste properties [12–14]. Metabolomics and lipidomics can provide more comprehensive insights of the metabolites and lipids of the meat [3]. LC–MS technology has been widely used in the metabolomic and lipidomic detection of pork because of its high sensitivity and wide detection range [15,16], such as the comparative study on PSE and red, firm, and non-exudative pork [9], and the identification of metabolomic changes in longissimus dorsi muscle of Finishing pigs following heat stress [17]. With the improvement of global living standards, meat quality has become a key factor influencing consumers' purchasing decisions [18]. However, the profiles of meat metabolomics and lipidomics of Shanghai native pig breeds have not been systematically elucidated. Therefore, through muscle untargeted metabolomics and lipidomics among Meishan Pig (MSZ), ShaWutou Pig (SWT), Shanghai White breed (SHB) [19], and commercial DLY breed, this study aimed to dissect differential metabolic biomarkers to characterize Shanghai local pig breeds in terms of meat quality, which will provide novel insights for the development and utilization of Shanghai local pig breeds.

2. Materials and Methods

2.1. Sample Collection

A total of three Shanghai native pig breeds (MSZ, SWT, and SHB) and one commercial pig breed (DLY) were selected for this study. MSZ was from Meishan Pig Breeding Center, Jiading District, Shanghai; SWT was from Shawutou Breeding Farm, Chongming District, Shanghai; and SHB was from Breeding Farm of Zhuanghang Comprehensive Experimental Base of Shanghai Academy of Agricultural Sciences (SAAS). DLY was from Wufeng Shangshi Foods Co. (Shanghai, China). Appearance pictures and feed nutrient levels of the pig breeds were supplemented in Supplementary Material S1. The pork samples were collected from these 4 breeds that had reached the standardized market slaughter weight (100 $\pm$ 5 kg). All slaughtering was performed by licensed commercial slaughtering companies, such as Wufeng Shangshi Food Co., and the researchers only handled the associated meat samples. Twelve heads of each breed, half male and half female, were collected from individual with two tissue parts including the longissimus dorsi muscle (L) and gluteal muscle (T). A total of 96 pork samples were obtained.

2.2. Measurement of Meat Quality Traits

Measurements were taken of 10 individuals per breed. Samples were taken from the thoracolumbar junction to the L muscle of the 3rd and 4th lumbar vertebrae for the measurement of meat quality traits. The meat sample was taken and quickly put into a self-sealing plastic bag, stored in a 4 °C refrigerator, and brought back to the laboratory for testing. Petroleum ether and other reagents required for the experiment were purchased from Sinopharm Chemical Reagent Co. (Shanghai, China). Meat quality traits, such as intramuscular fat (IMF), 24-h drip loss, meat color, and water holding capacity (WHC), were tested. Measurements were made with reference to the reported methods [11,20], and the instruments used were a pH meter (Thermo Star, Thermo Fisher Scientific, Waltham, MA, USA), color difference analyzer (SP60, X-rite, Grand Rapids, MI, USA), automatic fat analyzer (XT10, ANKOM, Macedon, NY, USA), and tenderness meter (C-LM3, Beijing Bullard Technology Development Co., Ltd., Beijing, China).

2.3. Metabolite and Lipid Extraction from Tissues

This study used 6 mm grinding beads to extract the metabolites from 50 mg of skinless muscle frozen tissue. An extraction solution of 400 μL methanol to water in a volume ratio of 4:1 was used. L-2-chlorophenylalanine (0.02 mg/mL) was added as an internal standard. Lipid extraction used 280 μL of methanol solution (2:5 volume ratio with water) and 400 μL of methyl tertiary butyl ether. The extract was ground for 6 min at 50 Hz at −10 °C, followed by low temperature extraction at 5 °C for 30 min (40 kHz), followed by standing at −20 °C for 30 min and centrifugation at 4 °C for 15 min (13,000× *g*).

2.4. LC–MS of Metabolites

The system used for the LC–MS analysis was a Thermo UHPLC-Q Exactive HF-X with an ACQUITY HSS T3 column (Waters, Milford, CT, USA). The mobile phases were 0.1% formic acid water: acetonitrile (volume ratio is 95:5) and 0.1% formic acid acetonitrile: isopropanol: water (volume ratio is 47.5:47.5). The column temperature was set at 40 °C, the flow rate was set at 0.40 mL/min, and the mass spectral signals of the samples were collected in positive and negative ion scanning modes. Quality control (QC) samples were prepared by mixing all samples in equal volumes. The experimental part of LC–MS was carried out by Shanghai Majorbio Company (Shanghai, China) (https://www.majorbio.com/ (accessed on 8 September 2023)) according to the established procedure [21,22].

2.5. LC–MS of Lipids

The same LC–MS analysis system as described above was used and equipped with an Accucore C30 column (Thermo Scientific, Waltham, MA, USA). The mobile phases consisted of 10 mM ammonium acetate in acetonitrile: H_2O (volume ratio is 1:1) (0.1% formic acid) and 2 mM ammonium acetate in acetonitrile: isopropanol: H_2O (volume ratio is 10:88:2) (0.02% formic acid). The standard sample injection parameters were as follows: 2 μL volume, flow rate set at 0.4 mL/min, column temperature set at 40 °C, and 20 min of total chromatographic separation. The QC samples were prepared as described previously.

2.6. Data Analysis

The LC–MS raw data were converted into the common format by Progenesis QI software 2.3 (Waters, Milford, CT, USA) [23] and Lipidsearch 4.2 (Thermo Fisher, San Diego, CA, USA). The metabolites were identified by searching the HMDB database (http://www.hmdb.ca/ (accessed on 12 September 2023)) [24] and Metlin (https://metlin.scripps.edu/ (accessed on 12 September 2023)) [25]. Metabolic and lipidomic features detected in at least 80% in any set of samples were retained [26,27]. Variables with a relative standard deviation (RSD) more than 30% of the QC samples were removed, and log10 logarithmization was performed [28]. The data were analyzed through the free online platform (https://Cloud.majorbio.com/ (accessed on 13 September 2023)) [29,30].

Variance analysis was performed on the matrix file after data preprocessing. Overall differences among the groups were analyzed by PLS-DA (partial least squares discriminant analysis) [31]. The number of times for the randomized permutation test was set to 200. The orthogonal least partial squares discriminant analysis (OPLS-DA) and 7-cycle interactive validation were used to evaluate the stability of the model. In addition, variable importance in the projection (VIP) [32] analysis obtained by the OPLS-DA model and the *p*-value of the Student's *t*-test, and the different expression metabolites (DEMs) with VIP score > 1 and *p*-value < 0.05 were considered as significant.

Venn plot (Venn) [33] is used for plots that show overlap of the number of the elements. One-way ANOVA [34] was used to compare the distribution of metabolites in three or more sample groups for significant differences, and then post hoc tests [35] compared two-by-two to detect sample groups that differed in multiple groups. Student's *t*-test [36] (unpaired) was used in the comparison of differential metabolites between Shanghai local pig breeds and DLY pig breeds, respectively. By default, the program screened for significant differences based on a *p*-value < 0.05.

Differential metabolites were mapped into their biochemical pathways based on a KEGG database search (http://www.genome.jp/kegg/ (accessed on 13 September 2023)) [37].

3. Results

3.1. Meat Quality Traits

This study compared differences in meat quality traits among three Shanghai local pig breeds and the commercial DLY breed. In Table 1, based on the *p*-value of the one-way ANOVA comparison, the results clearly indicate that, except for water content and shear force, which were not significantly different among the four breeds, all other meat quality traits were significantly different among the four breeds ($p < 0.05$). Intramuscular fat (IMF), protein content, lightness (L*), and yellowness (b*) were significantly higher in Shanghai local pigs than in the commercial DLY breed.

Table 1. Results of measurement of meat quality traits.

Meat Quality Traits	DLY	MSZ	SWT	SHB	*p*-Value
IMF (%)	3.22 ± 0.77	5.77 ± 0.51	4.74 ± 1.06	4.42 ± 2.29	**
Drip loss (%)	4.76 ± 1.73	3.36 ± 1.09	5.93 ± 0.26	4.66 ± 1.79	**
pH	5.78 ± 0.30	6.46 ± 0.27	5.61 ± 0.29	5.63 ± 0.26	****
Lightness (L*)	15.48 ± 4.20	44.14 ± 2.88	50.21 ± 6.87	48.44 ± 6.56	****
Redness (a*)	11.00 ± 1.52	3.81 ± 2.32	8.35 ± 2.64	6.98 ± 3.38	*
Yellowness (b*)	6.20 ± 1.71	9.39 ± 1.48	9.37 ± 2.67	10.67 ± 1.04	****
Water content (%)	69.08 ± 1.47	69.79 ± 3.62	65.68 ± 6.28	67.50 ± 2.06	ns
WHC (%)	85.21 ± 4.28	91.86 ± 5.02	91.25 ± 6.73	89.65 ± 4.69	*
Protein content (%)	11.19 ± 1.79	14.42 ± 5.34	18.63 ± 2.94	13.10 ± 2.54	***
Shear force (kgf)	3.35 ± 0.74	2.57 ± 0.31	4.79 ± 2.13	3.65 ± 0.42	ns

Note: IMF represents intramuscular fat, and WHC represents water holding capacity. "ns" means that the difference is not significant (*p*-value > 0.05); "*" means that the *p*-value < 0.05, "**" means that the *p*-value < 0.01, "***" means that the *p*-value < 0.001), and "****" means that the *p*-value < 0.0001.

The Pearson correlation results on these meat traits demonstrated that the protein content (%) showed a significant negative correlation with water content (%) ($r = -0.51$, *p*-value = 1.59×10^{-4}. The meat lightness (L) trait showed a significant positive correlation with the yellowness (b) trait ($r = 0.62$, *p*-value = 1.785×10^{-6}), protein content % ($r = 0.45$, *p*-value = 0.001), and IMF% ($r = 0.44$, *p*-value = 0.001).

3.2. Muscle Untargeted Metabolomic Analysis

A total of 786 (pos) and 343 (neg) metabolites were identified in this study. Significant differences are observed in the metabolites of interest, both when comparing the breeds and comparing the different parts of the pork (Figure 1A); it can be clearly observed that the metabolites of the samples from both MSZ (L and T) and DLY (L and T muscle) are clustered together individually. The permutation test model is to randomly disrupt the grouping labels (Y variables) of the experimental and control groups. The intercept of the Q2 regression line with the *y*-axis is less than 0.05 (Figure 1B), indicating that the model is robust and not overfitted. The results demonstrated a significant difference between the different breeds or parts of meat.

3.3. DEMs Identified between Breeds

The classification of HMDB compounds revealed that DEMs are 283 (28.16%) lipids and lipid-like molecules and 260 (25.87%) organic acids and their derivatives, which are the most numerous two classes of compounds (Figure 2A). Hundreds of DEMs were identified through comparison, respectively. For example, 357 DEMs were identified in the comparison of the L muscles of the DLY and MSZ groups (Figure 2B) for the one-way ANOVA analysis, revealing the significance of the 20 differential metabolites with the highest abundance among groups (Figure 2C).

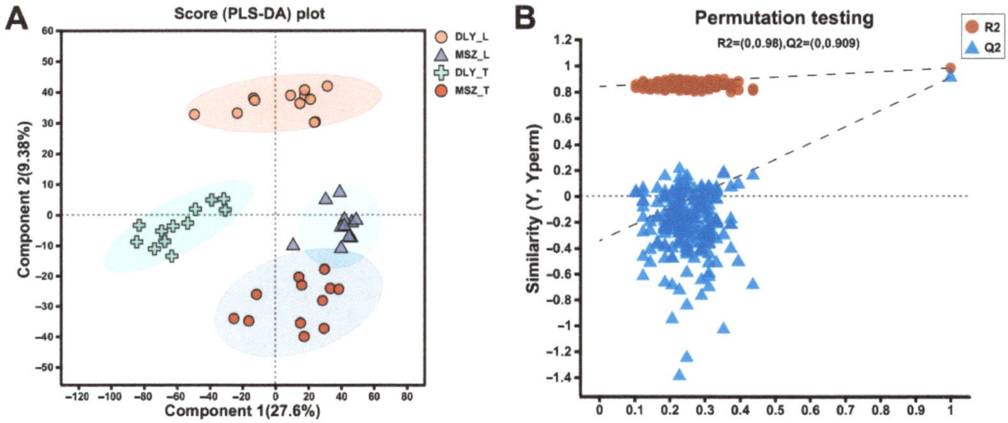

Figure 1. Partial least squares discriminant analysis (PLS-DA). (**A**) Metabolite PLS-DA plot of MSZ breed and DLY breed. The component 1 is the first principal component explanatory degree, and the component 2 is the second principal component explanatory degree. (**B**) Permutation test of PLS-DA model. The horizontal coordinates represent the replacement retention rate of the replacement test, the vertical coordinates represent the R2 (blue triangles) and Q2 (red dots) replacement test values, and the two dashed lines represent the regression lines for R2 and Q2, respectively.

For example, the abundance of L-carnitine (p-value = 7.322×10^{-11}) and inosine (7.985×10^{-17}) differed significantly between groups. The abundance of L-carnitine in L and T meat tissues was significantly lower (p-value < 0.001) in the commercial pig breed (DLY) than in the two Shanghai indigenous black pig breeds (MSZ and SWT) (Figure 2D). Similarly, there was also significant variability for L-acetylcarnitine (p-value = 1.53×10^{-9}). In contrast, the abundance of inosine in the two Shanghai indigenous pig breeds (MSZ and SWT) was significantly lower (p-value < 0.001) than that in the two white pig breeds (DLY and SHB) (Figure 2E). Detailed information on the DEM sets for each group was added in Supplementary Material S2.

Differential metabolite sets identified by comparing three Shanghai local pig breeds with the commercial pig breed DLY were analyzed by a Venn diagram. We identified 50 shared important differential metabolites in L muscle tissue (Figure 3A,B). It was found that the most significant differences remained for carnitines, such as decanoylcarnitine (Figure 3C), L-octanoylcarnitine (Figure 3D), and lauroylcarnitine (Figure 3E). Using the same method, we also analyzed T tissues and identified a total of 178 shared significant differential metabolites, among which 1-(beta-D-ribofuranosyl)-1,4-dihydronicotinamide (Figure 3F) had a significantly higher abundance in the T muscle of the commercial pig DLY than in the three local pig breeds in Shanghai. On the contrary, 10-formyldihydrofolate (Figure 3G) had a higher abundance in the T muscle of the Shanghai local pig breeds. On the other hand, 2-phenylethyl acetate had a lower abundance in the T muscle of the two indigenous black pig breeds (MSZ and SWT) in Shanghai than in the two white pig breeds (DLY and SHB) (Figure 3H).

3.4. DEMs Identified between L and T Meat Parts

In addition to comparing differences between breeds, this study also identified muscle differential metabolites in different parts of the same breed. By analyzing the DEMs of L and T muscles of each of the four breeds, 406, 290, 264, and 229 differential metabolites were identified in the DLY, SHB, MSZ, and SWT groups, respectively (Figure 4A). The Venn diagrams showed that there were 73 DEMs shared among them (Figure 4A), indicating that these 73 DEMs were significantly different metabolic collections between L and T muscle. We also performed a VIP analysis of these 73 DEMs and revealed the top 30 DEMs with the highest VIP scores (Figure 4B). The significant different abundance of the three highest-scoring metabolites, glutamylvaline, phenylacetylglutamine, and glucaric acid, were demonstrated among the groups (Figure 4C).

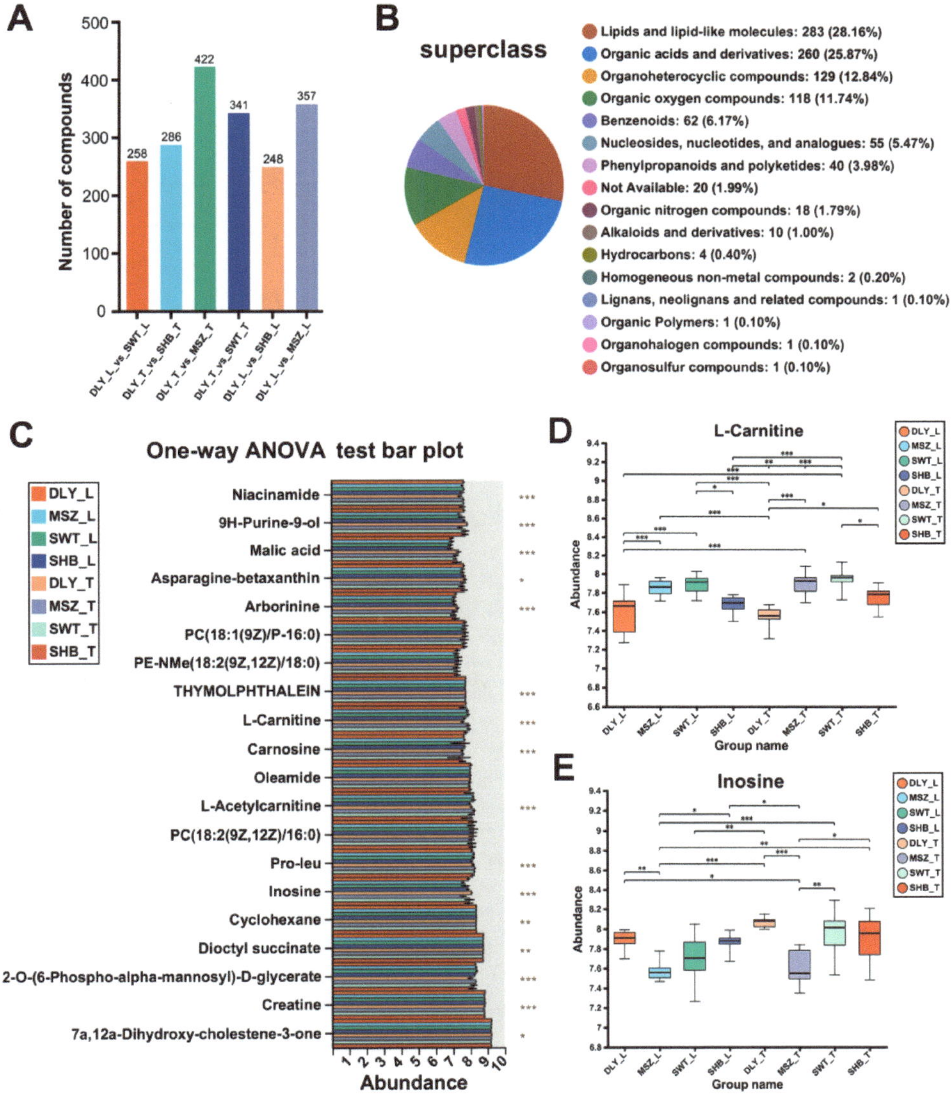

Figure 2. Differential metabolites between Shanghai local pig breeds and DLY breed. (**A**) Statistics of differential metabolites in L and T tissues of DLY versus three other breeds of pigs. (**B**) Compound classification statistics of differential metabolites using the Human Metabolome Database (HMDB). (**C**) One-way ANOVA analysis between groups and then post hoc tests. y-axis indicates the top 20 most abundant metabolite names, and x-axis indicates the average relative abundance of metabolites in different subgroups; p-value is shown on the rightmost side, * p-value < 0.05, ** 0.001 < p < 0.05, *** p-value < 0.001. (**D**) Abundance plot of L-carnitine between groups. (**E**) Abundance plot of inosine between groups.

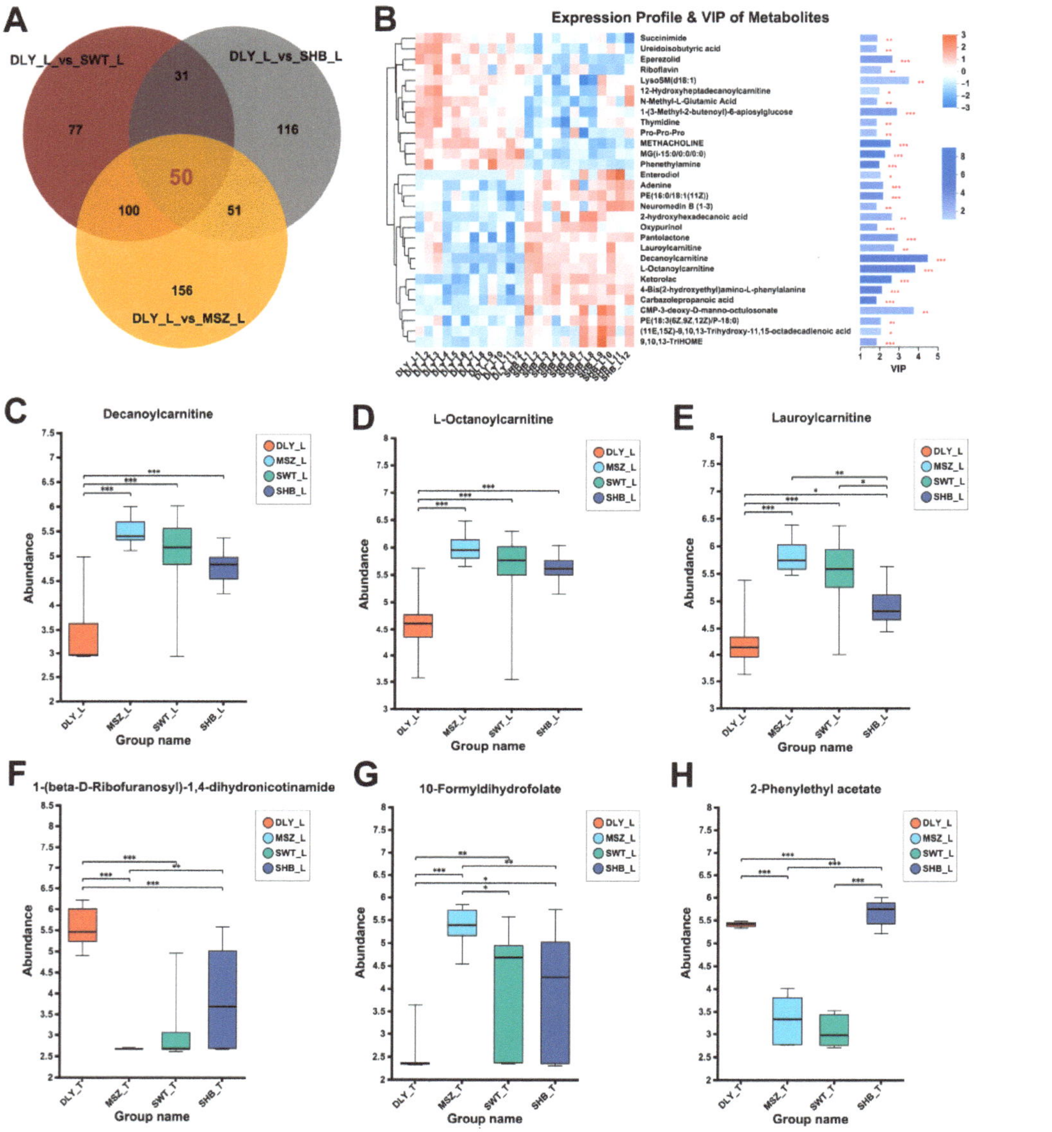

Figure 3. Shared DEMs between Shanghai local pig breeds and DLY breed. (**A**) Venn diagrams of the DEM sets identified between L groups. (**B**) VIP analysis of this collection of top 30 metabolites. On the left side is the metabolite clustering dendrogram, with the sample name at the bottom. On the right side, the bar length indicates the contribution of the metabolite to the difference. The bar color indicates the significance of the difference of the metabolite. The smaller the p-value, the larger $-\log10$ (p-value), the darker the color. (**C–E**) correspond to the abundance of the three significantly different metabolites in the L groups, respectively. (**F–H**) correspond to the abundance of the three significantly different metabolites in the T groups, respectively. * means p-value < 0.05, ** means p-value < 0.01, and *** means p-value < 0.001.

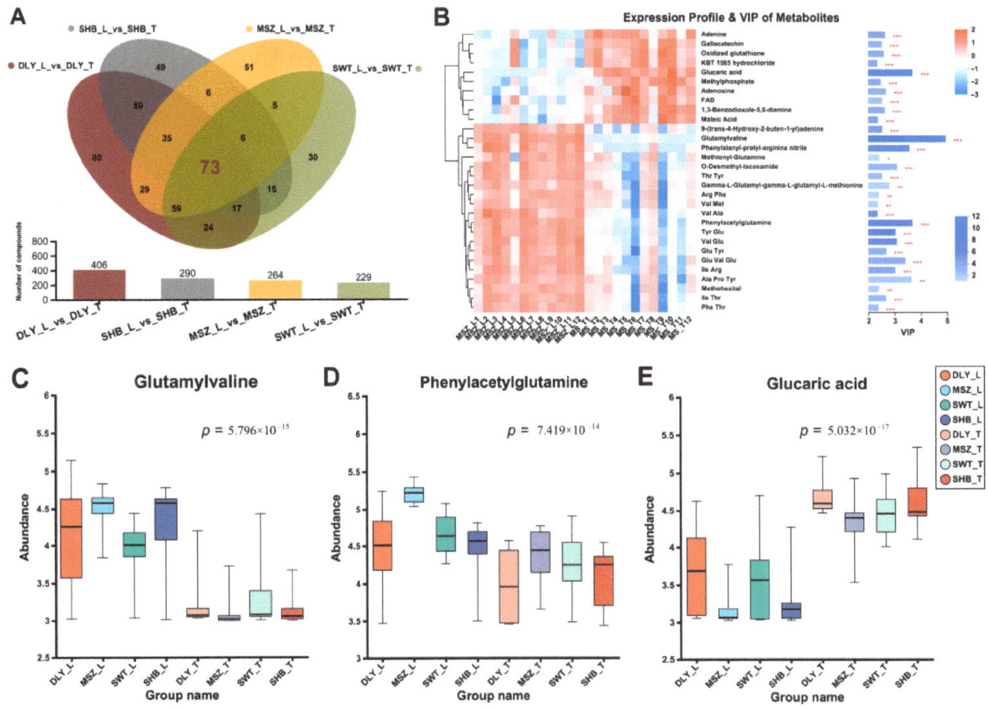

Figure 4. DEMs between longissimus dorsi (L) and gluteal muscle (T). (**A**) Venn diagrams of the metabolic sets identified in 4 breeds between L and T meat tissues. (**B**) Results of the VIP analysis of the 73 shared differential metabolite collections, with a graphical representation of the 30 DEMs with the top VIP values among them. (**C–E**) demonstrate the abundance of three DEMs with the top VIP values in each group. Significant *p*-values obtained from the ANOVA test are shown on the plot. * means *p*-value < 0.05, ** means *p*-value < 0.01, and *** means *p*-value < 0.001.

3.5. Lipidomic Analysis

Lipid metabolites can be categorized into different classes based on different classification methods to facilitate research for specific research directions. The lipid metabolites and pathways strategy [38] [project (LIPID MAPS)] classified lipid metabolites into eight major categories: fatty acids (FA), glycerolipids (GL), glycerophospholipids (GP), sphingolipids (SP), sterol lipids (ST), prenol lipids (PR), saccharolipids (SL), and polyketides (PK). A total of 483 (pos) and 273 (neg) lipid metabolites were identified in this study that consisted of four main categories (GP, GL, SP, and FA) (Figure 5A). An analysis of the KEGG pathway (Figure 5B) revealed that these differential expressed lipids (DELs) were mainly enriched in choline metabolism in cancer, fat digestion and absorption, and GnRH signaling pathways.

The Venn diagrams showed that 13 DELs were shared among the six groups in the comparison between Shanghai local and DLY pig breeds (Figure 5C). Most of them were mainly concentrated on glycerophospholipids (except GM3 belonged to Sphingolipids). Among them, VIP analysis showed that PEt (18:1e/20:4), PA (18:0/18:2), and GM3 (d16:0/20:1) had the highest VIP values in the differences between L muscles of MSZ and DLY (Figure 5D).

It is worth noting that the PA (18:0/18:2), chemical formula: C39H73O8P, is a phosphatidic acid and involved in several of the most prominent metabolism pathways mentioned above, such as fat digestion and absorption, glycerophospholipid metabolism, GnRH signaling pathway, etc. It is clearly shown that its abundance is significantly different (p-value = 6.207×10^{-20}) among the four pig breeds. In particular, in the commercial pig breed DLY, it is significantly lower (all $p < 0.001$) than in the two black pig breeds (MSZ and SWT) (Figure 5E).

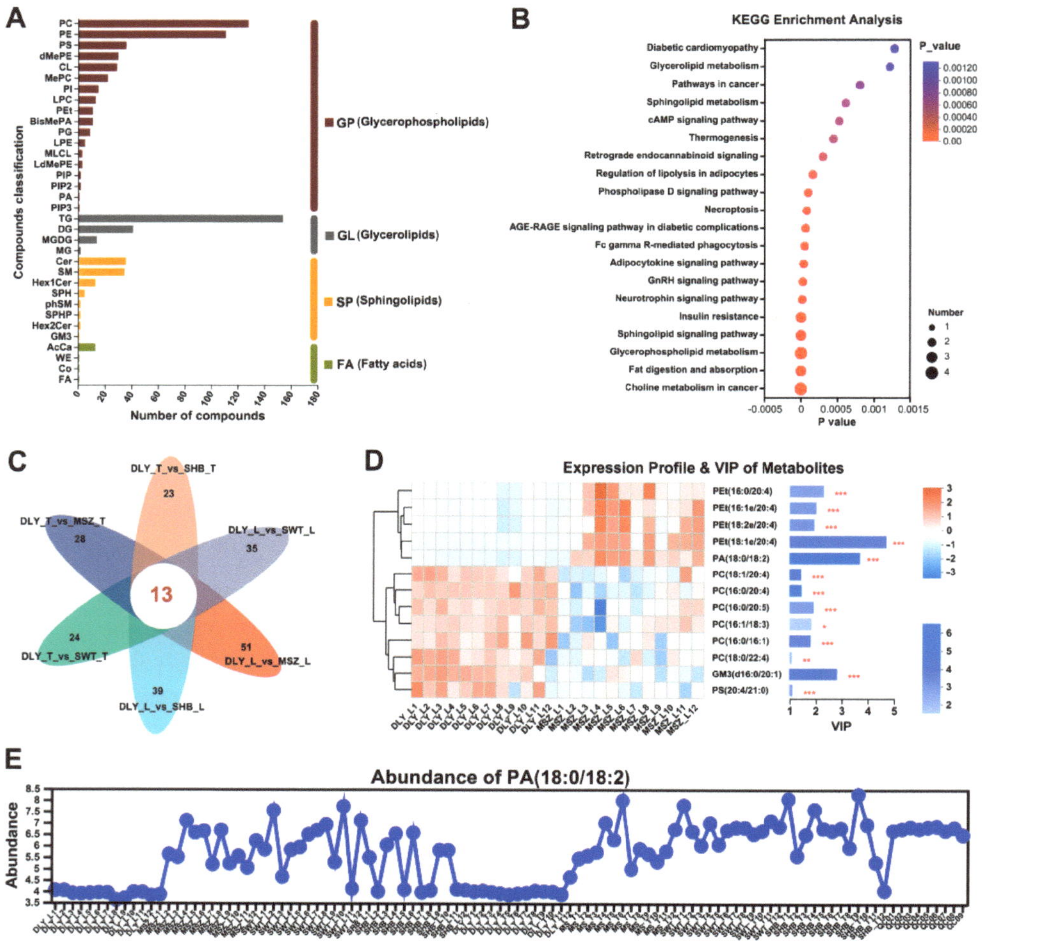

Figure 5. Lipid analysis of Shanghai local pig breeds compared with DLY breed. (**A**) Classification information of lipids in this study obtained based on LIPID MAPS. (**B**) Enrichment analysis of KEGG pathway for differentially metabolized lipids. (**C**) Venn diagrams of the metabolic sets identified in each group. (**D**) VIP analysis of 13 differential expressed lipids (DELs) shared by various groups. (**E**) The PA (18:0/18:2) abundance in all samples. * means p-value < 0.05, ** means p-value < 0.01, and *** means p-value < 0.001.

3.6. DELs Identified between L and T Meat Parts

The present study also identified muscle DELs in different meat tissues of the same breed by analyzing the DELs of L and T muscles in each breed. A total of 162, 128, 173, and 152 DELs were identified in the DLY, SHB, MSZ, and SWT groups, respectively (Figure 6A). On the other hand, the Venn diagrams showed 17 shared DELs between L and T muscles, respectively (Figure 6A). The VIP analysis of these 17 DELs presents those with the highest VIP scores (Figure 6B) and presents the abundance of the three highest scoring lipids: AcCa (22:0), CL (22:2/18:0/18:1/20:4), and BisMePA (18:0/20:5) (Figure 6C). Fifteen of these 17 DELs were also predominantly distributed in glycerophospholipids, while the other two, AcCa (22:0) and Co (Q9), belonged to fatty acids (FA).

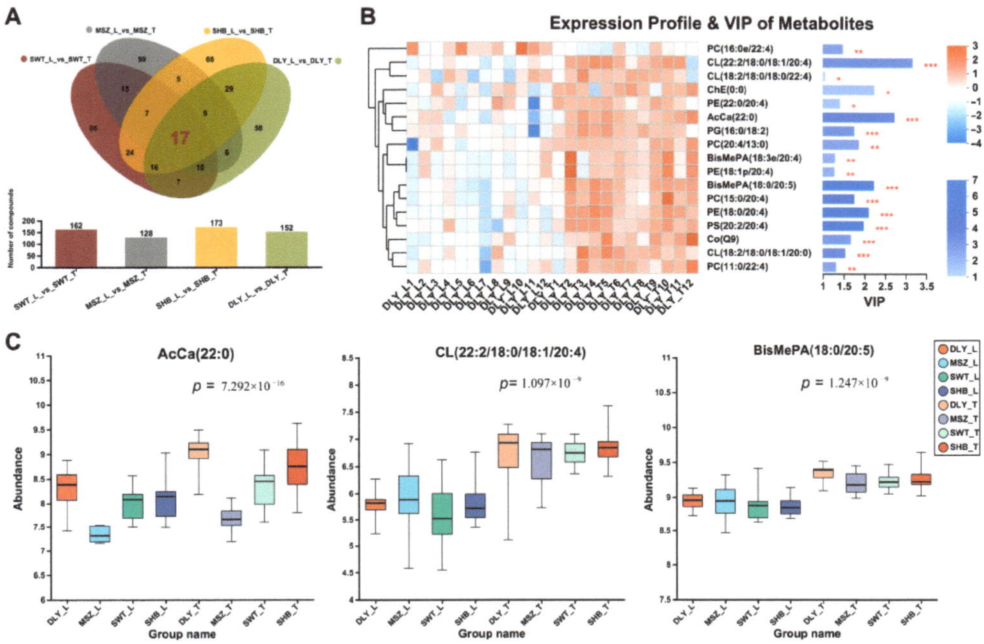

Figure 6. Differential analysis of lipids between L and T meat parts. (A) Venn diagrams of the differential expression lipids sets identified in 4 breeds between L and T type meat. (B) Results of the VIP analysis of the17 shared differential lipids collection. (C) Abundance of the three differential lipids with the highest VIP values in each group. * means p-value < 0.05, ** means p-value < 0.01, and *** means p-value < 0.001.

4. Discussion

Several studies have shown that Chinese indigenous pig breeds have relatively better meat quality (e.g., higher intermuscular fat content and better meat color) and flavor than the globalized hybrid commercial pig DLY, which is known for its growth rate, muscle deposition, and lean meat yield [5,39]. Similar results were also obtained in this study, for example, the IMF, protein content, and meat color of the Shanghai local breeds were significantly higher than those of the DLY breeds, indicating that the Shanghai local breeds also have relatively better meat quality. Among the 20 metabolic markers with the highest abundance in meat, we found that L-carnitine and L-acetylcarnitine were significantly higher in two meat tissues (L and T muscle) of two indigenous black pig breeds (MSZ and SWT) than in the commercial pig breed (DLY).

L-carnitine is an endogenous molecule involved in fatty acid metabolism and can be found in many foods, especially in red meat. L-carnitine transports fatty acid chains into the mitochondrial matrix, which allows for cells to break down fat and derive energy from stored fat reserves [40]. L-carnitine and its esters help reduce oxidative stress, exert a nutrigenomic effect via direct modulation of nuclear receptor signaling in adipocytes, demonstrating its nutritional impact, and have been proposed as treatments for many diseases [40,41]. The L-carnitine levels and storage stabilities of livestock products were reported to have a high correlation (r = 0.9764) with the redness values of a homogenized meat solution [42]. Gao et al. identified nine differential metabolites, including L-carnitine, that were involved in antioxidant function, lipid metabolism, and cell signal transduction, which may decrease postmortem meat quality and play important roles in anti-heat stress [17]. Comparisons of carcass traits, meat quality, and serum metabolomes between Chinese Shaziling pigs and Yorkshire pigs have also been reported, and higher serum L-carnitine levels have been found to be a promising indicator of improved meat quality [43]. From this point of view, the meat quality of Shanghai local pig breeds is also relatively better than that of commercial DLY breeds.

In a previous study, Piglets, especially those of low birth weight, may benefit from carnitine supplementation in the early postnatal period, which may mitigate the negative effects of low birth weight on body composition and market weight meat quality [44]. In addition, it has been shown that reconfiguration of carnitine metabolism using a Chinese obese Ningxiang pig-derived microbiota can promote muscle fatty acid deposition in lean DLY breed [45]. Therefore, the differences in carnitine levels between Shanghai local pig breeds and the DLY breed were confirmed in this study. Other distinctly different metabolites are also of interest, which may also have an important impact on meat flavor, such as the abundance content of inosine that was significantly lower in the two Shanghai indigenous black pig breeds than in the two white pig breeds (DLY and SHB). Inosine 5′-monophosphate (IMP) is an important metabolite that contributes to meat flavor [46]. The breakdown of inosine to form hypoxanthine along with free amino acids contributes to a bitter taste [47,48].

Differences in metabolites between different parts of pork may be an important factor contributing to variations in physicochemical properties, nutrient content, and flavor of different meats. Glutamylvaline is a peptide composed of glutamate and valine and is a proteolytic breakdown product of larger proteins. And there are reports that kokumi-active γ-glutamyl peptides could enhance the umami taste of monosodium glutamate [49]. Phenylacetylglutamine is a product of the conjugation of phenylacetic acid with glutamine. A study of dietary intake patterns demonstrated that *O*-acetylcarnitine and phenylacetylglutamine are positively correlated with the intake of red meat and vegetables, respectively [50]. The analysis of these metabolite differences helped to characterize the differences between L and T muscles at the level of small molecule metabolites.

A study comparing the lipidomics of the longest dorsal muscle of Chinese indigenous Luchuan pigs with that of DLY identified that the expression of 61 TGs and 20 DGs were significantly up-regulated in Luchuan pigs, and only three were down-regulated. The study concluded that this explains the better meat flavor of Luchuan pigs than Duroc pigs [15]. Similar findings were obtained in the present study, where GP (of which PC was the most abundant) and GL (including TG and DG) were the predominant differential lipids in terms of lipid group classification. In the comparison of the longissimus dorsi muscle of MSZ with those of DLY, GL lipids (TG+DG) generally showed an up-regulation in MSZ pigs. Both MSZ and Luchuan pigs belong to the local obese pig breeds in China, which are characterized by higher intermuscular fat content compared to the lean commercial pig DLY.

Through multi-group comparisons, we identified 13 DELs shared between the DLY breed with the Shanghai native breeds and 17 DELs between L and T muscles. Not many studies have been reported on these DELs. For example, PA (18:0/18:2) of the 13 DELs, chemical formula: C39H73O8P, is a phosphatidic acid. This study showed that its abundance level in meat was significantly lower in the DLY breed. And this lipid was involved in several lipid metabolic pathways. The major biosynthetic pathway of PA in animal tissues involves sequential acylation of α-glycerophosphate by acyl-CoA derivatives of fatty acids, making it a biologically active lipid that stimulates a variety of responses such as smooth muscle contraction. Few studies have been reported on this lipid, and in particular, its differences in meat quality among pig breeds have not been revealed. How it affects meat traits or flavor still remains to be elucidated.

5. Conclusions

In this study, we compared the meat quality traits of three Shanghai local pig breeds with the commercial crossbred pig breed DLY. IMF, protein content, and meat color traits reflect the better meat quality in Shanghai local pig breeds. Based on LC–MS analysis results, differences in the abundance of carnitine metabolites, such as L-carnitine, were the most significant muscle metabolism differentiating substances between Shanghai local pig breeds and DLY pig breeds. Characteristic differential lipids were also identified among the breeds, such as PA (18:0/18:2) abundance in the muscle of Shanghai local pigs, both of which were significantly higher than that of the DLY breed. Therefore, these muscle-related biomarkers identified in this study provided novel insights on characterization of pork metabolites in Shanghai local pig breeds.

Supplementary Materials: The following supporting information can be downloaded at: https://www.mdpi.com/article/10.3390/foods13152327/s1, Supplementary Material S1: Appearance pictures and feed nutrition levels of the four pig breeds in this study; Supplementary Material S2: Results of all data analysis generated by this study.

Author Contributions: J.G.: Writing and editing—original manuscript, methodology, data curation, visualization. L.S.: Sample collection and methodology. W.T.: Measurement of meat quality trait data. M.C.: Data statistics. S.Z.: Sample collection. J.X. and M.H.: Investigation. D.Z.: Project administration. C.W.: Funding acquisition. X.W.: Funding acquisition. J.D.: Project administration, Supervision, Funding acquisition, manuscript review and edit. All authors have read and agreed to the published version of the manuscript.

Funding: This study was funded by the National Key Research and Development Plan of China, grant number 2021YFD1200303 and 2021YFD1200305; Shanghai Agriculture Applied Technology Development Program, China (2021-02-08-00-12-F00768) and the Project of Developing Agriculture by Science and Technology in Shanghai (2022, No.1-1).

Institutional Review Board Statement: Not applicable.

Informed Consent Statement: Not applicable.

Data Availability Statement: The original contributions presented in the study are included in the article/Supplementary Materials, further inquiries can be directed to the corresponding authors.

Acknowledgments: We sincerely thank Shanghai Meishan Pig Breeding Center and other Shanghai local pig breeding units for their cooperation in this study.

Conflicts of Interest: The authors declare no conflicts of interest.

References

1. Liang, Y.; Cheng, Y.; Xu, Y.; Hua, G.; Zheng, Z.; Li, H.; Han, L. Consumer preferences for animal welfare in China: Optimization of pork production-marketing chains. *Animals* **2022**, *12*, 3051. [CrossRef] [PubMed]
2. Liu, Y.; Kong, X.; Jiang, G.; Tan, B.e.; Deng, J.; Yang, X.; Li, F.; Xiong, X.; Yin, Y. Effects of dietary protein/energy ratio on growth performance, carcass trait, meat quality, and plasma metabolites in pigs of different genotypes. *J. Anim. Sci. Biotechnol.* **2015**, *6*, 36. [CrossRef] [PubMed]
3. Hou, X.; Zhang, R.; Yang, M.; Niu, N.; Wu, J.; Shu, Z.; Zhang, P.; Shi, L.; Zhao, F.; Wang, L.; et al. Metabolomics and lipidomics profiles related to intramuscular fat content and flavor precursors between Laiwu and Yorkshire pigs. *Food Chem.* **2023**, *404*, 134699. [CrossRef] [PubMed]
4. Faucitano, L.; Ielo, M.C.; Ster, C.; Fiego, D.L.; Methot, S.; Saucier, L. Shelf life of pork from five different quality classes. *Meat Sci.* **2010**, *84*, 466–469. [CrossRef] [PubMed]
5. Huang, Y.; Zhou, L.; Zhang, J.; Liu, X.; Zhang, Y.; Cai, L.; Zhang, W.; Cui, L.; Yang, J.; Ji, J. A large-scale comparison of meat quality and intramuscular fatty acid composition among three Chinese indigenous pig breeds. *Meat Sci.* **2020**, *168*, 108182. [CrossRef] [PubMed]
6. Trevisan, L.; Brum, J.S. Incidence of pale, soft and exudative (PSE) pork meat in reason of extrinsic stress factors. *An. Acad. Bras. Ciências* **2020**, *92*, e20190086. [CrossRef] [PubMed]
7. Adzitey, F.; Nurul, H. Pale soft exudative (PSE) and dark firm dry (DFD) meats: Causes and measures to reduce these incidences—A mini review. *Int. Food Res. J.* **2011**, *18*, 11–20.
8. Choi, Y.; Kim, B. Muscle fiber characteristics, myofibrillar protein isoforms, and meat quality. *Livest. Sci.* **2009**, *122*, 105–118. [CrossRef]
9. Liu, R.; Wu, G.-Y.; Li, K.-Y.; Ge, Q.-F.; Wu, M.-G.; Yu, H.; Wu, S.-L.; Bao, W.-B. Comparative study on pale, soft and exudative (PSE) and red, firm and non-exudative (RFN) pork: Protein changes during aging and the differential protein expression of the myofibrillar fraction at 1 h postmortem. *Foods* **2021**, *10*, 733. [CrossRef] [PubMed]
10. Xu, Z.; Shao, Y.; Liu, G.; Xing, S.; Zhang, L.; Zhu, M.; Xu, Y.; Wang, Z. Proteomics analysis as an approach to understand the formation of pale, soft, and exudative (PSE) pork. *Meat Sci.* **2021**, *177*, 108353.

11. Liu, H.; He, J.; Yuan, Z.; Xie, K.; He, Z.; Zhou, X.; Wang, M.; He, J. Metabolomics Analysis Provides Novel Insights into the Difference in Meat Quality between Different Pig Breeds. *Foods* **2023**, *12*, 3476. [CrossRef] [PubMed]
12. Simopoulos, A.P. N−3 fatty acids and human health: Defining strategies for public policy. *Lipids* **2001**, *36*, S83–S89. [CrossRef] [PubMed]
13. Pereira, P.M.d.C.C.; Vicente, A.F.d.R.B. Meat nutritional composition and nutritive role in the human diet. *Meat Sci.* **2013**, *93*, 586–592. [CrossRef] [PubMed]
14. Yu, M.; Li, Z.; Rong, T.; Wang, G.; Liu, Z.; Chen, W.; Li, J.; Li, J.; Ma, X. Different dietary starch sources alter the carcass traits, meat quality, and the profile of muscle amino acid and fatty acid in finishing pigs. *J. Anim. Sci. Biotechnol.* **2020**, *11*, 78. [CrossRef] [PubMed]
15. Zhang, Z.; Liao, Q.; Sun, Y.; Pan, T.; Liu, S.; Miao, W.; Li, Y.; Zhou, L.; Xu, G. Lipidomic and transcriptomic analysis of the longissimus muscle of Luchuan and Duroc pigs. *Front. Nutr.* **2021**, *8*, 667622. [CrossRef] [PubMed]
16. Zhang, R.; Yang, M.; Hou, X.; Hou, R.; Wang, L.; Shi, L.; Zhao, F.; Liu, X.; Meng, Q.; Wang, L.; et al. Characterization and difference of lipids and metabolites from Jianhe White Xiang and Large White pork by high-performance liquid chromatography–tandem mass spectrometry. *Food Res. Int.* **2022**, *162*, 111946. [CrossRef] [PubMed]
17. Gao, J.; Yang, P.; Cui, Y.; Meng, Q.; Feng, Y.; Hao, Y.; Liu, J.; Piao, X.; Gu, X. Identification of metabonomics changes in longissimus dorsi muscle of finishing pigs following heat stress through lc-ms/ms-based metabonomics method. *Animals* **2020**, *10*, 129. [CrossRef] [PubMed]
18. Bai, M.; Liu, H.; Xu, K.; Yu, R.; Oso, A.O.; Deng, J.; Yin, Y. Effects of coated cysteamine hydrochloride on muscle fiber characteristics and amino acid composition of finishing pigs. *Asian-Australas. J. Anim. Sci.* **2019**, *32*, 1430. [CrossRef] [PubMed]
19. Gao, J.; Sun, L.; Pan, H.; Zhang, S.; Xu, J.; He, M.; Zhang, K.; Zhou, J.; Zhang, D.; Wu, C.; et al. Genomic Dissection through Whole-Genome Resequencing of Five Local Pig Breeds from Shanghai, China. *Animals* **2023**, *13*, 3727. [CrossRef] [PubMed]
20. Duan, H.; Lu, L.; Zhang, L.; Li, J.; Gu, X.; Li, J. Effects of Lactobacillus Lactis supplementation on growth performance, hematological parameters, meat quality and intestinal flora in growing-finishing pigs. *Animals* **2023**, *13*, 1247. [CrossRef] [PubMed]
21. Li, C.; Al-Dalali, S.; Zhou, H.; Xu, B. Influence of curing on the metabolite profile of water-boiled salted duck. *Food Chem.* **2022**, *397*, 133752. [CrossRef] [PubMed]
22. Wang, J.; Wen, X.; Zhang, Y.; Zou, P.; Cheng, L.; Gan, R.; Li, X.; Liu, D.; Geng, F. Quantitative proteomic and metabolomic analysis of Dictyophora indusiata fruiting bodies during post-harvest morphological development. *Food Chem.* **2021**, *339*, 127884. [CrossRef] [PubMed]
23. Zhang, J.; Yang, W.; Li, S.; Yao, S.; Qi, P.; Yang, Z.; Feng, Z.; Hou, J.; Cai, L.; Yang, M. An intelligentized strategy for endogenous small molecules characterization and quality evaluation of earthworm from two geographic origins by ultra-high performance HILIC/QTOF MS E and Progenesis QI. *Anal. Bioanal. Chem.* **2016**, *408*, 3881–3890. [CrossRef] [PubMed]
24. Wishart, D.S.; Guo, A.; Oler, E.; Wang, F.; Anjum, A.; Peters, H.; Dizon, R.; Sayeeda, Z.; Tian, S.; Lee, B.L. HMDB 5.0: The human metabolome database for 2022. *Nucleic Acids Res.* **2022**, *50*, D622–D631. [CrossRef]
25. Guijas, C.; Montenegro-Burke, J.R.; Domingo-Almenara, X.; Palermo, A.; Warth, B.; Hermann, G.; Koellensperger, G.; Huan, T.; Uritboonthai, W.; Aisporna, A.E. METLIN: A technology platform for identifying knowns and unknowns. *Anal. Chem.* **2018**, *90*, 3156–3164. [CrossRef] [PubMed]
26. Smilde, A.K.; Jansen, J.J.; Hoefsloot, H.C.; Lamers, R.-J.A.; Van Der Greef, J.; Timmerman, M.E. ANOVA-simultaneous component analysis (ASCA): A new tool for analyzing designed metabolomics data. *Bioinformatics* **2005**, *21*, 3043–3048. [CrossRef] [PubMed]
27. Hrydziuszko, O.; Viant, M.R. Missing values in mass spectrometry based metabolomics: An undervalued step in the data processing pipeline. *Metabolomics* **2012**, *8*, 161–174. [CrossRef]
28. Yang, J.; Zhao, X.; Lu, X.; Lin, X.; Xu, G. A data preprocessing strategy for metabolomics to reduce the mask effect in data analysis. *Front. Mol. Biosci.* **2015**, *2*, 4. [CrossRef] [PubMed]
29. Ren, Y.; Yu, G.; Shi, C.; Liu, L.; Guo, Q.; Han, C.; Zhang, D.; Zhang, L.; Liu, B.; Gao, H. Majorbio Cloud: A one-stop, comprehensive bioinformatic platform for multiomics analyses. *IMeta* **2022**, *1*, e12. [CrossRef] [PubMed]
30. Karaman, I. Preprocessing and pretreatment of metabolomics data for statistical analysis. In *Metabolomics: From Fundamentals to Clinical Applications*; Springer: Cham, Switzerland, 2017; pp. 145–161.
31. Lee, L.C.; Liong, C.-Y.; Jemain, A.A. Partial least squares-discriminant analysis (PLS-DA) for classification of high-dimensional (HD) data: A review of contemporary practice strategies and knowledge gaps. *Analyst* **2018**, *143*, 3526–3539. [CrossRef] [PubMed]
32. Galindo-Prieto, B.; Eriksson, L.; Trygg, J. Variable influence on projection (VIP) for OPLS models and its applicability in multivariate time series analysis. *Chemom. Intell. Lab. Syst.* **2015**, *146*, 297–304. [CrossRef]
33. Jia, A.; Xu, L.; Wang, Y. Venn diagrams in bioinformatics. *Brief. Bioinform.* **2021**, *22*, bbab108. [CrossRef] [PubMed]
34. St, L.; Wold, S. Analysis of variance (ANOVA). *Chemom. Intell. Lab. Syst.* **1989**, *6*, 259–272.
35. Kucuk, U.; Eyuboglu, M.; Kucuk, H.O.; Degirmencioglu, G. Importance of using proper post hoc test with ANOVA. *Int. J. Cardiol.* **2016**, *209*, 346. [CrossRef] [PubMed]
36. Mishra, P.; Singh, U.; Pandey, C.M.; Mishra, P.; Pandey, G. Application of Student's *t*-test, analysis of variance, and covariance. *Ann. Card. Anaesth.* **2019**, *22*, 407–411. [CrossRef] [PubMed]
37. Kanehisa, M.; Goto, S. KEGG: Kyoto encyclopedia of genes and genomes. *Nucleic Acids Res.* **2000**, *28*, 27–30. [CrossRef] [PubMed]

38. Fahy, E.; Sud, M.; Cotter, D.; Subramaniam, S. LIPID MAPS online tools for lipid research. *Nucleic Acids Res.* **2007**, *35*, W606–W612. [CrossRef] [PubMed]
39. Tu, T.; Wu, W.; Tang, X.; Ge, Q.; Zhan, J. Screening out important substances for distinguishing Chinese indigenous pork and hybrid pork and identifying different pork muscles by analyzing the fatty acid and nucleotide contents. *Food Chem.* **2021**, *350*, 129219. [CrossRef] [PubMed]
40. Pekala, J.; Patkowska-Sokola, B.; Bodkowski, R.; Jamroz, D.; Nowakowski, P.; Lochynski, S.; Librowski, T. L-carnitine-metabolic functions and meaning in humans life. *Curr. Drug Metab.* **2011**, *12*, 667–678. [CrossRef]
41. Förster, L.; Indra, D.; Rosenberger, K.; Zver, L.; Hofbauer, R. L-carnitine exerts a nutrigenomic effect via direct modulation of nuclear receptor signaling in adipocytes, hepatocytes and SKMC, demonstrating its nutritional impact. *Nutr. Res.* **2021**, *85*, 84–98. [CrossRef] [PubMed]
42. Fan, J.P.; Kim, D.Y.; Han, G.D. Correlation between meat color and L-carnitine content in livestock meats. *Food Sci. Biotechnol.* **2009**, *18*, 257–261.
43. Song, B.; Zheng, C.; Zheng, J.; Zhang, S.; Zhong, Y.; Guo, Q.; Li, F.; Long, C.; Xu, K.; Duan, Y.; et al. Comparisons of carcass traits, meat quality, and serum metabolome between Shaziling and Yorkshire pigs. *Anim. Nutr.* **2022**, *8*, 125–134. [CrossRef] [PubMed]
44. Losel, D.; Kalbe, C.; Rehfeldt, C. L-Carnitine supplementation during suckling intensifies the early postnatal skeletal myofiber formation in piglets of low birth weight. *J. Anim. Sci. Biotechnol.* **2009**, *87*, 2216–2226. [CrossRef] [PubMed]
45. Yin, J.; Li, Y.; Tian, Y.; Zhou, F.; Ma, J.; Xia, S.; Yang, T.; Ma, L.; Zeng, Q.; Liu, G.; et al. Obese Ningxiang pig-derived microbiota rewires carnitine metabolism to promote muscle fatty acid deposition in lean DLY pigs. *Innov.* **2023**, *4*, 100486. [CrossRef] [PubMed]
46. Kuchiba-Manabe, M.; Matoba, T.; Hasegawa, K. Sensory changes in umami taste of inosine 5′-monophosphate solution after heating. *J. Food Sci.* **1991**, *56*, 1429–1432. [CrossRef]
47. Subramaniyan, S.A.; Belal, S.A.; Jung, J.-H.; Jung, Y.-C.; Choi, Y.-I.; Shim, K.-S. Meat quality and physicochemical trait assessments of Berkshire and commercial 3-way crossbred pigs. *Korean J. Food Sci. Anim. Resour.* **2016**, *36*, 641. [CrossRef]
48. Tikk, M.; Tikk, K.; Tørngren, M.A.; Meinert, L.; Aaslyng, M.D.; Karlsson, A.H.; Andersen, H.J. Development of inosine monophosphate and its degradation products during aging of pork of different qualities in relation to basic taste and retronasal flavor perception of the meat. *J. Agric. Food Chem.* **2006**, *54*, 7769–7777. [CrossRef] [PubMed]
49. Yang, J.; Huang, Y.; Cui, C.; Dong, H.; Zeng, X.; Bai, W. Umami-enhancing effect of typical kokumi-active γ-glutamyl peptides evaluated via sensory analysis and molecular modeling approaches. *Food Chem.* **2021**, *338*, 128018. [CrossRef] [PubMed]
50. O'Sullivan, A.; Gibney, M.J.; Brennan, L. Dietary intake patterns are reflected in metabolomic profiles: Potential role in dietary assessment studies. *Am. J. Clin. Nutr.* **2011**, *93*, 314–321. [CrossRef] [PubMed]

foods

Article

Transcriptomic and Physiological Analysis Reveals the Possible Mechanism of Inhibiting Strawberry Aroma Changes by Ultrasound after Harvest

Yutong Li, Siyue Liu, Huiyu Kuang, Junyi Zhang, Bei Wang and Shaojia Wang *

Beijing Advanced Innovation Center for Food Nutrition and Human Health (BTBU), School of Food and Health, Beijing Higher Institution Engineering Research Center of Food Additives and Ingredients, Beijing Technology and Business University, Beijing 100048, China; 2130021015@st.btbu.edu.cn (Y.L.); friedb62y@163.com (S.L.); khy1128@126.com (H.K.); zjy2030202087@163.com (J.Z.); wangbei@th.btbu.edu.cn (B.W.)
* Correspondence: wangshaojia@btbu.edu.cn

Abstract: The volatile compounds in strawberries play a significant role in the formation of strawberry aroma. However, these compounds undergo continual changes during storage, resulting in a decline in quality. In this study, a total of 67 volatile organic compounds (VOCs) were identified in strawberries through quantitative analysis. At the end of the storage period, the VOC content in the ultrasonic group was 119.02 μg/kg higher than that in the control group. The results demonstrated that the ultrasonic treatment increased the contents of terpenes and esters at the end of storage. Among these, linalool increased from 67.09 to 91.41 μg/kg, while ethyl cinnamate increased from 92.22 to 106.79 μg/kg. Additionally, the expression of the key metabolic genes closely related to these substances was significantly up-regulated. The expression of the *FaNES* gene, related to terpene metabolism, was up-regulated by 2.8 times in the second day, while the expression of the *FaAAT* gene, related to ester metabolism, was up-regulated by 1.5 times. In summary, this study provides a theoretical basis for exploring the mechanism of ultrasonic effect on strawberry flavor and quality after harvest.

Keywords: strawberry; ultrasound; flavor; transcriptomic; volatile organic components; GC-MS

Citation: Li, Y.; Liu, S.; Kuang, H.; Zhang, J.; Wang, B.; Wang, S. Transcriptomic and Physiological Analysis Reveals the Possible Mechanism of Inhibiting Strawberry Aroma Changes by Ultrasound after Harvest. *Foods* **2024**, *13*, 2231. https://doi.org/10.3390/foods13142231

Academic Editor: Mehdi Rahimmalek

Received: 24 June 2024
Revised: 11 July 2024
Accepted: 14 July 2024
Published: 16 July 2024

1. Introduction

Strawberry (Fragaria × ananassa Duch) is an octaploid ($2n = 56$) perennial root herb of the rose family and is widely known as the "queen of fruits". It is characterized by a bright red appearance and a distinctive fruit shape, with a sweet taste and a tender and juicy texture [1]. Strawberries are rich in amino acids, vitamins, flavonoids and other nutrients, with a high level of consumer acceptance [2]. However, strawberries are prone to damage due to their high water content and absence of a hard peel, making them softened and infested with mold and other microorganisms, leading to quality deterioration after harvest [3–5]. Ultrasonic treatment can extend the shelf life of fruits by inhibiting the growth of harmful microorganisms and regulating the enzyme activity of key metabolic pathways. Feng et al. demonstrated that ultrasonic treatment could enhance the antioxidant capacity of grapes and improve the quality of fruits [6]. Previous studies have also proved that ultrasonic treatment could affect the transcriptome gene expression of strawberries and alleviate fruit softening [7].

The scientific name of strawberry, 'Fragaria', refers to the aromatic aspect, which is one of the important characteristics of fresh strawberries [8]. Typically, aroma is produced by volatile organic compounds (VOCs) and can be used as an indicator of fruit quality deterioration [9,10]. Although the total content of the VOCs only accounts for 0.001% to 0.01% of the fresh fruit mass, they exert a significant impact on the aroma and flavor of strawberries [3]. Previous studies have identified over 350 volatile substances in strawberries, with the majority belonging to the classes of esters, ketones, alcohols, terpenes, aldehydes, and acids [11,12]. The aroma of fruit is not only influenced by the type and content of VOCs, but also by their threshold. According to the aroma activity value OAV (content/threshold), VOCs can be classified into characteristic aroma components or non-characteristic aroma components [13]. Characteristic aroma components are aroma components representing the characteristic aroma of the fruit. As for strawberries, the characteristic aroma components include terpenoids, aldehydes, esters, and furanones. Terpenes are important secondary metabolites in plants and serve as the primary components affecting the floral flavor of many flowering plants. Of them, linalool and nerolidol have a greater effect on the aroma of strawberries and are responsible for the production of floral and citrus fragrances [14]. Aldehydes impart herbal aromas, with hexanal and hexenal being the most common in fruits [15,16].

The content of ester compounds in ripe strawberry fruits accounts for 25% to 90% of the total VOCs, which is an important part of the characteristic aromatic substances of strawberries [17,18]. The ester substances in strawberries mainly include methyl esters and ethyl esters. Among these, methyl butyrate, ethyl butyrate, methyl caproate, and ethyl caproate have been identified as the principal aromatic compounds of strawberry fruit, imparting a sweet, fruity aroma to the fruit [19]. Although the content of furanone in strawberries is relatively low, 4-Hydroxy-2,5-dimethyfuran-3(2H)-one (DMHF) and 4-Methoxy-2,5-dimethyfuran-3(2H)-one (DMMF) are referred as the most significant aromatic components. These components exert a significant impact on the aroma of strawberries, imparting caramel, fruit, and burnt pineapple aromas [20,21].

During storage, strawberries continue to metabolize, followed by the synthesis and metabolism of aroma compounds [15]. Previous studies have shown that both the type and content of the esters in strawberries increase during storage [3], especially ethyl acetate, imparting a distinctive fruity flavor. This is mainly related to the metabolism of fatty acids and amino acids in strawberry species [22–24]. Furthermore, some studies have found that storage can lead to a decrease in the alcohol content, but some specific alcohols, such as 1-propanol and 3-octanol, induce quality deterioration with their increased content during storage [22,25]. The consumption of other substances such as sugar [26], leading to the continued synthesis and increased contents of furanone metabolites [27,28]. In addition to the influence of storage time on aroma, many preservation treatments such as chilling, irradiation, different types of packaging, etc., have also been reported to have a significant influence on fruit aroma [29,30].

Ultrasonic treatment is an efficient method for maintaining the freshness of fruits. Additionally, it has been reported to alleviate the quality deterioration of fruits during storage [31,32]. Ultrasonic technology differs from other preservation technologies and is often used in auxiliary processing [33–35]. Previous studies have shown the effects of ultrasound on processing [36], cleaning [37], and sterilization [38]. Nevertheless, reports on the impact of ultrasound on the aroma of fruits, particularly strawberries, are scarce. Meanwhile, there are few studies on the influence of ultrasound on the transcriptome of fruits during storage. Consequently, this study analyzed the changes in transcriptomics and VOCs to investigate the impact of ultrasonic waves on the aroma metabolism of strawberries during storage. Overall, the experimental findings will provide a theoretical and experimental basis for future studies on the aroma changes of fruits and vegetables during storage and offer insights for the practical application of ultrasonic wave preservation.

2. Materials and Methods

2.1. Experimental Material

The octaploid "Red face" strawberry (Fragaria × ananassa Duch. cv. "Benihoppe") from Da Zi Ran farm (39.54° N, 116.23° E, Beijing, China) was selected as the experimental material. Strawberry plants were grown in a greenhouse containing a mixture of nutrient soil, vermiculite, and organic fertilizer (7:2:1, $v/v/v$). The seedlings were spaced 0.2 m within each row column and grown in a controlled environment with the following conditions: 25 °C:18 °C (day:night), 60% humidity, 12 h photoperiod. The samples were obtained by organic planting methods, and organic fertilizer made of fish protein was used. And insect infestations were prevented by physical techniques such as insect nets. Fresh strawberry fruits of similar ripening quality were transported to the laboratory immediately after picking. Strawberry fruits with similar size, color, maturity, and no obvious surface damage were pre-cooled at 5 °C for 24 h for use. The strawberries were randomly divided into a control group (CK) and an ultrasonic group (U) for treatment. Ultrasonic group (U): the strawberries were completely immersed in an ultrasonic tank filled with deionized water for ultrasonic treatment. The ultrasonic conditions were the same as in the previous study [7]: ultrasonic frequency was 28 kHz, ultrasonic power was 0.15 W/cm^2, and ultrasonic time was 3 min (to maintain a constant water temperature in the ultrasonic tank, the ultrasonic equipment (Beijing Jinxing Ultrasonic Equipment Technology Co., Ltd., Beijing, China) was connected to a low-temperature constant water bath, and the temperature was set at 15 °C). Control group (CK): the strawberries were completely immersed in deionized water for the same time as the ultrasonic group. After being treated, the surface water of the strawberries was removed with absorbent paper and then they were stored in a refrigerator at 5 °C and a humidity of 75–85%.

2.2. Determination of Volatile Compounds

Volatile compounds were collected by headspace solid-phase microextraction [39,40]. The strawberry samples from the ultrasonic group and the control group were randomly selected at 0 h, 3 h, 6 h, 9 h, 12 h, 1 d, 2 d, 3 d, 6 d, 9 d, 12 d, and 15 d during storage. After removing the leaf stalks, the strawberries were cut into parts weighing 5 g each and mixed with 5 mL saturated NaCl (>99%) solution. Then, 3 g of fruit pulp was placed in a 15 mL headspace bottle and the internal standard 2-methyl-3-heptanone (>99%) (0.816 mg/mL, 1 μL) was added. After the sealed headspace bottle was temperated in a water bath at 25 °C for 20 min, the SPME arrow extraction head was inserted into the headspace bottle and the extracted fiber was pushed out. After 30 min of headspace adsorption, the extracted fiber was retracted and the extraction head was pulled out. Each group was treated in triplicate per time point.

The volatile compounds were determined by gas chromatography–mass spectrometry (Agilent Technologies, Santa Clara, CA, USA) following the previously reported method with some modifications [41]. GC condition: DB-WAX column, 60 m × 0.25 mm × 0.25 μm, the carrier gas was helium at a flow rate of 1.2 mL/min. The temperature procedure of the column was as follows: the initial temperature of the column was 40 °C, then temperature increased to 75 °C at 7 °C/min; it was heated up to 150 °C at 2 °C/min; and the temperature was raised to 230 °C at 5 °C/min for 2 min. The diversion mode was adopted. MS conditions: electron ionization source, electron energy 70 eV, the inlet temperature was 250 °C, the ion source temperature was 230 °C, the quadrupole temperature was 150 °C, full scanning mode, and the quality scanning range was m/z 35~350.

2.3. Transcriptome Sequencing

The strawberry samples from the ultrasonic group and the control group were randomly selected at 0 h, 3 h, 6 h, 9 h, 12 h, 1 d, 2 d, 3 d, 6 d, and 9 d during storage. The strawberries were immediately beaten and mixed (5 s) after the leaf stem was removed and divided. About 10 g of strawberry pulp was weighed and packed in a 10 mL centrifuge tube, frozen with liquid nitrogen, and then refrigerated at −80 °C. For transcriptome sequencing, each treatment group was set up with 3 biological replicates per time point. The sequencing of strawberry samples, including total RNA extraction and quality inspection, cDNA library construction, transcriptome sequencing, and preliminary bioinformatics analysis, was performed by the Shanghai Meiji Biomedical Technology Co., Ltd. (Shanghai, China). The gene differential expression analysis was conducted using DESeg2 1.24.0 software [42]. The screening criteria for differential expression genes were $|\log_2 FC| > 1$ (difference multiple > 2) and p-adjust < 0.05.

All genes were functionally annotated in the GO, KEGG, COG, NR, Swiss-Prot, and Pfam databases. The gene function enrichment analysis mainly includes KEGG enrichment analysis to identify the genes involved in the metabolic pathways. The KEGG enrichment analysis was performed by Goatools 0.6.5 software, and the p-value was corrected by default. The KEGG pathway with p-adjust < 0.05 was significantly enriched in gene concentration.

2.4. Real-Time Fluorescence Quantitative PCR

The key genes related to the signal pathway of strawberry active aroma substances were selected and the cDNA samples of the ultrasonic group and the control group were used for real-time fluorescence quantitative PCR (qPCR) to verify the reliability of the transcriptome sequencing results. The octoploid strawberry Actin gene was used as the internal reference gene. The selected gene coding region used Primer Premier 5.0 software design primers through the Primer-BLAST (https://www.ncbi.nlm.nih.gov/tools/primer-blast (accessed on 15 January 2022)) for their early specific validation. The specificity of the primers was confirmed by the dissolution curve. The internal reference Actin primer was F: 5′-GGTGACGAGGCTCAATCCAA-3′, R: 5′-GGGCAACACGAAGCTCATTG-3′, and the primer sequence of the target gene is shown in Table 1.

Table 1. Primers used for real-time fluorescence quantitative PCR.

Gene Name		Sequences
FaAAT	F	GGGAGGACATCATGGATTG
	R	CTAGATTCACCCACGCTTC
FaOMT	F	CACCAGACACTAGCCTTGCC
	R	GGAATCCAGAACCCTTAGCC
FaQR	F	AACAATAGTAGGTCCAGCAA
	R	ACATACAAGGGAAAGGAAA
FaNES	F	TGGGTCGTATGTAAGGTGC
	R	TGAATGATGCTGGAAATGG

After the total RNA was qualified, cDNA with a concentration of 50 ng/uL was synthesized by reverse transcription and freeze-dried at −20 °C for further use. Later, the sample was diluted 10 times for use. The gene-specific primers were synthesized by Suzhou Hongxun Biotechnology Co., Ltd. (Suzhou, China), and real-time fluorescent quantitative PCR experiments were performed using the SuperReal Premix Plus (SYBR Green) kit (Tiangen Biochemical Technology Co., Ltd. Beijing, China). Real-time fluorescence quantitative PCR reaction system and the procedures were as follows: a 10 μL real-time fluorescence quantitative PCR reaction system: 2 × SuperReal Premix Plus (SYBR Green) 5 μL; Template cDNA 5 ng; 5 μM Primer F, 0.3 μL; 5 μM Primer R, 0.3 μL; ddH_2O was added to 10 μL. After adding a 10 μL real-time fluorescence quantitative PCR reaction system, the real-time fluorescence quantitative PCR reaction was performed using the CFX96 real-time fluorescence quantitative PCR instrument (BioRad, Hercules, CA, USA). The 10 μL real-time fluorescent quantitative PCR reaction procedure was as follows: predeformation stage 95 °C, 15 min; cycle reaction stage 95 °C 10 s, 55 °C 20 s, 72 °C 20 s, a total of 40 cycles; dissolution curve stage 65 °C 5 s, 95 °C 15 s.

Based on the expression level of each gene in the CK-0H samples, $2^{-\Delta\Delta CT}$ was used to calculate the relative expression level of the target genes [43]. Each gene in each sample was performed with 3 duplicate holes, and template-free amplification was used as a negative control.

2.5. Data Processing

The data significance and correlation analyses were performed using IBM SPSS Statistics 27. The data were analyzed and mapped using Excel 2021 and Origin 2022 software. Transcriptome-related data were analyzed and mapped using various online data processing and mapping tools provided by the Meggie Biocloud platform (http://www.majorbio.com/ (accessed on 19 January 2024)).

3. Results and Discussion

3.1. Analysis of Volatile Organic Compound Types of Strawberry during Storage

The types of volatile compounds in strawberries during storage mainly include esters, terpenes, alcohols, and aldehydes. As shown in Figure 1 and Table A1, ultrasonic treatment alleviated the reduction of VOC types during the storage period. A total of 67 volatile compounds were detected in strawberries by gas chromatography–mass spectrometry (attached table). Among them, esters accounted for the largest number (13 types), followed by terpenes (10 types), aldehydes (7 types), alcohols (11 types), acids (7 types), ketones (5 types), and other substances (14 types). A total of 53 aroma substances were identified in the ultrasonic group, including 10 terpenes, 10 alcohols, nine esters, four ketones, six aldehydes, five acids, and nine other substances. A total of 64 aroma substances were identified in the control group, including 10 terpenes, 11 alcohols, 12 esters, four ketones, six aldehydes, seven acids, and 14 other substances. This result was consistent with the findings of Ana et al. and Ulrich et al., reporting that strawberry aroma is mainly composed of esters, terpenes, and alcohols [40,44].

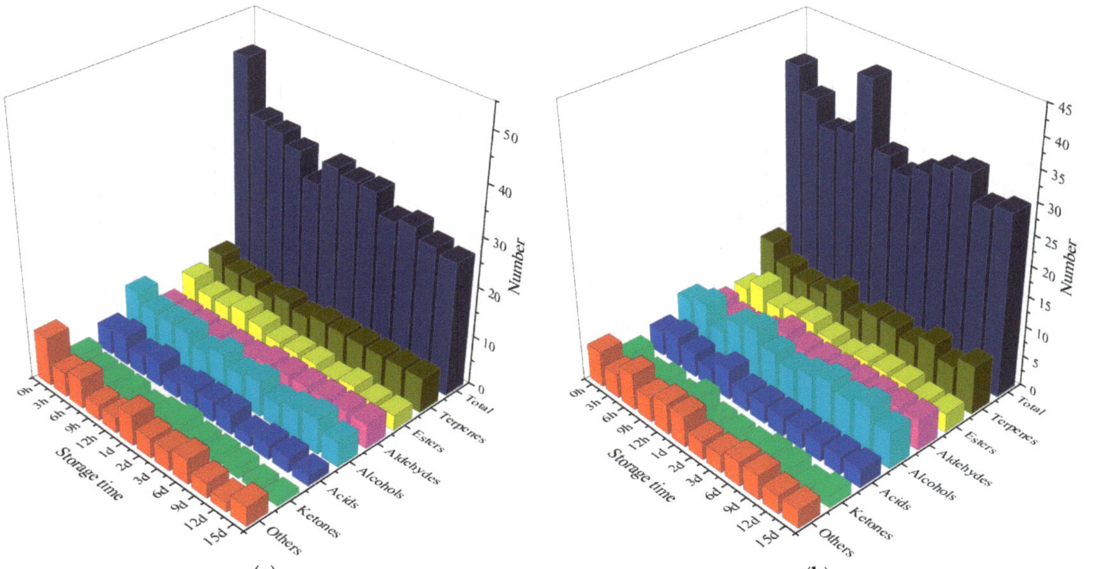

(**a**) (**b**)

Figure 1. Number of volatile organic components in strawberry fruit during storage. (**a**) Control group; (**b**) ultrasonic group.

As shown in Figure 1, a total of 48 and 38 volatile compounds were detected in the control and ultrasonic groups at the beginning of the storage period. This might be because the initial abiotic stress, such as ultrasound, inhibits the metabolism of VOCs [45]. Strawberries have a reduced range of volatiles during storage, similar to some fruits stored in cold stores [46]. Compared with the sample at 0 h, the types of volatile compounds in both groups decreased at 15 d, and the types of volatile compounds in the ultrasonic group were more than three compared to those in the control group at 15 d. These results indicated that ultrasound inhibited the reduction in volatile species during storage. Cai et al. and Caleb et al. found that preservation treatments such as heat treatment and air conditioning could also mitigate this trend [45,47].

Among the volatile compounds, there were more types of terpenoids and alcohols, and the number of alcohols in the control group decreased with the extension of storage time. The number of aldehydes was relatively stable during the storage period but increased after 1 d. Xu et al. also found that the number of aldehydes increased after 1 d storage [29]. During the period, the types of esters and other substances decreased with the extension of storage time. Compared to the decrease in species composition observed in the control group, the ultrasound group showed an increase in the abundance of terpenes and alcohols, which play a key role in species composition.

3.2. Volatile Organic Compounds Analysis of Strawberries during Storage

As shown in Figure 2 and Table A1, ultrasonic treatment increased the VOC contents during storage. The volatile compounds in the control group had little change after storage for 15 d compared that of storage for 0 h, which was consistent with the findings of Liu et al. [30]. The volatile compounds in the ultrasonic group showed a greater change, increasing from 356.79 μg/kg to 639.40 μg/kg. The total content of volatile compounds in the control group was higher after 9 h of storage, and the contents of VOCs in the ultrasonic group were higher than that in the control group at 12 h and the time points after 3 d. The total contents of VOCs were decreased by ultrasound at 0 h of storage, and increased by ultrasound at 12 h and after 3 d of storage. Similar to the results of hot water dipping treatment, it was found that the stress initially inhibits the metabolism and reduces the content of some substances. However, in this study, the metabolism was activated with increased storage time, increasing the content of VOCs in the treatment group [45].

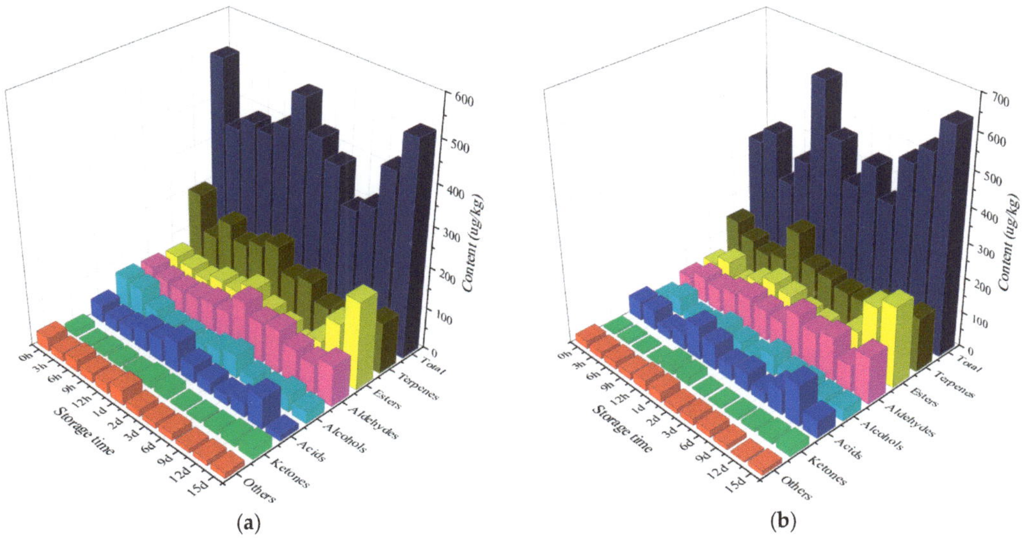

Figure 2. Content of volatile organic components in strawberry fruit during storage. (**a**) Control group; (**b**) ultrasonic group.

For various compounds, terpenoids accounted for a relatively high proportion of the total volatile compounds in the first 12 h of storage, among which the linalool content was the highest, reaching peaks around 12 h, followed by menthol and alpinene contents. Among the alcohols, hexanol and other C6 alcohols had higher contents. A similar content relationship was also found by Zhang et al. [20] and Xu et al. [48]. The contents of esters and aldehydes in both groups increased with the extension of storage time, which was consistent with the findings of Macario et al. [49]. Esters and aldehydes in the two groups increased with the extension of storage time. Joseana et al. and Liu et al. proved that UV-C irradiation and IPL treatment could increase the content of esters and aldehydes during storage [30,50]. Ethyl cinnamate, ethyl caproate, and methyl caproate accounted for most of the esters, and the levels of hexanal, 2-hexenal, and nonanaldehyde were higher among the aldehydes. Previous studies have shown that esters could enhance the fruity flavor of fresh strawberries, and esters and aldehydes contribute to the freshness of strawberries [3]. The higher levels of esters and aldehydes in the ultrasonic group at the end of 15 days of storage might be attributed to the fact that ultrasonic treatment could maintain the freshness of strawberries [21].

Therefore, it was speculated that ultrasonic treatment inhibited the metabolism of VOCs, and the content of VOCs was reduced under ultrasonic stress. However, after 12 h, ultrasonic treatment promoted the metabolism of esters and terpenoids, increased the contents of total volatile compounds, followed by an increase in the contents of linalool, hexanol, ethyl ester, hexaldehyde, terpenoids, alcohols, esters, and aldehydes.

3.3. Analysis of Key Volatile Compounds in Strawberry

Additionally, ultrasonic treatment increased the content of key aroma compounds and improved the fruity aroma of strawberries after 1 d of storage. The aroma activity value OAV (concentration/threshold value) can roughly represent the degree of influence of volatile compounds on the overall aroma of the sample. Compounds with OAV value greater than 1 contribute more to the aroma of the sample. In this study, eight key substances with OAV values greater than 1 were screened (Figure 3 and Table 2), including 1-hexanol, ethyl cinnamate, ethyl caproate, hexanal, nonanaldehyde, linalool, nerolidol, and 4-methoxy-2,5-dimethyl furanone. These compounds were identified as the key contributors to the aroma of the sample, implying that these substances play a pivotal role in the aroma of strawberries [51].

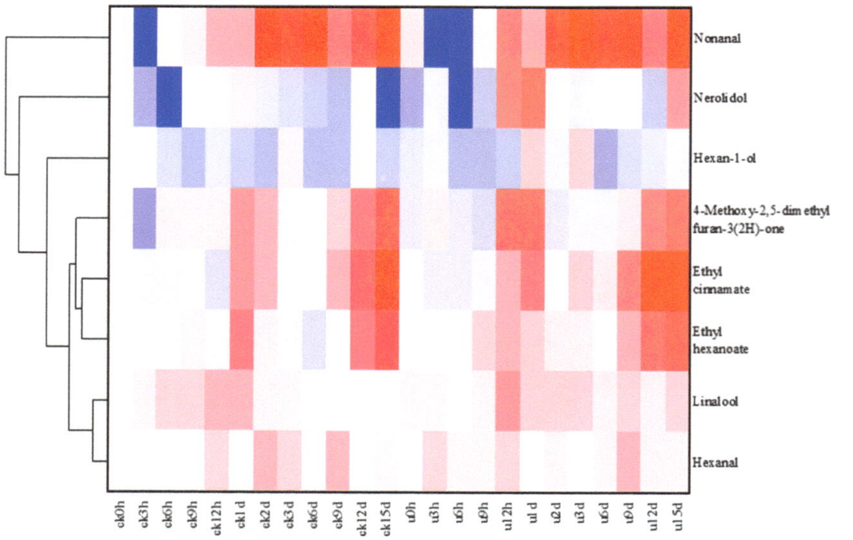

Figure 3. Heatmap of key volatile compounds of strawberry from control and ultrasonic groups. Red color represents the higher expression level of the substances in the sample compared to the sample of ck0h, and blue represents the lower expression level.

Table 2. OAV value of 8 key aroma substance in strawberries.

Name	Threshold	Storage Time											
Control Group		**0 h**	**3 h**	**6 h**	**9 h**	**12 h**	**1 d**	**2 d**	**3 d**	**6 d**	**9 d**	**12 d**	**15 d**
Hexanal	5	4.12	3.91	2.99	3.54	5.69	3.98	6.39	5.41	3.61	6.87	3.34	4.30
Ethyl hexanoate	5	5.13	4.31	4.19	5.80	2.79	13.05	6.11	3.05	1.98	4.88	13.89	18.59
Hexan-1-ol	5.6	1.80	1.65	0.75	0.46	0.84	0.62	0.42	2.22	0.43	0.51	1.29	0.53
Nonanal	8	0.54	0.00	0.38	0.63	0.85	0.95	4.88	3.06	4.75	1.38	2.19	4.41
Linalool	6	11.97	13.26	15.46	17.76	21.39	19.81	12.80	14.75	9.79	11.91	8.15	11.18
4-Methoxy-2,5-dimethyl furan-3(2H)-one	16	0.41	0.08	0.45	0.50	0.51	0.94	0.69	0.24	0.32	0.57	1.15	1.59
Nerolidol	10	0.67	0.15	0.00	0.59	0.39	0.79	0.29	0.25	0.24	0.16	0.67	0.00
Ethyl cinnamate	17	0.66	0.41	0.44	0.38	0.27	1.42	1.13	0.58	0.60	1.02	2.31	5.42
Ultrasonic Group		**0 h**	**3 h**	**6 h**	**9 h**	**12 h**	**1 d**	**2 d**	**3 d**	**6 d**	**9 d**	**12 d**	**15 d**
Hexanal	5	3.40	5.60	4.48	4.69	6.27	3.50	4.64	3.76	4.46	7.21	4.73	4.63
Ethyl hexanoate	5	5.00	4.84	4.75	7.84	8.47	7.40	5.91	5.72	3.70	8.25	16.57	20.74
Hexan-1-ol	5.6	0.80	1.17	0.48	0.50	0.62	2.62	1.04	2.27	0.37	0.75	0.80	1.11
Nonanal	8	0.55	0.00	0.00	0.40	1.46	0.93	2.75	6.67	3.25	4.33	1.30	5.20
Linalool	6	13.55	13.76	10.62	14.56	23.92	16.35	15.90	15.21	14.77	15.22	13.20	15.23
4-Methoxy-2,5-dimethyl furan-3(2H)-one	16	0.18	0.42	0.18	0.17	1.32	1.39	0.18	0.24	0.23	0.49	1.00	1.18
Nerolidol	10	0.15	0.41	0.00	0.20	1.92	2.03	0.47	0.44	0.47	0.60	0.23	1.49
Ethyl cinnamate	17	0.57	0.33	0.35	0.42	1.09	1.60	0.64	0.89	0.81	1.86	6.02	6.28

Among the key aroma substances, 4-methoxy-2,5-dimethylfuranone could affect the specific flavor of strawberries [52]. Strawberries continue to synthesize furanones even after the plant reaches full maturity, and the content of furanones increases after harvest [53]. A similar trend was observed in this study. In addition to their significant floral properties [54], terpenoids such as linalool have a series of physiological functions, especially their involvement in plant stress resistance [55]. It is reported that two key ester compounds, ethyl caproate and ethyl cinnamate, which increase with storage time, can impart a fruity and wine-like flavor [28,56]. Aldehydes are associated with the improvement of plant flavor, while hexanal has a grassy flavor. Du et al. also proved that hexanal is one of the characteristic aroma substances of strawberries [21]. Additionally, it is reported that hexanal has the potential to suppress postharvest diseases [57]. Li et al. showed that octanal and nonanal have the antifungal properties [58].

Compared with the control group, the content of 4-methoxy-2,5-dimethylfuranone in the ultrasonic group was lower than that in the control group at 0 h, and the concentration of 4-methoxy-2,5-dimethylfuranone was significantly increased at 12 h and 1 d after ultrasonic treatment. The levels of linalool and nerolidol were higher in the ultrasonic group at most times. The contents of key terpenoids increased gradually after 6 h, with the largest increase at 12 h, 1 d, 12 d, and 15 d. The ultrasonic group showed higher levels of these compounds at 12 h, 1 d, and 15 d. As shown in the accompanying table, ultrasonic treatment significantly increased the hexanol content at 1 d and 3 d. Similar to other substance treatments, ultrasonic treatment increased the contents of aldehydes, especially hexanal and nonanal in the two groups at 12 h, 1 d, 12 d, and 15 d.

These results showed that the contents of ethyl caproate, ethyl cinnamate, and 4-methoxy-2,5-dimethylfuranone were increased at 12 h, 1 d, 12 d, and 15 d compared with those at 0 h, indicating that the fruity flavor of strawberries was more intense after 12 h of storage. Similar to the results of other preservation treatments such as nano-selenium and high-CO_2 atmosphere [59,60], ultrasonic treatment increased the content of most key flavor compounds at 12 h, 1 d, 12 d, and 15 d.

3.4. Metabolic Pathway Analysis

The metabolic pathways of aroma substances such as linoleic acid, linolenic acid, and terpenes were activated by ultrasonic treatment, which enhanced the fruity aroma of strawberries during storage. Some studies have found that acoustic signals can induce changes in gene expression [61]. Sound stimulation can help plants defend against other abiotic (drought) or biological (insect) stresses [62]. Ultrasonic treatment can induce the antioxidant system, mainly in the early treatment response expression [63], and improve the fatty acid metabolism pathway. The changes in the key metabolic pathways were analyzed using transcriptome sequencing to explore the molecular mechanism of ultrasonic influence on strawberry aroma metabolism further. The transcriptome sequencing results showed that ultrasonic treatment could significantly affect the gene expression of strawberries. A total of 1905 genes were significantly differentially expressed at different times. In different storage time points, there were more differential genes at 0 h, 6 h, and 12 h (Figure 4a). The transcriptome mainly responds to ultrasonic treatment at the early stage of storage, and the response is mainly down-regulated. Further enrichment analysis of the metabolic pathways showed that the DEGs were more significant in the metabolic pathways such as plant–pathogen interaction and hormone response.

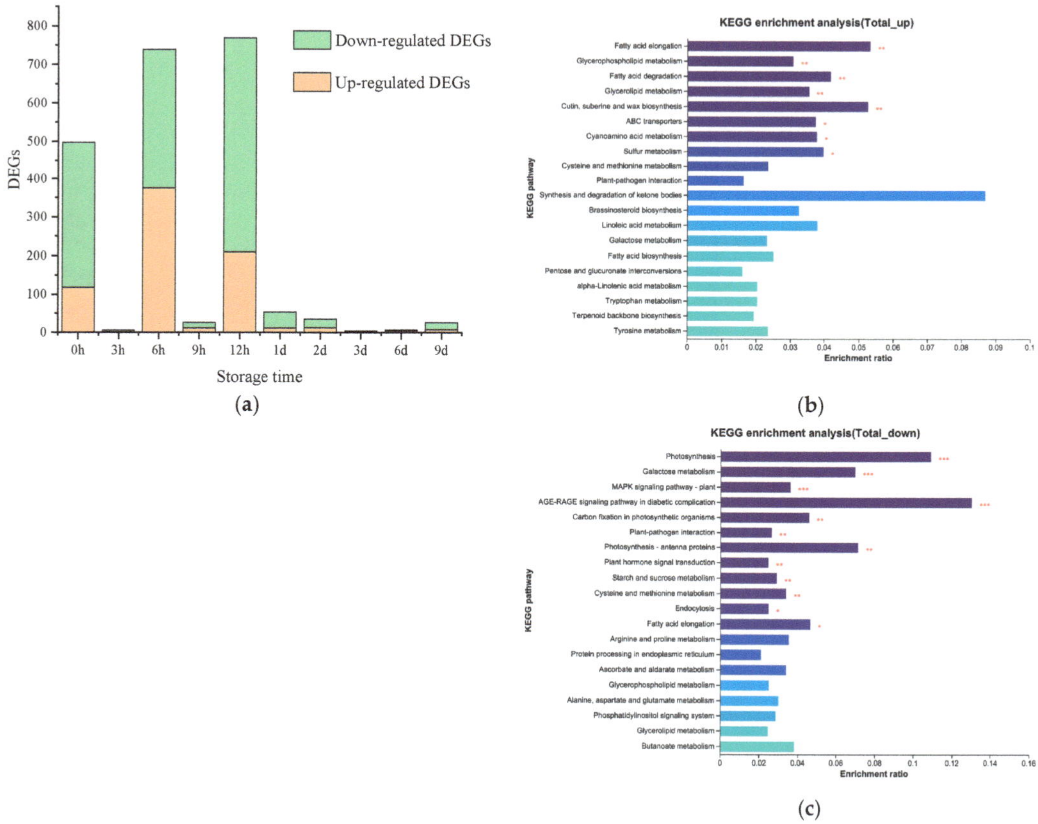

Figure 4. Enrichment analysis of differentially expressed genes of strawberry after treatments. (**a**) Numbers of DEGs in KEGG pathway enrichment analysis of DEGs ($|\log_2 FC| > 1$, p-adjust < 0.05) in strawberry fruit after ultrasound treatment. (**b**) Functional enrichment analysis of up-regulated DEGs. (**c**) Functional enrichment analysis of down-regulated DEGs. The column color gradient indicates the significance of enrichment, where Pvalue or Padjust < 0.001 is labeled ***, Pvalue or Padjust < 0.01 is labeled **, and Pvalue or Padjust < 0.05 is labeled *.

Previous studies have found that ultrasonic treatment could affect plant softening, plant hormone signal transduction, and other metabolic pathways. At the physiological level, ultrasonic treatment can alleviate the softening of strawberry fruits during storage and improve storage quality [7].

Concurrently, differential gene enrichment was observed in the linolenic acid and linoleic acid metabolic pathway, terpenoid skeleton metabolic pathway, and other pathways which are related to ester and terpenoid metabolism of key aroma substances of strawberry [64]. The KEGG enrichment results demonstrated that differential genes were predominantly up-regulated in these pathways, which was analogous to the impact of certain preservation techniques on the transcriptome. Some studies have found that pretreatment before storage can affect the transcriptome expression of fruits; for example, low-temperature storage can enrich DEGs in nectarine in the carbohydrate and terpenoid metabolic pathways [65]. Additionally, cold plasma can regulate the transcriptome of blueberry, and the enrichment pathway is related to the secondary metabolites [63].

3.5. Analysis of Key Gene Transcript Expression

Ultrasound treatment up-regulated the transcription expression levels of genes related to aroma metabolism at some time points. Esters, aldehydes, acids, and alcohols are formed mainly through the metabolism of fatty acids, amino acids and sugars [66,67]. In the fatty acid pathway, fatty acids continuously remove the C2 unit (acetyl-CoA) by β-oxidation to form different fatty acyl-CoA and finally convert to esters (Figure 5). In the amino acid pathway, aliphatic amino acids form branched-chain ketoacids through aminotransferase and then form acyl-CoA through dehydrogenase [68]. Acetyl-coenzyme A is catalyzed to produce acetic acid ester by alkyl transferase AAT or ethanol under the catalysis of reductase and then ethyl acid. Acetyl-CoA is involved in many metabolic pathways and is a common precursor of alcohols, acids, aldehydes, and esters [69]. The activation of the secondary metabolite pathway and an increase in the AAT transcription level to promote the increase in ester aroma were also observed in some genetically mutated mangoes [70].

In the terpenoid metabolic pathway, the nerolidol synthetase (NES) gene *FaNES1* is a key gene encoding the unique "flower fruit" substances of strawberry, while linalool and nerolidol can catalyze the production of linalool and nerolidol by Geranyl-PP or Farnesyl-PP [71]. In a previous study, the levels of NES transcriptional genes were regulated through pollination to regulate the content of terpenes [72]. The carbohydrate pathway mainly produces furanone compounds, representative products like 4-hydroxy-2,5-dimethyl-3(2H)-furanone (HDMF), and 4-methoxy-2,5-dimethyl-3(2H)-furanone (DMMF). FaQR has been shown to be involved in the final step in the synthesis of HDMF from 4-5-methyl-2-methylene-3 (2H) methanone [20]. HDMF is a methylation substrate catalyzed by OMT to produce a more stable DMMF during maturation [73].

The expression of many *FaAAT* transcripts in the ultrasound group was down-regulated compared with that in the control group at 9 h of storage, while some of the transcripts in the ultrasound group were up-regulated after 1 d. Some studies have found that jasmonate can induce the up-regulation of *FaAAT* transcripts in apples [74], and 1-MCP treatment can inhibit the expression level of *FaAAT* in peaches and bananas during storage [75,76], thus affecting the contents of ester substances. Compared with the control group, the expression of *FaNES1* in the ultrasound group was up-regulated at 3 h and 6 h, and no significant difference was observed between the two groups at other time points during storage. The expression of some *FaQR* transcripts in the ultrasound group was more up-regulated than that in the control group within 9 h, then more down-regulated than that in the control group after 1 d. The expression of *FaOMT* transcripts in the ultrasound group was partly up-regulated and partly down-regulated compared with that in the control group at 0 h, and the expression of *FaOMT* transcripts in the ultrasound group was up-regulated compared with that in the control group at 1 d of storage.

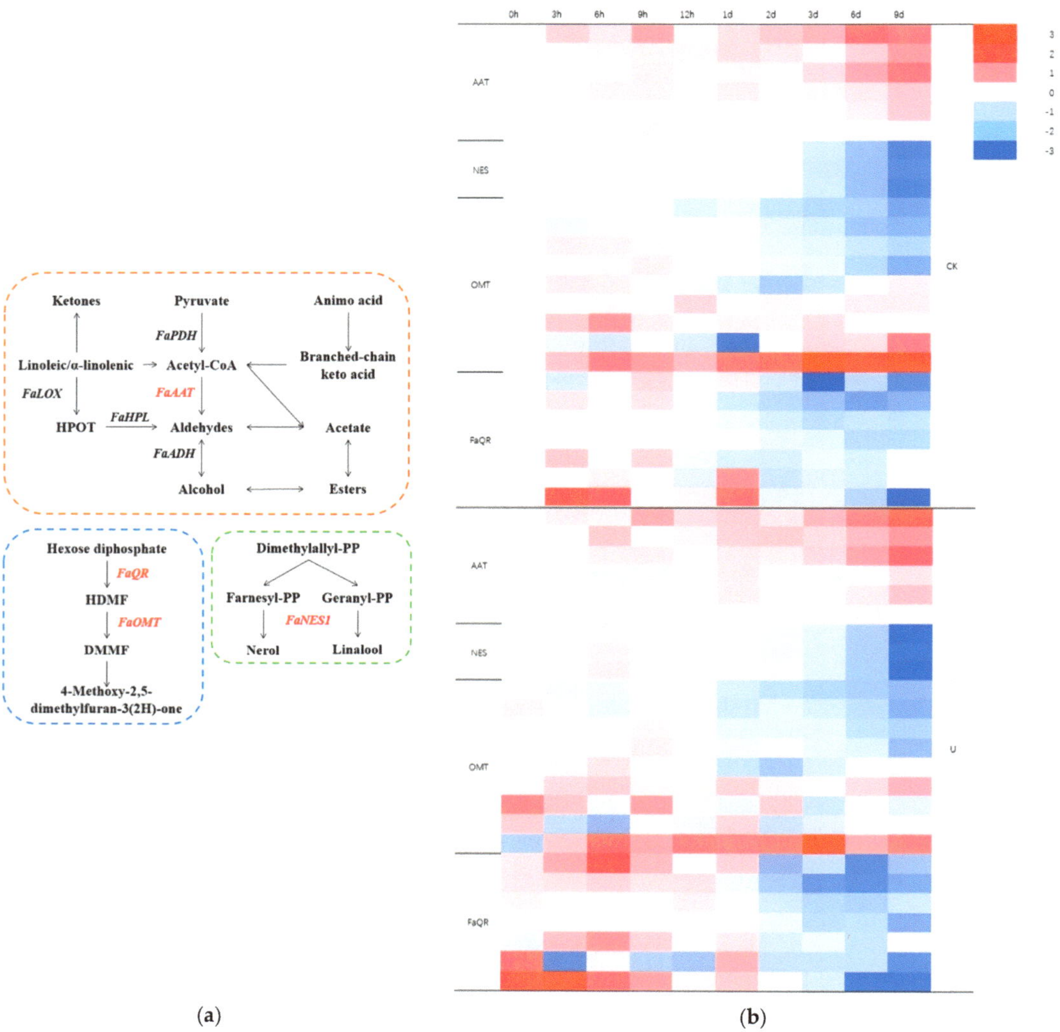

Figure 5. (**a**) Schematic diagram of the synthesis mechanism of strawberry fruit volatile compounds. (**b**) The relative expression of gene transcripts related to the synthesis of volatile compounds in strawberry fruit from control and ultrasonic treatment.

3.6. Analysis of Real-Time Quantitative Fluorescence

The data obtained by real-time fluorescence quantitative PCR was used to plot a calorimetric map of strawberry aroma gene expression, as shown in Figure 6. The gene expression of aroma metabolism was down-regulated at 0 h and up-regulated at 1 d after ultrasonic treatment. In the control group, the change was significant within 6 h after storage, and the expression level of related genes decreased significantly at 3 h, increased significantly at 6 h, and then decreased significantly at 2 d and 9 d of storage. According to Zhang et al., refrigeration could induce gene expression fluctuations [77]. Compared with the control group, gene expression in the ultrasound group was significantly down-regulated at 0 h and 6 h, which was closely related to the lower substance content in the ultrasound group at 9 h before storage. Gene expression in the ultrasonic group was significantly increased at 1 d and 2 d, and the metabolism of esters, terpenes, and furanones was accelerated, leading to an increase in the volatile compound contents in the ultrasound group after 1 d. The expression of *FaNES* was up-regulated within 6 h–2 d, and the biosynthesis of terpenes was accelerated, which was consistent with the trend of higher content of terpenes around 1 d. After 3–15 d of storage, the expression of the *FaAAT* gene was down-regulated compared with that of the control group at 0 h, but the content of esters still increased with the extension of storage time, indicating that ester catabolism was inhibited. Furthermore, other genes, such as *LOX* and *SAAT*, are also involved in the metabolism of ester substances [28]. Additionally, the enzyme activity of AAT is influenced by the transcription factors, including the MYB group [78]. These factors may influence the content of ester substances, which can be investigated in subsequent experiments.

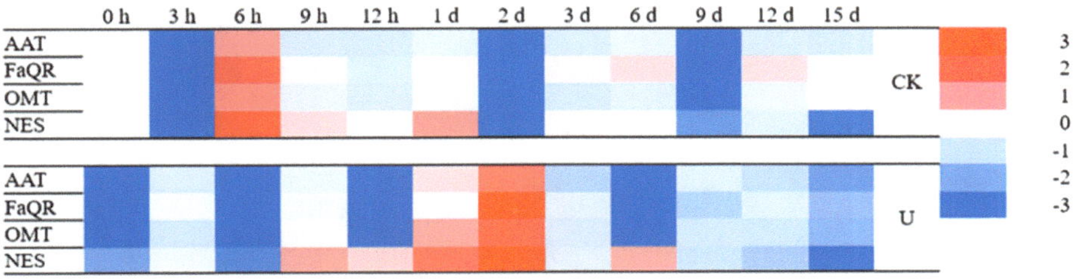

Figure 6. The relative expression of genes related to volatile synthesis of strawberry fruits from the control and ultrasound treatment.

4. Conclusions

In conclusion, this study offers insights into the effects of ultrasound on VOCs and key genes during the storage of strawberries. The results of this experiment showed that the types and contents of VOCs in the ultrasound group decreased at 0 h of storage. However, the contents of total VOCs, especially esters and terpenes, were increased by ultrasonic treatment at 1 d, 12 d, and 15 d of storage. According to the OAV value, eight key aroma compounds were selected, including 1-hexanol, ethyl cinnamate, ethyl caproate, hexanal, nonanaldehyde, linalool, nerolidol, and 4-methoxy-2,5-dimethyl furanone. The content of key aroma compounds such as linalool and ethyl caproate was increased by ultrasonic treatment after 1 d. The results of the transcriptome analysis showed that the metabolic pathways of fatty acids and terpenes were activated by ultrasonic treatment, and the expression levels of *FaAAT*, *FaNES1*, *FaQR*, and *FaOMT* in the ultrasonic group were decreased at 0 h of storage and then increased at 9 h and 1 d of storage, which was consistent with the changes in aroma substances. Therefore, ultrasound could activate the aroma metabolic pathway, improve gene expression, and alleviate the reduction in volatile compound content and variety.

Author Contributions: Y.L.: writing—original draft; S.L. and J.Z.: investigation; H.K.: data curation; B.W. and S.W.: writing—review & editing. All authors have read and agreed to the published version of the manuscript.

Funding: This research received no external funding.

Institutional Review Board Statement: Not applicable.

Informed Consent Statement: Not applicable.

Data Availability Statement: The original contributions presented in the study are included in the article; further inquiries can be directed to the corresponding author.

Conflicts of Interest: The authors declare no conflicts of interest.

Appendix A

Table A1. Volatile compounds in strawberries.

Compounds	RT	CAS	Storage Time	Content (μg/kg) 0 h	3 h	6 h	9 h	12 h	1 d	2 d	3 d	6 d	9 d	12 d	15 d
Esters															
Ethyl butyrate	7.545	105-54-4	ck	ND	ND	ND	ND	ND	ND	ND	ND	ND	ND	ND	ND
			u	ND	20.22 ± 2.85	ND	ND	ND	ND	ND	ND	ND	ND	ND	ND
Methyl hexoate	11.254	106-70-7	ck	17.25 ± 2.42	19.98 ± 3.24	19.21 ± 2.35	34.56 ± 4.21	50.47 ± 5.35	7.80 ± 1.84	19.50 ± 2.56	25.40 ± 4.21	9.88 ± 1.74	24.45 ± 2.45	27.63 ± 3.62	30.22 ± 5.32
			u	10.86 ± 2.01	28.43 ± 4.25	9.83 ± 2.10	27.37 ± 3.42	34.19 ± 2.46	24.25 ± 4.72	28.92 ± 3.62	18.86 ± 3.23	10.06 ± 1.72	26.50 ± 3.32	16.51 ± 2.32	22.73 ± 3.26
Ethyl hexanoate	12.773	123-66-0	ck	25.63 ± 4.23	21.55 ± 3.23	20.96 ± 3.12	29.00 ± 2.64	13.94 ± 1.24	65.27 ± 8.34	30.53 ± 5.63	15.26 ± 2.34	9.91 ± 1.64	24.40 ± 3.52	69.45 ± 7.35	92.97 ± 10.52
			u	25.00 ± 2.62	24.19 ± 2.73	23.74 ± 3.26	39.21 ± 5.25	42.37 ± 6.74	37.02 ± 4.26	29.55 ± 4.72	28.58 ± 3.72	18.52 ± 2.46	41.27 ± 6.34	82.86 ± 11.46	103.70 ± 12.63
Hexyl acetate	14.181	142-92-7	ck	3.98 ± 1.52	ND	ND	ND	ND	ND	ND	ND	ND	ND	ND	ND
			u	ND	ND	ND	ND	ND	ND	ND	ND	ND	ND	ND	ND
Trans-2-hexenyl acetate	16.638	2497-18-9	ck	9.09 ± 1.26	8.01 ± 2.01	ND	ND	ND	ND	ND	ND	ND	ND	ND	ND
			u	6.13 ± 2.13	ND	ND	ND	ND	ND	ND	ND	ND	ND	ND	ND
Cyclohexyl acetate	16.883	622-45-7	ck	ND	2.54 ± 0.23	ND	ND	ND	ND	ND	ND	ND	ND	ND	ND
			u	ND	2.38 ± 0.54	ND	ND	ND	ND	ND	ND	ND	ND	ND	ND
Methyl benzoate	30.819	93-58-3	ck	ND	ND	ND	0.58 ± 0.12	0.30 ± 0.06	ND	ND	ND	ND	ND	ND	ND
			u	ND	ND	0.87 ± 0.13	0.47 ± 0.10	ND	ND	ND	ND	ND	ND	ND	ND
Gamma-butyrolactone	31.299	96-48-0	ck	0.55 ± 0.08	ND	ND	ND	ND	ND	ND	ND	ND	ND	ND	ND
			u	ND	ND	ND	ND	ND	ND	ND	ND	ND	ND	ND	ND
Methyl salicylate	38.742	119-36-8	ck	ND	ND	ND	0.42 ± 0.14	ND	ND	ND	ND	ND	ND	ND	ND
			u	ND	ND	ND	ND	ND	ND	ND	ND	ND	ND	ND	ND
Methyl cinnamate	51.77	1754-62-7	ck	0.55 ± 0.11	ND	0.24 ± 0.04	0.29 ± 0.06	0.50 ± 0.14	ND	0.30 ± 0.06	0.49 ± 0.10	ND	0.64 ± 0.13	ND	ND
			u	ND	0.29 ± 0.05	ND	0.33 ± 0.04	0.70 ± 0.06	0.64 ± 0.15	ND	0.69 ± 0.21	0.47 ± 0.12	0.84 ± 0.25	ND	ND
Ethyl cinnamate	53.123	103-36-6	ck	11.22 ± 2.61	6.90 ± 1.26	7.42 ± 2.73	6.48 ± 2.71	4.61 ± 1.73	24.09 ± 4.71	19.16 ± 3.72	9.91 ± 2.26	10.19 ± 2.71	17.39 ± 1.28	39.33 ± 4.72	92.22 ± 10.27
			u	9.72 ± 1.83	5.63 ± 0.83	5.91 ± 1.61	7.18 ± 2.31	18.45 ± 3.72	27.24 ± 3.17	10.85 ± 2.72	15.21 ± 2.27	13.84 ± 2.21	31.61 ± 5.62	102.39 ± 17.36	106.79 ± 10.62
Gamma-Decanolactone	58.882	706-14-9	ck	ND	ND	ND	ND	ND	1.62 ± 0.15	ND	ND	ND	ND	ND	ND
			u	ND	ND	ND	ND	ND	ND	ND	ND	ND	ND	ND	ND
Diisobutyl phthalate	61.859	84-69-5	ck	ND	ND	3.05 ± 0.26	ND	ND	ND	ND	ND	ND	ND	ND	ND
			u	ND	ND	ND	ND	4.65 ± 0.45	ND	ND	ND	ND	ND	ND	ND
Alcohols															
Isoamylol	13.529	123-51-3	ck	6.41 ± 1.12	ND	7.84 ± 0.92	5.19 ± 0.63	4.29 ± 0.63	3.98 ± 0.42	5.23 ± 0.47	5.26 ± 1.02	ND	ND	3.75 ± 0.51	ND
			u	ND	ND	ND	4.85 ± 0.52	6.16 ± 0.72	ND	3.92 ± 0.42	3.97 ± 0.23	5.22 ± 0.41	4.80 ± 0.85	3.46 ± 0.45	ND
Pentanol	13.562	71-41-0	ck	7.77 ± 1.93	4.61 ± 0.32	9.76 ± 0.89	5.00 ± 0.46	4.43 ± 0.52	4.06 ± 0.35	4.70 ± 0.32	5.80 ± 0.56	ND	5.02 ± 0.67	3.95 ± 0.36	5.96 ± 0.63
			u	3.48 ± 0.43	7.36 ± 1.25	9.38 ± 1.56	ND	5.82 ± 1.31	2.16 ± 0.23	3.27 ± 0.34	3.89 ± 0.25	5.43 ± 0.43	5.00 ± 0.46	3.86 ± 0.25	3.61 ± 0.15
Cis-2-penten-1-ol	16.244	1576-95-0	ck	3.45 ± 0.35	2.72 ± 0.23	ND	3.60 ± 0.51	ND	ND	ND	ND	ND	ND	ND	ND
			u	2.02 ± 0.23	3.55 ± 0.42	ND	3.49 ± 0.14	4.12 ± 0.23	1.50 ± 0.06	ND	ND	ND	ND	2.55 ± 0.24	ND
Hexan-1-ol	17.703	111-27-3	ck	10.07 ± 1.45	9.25 ± 0.82	4.22 ± 0.24	2.56 ± 0.34	4.71 ± 0.45	3.49 ± 0.24	2.35 ± 0.15	12.45 ± 1.26	2.39 ± 0.42	2.85 ± 0.25	7.23 ± 0.92	2.95 ± 0.35
			u	4.48 ± 0.25	6.53 ± 0.45	2.66 ± 0.27	2.81 ± 0.29	3.48 ± 0.49	14.68 ± 1.52	5.83 ± 0.51	12.68 ± 1.53	2.06 ± 0.26	4.20 ± 0.42	4.49 ± 0.64	6.24 ± 0.52
Trans-2-Hexen-1-ol	20.141	928-95-0	ck	11.50 ± 1.64	19.30 ± 2.64	5.01 ± 0.56	ND	7.12 ± 0.62	2.46 ± 0.43	ND	2.13 ± 0.25	1.78 ± 0.15	2.52 ± 0.26	6.37 ± 0.53	1.09 ± 0.15
			u	8.83 ± 1.83	11.65 ± 0.92	ND	5.18 ± 0.63	5.10 ± 0.52	7.76 ± 0.61	1.49 ± 0.11	3.93 ± 0.46	1.56 ± 0.15	ND	ND	1.92 ± 0.29
Cyclohexanol	20.15	108-93-0	ck	15.98 ± 1.51	19.75 ± 2.31	ND	ND	ND	ND	ND	ND	ND	ND	ND	ND
			u	ND	ND	ND	ND	ND	ND	ND	ND	ND	ND	ND	ND
Cis-2-Hexen-1-ol	20.2	928-94-9	ck	13.24 ± 1.42	12.05 ± 2.21	3.19 ± 0.41	5.27 ± 0.45	7.26 ± 0.64	2.65 ± 00.25	ND	ND	ND	ND	ND	ND
			u	ND	12.25 ± 0.98	4.62 ± 0.52	4.79 ± 0.43	5.75 ± 1.06	2.19 ± 0.25	ND	ND	ND	3.01 ± 0.32	2.96 ± 0.31	2.41 ± 0.23
Oct-1-en-3-ol	22.344	3391-86-4	ck	ND	ND	5.15 ± 0.62	ND	ND	2.03 ± 0.43	9.48 ± 1.34	9.07 ± 0.82	7.30 ± 0.91	3.67 ± 0.43	4.21 ± 0.32	ND
			u	ND	4.45 ± 0.32	ND	ND	7.59 ± 0.94	ND	ND	12.13 ± 1.23	6.70 ± 0.42	5.72 ± 0.62	3.37 ± 0.31	ND
2-Ethylhexan-1-ol	24.237	104-76-7	ck	5.54 ± 0.41	3.68 ± 0.23	4.48 ± 0.45	2.36 ± 0.25	ND	4.56 ± 0.15	4.46 ± 0.25	5.01 ± 0.62	5.48 ± 0.56	6.58 ± 0.72	8.70 ± 1.02	11.13 ± 1.93
			u	5.13 ± 0.53	2.59 ± 0.20	3.68 ± 0.53	2.32 ± 0.30	5.53 ± 0.53	3.13 ± 0.24	4.10 ± 0.31	2.80 ± 0.45	6.39 ± 0.53	8.34 ± 0.75	4.97 ± 0.34	6.34 ± 0.25
Octan-1-ol	27.958	111-87-5	ck	2.95 ± 0.36	1.70 ± 0.16	3.57 ± 0.46	3.45 ± 0.26	2.27 ± 0.74	4.46 ± 0.53	17.03 ± 1.93	16.55 ± 0.37	8.73 ± 0.39	2.95 ± 0.32	5.31 ± 0.52	8.15 ± 0.89
			u	2.80 ± 0.34	3.62 ± 0.53	2.15 ± 0.32	2.21 ± 0.22	6.82 ± 0.34	3.60 ± 0.52	6.59 ± 0.63	17.78 ± 1.52	7.12 ± 0.82	8.67 ± 0.93	5.00 ± 0.52	11.96 ± 1.83
2-Phenylethanol	45.972	60-12-8	ck	0.65 ± 0.12	ND	ND	0.39 ± 0.07	ND	ND	ND	ND	ND	ND	ND	ND
			u	0.26 ± 0.12	ND	ND	0.41 ± 0.08	ND	ND	ND	ND	2.13 ± 0.24	ND	ND	ND

Table A1. *Cont.*

Compounds	RT	CAS	Storage Time	0 h	3 h	6 h	9 h	12 h	1 d	2 d	3 d	6 d	9 d	12 d	15 d
				Content (μg/kg)											
Esters															
Aldehydes															
Hexanal	8.452	66-25-1	ck	20.61 ± 3.18	19.56 ± 2.41	14.94 ± 2.12	17.72 ± 2.51	28.45 ± 3.52	19.92 ± 1.51	31.94 ± 3.24	27.04 ± 2.51	18.07 ± 1.72	34.33 ± 4.63	16.69 ± 2.52	21.50 ± 2.16
			u	16.99 ± 1.62	27.98 ± 2.32	22.39 ± 1.62	23.47 ± 2.63	31.34 ± 2.35	17.52 ± 2.15	23.20 ± 3.21	18.78 ± 2.61	22.32 ± 2.61	36.06 ± 3.12	23.65 ± 3.11	23.13 ± 2.62
Trans-2-hexenal	12.361	6728-26-3	ck	38.01 ± 4.61	33.23 ± 3.71	32.02 ± 2.46	30.99 ± 2.36	29.47 ± 3.72	45.00 ± 5.27	37.62 ± 3.61	35.72 ± 4.16	38.99 ± 4.26	27.63 ± 3.71	29.19 ± 3.77	29.19 ± 2.18
			u	19.92 ± 2.71	37.18 ± 4.17	37.18 ± 4.81	43.86 ± 4.16	35.94 ± 3.71	52.36 ± 5.16	28.28 ± 3.93	47.43 ± 4.12	44.59 ± 4.61	39.40 ± 3.36	22.70 ± 2.41	52.40 ± 6.27
Octanal	14.249	124-13-0	ck	ND	ND	ND	ND	ND	2.78 ± 0.31	6.70 ± 0.53	5.12 ± 0.66	4.84 ± 0.34	3.14 ± 0.23	3.30 ± 0.52	4.23 ± 0.33
			u	ND	ND	ND	ND	2.53 ± 0.21	1.79 ± 0.43	2.69 ± 0.22	3.80 ± 0.43	7.15 ± 0.56	3.72 ± 0.25	4.34 ± 0.36	3.52 ± 0.25
Nonanal	19.332	124-19-6	ck	4.36 ± 0.47	ND	3.01 ± 0.31	5.05 ± 0.46	6.81 ± 0.67	7.60 ± 0.83	39.02 ± 3.13	24.47 ± 2.41	38.01 ± 3.15	11.04 ± 1.23	17.55 ± 2.41	35.26 ± 3.15
			u	4.40 ± 0.32	ND	ND	3.23 ± 0.12	11.71 ± 2.31	7.43 ± 0.25	21.98 ± 3.51	53.40 ± 6.32	26.02 ± 3.23	34.65 ± 3.71	10.37 ± 2.01	41.62 ± 4.10
Furfural	22.628	98-01-1	ck	ND	0.94 ± 0.15	ND	ND	ND	ND	ND	ND	ND	ND	ND	ND
			u	ND	ND	ND	ND	ND	ND	ND	ND	ND	ND	ND	ND
Benzaldehyde	25.541	100-52-7	ck	9.60 ± 1.20	7.20 ± 1.23	9.19 ± 0.95	11.40 ± 1.21	9.86 ± 0.41	ND	10.94 ± 1.21	7.06 ± 0.57	7.22 ± 0.48	8.15 ± 1.02	7.59 ± 0.96	8.12 ± 0.83
			u	6.20 ± 0.73	8.62 ± 0.93	7.80 ± 0.94	13.04 ± 1.21	14.03 ± 2.56	9.00 ± 1.02	7.56 ± 0.53	10.59 ± 0.93	12.87 ± 2.10	14.71 ± 1.43	17.05 ± 1.20	13.75 ± 1.56
Undecan-4-olide	58.899	104-67-6	ck	ND	ND	ND	ND	ND	ND	ND	ND	ND	ND	ND	ND
			u	ND	ND	ND	ND	1.73 ± 0.53	ND	ND	ND	ND	ND	ND	ND
Acids															
Acetic acid	22.313	64-19-7	ck	2.99 ± 0.45	3.02 ± 0.32	3.83 ± 0.43	1.67 ± 0.34	2.98 ± 0.32	3.79 ± 0.44	ND	4.82 ± 0.45	ND	3.02 ± 0.22	5.91 ± 0.67	4.99 ± 0.43
			u	3.16 ± 0.32	3.43 ± 0.31	3.82 ± 0.43	3.83 ± 0.33	3.69 ± 0.43	8.37 ± 0.64	3.03 ± 0.52	ND	4.98 ± 0.64	4.13 ± 0.36	6.35 ± 0.67	3.84 ± 0.19
2-Methylbutyric acid	33.755	116-53-0	ck	ND	ND	ND	8.61 ± 1.38	ND	ND	ND	ND	ND	ND	ND	ND
			u	ND	2.51 ± 0.32	ND	ND	3.07 ± 0.71	ND	ND	ND	ND	ND	ND	ND
Pentanoic acid	37.311	109-52-4	ck	ND	0.25 ± 0.05	ND	ND	ND	ND	ND	ND	ND	ND	ND	ND
			u	ND	ND	ND	ND	ND	ND	ND	ND	ND	ND	ND	ND
Hexanoic acid	42.853	142-62-1	ck	24.59 ± 3.58	13.51 ± 1.42	24.32 ± 2.53	37.84 ± 4.23	49.69 ± 5.00	61.90 ± 5.19	34.64 ± 3.10	22.04 ± 2.84	20.42 ± 2.12	28.78 ± 3.82	64.36 ± 8.32	8.95 ± 1.49
			u	47.00 ± 4.19	43.27 ± 3.92	18.60 ± 2.94	40.18 ± 4.01	83.01 ± 9.91	56.30 ± 4.92	35.83 ± 3.01	46.75 ± 4.92	20.98 ± 2.27	60.94 ± 4.91	92.70 ± 10.93	43.69 ± 4.12
Octanoic acid	51.332	124-07-2	ck	3.34 ± 0.52	4.94 ± 0.23	2.36 ± 0.33	4.34 ± 0.41	2.05 ± 0.22	8.02 ± 0.84	3.38 ± 0.29	1.52 ± 0.19	1.35 ± 0.20	1.38 ± 0.10	4.81 ± 0.93	ND
			u	2.17 ± 0.23	3.01 ± 0.25	1.81 ± 0.21	3.06 ± 0.41	4.86 ± 0.51	9.57 ± 1.39	2.74 ± 0.42	4.96 ± 0.50	1.95 ± 0.29	5.49 ± 0.48	16.70 ± 1.94	2.20 ± 0.39
Nonanoic acid	54.164	112-05-0	ck	5.97 ± 0.39	3.10 ± 0.42	ND	ND	ND	ND	2.50 ± 0.28	1.37 ± 0.15	ND	ND	ND	ND
			u	ND	ND	ND	ND	ND	ND	ND	ND	ND	ND	ND	ND
Benzoic acid	60.014	65-85-0	ck	6.64 ± 1.03	4.46 ± 0.84	2.61 ± 0.42	ND	ND	2.82 ± 0.32	1.61 ± 0.15	ND	ND	ND	ND	ND
			u	2.29 ± 0.32	3.63 ± 0.35	1.81 ± 0.15	ND	2.35 ± 0.34	ND	ND	ND	ND	ND	ND	ND
Ketones															
6-Methylhept-5-en-2-one	17.057	110-93-0	ck	ND	ND	1.62 ± 0.30	ND	ND	3.21 ± 0.32	3.08 ± 0.43	ND	ND	ND	ND	ND
			u	ND	ND	ND	ND	1.27 ± 0.20	ND	ND	ND	ND	ND	ND	ND
4-Methoxy-2,5-dimethylfuran-3(2H)-one	29.556	4077-47-8	ck	6.51 ± 1.02	1.23 ± 0.29	7.15 ± 0.85	7.93 ± 0.95	8.13 ± 1.92	15.06 ± 1.72	11.10 ± 2.02	3.88 ± 0.83	5.13 ± 0.31	9.07 ± 1.09	18.34 ± 1.84	25.39 ± 3.92
			u	2.96 ± 0.42	6.77 ± 0.53	2.95 ± 0.32	2.68 ± 0.23	21.14 ± 0.21	22.30 ± 0.24	2.90 ± 0.37	3.78 ± 0.74	3.70 ± 0.46	7.80 ± 0.91	15.94 ± 1.38	18.96 ± 2.73
Acetophenone	32.305	98-86-2	ck	1.82 ± 0.31	1.58 ± 0.15	ND	1.35 ± 0.11	ND	ND	2.63 ± 0.27	1.37 ± 0.25	1.68 ± 0.15	1.21 ± 0.16	ND	ND
			u	1.63 ± 0.20	ND	1.33 ± 0.15	ND	3.70 ± 0.34	ND	ND	ND	2.75 ± 0.25	1.78 ± 0.14	ND	2.46 ± 0.24
1,5-Ditert-butyl-3,3-dimethylbicyclo[3.1.0]hexan-2-one	52.29	19377-95-8	ck	ND	ND	ND	ND	ND	ND	ND	ND	ND	ND	ND	ND
			u	1.03 ± 0.23	ND	ND	ND	ND	ND	ND	ND	ND	ND	ND	ND
Noroxoagarofuran	61.462	5986-25-4	ck	7.01 ± 0.91	ND	ND	ND	ND	ND	ND	ND	ND	ND	ND	ND
			u	ND	ND	ND	ND	ND	ND	ND	ND	ND	ND	ND	ND
Terpenes															
Alpha-pinene	10.507	80-56-8	ck	ND	ND	ND	ND	ND	ND	ND	ND	ND	ND	ND	ND
			u	3.08 ± 0.21	ND	ND	ND	ND	ND	ND	ND	ND	ND	ND	ND
Limonene	11.635	138-86-3	ck	3.77 ± 0.42	2.74 ± 0.23	2.23 ± 0.24	2.69 ± 0.41	2.99 ± 0.32	ND	2.70 ± 0.32	2.82 ± 0.44	2.20 ± 0.23	1.52 ± 0.18	1.78 ± 0.19	1.85 ± 0.21
			u	2.32 ± 0.21	1.98 ± 0.21	2.26 ± 0.22	2.17 ± 0.32	3.63 ± 0.34	ND	2.80 ± 0.26	2.38 ± 0.27	2.41 ± 0.21	1.97 ± 0.19	ND	2.85 ± 0.32
Cineole	11.84	470-82-6	ck	0.94 ± 0.19	ND	ND	ND	ND	ND	ND	ND	ND	ND	ND	ND
			u	0.67 ± 0.09	ND	ND	ND	ND	ND	ND	ND	ND	ND	ND	ND
(Z)-linalool oxide	23.287	5989-33-3	ck	4.78 ± 0.45	1.81 ± 0.19	2.04 ± 0.29	3.09 ± 0.30	3.25 ± 0.31	2.80 ± 0.25	2.28 ± 0.24	2.32 ± 0.22	1.59 ± 0.19	1.54 ± 0.23	1.24 ± 0.21	1.41 ± 0.16
			u	2.84 ± 0.23	3.11 ± 0.42	3.07 ± 0.63	2.25 ± 0.32	6.28 ± 1.21	3.02 ± 0.53	2.40 ± 0.25	1.78 ± 0.43	1.52 ± 0.22	2.02 ± 0.31	2.21 ± 0.34	2.30 ± 0.26
Linalool	27.136	78-70-6	ck	71.84 ± 8.39	79.58 ± 9.10	92.75 ± 8.38	106.58 ± 15.28	128.36 ± 10.38	118.83 ± 11.39	76.83 ± 9.38	88.51 ± 7.29	58.72 ± 6.29	71.46 ± 5.29	48.91 ± 5.10	67.09 ± 7.29
			u	81.28 ± 8.39	82.59 ± 8.03	63.70 ± 7.03	87.37 ± 9.38	143.52 ± 10.39	98.11 ± 9.48	95.42 ± 10.37	91.23 ± 9.57	88.64 ± 8.93	91.31 ± 10.39	79.19 ± 8.94	91.41 ± 9.38

Table A1. *Cont.*

Compounds	RT	CAS	Storage Time	Content (μg/kg) 0 h	3 h	6 h	9 h	12 h	1 d	2 d	3 d	6 d	9 d	12 d	15 d
Esters															
Terpinine-4-ol	30.014	562-74-3	ck	24.65 ± 2.93	12.68 ± 1.29	20.57 ± 3.02	6.16 ± 1.20	5.98 ± 0.84	19.19 ± 2.30	12.12 ± 1.20	14.24 ± 1.62	9.18 ± 0.84	9.48 ± 0.72	7.35 ± 0.82	16.40 ± 1.47
			u	20.84 ± 2.48	9.23 ± 1.29	10.78 ± 1.47	3.80 ± 0.48	16.04 ± 0.94	6.37 ± 0.73	9.87 ± 0.92	5.58 ± 0.56	12.79 ± 0.74	14.47 ± 1.37	5.67 ± 0.46	18.68 ± 2.48
Menthol	32.103	89-78-1	ck	66.39 ± 5.83	7.32 ± 0.78	36.04 ± 5.38	6.78 ± 1.36	5.06 ± 0.67	20.06 ± 2.04	14.03 ± 1.37	13.26 ± 0.95	10.61 ± 0.93	7.16 ± 0.45	5.02 ± 0.35	29.99 ± 3.47
			u	21.10 ± 2.02	7.77 ± 0.76	8.77 ± 1.02	8.56 ± 0.84	12.95 ± 1.27	ND	11.12 ± 1.40	5.00 ± 0.67	ND	12.22 ± 1.38	7.24 ± 0.85	15.50 ± 0.92
Terpineol	35.142	98-55-5	ck	10.56 ± 1.28	3.86 ± 0.45	7.14 ± 0.74	ND	2.27 ± 0.32	ND	3.24 ± 0.42	3.59 ± 0.32	3.08 ± 0.31	2.56 ± 0.30	2.95 ± 0.29	5.95 ± 0.65
			u	5.91 ± 0.76	2.76 ± 0.34	3.33 ± 0.32	ND	7.44 ± 0.67	3.83 ± 0.43	ND	2.58 ± 0.35	3.61 ± 0.36	4.30 ± 0.54	5.05 ± 0.58	4.64 ± 0.32
Piperitone	36.546	89-81-6	ck	11.06 ± 1.48	ND	4.20 ± 0.74	ND	ND	4.45 ± 0.43	ND	ND	ND	ND	ND	2.81 ± 0.53
			u	6.14 ± 0.74	ND	ND	ND	4.28 ± 0.32	ND	ND	ND	ND	3.30 ± 0.43	ND	5.34 ± 0.56
Nerolidol	50.741	7212-44-4	ck	6.69 ± 0.78	1.48 ± 0.32	ND	5.89 ± 1.02	3.95 ± 0.43	7.89 ± 0.84	2.94 ± 0.32	2.51 ± 0.32	2.35 ± 0.21	1.62 ± 0.18	6.69 ± 0.78	ND
			u	1.47 ± 0.17	4.12 ± 0.43	ND	2.04 ± 0.23	19.19 ± 2.01	20.30 ± 2.93	4.73 ± 0.43	4.44 ± 0.32	4.72 ± 0.56	6.02 ± 0.64	2.29 ± 0.32	14.93 ± 1.42
Others															
Toluene	7.575	108-88-3	ck	13.58 ± 1.24	15.54 ± 1.64	17.59 ± 2.83	19.46 ± 2.83	22.79 ± 2.735	23.04 ± 2.65	11.32 ± 1.27	9.55 ± 0.95	8.23 ± 0.75	7.49 ± 0.74	10.86 ± 0.91	8.26 ± 1.38
			u	15.85 ± 1.52	15.74 ± 1.38	15.84 ± 1.73	16.45 ± 1.82	20.57 ± 2.73	18.13 ± 2.10	11.73 ± 1.02	16.46 ± 1.28	10.77 ± 0.94	8.86 ± 0.83	ND	8.67 ± 0.93
Ethylbenzene	9.949	100-41-4	ck	2.90 ± 0.34	ND	ND	ND	ND	ND	ND	ND	ND	ND	ND	ND
			u	ND	ND	ND	ND	ND	ND	ND	ND	ND	ND	ND	ND
Styrene	13.093	100-42-5	ck	ND	ND	1.53 ± 0.20	ND	ND	ND	ND	ND	ND	ND	ND	ND
			u	ND	ND	ND	ND	ND	ND	ND	ND	ND	ND	1.16 ± 0.23	ND
M-cymene	13.318	535-77-3	ck	ND	ND	ND	ND	ND	ND	ND	ND	ND	1.44 ± 0.21	ND	ND
			u	1.55 ± 0.18	ND	ND	ND	ND	ND	ND	ND	ND	ND	ND	ND
P-cymene	14.011	99-87-6	ck	3.44 ± 0.45	2.57 ± 0.32	2.53 ± 0.23	1.92 ± 0.21	1.37 ± 0.19	2.52 ± 0.28	2.39 ± 0.29	1.52 ± 0.20	1.88 ± 0.31	ND	ND	1.70 ± 0.19
			u	ND	2.00 ± 0.29	1.63 ± 0.21	1.03 ± 0.19	1.99 ± 0.17	1.65 ± 0.15	1.92 ± 0.20	1.63 ± 0.18	1.97 ± 0.32	1.67 ± 0.27	2.57 ± 0.31	2.07 ± 0.29
1,3-Dimethyl-5-ethylbenzene	14.052	934-74-7	ck	ND	ND	2.08 ± 0.28	ND	ND	ND	ND	ND	ND	ND	ND	ND
			u	ND	ND	ND	ND	ND	ND	ND	ND	ND	ND	ND	ND
2,5-Dimethylpyrazine	16.858	123-32-0	ck	1.02 ± 0.09	ND	ND	ND	ND	ND	ND	ND	ND	ND	ND	ND
			u	ND	ND	ND	ND	ND	ND	ND	ND	ND	ND	ND	ND
2-(2-Ethoxyethoxy)ethanol	31.527	111-90-0	ck	2.42 ± 0.28	ND	ND	ND	ND	ND	ND	ND	ND	ND	ND	ND
			u	1.21 ± 0.17	ND	1.43 ± 0.18	0.92 ± 0.08	ND	ND	ND	ND	ND	ND	ND	ND
1-Methylnaphthalene	42.353	90-12-0	ck	0.34 ± 0.04	ND	ND	ND	ND	ND	ND	ND	ND	ND	ND	ND
			u	ND	ND	ND	ND	ND	ND	ND	ND	ND	ND	ND	ND
Butylated hydroxytoluene	45.532	128-37-0	ck	1.64 ± 0.26	ND	ND	ND	ND	4.72 ± 0.57	ND	ND	2.54 ± 0.21	ND	ND	ND
			u	ND	ND	ND	ND	3.28 ± 0.32	4.57 ± 0.65	ND	ND	2.79 ± 0.21	ND	ND	ND
Benzothiazole	47.414	95-16-9	ck	1.96 ± 0.19	1.10 ± 0.14	1.31 ± 0.15	0.85 ± 0.11	ND	1.89 ± 0.18	1.03 ± 0.21	1.37 ± 0.13	2.14 ± 0.25	1.12 ± 0.21	1.09 ± 0.10	1.74 ± 0.15
			u	1.33 ± 0.21	0.92 ± 0.09	0.92 ± 0.10	ND	2.05 ± 0.21	1.43 ± 0.24	0.89 ± 0.10	0.91 ± 0.08	2.10 ± 0.28	1.62 ± 0.16	1.39 ± 0.18	1.73 ± 0.13
M-Cresol	49.356	108-39-4	ck	ND	ND	ND	ND	ND	1.68 ± 0.19	ND	ND	ND	ND	ND	ND
			u	ND	ND	ND	ND	ND	ND	ND	ND	ND	ND	ND	ND
Phenol	49.396	108-95-2	ck	1.10 ± 0.19	0.63 ± 0.05	ND	ND	ND	ND	ND	ND	ND	ND	ND	ND
			u	0.57 ± 0.04	ND	0.75 ± 0.10	ND	ND	ND	ND	ND	ND	ND	ND	ND
2,4-Di-t-butylphenol	57.16	96-76-4	ck	4.55 ± 0.56	0.92 ± 0.19	3.88 ± 0.41	ND	0.34 ± 0.05	1.32 ± 0.21	1.52 ± 0.14	0.97 ± 0.19	1.43 ± 0.17	ND	ND	0.87 ± 0.11
			u	1.80 ± 0.21	0.64 ± 0.17	0.53 ± 0.08	0.42 ± 0.05	0.93 ± 0.09	1.27 ± 0.14	ND	ND	1.41 ± 0.17	0.51 ± 0.04	ND	ND

Ck: Control group, U: Ultrasonic group. ND means not detected.

References

1. Liu, L.; Ji, M.L.; Chen, M.; Sun, M.Y.; Fu, X.L.; Li, L.; Gao, D.S.; Zhu, C.Y. The flavor and nutritional characteristic of four strawberry varieties cultured in soilless system. *Food Sci. Nutr.* **2016**, *4*, 858–868. [CrossRef] [PubMed]
2. Zuñiga, P.E.; Castañeda, Y.; Arrey-Salas, O.; Fuentes, L.; Aburto, F.; Figueroa, C.R. Methyl Jasmonate Applications From Flowering to Ripe Fruit Stages of Strawberry (*Fragaria* × *ananassa* 'Camarosa') Reinforce the Fruit Antioxidant Response at Post-harvest. *Front. Plant Sci.* **2020**, *11*, 538. [CrossRef] [PubMed]
3. Yan, J.W.; Ban, Z.J.; Lu, H.Y.; Li, D.; Poverenov, E.; Luo, Z.S.; Li, L. The aroma volatile repertoire in strawberry fruit: A review. *J. Sci. Food Agric.* **2018**, *98*, 4395–4402. [CrossRef] [PubMed]
4. Feliziani, E.; Romanazzi, G. Postharvest decay of strawberry fruit: Etiology, epidemiology, and disease management. *J. Berry Res.* **2016**, *6*, 47–63. [CrossRef]
5. Shen, F.; Zhang, B.; Cao, C.J.; Jiang, X.S. On-line discrimination of storage shelf-life and prediction of post-harvest quality for strawberry fruit by visible and near infrared spectroscopy. *J. Food Process Eng.* **2018**, *41*, e12866. [CrossRef]
6. Feng, Y.B.; Suo, K.; Zhang, Y.; Yang, Z.F.; Zhou, C.S.; Shi, L.Y.; Chen, W.; Wang, J.C.; Wang, C.Y.; Zheng, Y.X. Ultrasound synergistic slightly acidic electrolyzed water treatment of grapes: Impacts on microbial loads, wettability, and postharvest storage quality. *Ultrason. Sonochem.* **2024**, *103*, 106751. [CrossRef]
7. Zhang, J.Y.; Jiang, H.; Li, Y.T.; Wang, S.J.; Wang, B.; Xiao, J.S.; Cao, Y.P. Transcriptomic and physiological analysis reveals the possible mechanism of ultrasound inhibiting strawberry (*Fragaria* × *ananassa* Duch.) postharvest softening. *Front. Nutr.* **2022**, *9*, 1066043. [CrossRef] [PubMed]
8. Lewers, K.S.; Newell, M.J.; Park, E.; Luo, Y.G. Consumer preference and physiochemical analyses of fresh strawberries from ten cultivars. *Small Fruits Rev.* **2020**, *20*, S733–S756. [CrossRef]
9. Vallone, S.; Sivertsen, H.; Anthon, G.E.; Barrett, D.M.; Mitcham, E.J.; Ebeler, S.E.; Zakharov, F. An integrated approach for flavour quality evaluation in muskmelon (*Cucumis melo* L. reticulatus group) during ripening. *Food Chem.* **2013**, *139*, 171–183. [CrossRef]
10. Kulkarni, R.S.; Chidley, H.G.; Pujari, K.H.; Giri, A.P.; Gupta, V.S. Geographic variation in the flavour volatiles of Alphonso mango. *Food Chem.* **2012**, *130*, 58–66. [CrossRef]
11. Vandendriessche, T.; Geerts, P.; Membrebe, B.N.; Keulemans, J.; NicolaI, B.M.; Hertog, M. Journeys through aroma space: A novel approach towards the selection of aroma-enriched strawberry cultivars in breeding programmes. *Plant Breed.* **2013**, *132*, 217–223. [CrossRef]
12. Oh, Y.; Barbey, C.R.; Chandra, S.; Bai, J.H.; Fan, Z.; Plotto, A.; Pillet, J.; Folta, K.M.; Whitaker, V.M.; Lee, S. Genomic Characterization of the Fruity Aroma Gene, FaFAD1, Reveals a Gene Dosage Effect on γ-Decalactone Production in Strawberry (*Fragaria* × *ananassa*). *Front Plant Sci.* **2021**, *12*, 639345. [CrossRef] [PubMed]
13. Wang, G.X.; Zhang, Y.T.; Dong, J.; Zhong, C.F.; Chang, L.L.; Wang, L.N. Study of volatile compound contents in a progeny issued from cross between 'Camarosa' and 'Benihoppe'. *Acta Hortic.* **2012**, *926*, 65–71. [CrossRef]
14. Aubert, C.; Bruaut, M.; Chalot, G.; Cottet, V. Impact of maturity stage at harvest on the main physicochemical characteristics, the levels of vitamin C, polyphenols and volatiles and the sensory quality of Gariguette strawberry. *Eur. Food Res. Technol.* **2021**, *247*, 37–49. [CrossRef]
15. Abouelenein, D.; Acquaticci, L.; Alessandroni, L.; Borsetta, G.; Caprioli, G.; Mannozzi, C.; Marconi, R.; Piatti, D.; Santanatoglia, A.; Sagratini, G.; et al. Volatile Profile of Strawberry Fruits and Influence of Different Drying Methods on Their Aroma and Flavor: A Review. *Molecules* **2023**, *28*, 5810. [CrossRef] [PubMed]
16. Rey Serra, P.; Mnejja, M.; Monfort, A. Inheritance of esters and other volatile compounds responsible for the fruity aroma in strawberry. *Front Plant Sci.* **2022**, *13*, 959155. [CrossRef] [PubMed]
17. Jetti, R.R.; Yang, E.; Kurnianta, A.; Finn, C.; Qian, M.C. Quantification of selected aroma-active compounds in strawberries by headspace solid-phase microextraction gas chromatography and correlation with sensory descriptive analysis. *J. Food Sci.* **2007**, *72*, S487–S496. [CrossRef] [PubMed]
18. Prat, L.; Espinoza, M.I.; Agosin, E.; Silva, H. Identification of volatile compounds associated with the aroma of white strawberries (*Fragaria chiloensis*). *J. Sci. Food Agric.* **2014**, *94*, 752–759. [CrossRef] [PubMed]
19. Lixia, S.; Yinan, N.; Jianwen, W.; Yue, C.; Hongsheng, G. Characteristic-Aroma-Component-Based Evaluation and Classification of Strawberry Varieties by Aroma Type. *Molecules* **2021**, *26*, 6219. [CrossRef]
20. Zhang, Y.Y.; Yin, X.R.; Xiao, Y.W.; Zhang, Z.Y.; Li, S.J.; Liu, X.F.; Zhang, B.; Yang, X.F.; Donald, G.; Jiang, G.H.; et al. An ethylene response factor-MYB transcription complex regulates furaneol biosynthesis by activating quinone oxidoreductase expression in strawberry. *Plant Physiol.* **2018**, *178*, 189–201. [CrossRef]
21. Du, X.F.; Anne, P.; Elizabeth, B.; Russell, R. Evaluation of volatiles from two subtropical strawberry cultivars using GC-olfactometry, GC-MS odor activity values, and sensory analysis. *J. Agric. Food Chem.* **2011**, *59*, 12569–12577. [CrossRef]
22. Gong, D.; Bi, Y.; Zong, Y.Y.; Li, Y.C.; Sionov, E.; Prusky, D. Characterization and sources of volatile organic compounds produced by postharvest pathogenic fungi colonized fruit. *Postharvest Biol. Technol.* **2022**, *188*, 111903. [CrossRef]
23. Hou, X.L.; Jiang, J.M.; Luo, C.Q.; Latifur, R.; Li, X.Y.; Xie, X. Advances in detecting fruit aroma compounds by combining chromatography and spectrometry. *J. Sci. Food Agric.* **2023**, *103*, 4755–4766. [CrossRef] [PubMed]
24. Pott, D.M.; Osorio, S.; Vallarino, J.G. From Central to Specialized Metabolism: An Overview of Some Secondary Compounds Derived From the Primary Metabolism for Their Role in Conferring Nutritional and Organoleptic Characteristics to Fruit. *Front Plant Sci.* **2019**, *10*, 835. [CrossRef]

25. Vandendriessche, T.; Keulemans, J.; Geeraerd, A.; Nicolai, B.M.; Hertog, M.L.A.T.M. Evaluation of fast volatile analysis for detection of Botrytis cinerea infections in strawberry. *Food Microbiol.* **2012**, *32*, 406–414. [CrossRef] [PubMed]
26. Aguiló-Aguayo, I.; Oms-Oliu, G.; Soliva-Fortuny, R.; Martín-Belloso, O. Flavour retention and related enzyme activities during storage of strawberry juices processed by high-intensity pulsed electric fields or heat. *Food Chem.* **2009**, *116*, 59–65. [CrossRef]
27. Zannou, O.; Kelebek, H.; Selli, S. Elucidation of key odorants in Beninese Roselle (*Hibiscus sabdariffa* L. infusions prepared by hot and cold brewing. *Food Res. Int.* **2020**, *133*, 109133. [CrossRef]
28. Parra-Palma, C.; Úbeda, C.; Gil, M.; Ramos, P.; Castro, R.I.; Morales-Quintana, L. Comparative study of the volatile organic compounds of four strawberry cultivars and it relation to alcohol acyltransferase enzymatic activity. *Sci. Hortic.* **2019**, *251*, 65–72. [CrossRef]
29. Xu, H.S.; Quan, Q.; Chang, X.; Ge, S.; Xu, S.Q.; Wang, R.R.; Xu, Y.Q.; Luo, Z.S.; Shan, Y.; Ding, S.H. Maintaining the balance of fungal community through active packaging film makes strawberry fruit pose pleasant flavor during storage. *Postharvest Biol. Technol.* **2024**, *211*, 112815. [CrossRef]
30. Liu, Q.; Ding, H.Z.; Zhang, T.T.; Zhou, D.D.; Zhu, T.; Pan, L.Q.; Ma, G.X.; Lan, W.J.; Zhao, S.Q.; Hu, Q.H.; et al. Effect on lipoxygenase pathway and flavor stabilization of postharvest strawberries by intense pulsed light treatment. *Postharvest Biol. Technol.* **2023**, *204*, 112472. [CrossRef]
31. Hou, W.F.; Ma, Y.; Zhang, C.H.; Zhao, W.T.; Zhao, S.; Wang, P.; Zhao, X.Y.; Wang, D. Investigation on the inactivation effect and mechanism of Listeria monocytogenes in fresh-cut cucumber during storage by ultrasound combined with sodium hypochlorite. *Ultrason. Sonochem.* **2023**, *101*, 106706. [CrossRef] [PubMed]
32. Adeboyejo, F.O.; Oyesanya, O.B. Ultrasound and autoclave-deacetylated Achatina fulica shell chitosan: Characterisation and effect on tomato and cucumber fruit qualities during storage. *Food Chem.* **2023**, *415*, 135750. [CrossRef] [PubMed]
33. Zhang, R.Y.; Xing, L.J.; Kang, D.C.; Zhou, L.; Wang, L.; Zhang, W.G. Effects of ultrasound-assisted vacuum tumbling on the oxidation and physicochemical properties of pork myofibrillar proteins. *Ultrason. Sonochem.* **2021**, *74*, 105582. [CrossRef]
34. Da Porto, C.; Decorti, D. Ultrasound-assisted extraction coupled with under vacuum distillation of flavour compounds from spearmint (carvone-rich) plants: Comparison with conventional hydrodistillation. *Ultrason. Sonochem.* **2009**, *16*, 795–799. [CrossRef] [PubMed]
35. Bruna-Maynou, F.J.; Castro, R.; Rodríguez-Dodero, M.C.; Barroso, C.G.; Durán-Guerrero, E. Flavored Sherry vinegar with citric notes: Characterization and effect of ultrasound in the maceration of orange peels. *Food Res. Int.* **2020**, *133*, 795–799. [CrossRef] [PubMed]
36. Singla, M.; Sit, N. Application of Ultrasound in Combination with Other Technologies in Food Processing: A Review. *Ultrason. Sonochem.* **2021**, *73*, 105506. [CrossRef] [PubMed]
37. Alenyorege, E.A.; Ma, H. Removal of pesticides from fresh-cut Chinese cabbage (*Brassica rapa* var. chinensis L.) by sequential multi-frequency ultrasound cleaning treatments. *Food Humanit.* **2023**, *1*, 958–965. [CrossRef]
38. Khandpur, P.; Gogate, P.R. Evaluation of ultrasound based sterilization approaches in terms of shelf life and quality parameters of fruit and vegetable juices. *Ultrason. Sonochem.* **2016**, *29*, 337–353. [CrossRef] [PubMed]
39. Li, L.; Song, J.; Kalt, W.; Forney, C.; Tsao, R.; Pinto, D.; Chisholm, K.; Campbell, L.; Fillmore, S.; Li, X.H. Quantitative proteomic investigation employing stable isotope labeling by peptide dimethylation on proteins of strawberry fruit at different ripening stages. *J. Proteom.* **2013**, *94*, 219–239. [CrossRef] [PubMed]
40. Pastrana, A.M.; Borrero, C.; Perez, A.G.; Aviles, M. Soilborne pathogens affect strawberry fruit flavor and quality. *Plant Sci.* **2022**, *326*, 111533. [CrossRef]
41. Wang, K.H.; Qi, J.; Jin, Y.; Li, F.; Wang, J.; Xu, H.D. Influence of fruit maturity and lactic fermentation on physicochemical properties, phenolics, volatiles, and sensory of mulberry juice. *Food Biosci.* **2022**, *48*, 101782. [CrossRef]
42. Love, M.I.; Huber, W.; Anders, S. Moderated estimation of fold change and dispersion for RNA-seq data with DESeq2. *Genome Biol.* **2014**, *15*, 1–21. [CrossRef] [PubMed]
43. Pfaffl, M.W. A new mathematical model for relative quantification in real-time RT-PCR. *Nucleic Acids Res.* **2001**, *29*, 2002–2007. [CrossRef] [PubMed]
44. Ulrich, D.; Kecke, S.; Olbricht, K. What Do We Know about the Chemistry of Strawberry Aroma? *J. Agric. Food Chem.* **2018**, *66*, 3291–3301. [CrossRef] [PubMed]
45. Caleb, O.J.; Ilte, K.; Herppich, W.B.; Geyer, M.; Mahajan, P.V. Impacts of minimal processing and hot water dipping of 'Sonata' strawberries on volatiles emitted during storage. *Sci. Hortic.* **2019**, *243*, 385–391. [CrossRef]
46. Waghmode, B.; Masoodi, L.; Kushwaha, K.; Iqbal Mir, J.; Sircar, D. Volatile components are non-invasive biomarkers to track shelf-life and nutritional changes in apple cv. 'Golden Delicious' during low temperature postharvest storage. *J. Food Compos. Anal.* **2021**, *102*, 104075. [CrossRef]
47. Cai, H.F.; Han, S.; Jiang, L.; Yu, M.L.; Ma, R.J.; Yu, Z.F. Exogenous nitric oxide fumigation promoted the emission of volatile organic compounds in peach fruit during shelf life after long-term cold storage. *Food Res. Int.* **2020**, *133*, 109135. [CrossRef] [PubMed]
48. Xu, Y.Q.; Luo, Z.S.; Charles, M.T.; Rolland, D.; Roussel, D. Pre-harvest UV-C irradiation triggers VOCs accumulation with alteration of antioxidant enzymes and phytohormones in strawberry leaves. *J. Plant Physiol.* **2017**, *218*, 265–274. [CrossRef]

49. Padilla-Jimenez, S.M.; Angoa-Perez, M.V.; Mena-Violante, H.G.; Oyoque-Salcedo, G.; Montanez-Soto, J.L.; Oregel-Zamudio, E. Identification of Organic Volatile Markers Associated with Aroma during Maturation of Strawberry Fruits. *Molecules* **2021**, *26*, 504. [CrossRef]
50. Severo, J.; de Oliveira, I.R.; Bott, R.; Le Bourvellec, C.; Renard, C.M.G.C.; Page, D.; Chaves, F.C.; Rombaldi, C.V. Preharvest UV-C radiation impacts strawberry metabolite content and volatile organic compound production. *LWT-Food Sci. Technol.* **2016**, *85*, 390–393. [CrossRef]
51. Yang, W.B.; Liu, J.C.; Liu, H.; Zhang, Q.; Lv, Z.Z.; Jiao, Z.G. Characterization of strawberry purees fermented by Lactobacillus spp. based on nutrition and flavor profiles using LC-TOF/MS, HS-SPME-GC/MS and E-nose. *LWT-Food Sci. Technol.* **2023**, *189*, 115457. [CrossRef]
52. Schiefner, A.; Sinz, Q.; Neumaier, I.; Schwab, W.; Skerra, A. Structural basis for the enzymatic formation of the key strawberry flavor compound 4-hydroxy-2,5-dimethyl-3(2H)-furanone. *J. Biol. Chem.* **2013**, *288*, 16815–16826. [CrossRef] [PubMed]
53. Li, H.; Brouwer, B.; Oud, N.; Verdonk, J.C.; Tikunov, Y.; Woltering, E.; Schouten, R.; da Silva, F.P. Sensory, GC-MS and PTR-ToF-MS profiling of strawberries varying in maturity at harvest with subsequent cold storage. *Postharvest Biol. Technol.* **2021**, *182*, 111719. [CrossRef]
54. Knudsen, J.T.; Eriksson, R.; Gershenzon, J.; Stahl, B. Diversity and Distribution of Floral Scent. *Bot. Rev.* **2006**, *72*, 1–120. [CrossRef]
55. Gao, L.Y.; Chen, H.M.; Chen, W.X.; Chen, W.J.; Jian, H.Y.; Zhong, Q.P.; Zhang, M. Linalool against Hafnia alvei, its antibacterial mechanism revealed by metabolomic analyses. *Food Biosci.* **2023**, *51*, 102316. [CrossRef]
56. Do, E.; Kim, M.; Ko, D.Y.; Lee, M.; Lee, C.; Ku, K.M. Machine learning for storage duration based on volatile organic compounds emitted from 'Jukhyang' and 'Merry Queen' strawberries during post-harvest storage. *Postharvest Biol. Technol.* **2024**, *211*, 112808. [CrossRef]
57. Öz, A.T.; Ali, A. Retaining overall quality of fresh figs by postharvest hexanal vapor treatment during cold storage. *Postharvest Biol. Technol.* **2023**, *205*, 112539. [CrossRef]
58. Qian, L.; Xiaoman, Z.; Yanli, X.; Jingmeng, L. Antifungal properties and mechanisms of three volatile aldehydes (octanal, nonanal and decanal) on Aspergillus flavus. *Grain Oil Sci. Technol.* **2021**, *4*, 131–140.
59. Pelayo-Zaldivar, C.; Ben Abda, J.; Ebeler, S.E.; Kader, A.A. Quality and chemical changes associated with flavor of 'Camarosa' strawberries in response to a CO2-enriched atmosphere. *HortScience* **2007**, *42*, 299–303. [CrossRef]
60. Liu, Y.P.; Liu, R.; Li, F.F.; Yu, S.M.; Nie, Y.F.; Li, J.Q.; Pan, C.P.; Zhu, W.T.; Zhou, Z.Q.; Diao, J.L. Nano-selenium repaired the damage caused by fungicides on strawberry flavor quality and antioxidant capacity by regulating ABA biosynthesis and ripening-related transcription factors. *Pestic. Biochem. Physiol.* **2024**, *198*, 105753. [CrossRef]
61. Teixeira da Silva, J.A.; Hidvegi, N.; Gulyas, A.; Toth, B.; Dobranszki, J. Transcriptomic Response of In Vitro Potato (*Solanum tuberosum* L.) to Piezoelectric Ultrasound. *Plant Mol. Biol. Rep.* **2020**, *38*, 404–418. [CrossRef]
62. Mishra, R.C.; Ghosh, R.; Bae, H. Plant acoustics: In the search of a sound mechanism for sound signaling in plants. *J. Exp. Bot.* **2016**, *67*, 4483–4494. [CrossRef] [PubMed]
63. Dobranszki, J.; Hidvegi, N.; Gulyas, A.; Toth, B.; Teixeira da Silva, J.A. Abiotic stress elements in in vitro potato (*Solanum tuberosum* L.) exposed to air-based and liquid-based ultrasound: A comparative transcriptomic assessment. *Prog. Biophys. Mol. Biol.* **2020**, *158*, 47–56. [CrossRef] [PubMed]
64. Song, F.Y.; Huangfu, Z.Q.; Han, Y.R.; Li, H.M.; Wang, Z.P.; Jin, X.W.; Chen, J.L. Nitric oxide fumigation can affect the metabolism of volatile compounds derived from analyses of fatty acids and amino acids in post-harvest flat peach during cold storage. *Food Control* **2024**, *159*, 110258. [CrossRef]
65. Wang, Y.; Sun, Y.J.; Zhou, D.D.; Zhang, Q.; Pan, L.Q.; Tu, K. Transcriptomics analysis provides insights into metabolisms of sugars and carotenoids of nectarine fruit subjected to different temperature storage. *Sci. Hortic.* **2022**, *304*, 111262. [CrossRef]
66. Schwab, W.; Davidovich-Rikanati, R.; Lewinsohn, E. Biosynthesis of plant-derived flavor compounds. *Plant J.* **2008**, *54*, 712–732. [CrossRef] [PubMed]
67. Magalhaes, H.C.R.; Garruti, D.d.S.; Gandra, E.A.; Purgatto, E. Effect of Postharvest Treatments on the Biosynthesis of Fruit Volatile Compounds: A Literature Review. *Curr. Nutr. Food Sci.* **2023**, *19*, 246–261. [CrossRef]
68. Aharoni, A.; Keizer, L.C.P.; Bouwmeester, H.J.; Sun, Z.K.; Alvarez-Huerta, M.; Verhoeven, H.A.; Blaas, J.; van Houwelingen, A.M.M.L.; De Vos, R.C.H.; van der Voet, H. Identification of the SAAT gene involved in strawberry flavor biogenesis by use of DNA microarrays. *Plant Cell* **2000**, *12*, 647–661. [CrossRef] [PubMed]
69. Saez, D.; Rodríguez-Arriaza, F.; Urra, G.; Fabi, J.P.; Hormazábal-Abarza, F.; Méndez-Yáñez, A.; Castro, E.; Bustos, D.; Ramos, P.; Morales-Quintana, L. Unraveling the Key step in the Aroma Puzzle: Insights into Alcohol Acyltransferases in Strawberries. *Plant Physiol. Biochem.* **2024**, *28*, 108668. [CrossRef] [PubMed]
70. Wang, Z.R.; He, D.N.; Gao, W.K.; Li, M.H.; Wu, X.E.; Lv, J.H. Integrated transcriptomic and metabolomic analyses of 'Guifei' mango fruit flavor in an endospermic genotype and a mutated genotype without endosperm. *Sci. Hortic.* **2022**, *303*, 111189. [CrossRef]
71. Aharoni, A.; Verstappen, F.W.; Schwab, W.; Giri, A.P.; Bouwmeester, H.J. Gain and loss of fruit flavor compounds produced by wild and cultivated strawberry species. *Abstr. Pap. Am. Chem. Soc.* **2004**, *228*, U51. [CrossRef] [PubMed]
72. Liu, W.J.; Zhang, Y.; Wang, L.; Bilal, A.; Shi, X.X.; Ren, Y.; Liang, C.; Zhang, X.K.; Zhang, Y.X.; Du, G.Q. Integrated transcriptome and metabolome analysis unveiled the mechanisms of xenia effect and the role of different pollens on aroma formation in 'Yali' pear (*Pyrus bretschneideri* Rehd). *Sci. Hortic.* **2023**, *307*, 111503. [CrossRef]

73. Lunkenbein, S.; Salentijn, E.M.J.; Coiner, H.A.; Boone, M.J.; Krens, F.A.; Schwab, W. Up- and down-regulation of Fragaria x ananassa O-methyltransferase: Impacts on furanone and phenylpropanoid metabolism. *J. Exp. Bot.* **2006**, *57*, 2445–2453. [CrossRef] [PubMed]
74. Liu, X.; Feng, Y.F.; Li, S.S.; Li, D.M.; Yu, J.; Zhao, Z.Y. Jasmonate-induced MdMYC2 improves fruit aroma during storage of 'Ruixue' apple based on transcriptomic, metabolic and functional analyses. *LWT-Food Sci. Technol.* **2023**, *185*, 115168. [CrossRef]
75. Cai, H.F.; Han, S.; Jiang, L.; Yu, M.L.; Ma, R.J.; Yu, Z.F. 1-MCP treatment affects peach fruit aroma metabolism as revealed by transcriptomics and metabolite analyses. *Food Res. Int.* **2019**, *122*, 573–584. [CrossRef] [PubMed]
76. Zhu, X.Y.; Song, Z.Y.; Li, Q.M.; Li, J.; Chen, W.X.; Li, X.P. Physiological and transcriptomic analysis reveals the roles of 1-MCP in the ripening and fruit aroma quality of banana fruit (Fenjiao). *Food Res. Int.* **2020**, *130*, 108968. [CrossRef] [PubMed]
77. Xi, W.P.; Zhang, B.; Shen, J.Y.; Sun, C.D.; Xu, C.J.; Chen, K.S. Intermittent warming alleviated the loss of peach fruit aroma-related esters by regulation of AAT during cold storage. *Postharvest Biol. Technol.* **2012**, *74*, 42–48. [CrossRef]
78. Lu, H.Y.; Luo, Z.S.; Li, D.; Jiang, Y.H.; Li, L. FaMYB11 promotes the accumulation of volatile esters by regulating FaLOX5 during strawberry (*Fragaria* × *ananassa*) ripening. *Postharvest Biol. Technol.* **2021**, *178*, 111560. [CrossRef]

molecules

MDPI

Article

Development and Validation of Liquid Chromatographic Method for Fast Determination of Lincomycin, Polymyxin and Vancomycin in Preservation Solution for Transplants

Qi Lin, Tam Nguyen, Chiara Staffieri, Ann Van Schepdael and Erwin Adams *

Department of Pharmaceutical and Pharmacological Sciences, Pharmaceutical Analysis, KU Leuven, Herestraat 49, O&N2, PB 923, 3000 Leuven, Belgium; qi.lin923@gmail.com (Q.L.); thiminhtam.nguyen@kuleuven.be (T.N.); staffieri.chiara@gmail.com (C.S.); ann.vanschepdael@kuleuven.be (A.V.S.)

* Correspondence: erwin.adams@kuleuven.be

Abstract: In this study, a liquid chromatographic method was developed for the fast determination of lincomycin, polymyxin and vancomycin in a preservation solution for transplants. A Kinetex EVO C18 (150 × 4.6 mm, 2.6 μm) column was utilized at 45 °C. Gradient elution was applied using a mixture of mobile phases A and B, both including 30 mM phosphate buffer at pH 2.0 and acetonitrile, at a ratio of 95:5 (*v*/*v*) for A and 50:50 (*v*/*v*) for B. A flow rate of 1.0 mL/min, an injection volume of 20 μL and UV detection at 210 nm were used. A degradation study treating the three antibiotics with 0.5 M hydrochloric acid, 0.5 M sodium hydroxide and 3% H_2O_2 indicated that the developed method was selective toward lincomycin, polymyxin, vancomycin and their degradation products. Other ingredients of the preservation solution, like those from the cell culture medium, did not interfere. The method was validated with good sensitivity, linearity, precision and accuracy. Furthermore, lincomycin, polymyxin and vancomycin were found to be stable in this preservation solution for 4 weeks when stored at −20 °C.

Keywords: transplant preservation solution; lincomycin; polymyxin; vancomycin; superficially porous particles; liquid chromatography; validation

Citation: Lin, Q.; Nguyen, T.; Staffieri, C.; Van Schepdael, A.; Adams, E. Development and Validation of Liquid Chromatographic Method for Fast Determination of Lincomycin, Polymyxin and Vancomycin in Preservation Solution for Transplants. *Molecules* **2024**, *29*, 3166. https://doi.org/10.3390/molecules29133166

Academic Editor: Angelo Antonio D'Archivio

Received: 27 May 2024
Revised: 25 June 2024
Accepted: 27 June 2024
Published: 3 July 2024

1. Introduction

Antibiotics are added to solutions to preserve transplants for different reasons, for example, to prevent bacterial growth before and immediately after transplantation [1,2], to promote the proliferation of transplanted cells [3] and to instruct and develop the immune system of the recipients [4]. Lincomycin is a lincosamide used for treating severe bacterial infections. It has a narrow antibiotic spectrum showing antibacterial activity against Gram-positive and pathogenic species like *Streptococcus*, *Staphylococcus* and *Mycoplasma* [4]. Polymyxin is a biosynthesized complex consisting of 10 amino acid residuals which is mainly active against Gram-negative bacteria [5]. Vancomycin is generated from the *Amycolatopsis orientalis* strain. It is suggested to be used after the failure of other antibiotics or when treating serious or life-threatening infections [6].

A preservation solution containing these three antibiotics in a matrix of RPMI 1640 Medium (which includes inorganic salts, amino acids, vitamins, nutrients like glucose, pH indicator, etc.) is generally prepared by hospitals involved in organ transplants. However, the solution should be controlled before use to check if it was properly prepared and that the assay values comply with the limits. Moreover, the stability of the antibiotics in this preservation solution is not clear; it is not known for how long and under which conditions it can best be stored. Hence, a sensitive and reliable analytical method will be helpful to determine the three antibiotics.

Liquid chromatography (LC), as an efficient analytical tool to separate complex samples, has been employed. Several LC methods have been described for the analysis of lincomycin, polymyxin and vancomycin separately, but not in a mixture of all three. For example, the LC separation of lincomycin from its related substances has been described by Orwa et al. [7]. The separation of vancomycin from its impurities was published by Diana et al. [8]. In addition, Mathew and Das Gupta [9] and Orwa et al. [10] dedicated two LC methods to assess the stability of vancomycin and polymyxin, respectively. The European Pharmacopoeia (Ph. Eur.) also adopted LC methods in the respective monographs of the three antibiotics [11]. Only Cheng et al. [12] published an LC-MS method for the determination of polymyxin and vancomycin in rat plasma. However, after checking, their method did not chromatographically separate polymyxin and vancomycin.

Since no chromatographic method has been described in the literature to separate lincomycin, polymyxin and vancomycin in a single run, the aim of this study was to develop and validate an LC-UV method for the determination of the three antibiotics simultaneously. The separation was complicated by RPMI 1640 Medium, which contains numerous compounds that are also present in the sample. The method should also be able to evaluate the stability of the preservation solution. Although lincomycin and polymyxin are mixtures of active compounds, no efforts will be made to separate all of these components, but they will be analyzed as the respective cluster if they are not separated.

2. Results and Discussion

2.1. Method Development

In this study, an LC-UV method was developed to determine lincomycin hydrochloride (120 μg/mL), polymyxin sulfate (100 μg/mL) and vancomycin hydrochloride (50 μg/mL) in an RPMI 1640 matrix used as a preservation solution for transplants. The three antibiotics were combined to obtain antibacterial activity against a broad range of micro-organisms. Since it is the intention to use the method routinely in quality control laboratories, LC-UV was preferred over LC-MS because of its robustness and lower operational and purchase costs.

First, a comparison was made of the LC methods prescribed in the related substance test of the respective monographs in the Ph. Eur. [11] to set the starting point of the LC conditions on the Kinetex EVO C18 (150 × 4.6 mm, 2.6 μm) column. It can be derived from the overview in Table 1 that for simultaneous analyses of the three antibiotics, the following conditions should be beneficial: (1) a reversed-phase column, preferably end-capped and base-deactivated, which is the case for the Kinetex EVO C18 (150 × 4.6 mm, 2.6 μm) column; (2) a moderate column temperature avoiding on-column degradation of the analytes; (3) a mobile phase containing acetonitrile (ACN) as an organic modifier and a buffer solution with a pH between 2 and 7; (4) a flow rate of 1 mL/min; (5) UV detection at low wavelengths since lincomycin and polymyxin have no strong UV absorbing chromophore.

Table 1. An overview of the LC methods prescribed in the related substance test of the respective monographs in the Ph. Eur. [11].

	Lincomycin Hydrochloride	Polymyxin B Sulfate	Vancomycin Hydrochloride
Stationary phase	End-capped base-deactivated octylsilyl silica gel (250 × 4.6 mm, 5 μm)	End-capped octadecylsilyl silica gel (250 × 4.6 mm, 5 μm)	Octadecylsilyl silica gel (250 × 4.6 mm, 5 μm)
Column temperature	50 °C	30 °C	Room temperature
Mobile phase	MeOH-ACN–phosphate buffer (0.15 M, pH 6.1), 8:17:75 (*v*/*v*/*v*)	ACN—4.46 g/L of Na_2SO_4 (to pH 2.3 with 31.4 mM H_3PO_4 solution), 20:80 (*v*/*v*)	4 mL/L of Triethylamine (to pH 3.2 with H_3PO_4 solution)-THF-ACN, gradient with A) 920:10:70 (*v*/*v*/*v*) and B) 700:10:290 (*v*/*v*/*v*)
Flow rate	1.0 mL/min	1.0 mL/min	1.0 mL/min
UV detection	210 nm	215 nm	280 nm

The initial column temperature was set at 30 °C. The mobile phase was a gradient mixture from solution A that was composed of 90 volumes of phosphate buffer (30 mM, pH 3.0) and 10 volumes of ACN, and solution B that consisted of the same components, but at a ratio of 50:50, *v*/*v*. A flow rate of 1.0 mL/min and a UV detection wavelength at

210 nm were selected. Further optimization was implemented for the pH, concentration of phosphate buffer, amount of ACN in solution A, column temperature and UV wavelength.

The influence of buffer pH was investigated in the range from pH 2.0 to 7.0. It was observed that the retention of lincomycin was higher with increasing buffer pH across the range examined. For vancomycin, its retention showed a U-shaped relationship with pH. The retention reached a minimum at pH 5.0. Furthermore, a reversed U-shaped relationship was found for buffer pH and the retention of polymyxin, which showed maximum peak retention at pH 5.0. In addition, the UV response of polymyxin became weaker and no good peak shape could be obtained when buffers above pH 5.0 were used. Moreover, in a stability study published by Orwa et al. [10], it is mentioned that the degradation of polymyxin easily happened in solutions with pH values above 5 and that it is more stable in acidic media. Concerning vancomycin, it has been reported by Mathew and Das Gupta [9] that higher concentrations of phosphate ions accelerate the degradation of the antibiotic and that this is more pronounced at about pH 7 than pH 4. So, a buffer with pH 2.0 was selected, since this ensured the buffer capacity of phosphate and led to the best resolution between lincomycin, polymyxin and vancomycin, as well as the components of RPMI 1640 Medium.

The performance of different concentrations of phosphate buffer was assessed at 20 and 30 mM. Vancomycin and polymyxin showed a better peak shape (peak symmetry ≤ 1.5) when the higher concentration was used. The concentration was not further increased in order to not compromise the stability of vancomycin. The shape of the lincomycin peak was not influenced by either concentration. The retention times of the three antibiotics hardly changed between using 20 and 30 mM of phosphate buffer. So, 30 mM of buffer was preferred.

Reducing the amount of ACN from 10% to 5% in solution A resulted in an increase in retention for lincomycin, polymyxin and vancomycin. Vancomycin was the most affected, followed by lincomycin and then polymyxin.

The column temperature had an effect on the retention times as could be expected. Among the investigated temperatures from 30 °C to 55 °C, shorter retention times of several minutes were found with raising column temperature. The best separation of the three antibiotics and components of RPMI 1640 Medium within a reasonable analysis time was observed at a column temperature of 45 °C.

Next, UV detection at 205 nm, 210 nm, 215 nm and 220 nm was evaluated concerning the response intensity versus the noise. The most intense response was found at 205 nm, and it was reduced with increasing wavelengths. However, the lowest wavelength also showed the highest baseline noise. The best signal-to-noise ratio and baseline stability were obtained at 210 nm; so, this wavelength was selected for the final method.

Furthermore, it was observed that regarding the preparation of the reference solution, polymyxin was more stable when it was dissolved in the mobile phase than in water, while lincomycin and vancomycin were stable in both solvents. For the sample solution, no diluent was chosen since the preservation solution was injected as such, owing to the low concentration of the three antibiotics. It was also well buffered by RPMI 1640 Medium.

A chromatogram obtained with the optimized LC-UV method is illustrated in Figure 1. Both lincomycin and polymyxin are each eluted as a single peak and not as a cluster of peaks corresponding to the different compounds. This simplifies the integration and is sufficient to check the total content of the three antibiotics.

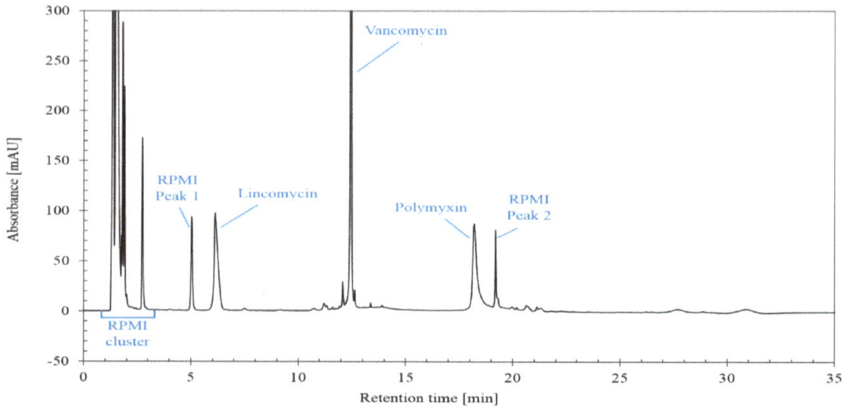

Figure 1. Chromatogram obtained with optimized LC-UV conditions for mixture consisting of lincomycin, polymyxin, vancomycin and RPMI Medium 1640.

2.2. Method Validation

2.2.1. Selectivity

From the chromatograms shown in Figure 2 that are corresponding to injections of degraded lincomycin, polymyxin and vancomycin after three ways of forced degradation, the RPMI Medium 1640 and the mobile phase, it can be derived that no interferences were found in the main peaks of lincomycin and vancomycin.

2.2.2. Sensitivity

The limit of detection (LOD) of lincomycin, polymyxin and vancomycin was found to be 24 ng, 24 ng and 3 ng, respectively, expressed as the amount injected on the column corresponding to a signal-to-noise ratio of 3. Amounts of 80 ng for lincomycin, 80 ng for polymyxin and 10 ng for vancomycin were demonstrated to be the limit of quantification (LOQ), determined at a signal-to-noise ratio of 10.

2.2.3. Linearity

Three regression equations were calculated as follows: $y = 0.1694\,x + 0.1100$ ($R^2 = 0.9998$) for lincomycin; $y = 0.2280\,x - 0.1391$ ($R^2 = 0.9995$) for polymyxin; and $y = 1.4617\,x + 0.0946$ ($R^2 > 0.9999$) for vancomycin. So, the R^2 values were clearly above the acceptance value of 0.995. Further, it was confirmed that zero was included in the 95% confidence interval (CI) of the y intercepts in all three equations, and their residual plots were randomly distributed. In addition, a lack-of-fit test for the linearity model revealed p-values above 0.05 for lincomycin, polymyxin and vancomycin. It may be concluded that linearity was established for the three antibiotics in the examined range.

2.2.4. Accuracy

Good accuracy of this method was demonstrated by the recoveries, which were each determined in triplicate and amounted at the 80, 100 and 120% level to, respectively, 101.9% (RSD: 0.5%), 101.8% (RSD: 0.5%) and 101.6% (RSD: 0.7%) for lincomycin; 99.9% (RSD: 0.6%), 99.8% (RSD: 1.2%) and 98.5% (RSD: 1.3%) for polymyxin; and 99.8% (RSD: 0.2%), 99.1% (RSD: 0.4%) and 100.1% (RSD: 0.3%) for vancomycin. All of these values fell within the acceptance range of 98–102%.

2.2.5. Precision

The precision results were expressed as relative standard deviation (RSD). The intra-day precision found on each day was not more than 1.4% for lincomycin, 1.2% for polymyxin and 1.0% for vancomycin. The inter-day precision investigated over 18 injections in three days was 1.2%, 1.5% and 1.2% for lincomycin, polymyxin and vancomycin, respectively. Since none of the values exceeded 2%, the precision was found to be acceptable.

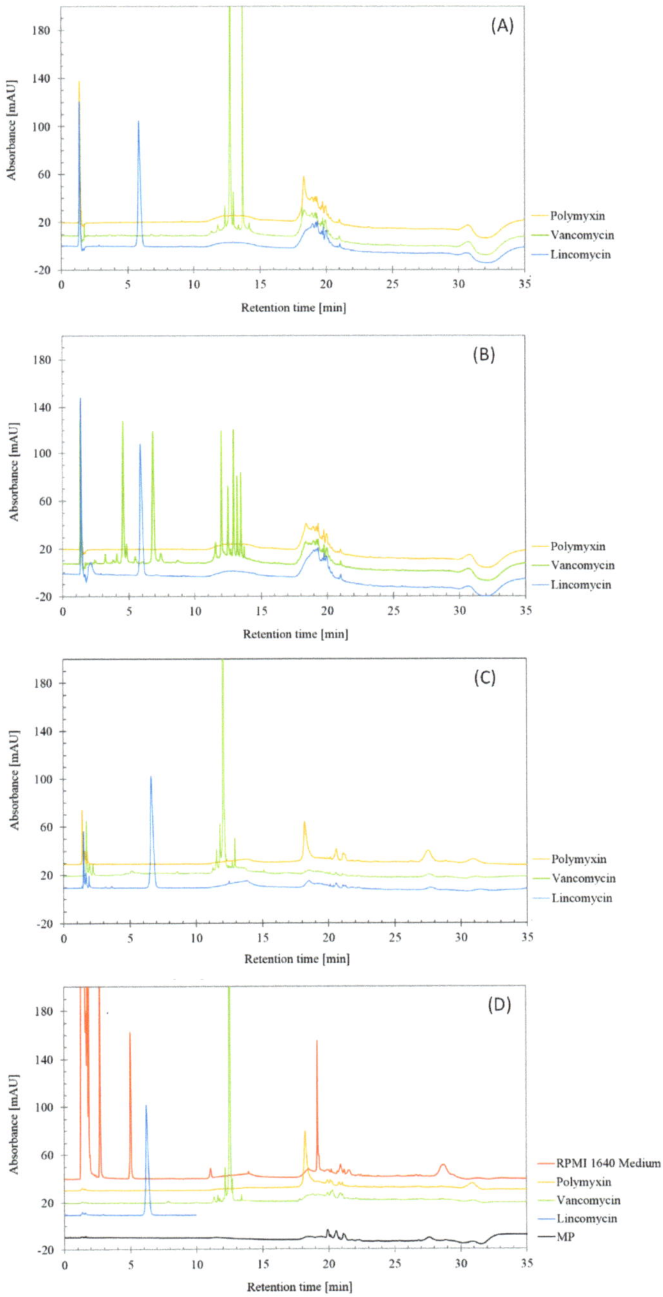

Figure 2. Chromatograms obtained from forced degradation studies of lincomycin, polymyxin and vancomycin treated by (**A**) 0.5 M HCl, (**B**) 0.5 M NaOH, (**C**) 3% H_2O_2 and (**D**) non-treated antibiotics, RPMI 1640 Medium and mobile phase (MP).

2.3. *Stability of Sample Solution*

At release (0 h), the three antibiotics complied with the 95–105% limits. More detailed results are shown in Table 2. After 4 weeks, the content of lincomycin and polymyxin remained within these limits under both storage conditions. However, the vancomycin content was observed to decrease. When stored at −20 °C for 4 weeks, around 93% of vancomycin was left, which is still within the end of the shelf-life specifications (90–110%). When stored at 4 °C, the sample could only be used for up to 2 weeks.

Table 2. Contents (RSD, n = 3) of lincomycin, polymyxin and vancomycin at different durations and conditions of storage.

	Lincomycin		**Polymyxin**		**Vancomycin**	
	4 °C	**−20 °C**	**4 °C**	**−20 °C**	**4 °C**	**−20 °C**
0 h	101.8% (0.5%)		99.4% (1.2%)		99.6% (1.3%)	
1 week	99.7% (0.6%)	99.6% (1.0%)	99.4% (1.6%)	101.0% (0.6%)	95.2% (1.8%)	97.9% (2.0%)
2 weeks	98.3% (0.2%)	97.6% (0.2%)	98.6% (1.6%)	99.6% (0.2%)	92.8% (0.1%)	96.1% (0.2%)
3 weeks	98.2% (0.3%)	98.5% (0.3%)	97.5% (1.4%)	101.8% (0.7%)	88.1% (0.2%)	93.5% (0.2%)
4 weeks	99.4% (0.2%)	98.3% (0.1%)	99.4% (1.7%)	101.0% (1.0%)	86.4% (0.2%)	93.3% (0.4%)

3. Materials and Methods

3.1. *Chemicals*

ACN (HPLC grade) was purchased from Fisher Scientific (Leicestershire, UK). Dipotassium hydrogen phosphate was provided by Acros Organics (Geel, Belgium). Potassium dihydrogen phosphate was obtained from Merck Millipore (Darmstadt, Germany) and 85% (*w/w*) phosphoric acid from Chem-Lab (Zedelgem, Belgium). RPMI 1640 Medium was obtained from Thermo Fisher (Paisley, UK). A Milli-Q water purification system from Millipore (Bedford, MA, USA) produced the purified water used in this study.

Lincomycin hydrochloride and polymyxin sulfate from EDQM (Strasbourg, France) and vancomycin hydrochloride from Xellia (Copenhagen, Denmark) that were used to prepare the preservation solution were also used as reference substances.

3.2. *Instrumentation and Chromatographic Conditions*

The LC-UV system used for method development and validation consisted of an LPG-3400A pump and an ASI-100 autosampler from Dionex (Sunnyvale, CA, USA), and a Spectra LINEAR UVIS 200 detector from Thermo Separation Products (San Jose, CA, USA). Chromeleon 6.8 software from Dionex was utilized for LC system operation and data acquisition. A Julabo ED heating thermostat (Seelbach, Germany) was immersed in a water bath to maintain the column temperature.

The separation was performed on a Kinetex EVO C18 (150 × 4.6 mm, 2.6 μm) column from Phenomenex (Torrance, CA, USA). The final mobile phase was a gradient mixture of mobile phases A and B that were both composed of ACN and 30 mM phosphate buffer with pH 2.0, but with different composition ratios, 5:95 (v/v) for A and 50:50 (v/v) for B. The gradient started with 100% of mobile phase A and remained the same for 8 min. In the next 3 min, mobile phase B increased linearly to 15%. This ratio was kept for 4 min before mobile phase B quickly reached 100% in 2 min. The mobile phase was isocratic for the next 10 min and took 1.5 min to return to the initial composition followed by the re-equilibration time. The flow rate was 1.0 mL/min, the column temperature was 45 °C, the injection volume was 20 μL and the detection wavelength was 210 nm.

3.3. Solution Preparation

The preservation solution was prepared by adding 100 μL of 300 mg/mL lincomycin hydrochloride, 250 μL of 100 mg/mL polymyxin sulfate, and 250 μL of 50 mg/mL vancomycin hydrochloride, all in water, to 250 mL of RPMI 1640 Medium. So, the solution consisted of 120 μg/mL of lincomycin hydrochloride, 100 μg/mL of polymyxin sulfate and 50 μg/mL of vancomycin hydrochloride in the matrix. It was injected as such.

Reference solutions were prepared at the same concentrations as the sample solution using mobile phase A. For the analysis of samples and validation of the linearity, the reference solution was prepared as a mixture of the 3 antibiotics. For verification of the accuracy, the antibiotics were injected separately.

3.4. Validation

The developed method was validated in accordance with the ICH Q2 guideline [13], including selectivity, sensitivity, linearity, precision and accuracy, as described below.

- Selectivity

The selectivity of the method was assessed by injecting the blank solution (mobile phase A), the placebo matrix solution (RPMI 1640 solution) and the forced degradation solutions in order to verify the selectivity among the major compounds, degradation products and RPMI 1640 components. The forced degradation solutions were prepared by separately treating lincomycin, polymyxin and vancomycin with 0.5 M HCl for 6 h at ambient temperature, 0.5 M NaOH for 6 h at ambient temperature and 3% H_2O_2 for 6 h at 60 °C. The chromatograms were examined to confirm that there was no interference at the retention time of the active compounds.

- Sensitivity

The LOD and LOQ were evaluated by separately injecting the solutions of each antibiotic, which were diluted until the achievement of signal-to-noise ratios of 3 and 10, respectively.

- Linearity

The calibration curves, implying the relationship between sample concentration (x) and UV response (y), were evaluated for lincomycin, polymyxin and vancomycin in the ranges from 30 to 150 μg/mL, from 25 to 125 μg/mL, and from 12.5 to 62.5 μg/mL, respectively. The UV signals generated from five concentrations were determined in triplicate for each antibiotic. The slope, intercept and determination coefficient (R^2) from the calibration curve were determined. R^2 should not be less than 0.995. The residual plots should be randomly distributed with zero included in the 95% confidence interval (CI) of the y intercepts.

- Accuracy

Since the sample was a preservation solution consisting of lincomycin, polymyxin, vancomycin and RPMI 1640 Medium, a comparison of responses measured from an RPMI 1640 solution spiked with the individual drug substances was made to assess the recovery of each antibiotic when using the developed method. The solutions were prepared with known amounts of lincomycin, polymyxin and vancomycin at concentrations of 80%, 100% and 120% of these in the sample, namely 96, 120 and 144 μg/mL, 80, 100 and 120 μg/mL, and 40, 50 and 60 μg/mL, respectively. Corresponding amounts of the three drug components were added to RPMI 1640 Medium to reach concentrations from 80% to 120%, as prescribed in the preservation solution. The solutions were injected in triplicate and the recovery at each concentration level was calculated. The recovery values should be between 98% and 102%.

- Precision

Both intra- and inter-day precisions were examined using the UV responses obtained for 120 μg/mL of lincomycin, 100 μg/mL of polymyxin and 50 μg/mL of vancomycin. Six consecutive injections were measured at each of the three days. RSD values of less than 2% should be obtained.

3.5. Stability of Sample Solution

The stability of the preservation solution was evaluated for 4 weeks after preparation and stored at 4 °C in a refrigerator and at −20 °C in a freezer. The preservation solution was injected in triplicate immediately after preparation (0 h) and at 1 week, 2 weeks, 3 weeks and 4 weeks. Reference solutions were freshly prepared at each time point.

4. Conclusions

An LC-UV method for the simultaneous determination of lincomycin, polymyxin and vancomycin, in the presence of a complex matrix, was developed on a Kinetex EVO C18 column. This is the first time that a method has been described to separate those three antibiotics. Since no sample pretreatment is necessary and a conventional LC-UV system can be used, the method is simple, which is an advantage when applied in quality control laboratories. Good validation data were obtained and demonstrated that this method is selective, sensitive, linear, precise and accurate. This method is useful for quality control of the content and stability of the three antibiotics in the preservation solution. At the concentrations studied, the solution was stable for up to 4 weeks at −20 °C. However, if the solution was kept at 4 °C, a stability of 2 weeks should be taken into account.

Author Contributions: Conceptualization: E.A.; methodology: Q.L. and E.A.; validation: Q.L.; formal analysis: Q.L., C.S. and E.A.; investigation: Q.L. and C.S.; resources: E.A.; data curation: Q.L.; writing—original draft preparation: Q.L.; writing—review and editing: E.A., T.N. and A.V.S.; supervision: E.A. and A.V.S.; project administration: E.A.; funding acquisitition: E.A. and A.V.S. All authors have read and agreed to the published version of the manuscript.

Funding: Qi Lin gratefully thanks the Chinese Scholarship Council (CSC) for her grant. Tam Nguyen gratefully thanks Internal Funds—KU Leuven for her grant (ZB/20/033).

Institutional Review Board Statement: Not applicable.

Informed Consent Statement: Not applicable.

Data Availability Statement: According to the policy of KU Leuven, the authors may be contacted for more details concerning the data supporting the reported results.

Conflicts of Interest: The authors declare that they have no conflicts of interest.

References

1. Fagiuoli, K.; Colli, A.; Bruno, R.; Craxì, A.; Gaeta, G.B.; Grossi, P.; Mondelli, M.U.; Puoti, M.; Sagnelli, E.; Stefani, S.; et al. Management of infections pre- and post-liver transplantation: Report of an AISF consensus conference. *J. Hepatol.* **2014**, *60*, 1075–1089. [CrossRef] [PubMed]
2. Allen, K.J.; Cheah, D.M.Y.; Lee, X.L.; Pettigrew-Buck, N.E.; Vadolas, J.; Mercer, J.F.B.; Ioannou, P.A.; Williamson, R. The Potential of Bone Marrow Stem Cells to Correct Liver Dysfunction in a Mouse Model of Wilson's Disease 2004. Available online: www.cognizantcommunication.com (accessed on 29 May 2019).
3. Skubis, A.; Gola, J.; Sikora, B.; Hybiak, J.; Paul-Samojedny, M.; Mazurek, U.; Łos, M.J. Impact of Antibiotics on the Proliferation and Differentiation of Human Adipose-Derived Mesenchymal Stem Cells. *Int. J. Mol. Sci.* **2017**, *18*, 2522. [CrossRef] [PubMed]
4. Khoruts, A.; Hippen, K.L.; Lemire, A.M.; Holtan, S.G.; Knights, D.; Young, J.H. Toward revision of antimicrobial therapies in hematopoietic stem cell transplantation: Target the pathogens, but protect the indigenous microbiota. *Transl. Res.* **2017**, *179*, 116–125. [CrossRef] [PubMed]
5. Velkov, T.; Roberts, K.D.; Nation, R.L.; Thompson, P.E.; Li, J. Pharmacology of polymyxins: New insights into an "old" class of antibiotics. *Future Microbiol.* **2013**, *8*, 711–724. [CrossRef] [PubMed]
6. Paterson, D.L.; Singh, N. Interactions between Tacrolimus and Antimicrobial Agents. *Clin. Infect. Dis.* **1997**, *25*, 1430–1440. [CrossRef] [PubMed]
7. Orwa, J.A.; Bosmans, F.; Depuydt, S.; Roets, E.; Hoogmartens, J. Liquid chromatographic method for separation of lincomycin from its related substances. *J. Chromatogr. A* **1998**, *829*, 161–166. [CrossRef]

8. Diana, J.; Visky, D.; Roets, E.; Hoogmartens, J. Development and validation of an improved method for the analysis of vancomycin by liquid chromatography: Selectivity of reversed-phase columns towards vancomycin components. *J. Chromatogr. A* **2003**, *996*, 115–131. [CrossRef] [PubMed]
9. Mathew, M.; Das Gupta, V. Stability of vancomycin hydrochloride solutions at various pH values as determined by high-performance liquid chromatography. *Drug Dev. Ind. Pharm.* **1995**, *21*, 257–264. [CrossRef]
10. Orwa, J.A.; Govaerts, C.; Gevers, K.; Roets, E.; Van Schepdael, A.; Hoogmartens, J. Study of the stability of polymyxins B1, E1 and E2 in aqueous solution using liquid chromatography and mass spectrometry. *J. Pharm. Biomed. Anal.* **2002**, *29*, 203–212. [CrossRef]
11. EDQM Council of Europe. *European Pharmacopoeia*, 11th ed.; EDQM Council of Europe: Strasbourg, France, 2023.
12. Cheng, C.; Liu, S.; Xiao, D.; Hollembaek, J.; Yao, L.; Lin, J.; Hansel, S. LC-MS/MS method development and validation for the determination of polymyxins and vancomycin in rat plasma. *J. Chromatogr. B* **2010**, *878*, 2831–2838. [CrossRef] [PubMed]
13. ICH. *ICH Q2 Guideline on Validation of Analytical Procedures;* International Council for Harmonisation of Technical Requirements for Pharmaceuticals for Human Use (ICH): Geneva, Switzerland, 2005.

molecules

Article

A Support Vector Machine-Assisted Metabolomics Approach for Non-Targeted Screening of Multi-Class Pesticides and Veterinary Drugs in Maize

Weifeng Xue *, Fang Li, Xuemei Li and Ying Liu *

Technology Centre of Dalian Customs, Dalian 116000, China; fanglee_4125@163.com (F.L.); lxmciq@163.com (X.L.)
* Correspondence: xwf526@163.com (W.X.); lnciqly@126.com (Y.L.)

Abstract: The contamination risks of plant-derived foods due to the co-existence of pesticides and veterinary drugs (P&VDs) have not been fully understood. With an increasing number of unexpected P&VDs illegally added to foods, it is essential to develop a non-targeted screening method for P&VDs for their comprehensive risk assessment. In this study, a modified support vector machine (SVM)-assisted metabolomics approach by screening eligible variables to represent marker compounds of 124 multi-class P&VDs in maize was developed based on the results of high-performance liquid chromatography–tandem mass spectrometry. Principal component analysis and orthogonal partial least squares discriminant analysis indicate the existence of variables with obvious inter-group differences, which were further investigated by S-plot plots, permutation tests, and variable importance in projection to obtain eligible variables. Meanwhile, SVM recursive feature elimination under the radial basis function was employed to obtain the weight-squared values of all the variables ranging from large to small for the screening of eligible variables as well. Pairwise *t*-tests and fold changes of concentration were further employed to confirm these eligible variables to represent marker compounds. The results indicate that 120 out of 124 P&VDs can be identified by the SVM-assisted metabolomics method, while only 109 P&VDs can be found by the metabolomics method alone, implying that SVM can promote the screening accuracy of the metabolomics method. In addition, the method's practicability was validated by the real contaminated maize samples, which provide a bright application prospect in non-targeted screening of contaminants. The limits of detection for 120 P&VDs in maize samples were calculated to be 0.3~1.5 μg/kg.

Keywords: maize; marker compounds; metabolomics; non-targeted screening; pesticides and veterinary drugs; support vector machine

Citation: Xue, W.; Li, F.; Li, X.; Liu, Y. A Support Vector Machine-Assisted Metabolomics Approach for Non-Targeted Screening of Multi-Class Pesticides and Veterinary Drugs in Maize. *Molecules* **2024**, *29*, 3026. https://doi.org/10.3390/molecules29133026

Academic Editor: Davide Bertelli

Received: 8 May 2024
Revised: 15 June 2024
Accepted: 19 June 2024
Published: 26 June 2024

1. Introduction

It has always been a headache to resolve pesticide contamination in plant-derived foods, which poses a considerable threat to global food safety. Some national governments (e.g., China [1], the United States [2], and Japan [3]) and international authorities (e.g., the European Union [4]) have issued a series of formal regulatory documents on the maximum residue limits (MRLs) of pesticides in plant-derived foods. With the deepening of research, more and more evidence has shown that plant-derived foods are also suffering from serious contamination by veterinary drugs [5–13]. Previous studies [5–13] have shown that veterinary drugs excreted by livestock and poultry are commonly used as fertilizers on farmland, and some plant-derived foods (e.g., maize, potatoes, cucumbers, and lettuce) are easy to absorb from the soil, resulting in a variety of veterinary drugs (e.g., tetracycline, quinolones, and sulfonamides) accumulating in the foods, with a total concentration up to several mg/kg levels [8,14–17]. Due to the lack of risk evaluation standards for veterinary drugs in plant-derived foods at home and abroad, it is difficult to directly judge whether the concentration of veterinary drugs will cause adverse effects. Referring to the regulatory documents of MRLs of veterinary drugs in animal-derived foods [18,19], the concentration of 10 μg/kg was deemed the safety threshold for most veterinary drugs, and it is inferred that the reported concentration level of veterinary drugs in some plant-derived foods is likely to cause serious food safety incidents. For example, some organophosphorus pesticides can damage the nervous system, while sulfonamide antibiotics can cause allergic reactions after entering the human body. To sum up, resolving the food safety issue caused by the co-existence of pesticides and veterinary drugs (P&VDs) is urgent, and developing an effective screening method is a feasible way to prevent the occurrence of potential contamination risks.

Traditional methods to screen contaminants in foods are mostly based on the establishment of a set of databases containing a variety of P&VDs [20–22]. These methods have proved to be successful in discerning contaminants in the database, but they are helpless for contaminant identification outside the database, which casts a shadow on food safety. For example, some infamous food safety incidents took place in the 21st century, including melamine milk powder and fipronil eggs, which arose from some enterprises or individuals illegally adding unexpected contaminants not in the routine test scope of foods. Even if the concentrations of these contaminants were highly above the red line of food safety, these defective products were also considered to be qualified. As a result, they have gradually developed into extremely serious food safety incidents. It can be seen that the constant establishment of new databases of contaminants to screen unexpected P&VDs in plant-derived foods is a passive way that consumes a lot of human and material resources and is also unable to meet the increasing demands for the detection of unexpected P&VDs in foods. Therefore, it has become a popular trend to develop highly efficient non-targeted screening methods for contaminants, as the European Commission NORMAN network proposed [23,24].

Metabolomics has shown great promise in screening unexpected contaminants in the food safety field due to its advantages in handling massive data with complicated characteristics, such as small sample sizes, mass interferents, and high noise [25–29]. Marker compounds on behalf of unexpected contaminants in foods were screened and identified by in-house or network databases (e.g., Massbank, SciFinder, ChemSpider, PubChem, and Metlin) during metabolomics analysis in order to achieve non-targeted screening of contaminants [28,29].

As we know, metabolomics data are composed of fewer samples but massive features (i.e., variables). How to extract meaningful information and find the eligible variables on behalf of marker compounds is a primary issue for researchers to resolve. Generally speaking, the distinguishable factors for metabolomics data between any two groups (e.g., experiment vs. control) are a combination of variables rather than a single one. Therefore, several chemometric approaches, such as principal component analysis (PCA) [30–32], orthogonal partial least squares discriminant analysis (OPLS-DA) [30–32], random forest (RF) [33,34], and support vector machine (SVM) [33,35,36], have widely been used to deal with these metabolomics data. Among all these modeling methods, SVM is gaining popularity in a wide variety of metabolomics studies due to its prediction performance. SVM is known to have excellent generalization ability and can be applied to nonlinear cases with the assistance of kernels relative to PCA and OPLS-DA, with only an assumption of linearity. SVM recursive feature elimination (SVMRFE) is a wrapper approach that adopts the rule of weight (w) to rank the variables and works well when the number of samples is small but the number of variables is large. Due to this feature of SVMRFE, which is similar to that of metabolomics data, SVMRFE is often used in metabolomics studies [37–40]. In addition, SVMRFE has shown better predictability than OPLS-DA. To date, SVMRFE-involved metabolomics studies are mainly applied to seek eligible biomarkers for disease diagnosis, but they remain insufficient to complete the non-targeted screening of contaminants in foods. In this work, we try to apply SVM to metabolomics analysis to better screen and identify marker compounds on behalf of 124 multi-class P&VDs in maize so as to realize non-targeted screening of residual contaminants in plant-derived foods.

2. Results and Discussion

2.1. Data Preprocessing

The total ion chromatograms containing 124 P&VDs in three concentration groups are shown in Figure 1. The matrix complexity of maize may cause the recoveries of each contaminant in different samples to be significantly discrepant. As a result, no evident correlation between peak intensities and concentration levels was observed, as shown in Figure 1. To eliminate the peak intensity errors of variables arising from different recoveries of P&VDs during the pretreatment process, enrofloxacin-d5 (parent ion m/z 365.21092; fragment ions m/z 348.19692, 322.21804, and 246.11012; retention time 7.01 min) and atrazine-d5 (parent ion m/z 221.14017; fragment ions m/z 179.08602, 137.06393, and 101.08703; retention time 5.77 min) were jointly spiked for recovery calibration. Blank maize extract solutions were employed to prepare the standard curves (5, 10, 25, 50, and 100 ng/mL) of enrofloxacin-d5 and atrazine-d5 to calculate the recoveries of these two deuterated compounds in maize samples, with the recoveries to be 75.6%~93.1%, 72.9%~93.4%, and 74.9%~93.6% for enrofloxacin-d5 and 73.6%~95.2%, 81.2%~94.7%, and 72.9%~96.4% for atrazine-d5 in the 20, 50, and 100 ng/mL groups, respectively (Table S1, Supplementary Materials).

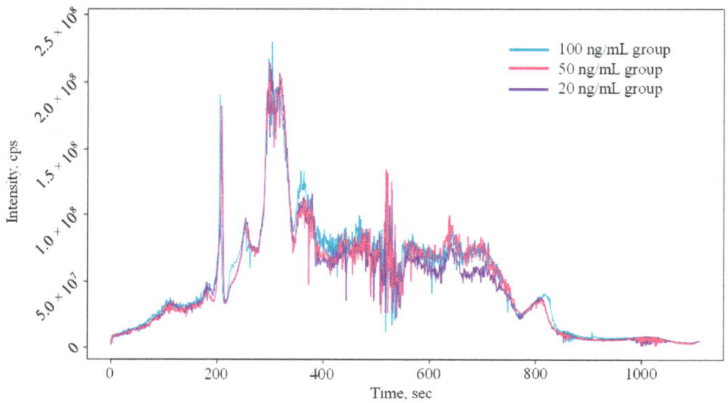

Figure 1. Total ion chromatograms of spiked maize sample groups on the W4M platform.

For each sample, we adopted the formula of 2 × 100%/((100%/recovery of enrofloxacin-d5) + (100%/recovery of atrazine-d5)) to calculate the final recovery of internal standards (Table S1), whose average value from the same concentration group needs to multiply a calibration coefficient to obtain 100% recovery. Peak intensities for internal standards and all other variables were also calibrated after multiplying the coefficient. After this, eligible variables with a relative standard deviation of peak intensity less than 30% in QC and three concentration groups were obtained to establish a 1447 × 39 data matrix for further analysis [41].

2.2. Multivariate Analysis

2.2.1. PCA Results

As indicated in Figure 2, all samples showed their credibility within a 95% confidence level. QC group samples have a close gathering, implying that the high quality of the data was available and deserved further analysis [42]. Nine samples from the same concentration group clustered closely, indicating the good intra-group similarity of these samples. Inter-group samples separated obviously on the first principal component axis, implying distinguishable differences in variables existed among these groups, which provided the possibility to hunt for marker compounds among concentration groups.

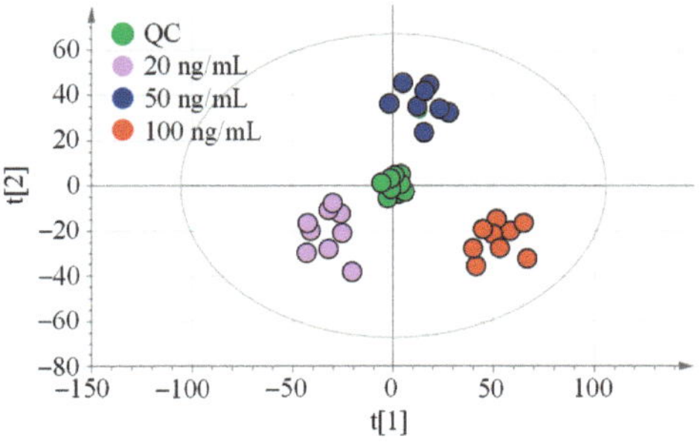

Figure 2. PCA score plot of spiked maize sample groups.

2.2.2. Cluster Analysis Results

Cluster analysis can directly describe the similarity level of samples, meaning that samples with the highest similarity gather together preferentially, level by level, until all the samples finish their aggregation. As observed in Figure S1 (Supplementary Materials), the samples from the same concentration group were obviously separated from others, showing their high intra-group congeniality. The evident inter-group separation can be attributed to the differences in peak intensities of variables. The findings of cluster analysis were consistent with those of PCA, and both supported the existence of variables with significant differences.

2.2.3. OPLS-DA Results

As we know, OPLS-DA is a two-class classification model; the size imbalance of two classes may introduce a bias in the computation of the decision rule, which can result in the unreliability of the OPLS-DA model [43]. To resolve this issue, we adopted the synthetic minority over-sampling technique (SMOTE) [44] to increase the sample size of the class with fewer samples. Herein, when the 20 ng/mL concentration group was designed as Class 1 (only 9 samples), and the 50 and 100 ng/mL concentration groups as a whole (18 samples in total) were designated Class 2, it needed to increase the size of Class 1 from 9 to 18 samples (Figure 3a). Similarly, the 100 ng/mL concentration group in Class 1 also needed to increase the class size to match that of the 20 and 50 ng/mL concentration groups in Class 2 (Figure 3b). As indicated in Figure 3, the obvious separation between Class 1 and Class 2 on the first principal component axis in OPLS-DA score plots was observed, implying that the differences in variables were significant between the two classes. R^2Y and Q^2 are two key parameters in the OPLS-DA model to evaluate the interpretability along the Y-axis and the predictability of the model, respectively [45,46]. When R^2Y and Q^2 values are close to 1, it means that the OPLS-DA model has high reliability and predictability. In Figure 3, the R^2Y and Q^2 values were both equal to 1, which favored the robustness of these OPLS-DA models.

A series of data points to represent the variables were shown in the S-plot plots (Figure 4), in which the points near the two tips of the 'S' plots signified the larger contribution and higher confidence levels of the corresponding variables in differentiating two classes of the OPLS-DA model. These variables were qualified to be marker compound candidates. Herein, marker compounds in the significantly low (20 ng/mL) and high (100 ng/mL) concentration groups should be sought at the right end of Figure 4a and the left end of Figure 4b, respectively.

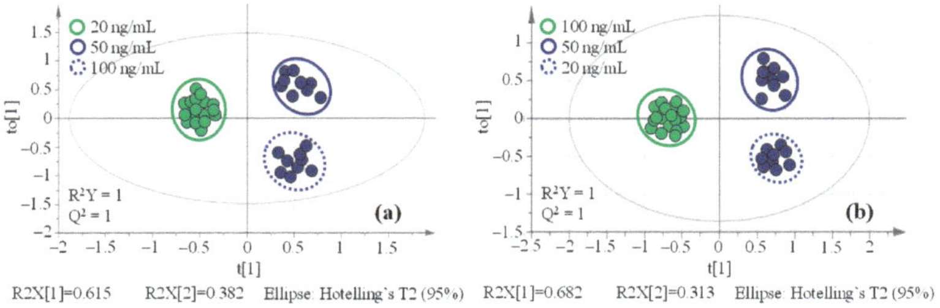

Figure 3. OPLS-DA score plots of spiked maize sample groups ((**a**): 20 ng/mL vs. 50 & 100 ng/mL; (**b**): 100 ng/mL vs. 20 & 50 ng/mL).

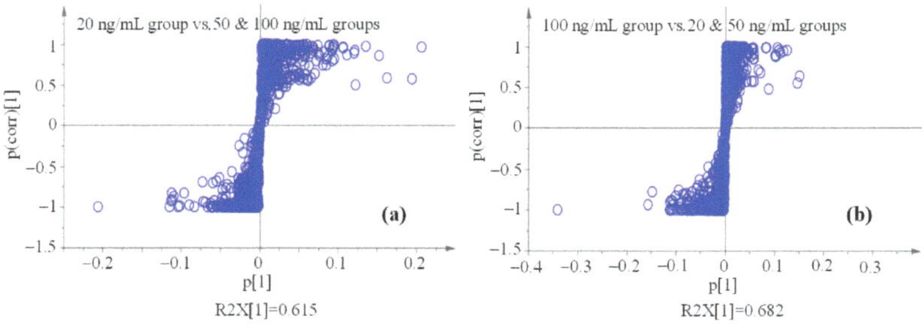

Figure 4. S-plot plots of spiked maize sample groups ((**a**): 20 ng/mL vs. 50 & 100 ng/mL; (**b**): 100 ng/mL vs. 20 & 50 ng/mL).

As we know, the Y-axis in the permutation test plot is on behalf of R^2Y and Q^2Y for the model, and the X-axis represents the correlation coefficient between original and permuted response data. It is desirable to summarize the results of the permutation test in a quantitative manner. One way to do this is to conduct conventional regression analysis on the two sets of points, i.e., one regression line is fitted among the R^2Y points (green circles) and another line among the Q^2Y points (blue squares), as shown in Figure 5. The intercepts of the resulting regression lines are interpretable as measures of 'background' R^2Y and Q^2Y obtained by fit to random data. Experience shows that the R^2Y-intercept should not exceed 0.3~0.4 and that the O^2Y-intercept should not exceed 0.05. Intercepts below these limits indicate valid models. A 200-iteration permutation test was a common method to evaluate whether the OPLS-DA model underwent over-fitting by the significance test (*p*-value) of Q^2Y metrics [47]. If $p < 0.05$, it meant that the OPLS-DA models were free of over-fitting, as reflected by the Q^2Y-intercept being less than 0.05 [47]. Figure 5 clearly points out that all OPLS-DA models were immune to over-fitting under the principles mentioned above.

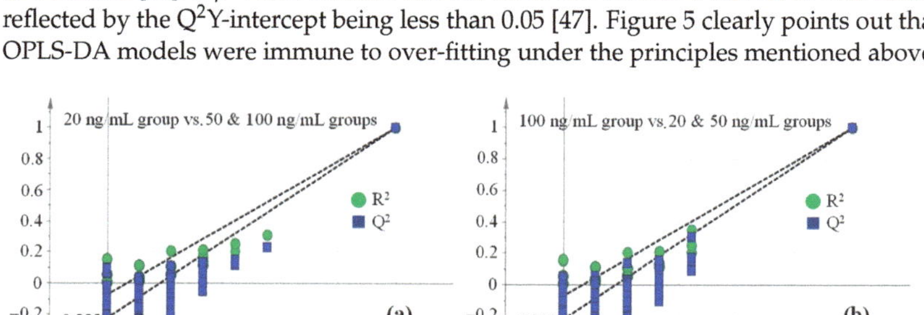

Figure 5. Permutation test plots of spiked maize sample groups ((**a**): 20 ng/mL vs. 50 & 100 ng/mL; (**b**): 100 ng/mL vs. 20 & 50 ng/mL).

VIP, as a measure of the importance of the variable in the OPLS-DA model, was computed for each extracted variable. Usually, VIP > 1 was considered to be the crucial threshold to pick eligible variables on behalf of marker compounds. In this study, a total of 228 variables with VIP > 1 from 20 concentration groups vs. 50 and 100 concentration groups (20 vs. 50 and 100 in abbreviation applied below) were chosen to be the marker compound candidates, but only 188 variables were eligible from 100 concentration groups vs. 20 and 50 concentration groups (100 vs. 20 and 50 in abbreviation applied below). One hundred and thirty-nine variables were observed to be overlapped between two VIP lists.

2.2.4. SVM Results

After running SVM codes, all the classification accuracy of the 10-fold cross-validation was computed to be 100%, in favor of the models with good generalization ability. The best penalty factor (c) and best kernel function parameter (g) from 3-fold cross-validation at 100% accuracy rate were computed to be 0.0039 and 0.0039 for 20 vs. 50 and 100, and 0.0037 and 0.0037 for 100 vs. 20 and 50. Two hundred and twenty-eight variables with VIP > 1 obtained from 20 vs. 50 and 100 were included in the weight table of two hundred and fifty-three variables. Similarly, 188 variables with VIP > 1 corresponded to those in the weight table, which contained an additional 26 variables with VIP < 1 from 100 vs. 20 and 50. From the above, we can see that the overlap ratio of variables obtained between VIP and weight ranking methods was 90.1% (i.e., 228 × 100%/253) from 20 vs. 50 and 100, while this ratio fell to 87.8% (i.e., 188 × 100%/214) from 100 vs. 20 and 50, indicating that the variables sought by the two methods were of high consistency, which laid a solid foundation to accurately seek eligible variables to represent marker compounds. One hundred and fifty-four overlapped variables from the two weight lists mentioned above were obtained for univariate analysis to further confirm the validity of those variables.

2.3. *Univariate Analysis*

Pairwise *t*-tests [47–49] and fold changes of concentration [27,50] have proved to be appropriate for univariate analysis of the metabolomics data. A pairwise *t*-test focuses on the significant differences of variables in peak intensity between two concentration groups, while a fold change of concentration considers the peak intensity ratio of variables between the high and low concentration groups. In general, the variables with a significance level (p) of pairwise *t*-tests below 0.05 and fold change (FC) of concentration above 2 are deemed eligible. In this study, p values of 154 variables between 20 and 50, 20 and 100, as well as 50 and 100 ng/mL concentration groups, i.e., $p_{20vs.50}$, $p_{20vs.100}$, and $p_{50vs.100}$, were calculated to be lower than 0.05 in a pairwise *t*-test (Table S2), indicating that the significant inter-group differences of 154 variables indeed existed. In addition, the fold changes of concentration between 50 and 20 ($FC_{50vs.20}$), as well as 100 and 20 ng/mL groups ($FC_{100vs.20}$), were also calculated to be 2.333~2.671 and 4.555~5.332 (Table S3), which were above 2 to further support the eligibility of 154 variables as marker compound candidates.

As shown in Table 1, 120 out of 154 variables were confirmed as marker compounds (124 in total) by the weight ranking method, with a screening rate of over 96% (i.e., 120 × 100%/124), while the VIP method can only find 109 marker compounds with a screening rate of about 88% (i.e., 109 × 100%/124), showing that the weight ranking method expressed better performance in selecting eligible variables to represent marker compounds than the VIP method. Fleroxacin, lincomycin, clindamycin, and mebendazole, as four veterinary drugs, have no variables equal to their identities as marker compounds in weight and VIP lists. To be different with these four contaminants, three pesticides (i.e., pyridaben, hexazinone, and phosmet) and another eight veterinary drugs, including sulfaphenazole, sulfamethoxazole, sulfamethoxypyridazine, lomefloxacin, fenthion, danofloxacin, sparfloxacin, and tilmicosin, played the role of unique marker compounds, which corresponded to the variables with VIP < 1 from 20 vs. 50 and 100, but were also included in the weight lists (Table 1). Thirty-four in one hundred and fifty-four variables not identified as marker compounds in weight lists showed more complicated and erratic

LC–MS/MS and metabolomics information, which needed more in-depth studies to discern their origins from a matrix or foreign impurity.

Table 1. Marker compounds were screened in maize sample groups.

No.	Var ID (Primary)	Marker Compounds	VIP Pred [a]	(Weight Squared Value × 10^5) [b]	m/z [c]	t (min) [d]	Mass Error (ppm) [e]	LOD (μg/kg)
1	M406T575	Difenoconazole	7.378/12.942	56,490/23,814	406.07130	9.58	−1.654	0.5
2	M326T524	Benalaxyl	5.798/5.975	9548/20,867	326.17516	8.73	0.279	1.1
3	M336T541	Zoxamide	4.658/5.732	11,861/11,744	336.03283	9.02	2.657	0.7
4	M218T216	Pymetrozine	7.828/5.649	25,040/13,579	218.10371	3.60	0.337	0.7
5	M372T574	Profenofos	6.189/5.531	26,626/4524	372.94268	9.57	0.699	0.9
6	M353T444	Tebufenozide	4.262/4.761	3052/4239	353.22383	7.40	4.178	1.0
7	M365T664	Pyridaben	0.514/4.492	73/3656	365.14408	11.07	−2.218	0.9
8	M256T210	Trichlorfon	2.470/4.263	689/3233	256.92941	3.50	−1.753	0.5
9	M218T300	Propanil	2.321/4.243	599/2611	218.01315	5.00	−1.160	1.3
10	M220T226	Dichlorvos	4.161/4.225	2493/2742	220.95367	3.77	2.220	0.9
11	M304T410	Fenpropimorph	3.241/4.191	1248/2663	304.26361	6.83	0.386	1.5
12	M223T241	Acetamiprid	4.561/4.170	3386/2965	223.07489	4.02	1.761	0.5
13	M349T621	Clorpyrifos	3.678/4.141	1708/2579	349.93393	10.35	1.049	0.5
14	M192T245	Carbendazim	2.651/4.094	746/2573	192.07637	4.08	−1.957	1.1
15	M732T421	Spinosad	4.227/4.059	2608/2462	732.46855	7.02	0.588	1.0
16	M294T344	Paclobutrazol	2.783/4.041	815/2289	294.13689	5.73	0.420	0.9
17	M345T650	Oxadiazon	3.066/3.997	1010/2326	345.07613	10.83	−1.724	1.0
18	M279T211	Oxadixyl	3.629/3.980	1621/2292	279.13363	3.52	−1.063	0.9
19	M276T326	Dimethenamid	3.496/3.968	1471/2338	276.08142	5.43	−1.904	0.8
20	M282T660	Pendimethalin	4.213/3.873	2396/2265	282.14480	11.00	−0.092	0.7
21	M305T541	Diazinon	4.289/3.844	2528/2250	305.10830	9.02	−0.114	0.4
22	M302T228	Flutriafol	4.602/3.813	3265/2173	302.11046	3.80	1.695	0.4
23	M330T429	Epoxiconazol	2.132/3.739	394/2123	330.08077	7.15	1.141	0.4
24	M324T499	Flutolanil	5.143/3.569	4610/1575	324.11992	8.32	−2.068	0.3
25	M292T384	Cyproconazol	3.236/3.488	1016/1409	292.12131	6.40	0.645	0.5
26	M253T239	Hexazinone	0.434/3.446	44/1392	253.16520	3.98	−2.774	0.8
27	M319T370	Pyriftalid	2.171/3.416	347/1241	319.07443	6.17	−0.885	0.9
28	M321T561	Chlorpyrifos-methyl	3.301/3.349	1028/1237	321.90281	9.35	1.702	1.1
29	M343T308	Thiophanate-methyl	1.896/3.345	249/1184	343.05204	5.13	−2.599	1.1
30	M306T540	Pirimiphos-methyl	7.863/3.343	24,423/1100	306.10351	9.00	−0.217	0.6
31	M888T502	Emamectin benzoate	1.008/3.257	106/1067	888.54603	8.37	−0.830	1.1
32	M337T436	Fenbuconazole	1.937/3.235	245/1010	337.12103	7.27	−1.260	0.9
33	M289T354	Myclobutanil	2.280/3.197	347/895	289.12143	5.90	−0.082	0.7
34	M224T228	Monocrotophos	4.047/3.145	1844/885	224.06760	3.80	−2.848	0.8
35	M222T196	Carbofuran	2.358/3.126	362/856	222.11230	3.27	−0.773	0.5
36	M304T200	Fenamiphos	1.974/3.103	237/790	304.11316	3.33	0.267	0.9
37	M299T446	Quinalphos	4.042/3.029	1806/729	299.06183	7.43	1.514	1.2
38	M338T555	Bitertanol	2.887/2.982	588/752	338.18671	9.25	1.223	0.9
39	M300T306	Phosphamidon	1.897/2.956	199/789	300.07668	5.10	1.536	0.8
40	M256T277	Phosfolan	1.855/2.946	188/728	256.02283	4.62	1.078	1.1
41	M271T540	Cadusafos	3.111/2.930	729/640	271.09408	9.00	−3.356	1.1
42	M376T359	Prochloraz	1.725/2.915	158/619	376.03841	5.98	0.850	0.5
43	M226T450	Cyprodinil	3.535/2.786	1080/608	226.13315	7.50	−3.196	0.5
44	M243T411	Mocap	1.673/2.746	132/591	243.06367	6.85	−0.080	0.9
45	M215T432	Metribuzin	2.828/2.720	507/396	215.09661	7.20	2.312	0.9
46	M436T523	Fipronil	4.356/2.698	2316/545	436.94589	8.72	−0.199	0.9
47	M250T234	Clothianidin	1.491/2.630	95/531	250.01693	3.90	3.708	0.9
48	M253T261	Thiacloprid	1.262/2.621	68/460	253.03037	4.35	−2.158	0.5
49	M292T214	Thiamethoxam	1.650/2.596	116/582	292.02655	3.57	−0.066	0.4
50	M294T360	Triadimefon	3.591/2.572	1107/393	294.10089	6.00	1.746	0.5
51	M302T266	Methidathion	3.877/2.538	1467/386	302.96913	4.43	−0.049	0.9
52	M368T509	Anilofos	1.299/2.491	65/377	368.03063	8.48	0.281	0.9
53	M299T346	Phoxim	1.463/2.430	79/367	299.06142	5.77	0.149	1.0

Table 1. *Cont.*

No.	Var ID (Primary)	Marker Compounds	VIP Pred [a]	(Weight Squared Value × 10^5) [b]	m/z [c]	t (min) [d]	Mass Error (ppm) [e]	LOD (μg/kg)
54	M318T521	Phosmet	0.495/2.374	31/326	318.00180	8.68	−0.050	0.6
55	M293T309	Etrimfos	2.431/2.362	279/281	293.07130	5.15	−2.200	0.8
56	M284T503	Metolachlor	1.320/2.289	57/312	284.14146	8.38	0.993	0.7
57	M330T488	Iprodione	1.610/2.279	88/299	330.04077	8.13	0.293	0.6
58	M216T346	Atrazine	2.591/2.276	334/284	216.10192	5.77	4.024	0.5
59	M307T499	Sulfotep	2.147/2.267	186/321	307.05231	8.32	−1.796	0.6
60	M208T297	Fenobucarb	1.281/2.247	51/261	208.13320	4.95	−0.059	1.0
61	M256T248	Imidacloprid	3.372/2.230	845/260	256.05943	4.13	−0.595	1.0
62	M203T178	Dinotefuran	2.636/2.201	345/241	203.11381	2.97	−0.305	1.1
63	M249T290	Linuron	2.276/2.189	521/222	249.01904	4.83	−0.689	0.8
64	M242T356	Prometryn	1.801/2.173	108/198	242.14414	5.93	3.079	0.5
65	M230T309	Terbutylazine	2.165/2.136	178/220	230.11703	5.15	1.443	0.8
66	M329T323	Pencycuron	1.550/2.124	71/221	329.14203	5.38	1.535	0.7
67	M318T276	Azinphos-methyl	1.632/1.675	60/94	318.01430	4.60	3.916	0.3
68	M279T490	Fenthion	0.372/1.438	8/54	279.02730	8.17	−0.006	0.9
69	M214T317	Simetryn	3.354/2.111	815/190	214.11264	5.28	2.581	0.5
70	M313T320	Praziquantel	1.624/2.045	76/183	313.19130	5.33	0.793	1.0
71	M479T347	Chlortetracycline	3.939/2.020	1530/182	479.12216	5.78	1.234	1.1
72	M463T377	Tetracycline	2.347/2.005	221/176	463.17183	6.28	1.560	0.3
73	M445T384	Doxycycline	3.592/1.990	1061/167	445.16071	6.40	0.390	0.5
74	M277T489	Sulfabenzamide	2.596/1.981	310/162	277.06427	8.15	0.462	0.4
75	M275T541	Ormetoprim	1.289/1.980	41/159	275.15083	9.02	2.093	0.6
76	M281T212	Sulfamonomethoxine	2.221/1.979	181/147	281.07014	3.53	−0.519	0.9
77	M315T378	Sulfaphenazole	0.518/1.908	28/144	315.09129	6.30	0.870	0.9
78	M215T306	Sulfacetamide	1.919/1.893	112/125	215.04915	5.10	3.056	0.9
79	M281T361	Sulfameter	2.102/1.822	142/112	281.07037	6.02	0.274	0.8
80	M279T225	Sulfamethazine	2.008/1.818	123/111	279.09161	3.75	2.104	0.9
81	M254T427	Sulfamethoxazole	1.564/1.815	58/118	254.05989	7.12	1.979	1.0
82	M265T308	Sulfamerazine	1.827/1.813	91/106	265.07593	5.13	2.102	0.8
83	M254T429	Sulfamethoxazole	0.616/1.811	16/105	254.05937	7.15	−0.063	0.8
84	M311T666	Sulfamethazine	1.671/1.800	71/103	311.08055	11.10	−0.962	0.8
85	M301T301	Sulfaquinoxaline	1.493/1.798	51/102	301.07589	5.02	1.739	1.0
86	M285T412	sulfachloropyridazine	1.001/1.758	19/95	285.02113	6.87	1.316	0.8
87	M215T293	Sulfaguanidine	1.926/1.734	103/97	215.05963	4.88	−0.403	0.9
88	M251T288	Sulfadiazine	2.213/1.692	164/96	251.05942	4.80	−1.177	0.4
89	M256T275	Sulfathiazole	2.800/1.687	396/99	256.02180	4.58	3.531	0.4
90	M279T377	Sulfisomidine	2.312/1.671	192/77	279.09146	6.28	1.584	0.4
91	M268T208	Sulfafurazole	1.659/1.668	64/87	268.07577	3.47	2.711	1.0
92	M291T365	Trimethoprim	1.707/1.652	70/77	291.14592	6.08	2.575	0.9
93	M281T380	Sulfamethoxypyri-dazine	0.735/1.636	26/89	281.07031	6.33	0.066	0.8
94	M360T414	Enrofloxacin	3.164/1.616	637/75	360.17198	6.90	0.491	0.8
95	M320T382	Norfloxacin	1.198/1.601	24/75	320.14194	6.37	4.507	0.7
96	M334T383	Pefloxacin	2.511/1.588	260/73	334.15581	6.38	−1.023	0.9
97	M332T404	Ciprofloxacin	1.473/1.573	37/70	332.14104	6.73	1.621	0.7
98	M362T385	Ofloxacin	1.266/1.535	27/68	362.15151	6.42	1.252	0.9
99	M386T456	Sarafloxacin	1.148/1.507	47/67	386.13203	7.60	2.492	0.4
100	M352T427	Lomefloxacin	0.728/1.451	7/62	352.14603	7.12	−1.973	0.3
101	M233T608	Nalidixic acid	1.036/1.449	15/56	233.09243	10.13	1.528	0.3
102	M262T619	Flumequine	2.101/1.441	128/55	262.08760	10.32	0.770	0.5
103	M358T411	Danofloxacin	0.814/1.415	7/52	358.15716	6.85	2.823	0.8
104	M400T442	Difloxacin	1.830/1.414	77/50	400.14683	7.37	0.282	0.8
105	M396T433	Orbifloxacin	1.500/1.395	38/48	396.15371	7.22	1.927	1.0
106	M393T494	Sparfloxacin	0.671/1.393	10/48	393.17268	8.23	−1.498	0.9
107	M734T616	Erythromycin	1.106/1.365	16/46	734.46808	10.27	−0.599	0.9

Table 1. *Cont.*

No.	Var ID (Primary)	Marker Compounds	VIP Pred [a]	(Weight Squared Value × 10^5) [b]	*m/z* [c]	*t* (min) [d]	Mass Error (ppm) [e]	LOD (μg/kg)
108	M837T659	Roxithromycin	1.236/1.354	23/38	837.53167	10.98	−0.214	1.0
109	M828T630	Josamycin	1.349/1.351	68/36	828.47461	10.50	0.733	0.8
110	M916T611	Tylosin	1.039/1.348	13/36	916.52689	10.18	0.505	0.9
111	M702T622	Kitasamycin	1.758/1.345	65/36	702.40593	10.37	0.010	0.8
112	M869T566	Tilmicosin	0.491/1.343	7/35	869.57374	9.43	0.484	1.0
113	M128T254	2-methyl-5-nitroimidazole	2.095/1.336	269/35	128.04551	4.23	0.444	0.9
114	M114T265	4-nitroimidazole	1.036/1.333	14/32	114.02989	4.42	0.821	1.0
115	M164T246	5-nitrobenzimidazole	1.122/1.319	16/32	164.04613	4.10	4.115	1.3
116	M142T207	Dimetridazole	1.492/1.311	36/32	142.06163	3.45	3.755	0.5
117	M172T218	Metronidazole	2.251/1.281	164/31	172.07180	3.63	0.779	1.1
118	M201T240	Ronidazole	1.356/1.280	26/31	201.06230	4.00	2.315	0.4
119	M158T329	Hydroxy dimetridazole	2.126/1.262	131/31	158.05646	5.48	2.797	0.4
120	M170T362	Ipronidazole	1.305/1.258	23/31	170.09277	6.03	2.156	0.9

Note: [a] two VIP values from 20 vs. 50 and 100 and 100 vs. 20 and 50, respectively; [b] two weight squared values (×10^5) from 20 vs. 50 and 100 and 100 vs. 20 and 50, respectively; [c] *m/z* represents extracted molecular weight from W4M platform; [d] *t* represents retention time from W4M platform; [e] Mass error (ppm) = (extracted molecular weight from W4M platform − extracted molecular weight from LC–MS/MS) × 10^6/extracted molecular weight from LC–MS/MS.

A blank maize sample was also investigated by the SVM-assisted metabolomics method with no corresponding variables of 124 P&VDs found, eliminating the inherent (rather than spiked) interference of these contaminants in maize to screen marker compounds and the existence of false positives.

2.4. Limits of Detection

According to the method proposed by the US Environmental Protection Agency to calculate the limits of detection (LODs) of 120 P&VDs [51], we first prepared a 2.0 g blank maize sample for the same pretreatment as introduced above to obtain a 1 mL extract solution. In this extract, we spiked a mixed solution (1 μg/mL, 20 μL) containing 120 P&VDs, resulting in a final concentration of 20 ng/mL. This process was repeated to gain seven replicates, which were also subjected to metabolomics analysis to obtain peak intensities of 120 P&VDs. For each P&VD, the concentration level of 20 ng/mL corresponds to the peak intensity average value of seven replicates. Therefore, the concentration (ng/mL) of each P&VD can be calculated by the formula for the respective peak intensity × 20/average peak intensity. As shown in Table 1, the LOD results for 120 P&VDs varied between 0.3 and 1.5 μg/kg.

The metabolomics method has successfully been applied to non-targeted contaminant screening in plant-derived foods such as tea [27], lettuce [29], maize [29], and orange juice [33]. In contrast to these studies, which only focused on pesticides or veterinary drugs, our study expands the domain of target compounds to simultaneously screen multi-class P&VDs. This is the first time to prove the introduction of SVM as an effective tool for promoting the screening rate of the metabolomics approach. In addition, method LODs were calculated to be significantly below 10 μg/kg for 120 P&VDs, which even have the upper hand over some targeted multi-residue methods [20–22].

2.5. Practicability Test

For plant-derived foods, the contamination pathways of pesticides are almost explicit; however, the contamination sources of veterinary drugs have not been fully understood. As reported in previous studies [52–54], the application of fertilizer and irrigation water contaminated with veterinary drugs onto farmland is a primary reason to cause the contamination of plant-derived foods; we hypothesized that some plant-derived foods from rural areas exposed to the abuse of veterinary drugs in livestock farming may suffer higher contamination risks. To investigate the practicability of our proposed SVM-assisted metabolomics screening method, we adopted maize samples from Qili and Bali villages as a case study. These two villages in Jinpu New Area, affiliated with Dalian City, are not only crucial maize suppliers but also pivotal providers of fresh livestock and poultry meat, which makes the maize samples highly susceptible to contamination by veterinary drugs.

Maize samples were collected after the maturity period in late September 2023. After metabolomics analysis, four contaminants, including norfloxacin, enrofloxacin, imidacloprid, and carbendazim, were all detected in two villages, with their concentrations being 10.7~17.7 μg/kg (Table S4). The abundance of norfloxacin and enrofloxacin in maize samples can be attributed to their extensive use as medicines and feed additives in animal husbandry [55], while the residual imidacloprid and carbendazim in maize samples with relatively high concentrations can be due to their widespread application to combat aphid [56] and bacterial wilt [56], respectively. As indicated in Table S4, the contamination of pesticides remained predominant over that of veterinary drugs, and imidacloprid presented the highest concentration among the four contaminants in both villages. In addition, Qili village suffered more serious overall contamination from four contaminants than Bali village. The test results supported the practicability of our proposed SVM-assisted metabolomics method in non-targeted screening of contaminants. It is worth noting that all four contaminants identified by the method presented a concentration above 10 μg/kg, which was understandable to show the feasibility of the method. But for those P&VDs with concentrations below 10 μg/kg, it is still unclear how to prove the applicability of the method, which probably led to those contaminants being unable to be detected. That is to say, the actual contamination of maize by P&VDs in the Jinpu New Area can be even worse than what we expected, and it deserves more attention to defuse the underlying risks. Food safety has no borders. In view of some food issues that initially occurred in local areas on a small scale, it is still essential for any country and international organization to come to notice and to strengthen cooperation in the prevention and control of the issues, such as through information sharing, experience exchange, and even the publication of standardized documents to ensure food safety from the source.

3. Materials and Methods

3.1. Chemicals and Materials

Maize samples were purchased from a local farmer's market in Dalian City, China. Methanol and acetonitrile (HPLC grade) were purchased from Merck Corporation (Darmstadt, Germany). Formic acid and acetic acid (HPLC grade) were provided by China Shanghai ANPEL Laboratory Technologies Inc. (Shanghai, China). Ammonium formate, anhydrous sodium sulfate (Na_2SO_4), anhydrous sodium acetate (NaAC), and primary secondary amine (PSA) sorbent were obtained from China Sinopharm Chemical Reagent Co., Ltd. (Shanghai, China). Filter membrane (0.22 μm) was purchased from Agilent Technologies (Santa Clara, CA, USA). Ultrapure water was produced by the Milli-Q ultrapure water system from Merck Corporation (Darmstadt, Germany). Enrofloxacin-d5 and atrazine-d5 (100 μg/mL in methanol) for recovery calibration were provided by First Standard (Worcester, MA, USA). Sixty-nine pesticides and fifty-five veterinary drugs were purchased from the corporations First Standard (Worcester, MA, USA), Sigma (St. Louis, MO, USA), Dr. Ehrenstorfer (Wesel, Germany), and TRC (Montréal, QC, Canada), with a purity greater than 98%. The detailed information on these P&VDs is shown in Table 2.

Table 2. Basic information on 69 pesticides (No. **1~69**) and 55 veterinary drugs (No. **70~124**).

No.	Compounds	Category	CAS No.	Extracted Molecular Weight	Retention Time (min)
1	Dinotefuran	insecticide	165252-70-0	203.11387	3.06
2	Carbofuran	insecticide	1563-66-2	222.11247	3.21

Table 2. *Cont.*

No.	Compounds	Category	CAS No.	Extracted Molecular Weight	Retention Time (min)
3	Fenamiphos	insecticide	22224-92-6	304.11308	3.44
4	Trichlorfon	insecticide	52-68-6	256.92986	3.47
5	Thiamethoxam	insecticide	153719-23-4	292.02657	3.54
6	Pymetrozine	insecticide	123312-89-0	218.10364	3.61
7	Dichlorvos	insecticide	62-73-7	220.95318	3.78
8	Monocrotophos	insecticide	2157-98-4	224.06824	3.81
9	Clothianidin	insecticide	210880-92-5	250.01600	3.95
10	Acetamiprid	insecticide	135410-20-7	223.07450	4.03
11	Imidacloprid	insecticide	105827-78-9	256.05958	4.09
12	Thiacloprid	insecticide	111988-49-9	253.03092	4.33
13	Methidathion	insecticide	950-37-8	302.96914	4.37
14	Phosfolan	insecticide	947-02-4	256.02255	4.54
15	Azinphos-methyl	insecticide	86-50-0	318.01305	4.59
16	Fenobucarb	insecticide	3766-81-2	208.13321	4.96
17	Phosphamidon	insecticide	13171-21-6	300.07622	5.05
18	Etrimfos	insecticide	38260-54-7	293.07194	5.16
19	Phoxim	insecticide	14816-18-3	299.06138	5.71
20	Mocap	insecticide	13194-48-4	243.06369	6.89
21	Spinosad	insecticide	131929-60-7	732.46812	7.15
22	Tebufenozide	insecticide	112410-23-8	353.22235	7.35
23	Quinalphos	insecticide	13593-03-8	299.06138	7.41
24	Fenthion	insecticide	55-38-9	279.02730	8.07
25	Sulfotep	insecticide	3689-24-5	307.05286	8.26
26	Emamectin benzoate	insecticide	155569-91-8	888.54677	8.54
27	Phosmet	insecticide	732-11-6	318.00182	8.57
28	Fipronil	insecticide	120068-37-3	436.94598	8.73
29	Cadusafos	insecticide	95465-99-9	271.09499	8.95
30	Diazinon	insecticide	333-41-5	305.10833	8.96
31	Pirimiphos-methyl	insecticide	29232-93-7	306.10358	8.96
32	Chlorpyrifos-methyl	insecticide	5598-13-0	321.90226	9.32
33	Profenofos	insecticide	41198-08-7	372.94242	9.49
34	Clorpyrifos	insecticide	2921-88-2	349.93356	10.26
35	Pyridaben	insecticide	96489-71-3	365.14489	11.06
36	Oxadixyl	bactericide	77732-09-3	279.13393	3.49
37	Flutriafol	bactericide	76674-21-0	302.10995	3.79
38	Carbendazim	bactericide	10605-21-7	192.07675	4.11
39	Thiophanate-methyl	bactericide	23564-05-8	343.05293	5.14
40	Pencycuron	bactericide	66063-05-6	329.14152	5.47
41	Prochloraz	bactericide	67747-09-5	376.03809	5.89
42	Myclobutanil	bactericide	88671-89-0	289.12145	6.06
43	Triadimefon	bactericide	43121-43-3	294.10038	6.11
44	Cyproconazol	bactericide	113096-99-4	292.12112	6.36
45	Fenpropimorph	bactericide	67306-03-0	304.26349	6.72
46	Epoxiconazol	bactericide	106325-08-0	330.08039	7.11
47	Fenbuconazole	bactericide	114369-43-6	337.12145	7.33
48	Cyprodinil	bactericide	121552-61-2	226.13387	7.64
49	Iprodione	bactericide	36734-19-7	330.04067	8.13
50	Flutolanil	bactericide	66332-96-5	324.12059	8.23
51	Benalaxyl	bactericide	71626-11-4	326.17507	8.75
52	Zoxamide	bactericide	156052-68-5	336.03194	8.99
53	Bitertanol	bactericide	55179-31-2	338.18630	9.17
54	Difenoconazole	bactericide	119446-68-3	406.07197	9.64
55	Hexazinone	herbicide	51235-04-2	253.16590	3.94
56	Linuron	herbicide	330-55-2	249.01921	4.92
57	Propanil	herbicide	709-98-8	218.01340	4.98
58	Terbutylazine	herbicide	5915-41-3	230.11670	5.07

Table 2. *Cont.*

No.	Compounds	Category	CAS No.	Extracted Molecular Weight	Retention Time (min)
59	Simetryn	herbicide	1014-70-6	214.11209	5.32
60	Dimethenamid	herbicide	87674-68-8	276.08195	5.41
61	Atrazine	herbicide	102029-43-6	216.10105	5.77
62	Prometryn	herbicide	7287-19-6	242.14339	5.97
63	Pyriftalid	herbicide	135186-78-6	319.07471	6.02
64	Metribuzin	herbicide	21087-64-9	215.09611	7.14
65	Metolachlor	herbicide	51218-45-2	284.14118	8.34
66	Anilofos	herbicide	64249-01-0	368.03053	8.49
67	Oxadiazon	herbicide	19666-30-9	345.07672	10.77
68	Pendimethalin	herbicide	40487-42-1	282.14483	10.92
69	Paclobutrazol	growth regulator	76738-62-0	294.13677	5.77
70	Sulfafurazole	sulfamido	127-69-5	268.07504	3.38
71	Sulfamonomethoxine	sulfamido	1220-83-3	281.07029	3.54
72	Sulfamethazine	sulfamido	57-68-1	279.09102	3.76
73	Sulfathiazole	sulfamido	72-14-0	256.02090	4.54
74	Sulfadiazine	sulfamido	68-35-9	251.05972	4.73
75	Sulfaguanidine	sulfamido	57-67-0	215.05972	4.85
76	Sulfaquinoxaline	sulfamido	59-40-5	301.07537	5.05
77	Sulfacetamide	sulfamido	144-80-9	215.04849	5.06
78	Sulfamerazine	sulfamido	127-79-7	265.07537	5.11
79	Sulfameter	sulfamido	651-06-9	281.07029	5.95
80	Trimethoprim	sulfamido	738-70-5	291.14517	6.01
81	Sulfaphenazole	sulfamido	526-08-9	315.09102	6.21
82	Sulfisomidine	sulfamido	515-64-0	279.09102	6.24
83	Sulfamethoxypyridazine	sulfamido	80-35-3	281.07029	6.51
84	Sulfachloropyridazine	sulfamido	80-32-0	285.02075	6.87
85	Sulfamethoxazole	sulfamido	723-46-6	254.05939	7.09
86	Sulfamethoxazole	sulfamido	723-46-6	254.05939	7.09
87	Sulfabenzamide	sulfamido	127-71-9	277.06414	8.05
88	Ormetoprim	sulfamido	6981-18-6	275.15025	8.94
89	Sulfamethazine	sulfamido	57-68-1	311.08085	11.05
90	Dimetridazole	nitroimidazoles	551-92-8	142.06110	3.46
91	Metronidazole	nitroimidazoles	443-48-1	172.07167	3.58
92	Ronidazole	nitroimidazoles	7681-76-7	201.06183	3.96
93	5-nitrobenzimidazole	nitroimidazoles	94-52-0	164.04545	4.05
94	2-methyl-5-nitroimidazole	nitroimidazoles	88054-22-2	128.04545	4.17
95	4-nitroimidazole	nitroimidazoles	3034-38-6	114.02980	4.41
96	Hydroxy dimetridazole	nitroimidazoles	936-05-0	158.05602	5.55
97	Ipronidazole	nitroimidazoles	14885-29-1	170.09240	6.05
98	Mebendazole	nitroimidazoles	31431-39-7	296.10297	7.74
99	Fleroxacin	quinolones	79660-72-3	370.13730	5.99
100	Ofloxacin	quinolones	82419-36-1	362.15106	6.39
101	Pefloxacin	quinolones	70458-92-3	334.15615	6.41
102	Norfloxacin	quinolones	70458-96-7	320.14050	6.51
103	Ciprofloxacin	quinolones	85721-33-1	332.14050	6.73
104	Enrofloxacin	quinolones	93106-60-6	360.17180	7.01
105	Danofloxacin	quinolones	112398-08-0	358.15615	7.02
106	Lomefloxacin	quinolones	98079-51-7	352.14672	7.12
107	Orbifloxacin	quinolones	113617-63-3	396.15295	7.27
108	Difloxacin	quinolones	98106-17-3	400.14672	7.42
109	Sarafloxacin	quinolones	98105-99-8	386.13107	7.64
110	Sparfloxacin	quinolones	110871-86-8	393.17327	8.34

Table 2. *Cont.*

No.	Compounds	Category	CAS No.	Extracted Molecular Weight	Retention Time (min)
111	Nalidixic acid	quinolones	389-08-2	233.09207	10.01
112	Flumequine	quinolones	42835-25-6	262.08740	10.15
113	Praziquantel	vermifuge	55268-74-1	313.19105	5.31
114	Chlortetracycline	tetracyclines	57-62-5	479.12157	5.68
115	Tetracycline	tetracyclines	60-54-8	463.17111	6.21
116	Doxycycline	tetracyclines	564-25-0	445.16054	6.34
117	Lincomycin	macrolides	154-21-2	407.22104	9.41
118	Tilmicosin	macrolides	108050-54-0	869.57332	9.49
119	Clindamycin	macrolides	18323-44-9	425.18715	9.68
120	Tylosin	macrolides	1401-69-0	916.52643	10.23
121	Erythromycin	macrolides	114-07-8	734.46852	10.25
122	Kitasamycin	macrolides	1392-21-8	702.40592	10.48
123	Josamycin	macrolides	16846-24-5	828.47400	10.65
124	Roxithromycin	macrolides	80214-83-1	837.53185	10.84
125	Atrazine-d5		163165-75-1	221.14017	5.77
126	Enrofloxacin-d5		1173021-92-5	365.21092	7.01

3.2. Solution Preparation

A total of 124 P&VDs was separately prepared in methanol with a concentration of 100 μg/mL as the stock solution, 1 mL of which was taken, mixed together, and diluted with methanol to prepare a 1 μg/mL working solution. A 100 ng/mL mixed working solution of enrofloxacin-d5 and atrazine-d5 was prepared by diluting their 100 μg/mL stock solution with methanol.

3.3. Sample Preparation and Pretreatment Process

(a) The maize sample was ground into powder by a grinder; (b) 2.0, 5.0, and 10.0 g of maize powder, together with corresponding 20, 50, and 100 μL of 124 P&VDs mixed solutions (1 μg/mL), were poured into 50 mL polypropylene centrifuge tubes, respectively. To calibrate the recovery during the sample pretreatment process, a mixed solution (0.5 mL, 100 ng/mL) of enrofloxacin-d5 and atrazine-d5 was further added as recovery internal standards; (c) a total of 20 mL of acetonitrile/water (80/20, *v*/*v*) with 1% acetic acid was dumped into the tube by vortex and ultrasonic extraction for 1 and 30 min, respectively. A total of 6.0 g anhydrous Na_2SO_4 and 1.5 g NaAC were added, shocking violently for 1 min and centrifuging at 6000 r/min for 5 min; (d) all the supernatant solutions were taken into a new tube containing 2.0 g anhydrous Na_2SO_4 and 0.3 g PSA, centrifuging at 6000 r/min for 2 min; (e) all the solutions were extracted, dried with nitrogen gas flow, and redissolved with 1 mL 40% (*v*/*v*) methanol–4 mmol/L ammonium formate buffer solution, vortexing for 1 min; (f) filtered with a 0.22 μm filter membrane, the sample solutions of 124 P&VDs at the theoretical concentrations of 20, 50, and 100 ng/mL were prepared. Each concentration experiment was completed in nonuplicate to obtain replicate samples.

3.4. Sample Grouping and Naming

Twenty-seven samples named 20 ng/mL−1~20 ng/mL−9, 50 ng/mL−1~50 ng/mL−9, and 100 ng/mL−1~100 ng/mL−9 were obtained from three concentration groups. A quality control (QC) sample was prepared by mixing 30 μL of each sample from the above-mentioned concentration groups [27,57,58] and underwent three repeated injections into the LC–MS/MS system before and after the injection of each concentration group. Twelve injections of QC samples were named after QC-1, QC-2, . . ., and QC-12 to evaluate the stability of LC–MS/MS.

3.5. Analytical Method

A quadrupole/electrostatic field orbitrap LC–MS/MS system (Q Exactive Plus, Thermo Fisher Scientific Inc., Waltham, MA, USA) equipped with a heatable electrospray ion (ESI) source was used to analyze all pesticides, veterinary drugs, enrofloxacin-d5, and atrazine-d5. The ESI-positive mode was adopted to obtain M+H adducts for metabolomics analysis. An Accucore RP-MS column (100 mm $\times$ 2.1 mm, 2.6 μm particle diameter) purchased from Thermo Fisher Corporation (Waltham, MA, USA) was employed to separate components after an injection of 10 μL sample solution. The mobile phases were composed of 0.1% (v/v) formic acid and ammonium formate at 4 mmol/L in water (eluent A) and 0.1% (v/v) formic acid and ammonium formate at 4 mmol/L in methanol (eluent B). The flow rate was kept at 0.3 mL/min. The elution program was as follows: 100% A (0 min), 100% A (2 min), 0% A (8 min), 0% A (13 min), and 100% A (14 min) until the end of the run. The oven temperature was kept at 40 °C. The Q Exactive Plus parameter settings include (a) heating and capillary temperature at 320 °C; (b) sheath and auxiliary gas from N_2, with flow rates of 40 and 10 arb, respectively; (c) lens and spray voltage at 50 and 3200 V, respectively; (d) scan mode to be full-scan/data-dependent two-stage scanning; (e) MS parameters to be full-scan resolution at 70,000, AGC target at 1×10^6, maximum dwell time at 100 ms, and scan range of m/z between 100 and 1000; (f) MS/MS parameters to be resolution at 17,500, AGC target at 2×10^5, and maximum dwell time at 50 ms.

Trace Finder software (Version 3.3) installed on the LC–MS/MS system was employed for residue analysis, with specific screening conditions as follows: (a) for primary parent ions, the response intensity threshold at 10,000, the ratio of signal to noise at 5.0, and mass error at 5 ppm; (b) for secondary fragment ions, the minimum number of matching ions at 1, the response intensity threshold at 10,000, and mass error at 5 ppm. The quantification of enrofloxacin-d5 and atrazine-d5 was finished by standard curves on the basis of the peak areas of primary parent ions for recovery calibration.

3.6. Metabolomics Data Processing

LC–MS/MS output files cannot be directly employed for metabolomics analysis and need to be converted from .RAW format to .mzXML format [59]. The new-formatted files were uploaded onto the Workflow4Metabolomics (W4M) platform (https://workflow4metabolomics.usegalaxy.fr/, accessed on 19 October 2022) for further analysis [60]. Chromatographic peak detection, alignment, and retention time calibration were performed on the platform, with further operations including normalization, centralizing, scaling, and data transformation of peak intensity to acquire the data matrix, in which variable and sample names were designated abscissa and ordinate, respectively [60,61]. PCA [62–64] and OPLS-DA [65,66] embedded in SIMCA software (Version 14.1) [67], together with cluster analysis in Heml software (Version 1.0.3.7), were performed for multivariate analysis of the data matrix. Over-fitting of the OPLS-DA model was evaluated by permutation tests [67,68]. Variable importance in projection (VIP) > 1 is the threshold to screen variables as marker compound candidates [67–69]. Under this principle, eligible variables from three concentration groups were included in the VIP lists.

SVM can perform nonlinear classification by mapping the input data into a high-dimensional feature space, the process of which is known to be the kernel trick [70]. Many kernel functions are available, but the most commonly used one is the radial basis function (RBF). Therefore, we chose SVMRFE under RBF to run the program in MATLAB software (Version R2019b). A 10-fold cross-validation was performed by randomly dividing the dataset into 10 subsets and computing the mean value of 10 accuracy values. This process was repeated 10 times, and the mean accuracy was obtained as the final classification accuracy. Kernel parameters were chosen through a 3-fold cross-validation approach, which divided the training dataset randomly into three subsets. The classifier was trained on either of the two subsets and tested on the third one. A set of parameters that provided the best cross-validation accuracy was employed for further analysis using the LIBSVM interface [71]. The kernel parameters with the best cross-validation accuracy were chosen to perform the actual classification. Under the conditions described above, weight values (w) and weight squared values (w^2) of variables were obtained, with the latter ones ranging from large to small to form a variable weight table for further analysis. The MATLAB codes used to generate the results were provided in the Supplementary Materials.

All VIP > 1 variables filled their corresponding vacancies in the variable weight table. For those variables, their sequence in the weight table was sequential, implying that two methods, including VIP and weight ranking, to seek eligible variables have completely consistent findings. But for those variables, their VIP > 1 sequence in the weight table was not consecutive, implying some variables with VIP < 1 were also listed in the weight table. As mentioned above, SVMRFE has higher predictability than OPLS-DA in general, i.e., the weight ranking method is more reliable than the VIP method in picking eligible variables. Therefore, we still kept those variables with VIP < 1 in the weight table to prevent the possible loss of some valid variables during VIP computation for further analysis. Overlapped variables in 20 and 100 ng/mL groups, on behalf of their significantly low and high concentrations, underwent further verification by pairwise *t*-test [47–49] performed in SPSS Statistics software (Version 17.0) and fold change calculation of concentration during univariate analysis. Finally, eligible variables were confirmed to represent the marker compounds by comparing the retention time and precise molecular weight (absolute value of error less than 5 ppm) of 124 P&VDs (Table 2).

4. Conclusions

This study developed an SVM-assisted metabolomics method to screen the marker compounds of 124 P&VDs in maize samples. One hundred and twenty out of one hundred and twenty-four P&VDs can be identified by our proposed method, while only one hundred and nine P&VDs can be found by the metabolomics method alone, implying that SVM can promote the screening accuracy of the metabolomics method. It is the first time to apply the SVM-assisted metabolomics method to non-targeted screening of P&VDs in plant-derived foods. Method LODs were calculated for 120 P&VDs, with values significantly below 10 μg/kg for all of them, which were favorably comparable with a targeted multi-residue method. Our approach, developed on a simple and self-designed case, has been validated on a more complicated and realistic condition. This study promoted insight into the development of truly non-targeted screening approaches on the basis of metabolomics (particularly LC–MS/MS and chemometrics) for food safety assessment, which may be in need of non-targeted methods as a complement to targeted methods in view of screening potentially contaminated food products in the near future. Since our proposed method was rather generic, it was applicable with only a few modifications to any other LC–MS/MS dataset or even to other fields like authenticity or origin issues to complement existing methods. In conclusion, there are reasons to believe that these findings will encourage more breakthroughs in analytical issues, both in the methods and the tools.

Supplementary Materials: The following supporting information can be downloaded at: https://www.mdpi.com/article/10.3390/molecules29133026/s1, MATLAB codes to run SVM; Table S1: Recovery of internal standards in maize sample groups; Table S2: Pairwise *t*-test results of 154 variables; Table S3: Fold change results of 154 variables; Table S4: Concentration of P&VDs in maize samples from Jinpu New Area. Figure S1. Cluster analysis plot of spiked maize sample groups.

Author Contributions: Conceptualization, W.X.; data curation, W.X.; investigation, W.X., F.L. and X.L.; methodology, W.X.; writing—original draft, W.X.; writing—review and editing, W.X. and Y.L. All authors have read and agreed to the published version of the manuscript.

Funding: This work was funded by the Natural Science Foundation of Liaoning Province of China (No. 2023-MS-349) and the Science and Technology Project of the General Administration of Customs of the PRC (No. 2023HK124).

Institutional Review Board Statement: Not applicable.

Informed Consent Statement: Not applicable.

Data Availability Statement: The data presented in this study are available in this article and Supplementary Materials.

Conflicts of Interest: The authors declare no conflicts of interest.

References

1. *GB 2763-2021;* National Food Safety Standard-Maximum Residue Limits for Pesticides in Food. Ministry of Agriculture and Rural Affairs of the People's Republic of China (MARAPRC): Beijing, China, 2021.
2. National Archives and Records Administration (NARA). Available online: https://www.ecfr.gov/current/title-40/chapter-I/subchapter-E/part-180 (accessed on 15 October 2022).
3. The Japan Food Chemical Research Foundation (JFCRF). Available online: http://db.ffcr.or.jp/front/ (accessed on 15 October 2022).
4. European Union (EU). Commission Regulation (EU) No 396/2005 of 23 February 2005. 2005. Available online: https://eur-lex.europa.eu/eli/reg/2005/396/oj (accessed on 15 October 2022).
5. Ahmed, M.B.M.; Rajapaksha, A.U.; Lim, J.E.; Vu, N.T.; Kim, S.; Kang, H.M.; Lee, S.S.; Ok, Y.S. Distribution and accumulative pattern of tetracyclines and sulfonamides in edible vegetables of cucumber, tomato and lettuce. *J. Agric. Food Chem.* **2015**, *63*, 398–405. [CrossRef] [PubMed]
6. Boxall, A.B.A.; Johnson, P.; Smith, E.J.; Sinclair, C.J.; Stutt, E.; Levy, L.S. Uptake of veterinary medicines from soils into plants. *J. Agric. Food Chem.* **2006**, *54*, 2288–2297. [CrossRef] [PubMed]
7. Hawker, D.W.; Cropp, R.; Boonsaner, M. Uptake of zwitterionic antibiotics by rice (*Oryza sativa* L.) in contaminated soil. *J. Hazard. Mater.* **2013**, *263*, 458–466. [CrossRef] [PubMed]
8. Hu, X.; Zhou, Q.; Luo, Y. Occurrence and source analysis of typical veterinary antibiotics in manure, soil, vegetables and groundwater from organic vegetable bases, northern China. *Environ. Pollut.* **2010**, *158*, 2992–2998. [CrossRef] [PubMed]
9. Jones-Lepp, T.L.; Sanchez, C.A.; Moy, T.; Kazemi, R. Method development and application to determine potential plant uptake of antibiotics and other drugs in irrigated crop production systems. *J. Agric. Food Chem.* **2010**, *58*, 11568–11573. [CrossRef] [PubMed]
10. Tanoue, R.; Sato, Y.; Motoyama, M.; Nakagawa, S.; Shinohara, R.; Nomiyama, K. Plant uptake of pharmaceutical chemicals detected in recycled organic manure and reclaimed wastewater. *J. Agric. Food Chem.* **2012**, *60*, 10203–10211. [CrossRef] [PubMed]
11. Tasho, R.P.; Cho, J.Y. Veterinary antibiotics in animal waste, its distribution in soil and uptake by plants: A review. *Sci. Total Environ.* **2016**, *563–564*, 366–376. [CrossRef] [PubMed]
12. Wang, C.; Luo, Y.; Mao, D. Sources, fate, ecological risks and mitigation strategies of antibiotics in the soil environment. *Environ. Chem.* **2014**, *33*, 19–29.
13. Zhang, M.; Xu, Q. Review on pollution and behavior of veterinary antibiotics in agricultural system. *Acta Agric. Zhejiang* **2013**, *25*, 416–424. (In Chinese)
14. Bao, Y.; Li, Y.; Mo, C.; Yao, Y.; Tai, Y.; Wu, X.; Zhang, Y. Determination of six sulfonamide antibiotics in vegetables by solid phase extraction and high performance liquid chromatography. *Environ. Chem.* **2010**, *29*, 513–518. (In Chinese)
15. Wu, X.; Xiang, L.; Mo, C.; Li, Y.; Jiang, Y.; Yan, Q.; Lv, X.; Huang, X. Determination of quinolones in vegetables using ultra performance liquid chromatography-electrospray ionization tandem mass spectrometry. *Chin. J. Anal. Chem.* **2013**, *41*, 876–881. (In Chinese) [CrossRef]
16. Wu, X.; Bao, Y.; Xiang, L.; Yan, Q.; Jiang, Y.; Li, Y.; Huang, X.; Mo, C. Simultaneous sulfonamide antibiotics in vegetables using solid phase extraction and high performance liquid chromatography coupled with mass spectrometry. *Environ. Chem.* **2013**, *32*, 1038–1044. (In Chinese)
17. Yao, Y.; Mo, C.; Li, Y.; Bao, Y.; Tai, Y.; Wu, X. Determination of tetracyclines in vegetables using solid phase extraction and HPLC with fluorescence detection. *Environ. Chem.* **2010**, *29*, 536–541. (In Chinese)
18. European Union (EU). Commission Regulation (EU) No 37/2010 of 22 December 2009. 2010. Available online: https://eur-lex.europa.eu/legal-content/EN/TXT/?uri=CELEX:32010R0037 (accessed on 15 October 2022).
19. *GB 31650-2019;* National Food Safety Standard-Maximum Residue Limits for Veterinary Drugs in Foods. Ministry of Agriculture and Rural Affairs of the People's Republic of China (MARAPRC): Beijing, China, 2019.
20. Jadhav, M.R.; Pudale, A.; Raut, P.; Utture, S.; Shabeer, T.P.A.; Banerjee, K. A unified approach for high-throughput quantitative analysis of the residues of multi-class veterinary drugs and pesticides in bovine milk using LC-MS/MS and GC-MS/MS. *Food Chem.* **2019**, *272*, 292–305. [CrossRef] [PubMed]
21. Jia, W.; Chu, X.; Ling, Y.; Huang, J.; Chang, J. High-throughput screening of pesticide and veterinary drug residues in baby food by liquid chromatography coupled to quadrupole Orbitrap mass spectrometry. *J. Chromatogr. A* **2014**, *1347*, 122–128. [CrossRef] [PubMed]
22. Gao, F.; Zhao, Y.; Shao, B.; Zhang, J. Determination of residues of pesticides and veterinary drugs in milk by ultra performance liquid chromatography coupled with quadrupole-time of flight mass spectrometry. *Chin. J. Chromatogr.* **2012**, *30*, 560–567. (In Chinese) [CrossRef] [PubMed]
23. Brack, W.; Dulio, V.; Slobodnik, J. The NORMAN network and its activities on emerging environmental substances with a focus on effect-directed analysis of complex environmental contamination. *Environ. Sci. Eur.* **2012**, *24*, 29–33. [CrossRef]

24. Fu, Y.; Zhou, Z.; Kong, H.; Lu, X.; Zhao, X.; Chen, Y.; Chen, J.; Wu, Z.; Xu, Z.; Zhao, C.; et al. Nontargeted screening method for illegal additives based on ultrahigh-performance liquid chromatography-high-resolution mass spectrometry. *Anal. Chem.* **2016**, *88*, 8870–8877. [CrossRef] [PubMed]
25. Kunzelmann, M.; Winter, M.; Aberg, M.; Hellenäs, K.E.; Rosén, J. Non-targeted analysis of unexpected food contaminants using LC-HRMS. *Anal. Bioanal. Chem.* **2018**, *410*, 5593–5602. [CrossRef]
26. Tengstrand, E.; Rosén, J.; Hellenäs, K.E.; Aberg, K.M. A concept study on non-targeted screening for chemical contaminants in food using liquid chromatography-mass spectrometry in combination with a metabolomics approach. *Anal. Bioanal. Chem.* **2013**, *405*, 1237–1243. [CrossRef]
27. Delaporte, G.; Cladière, M.; Jouan-Rimbaud Bouveresse, D.; Camel, V. Untargeted food contaminant detection using UHPLC-HRMS combined with multivariate analysis: Feasibility study on tea. *Food Chem.* **2019**, *277*, 54–62. [CrossRef] [PubMed]
28. Xue, W.; Zhang, H.; Wang, M.; Liu, Y.; Liu, M.; Shen, B. Metabolomics-based non-targeted screening analysis of 34 PPCPs in bovine and piscine muscles. *Anal. Methods* **2022**, *14*, 233–240. [CrossRef] [PubMed]
29. Xue, W.; Yang, C.; Liu, M.; Lin, X.; Wang, M.; Wang, X. Metabolomics approach on non-targeted screening of 50 PPCPs in lettuce and maize. *Molecules* **2022**, *27*, 4711. [CrossRef] [PubMed]
30. Rivera-Pérez, A.; Romero-González, R.; Frenich, A.G. A metabolomics approach based on 1H NMR fingerprinting and chemometrics for quality control and geographical discrimination of black pepper. *J. Food Compos. Anal.* **2022**, *105*, 104235–104244. [CrossRef]
31. Azizan, K.A.; Zamani, A.I.; Muhammad, N.A.N.; Khairudin, K.; Yusoff, N.; Nawawi, M.F. Dose-dependent effect of wedelia trilobata essential oil (EO) on lettuce (*Lactuca sativa L.*) with multivariate analysis. *Chem. Biodivers.* **2022**, *19*, e202100833. [CrossRef] [PubMed]
32. Cui, Y.; Ge, L.; Lu, W.; Wang, S.; Li, Y.; Wang, H.; Huang, M.; Xie, H.; Liao, J.; Tao, Y. Real-time profiling and distinction of lipids from different mammalian milks using rapid evaporative ionization mass spectrometry combined with chemometric analysis. *J. Agric. Food Chem.* **2022**, *70*, 7786–7795. [CrossRef] [PubMed]
33. Sun, R.; Xing, R.; Zhang, J.; Yu, N.; Ge, Y.; Zhang, W.; Chen, Y. UPLC-QTOF-MS coupled with machine learning to discriminate between NFC and FC orange juice. *Food Control.* **2023**, *145*, 109487. [CrossRef]
34. Yang, S.; Yang, H.; Huang, Y.; Chen, G.; Shen, C.; Zheng, S. Serum metabolomic signatures of hirschsprung's disease based on GC–MS and LC–MS. *J. Proteome Res.* **2023**, *22*, 2307–2318. [CrossRef] [PubMed]
35. Zheng, J.; Zheng, Y.; Li, W.; Zhi, J.; Huang, X.; Zhu, W.; Liu, Z.; Gong, L. Combined metabolomics with transcriptomics reveals potential plasma biomarkers correlated with non-small-cell lung cancer proliferation through the Akt pathway. *Clin. Chim. Acta* **2022**, *530*, 66–73. [CrossRef]
36. Lu, Y.; Yao, G.; Wang, X.; Zhang, Y.; Zhao, J.; Yu, Y.J.; Wang, H. Chemometric discrimination of the geographical origin of licorice in China by untargeted metabolomics. *Food Chem.* **2022**, *380*, 132235. [CrossRef]
37. Lin, X.; Yang, F.; Zhou, L.; Yin, P.; Kong, H.; Xing, W.; Lu, X.; Jia, L.; Wang, Q.; Xu, G. A support vector machine-recursive feature elimination feature selection method based on artificial contrast variables and mutual information. *J. Chromatogr. B* **2012**, *910*, 149–155. [CrossRef]
38. Huang, X.; Lin, X.; Zeng, J.; Wang, L.; Yin, P.; Zhou, L.; Hu, C.; Yao, W. A computational method of defining potential biomarkers based on differential sub-networks. *Sci. Rep.* **2017**, *7*, 14339. [CrossRef] [PubMed]
39. Smolinska, A.; Blanchet, L.; Coulier, L.; Ampt, K.A.M.; Luider, T.; Hintzen, R.Q.; Wijmenga, S.S.; Buydens, L.M.C. Interpretation and visualization of non-linear data fusion in kernel space: Study on metabolomic characterization of progression of multiple sclerosis. *PLoS ONE* **2012**, *7*, e38163. [CrossRef] [PubMed]
40. Li, N.; Xiong, Z.; Deng, T.; Ren, K. Automated brain tumor segmentation from multimodal MRI data based on Tamura texture feature and an ensemble SVM classifier. *Int. J. Intel. Comput. Cyber.* **2019**, *12*, 466–480.
41. Chen, Q.; Dai, W.; Lin, Z.; Xie, D.; Lv, M.; Lin, Z. Effects of shading on main quality components in tea (*Camellia Sinensis* (L) *O. Kuntze*) leaves based on metabolomics analysis. *Sci. Agric. Sin.* **2019**, *52*, 1066–1077. (In Chinese)
42. Pasikanti, K.K.; Esuvaranathan, K.; Ho, P.C.; Mahendran, R.; Kamaraj, R.; Wu, Q.H.; Chiong, E.; Chan, E.C.Y. Noninvasive urinary metabonomic diagnosis of human bladder cancer. *J. Proteome Res.* **2010**, *9*, 2988–2995. [CrossRef]
43. Brereton, R.G.; Lloyd, G.R. Partial least squares discriminant analysis: Taking the magic away. *J. Chemom.* **2014**, *28*, 213–225. [CrossRef]
44. Chawla, N.V.; Bowyer, K.W.; Hall, L.O.; Kegelmeyer, W.P. 'SMOTE: Synthetic minority over-sampling technique'. *J. Artif. Intell. Res.* **2002**, *16*, 321–357. [CrossRef]
45. Broadhurst, D.I.; Kella, D.B. Statistical strategies for avoiding false discoveries in metabolomics and related experiments. *Metabolomics* **2006**, *2*, 171–196. [CrossRef]
46. Slupsky, C.M.; Steed, H.; Wells, T.H.; Dabbs, K.; Schepansky, A.; Capstick, V.; Faught, W.; Sawyer, M.B. Urine metabolite analysis offers potential early diagnosis of ovarian and breast cancers. *Clin. Cancer Res.* **2010**, *16*, 5835–5841. [CrossRef]
47. Thévenot, E.A.; Roux, A.; Xu, Y.; Ezan, E.; Junot, C. Analysis of the human adult urinary metabolome variations with age, body mass index, and gender by implementing a comprehensive workflow for univariate and OPLS statistical analyses. *J. Proteome Res.* **2015**, *14*, 3322–3335. [CrossRef]
48. Gika, H.G.; Theodoridis, G.A.; Plumb, R.S.; Wilson, D. Current practice of liquid chromatography-mass spectrometry in metabolomics and metabonomics. *J. Pharm. Biomed. Anal.* **2014**, *87*, 12–25. [CrossRef]

49. Rubert, J.; Righetti, L.; Stranska-Zachariasova, M.; Dzuman, Z.; Chrpova, J.; Dall'Asta, C.; Hajslova, J. Untargeted metabolomics based on ultra-high-performance liquid chromatography-high-resolution mass spectrometry merged with chemometrics: A new predictable tool for an early detection of mycotoxins. *Food Chem.* **2017**, *224*, 423–431. [CrossRef] [PubMed]
50. Ortmayr, K.; Charwat, V.; Kasper, C.; Hann, S.; Koellensperger, G. Uncertainty budgeting in fold change determination and implications for non-targeted metabolomics studies in model systems. *Analyst* **2017**, *142*, 80–90. [CrossRef]
51. Environmental Protection Agency (EPA). *Method 8061A Phthalate Esters by Gas Chromatography with Electron Capture Detection (GC/ECD)*; Environmental Protection Agency: Washington, DC, USA, 1996.
52. Tai, Y.; Mo, C.; Li, Y.; Wu, X.; Wang, J.; Su, Q. Concentration and distribution of tetracycline antibiotics in soils from vegetable fields of Dongguan City. *China Environ. Sci.* **2011**, *31*, 90–95. (In Chinese)
53. Tang, X.; Lou, C.; Wang, S.; Lu, Y.; Liu, M.; Hashmi, M.Z.; Liang, X.; Li, Z.; Liao, Y.; Qin, W.; et al. Effects of long-term manure applications on the occurrence of antibiotics and antibiotic resistance genes (ARGs) in paddy soils: Evidence from four field experiments in south of China. *Soil Biol. Biochem.* **2015**, *90*, 179–187. [CrossRef]
54. Zhu, X.; Zeng, Q.; Xie, Q.; Ding, D.; Ru, S. Residues and risk assessment of tetracycline antibiotics in vegetable-growing soils from suburban areas of northern Guangzhou. *J. Agro-Environ. Sci.* **2017**, *36*, 2257–2266. (In Chinese)
55. Hang, L.; Zhao, Y.; Liu, C.; Yu, Y.; He, Y.; Xu, J.; Lu, Z. Determine multiple classes of veterinary antibiotics in soil: Comparing dispersive and solid-phase extraction for sample cleanup. *Chromatographia* **2021**, *84*, 833–844. [CrossRef]
56. Lian, Y.L. Control of maize pests and diseases and measures of rational use of pesticides. *World Trop. Agric. Info.* **2023**, *10*, 33–34. (In Chinese)
57. Want, E.J.; Wilson, I.D.; Gika, H.; Theodoridis, G.; Plumb, R.S.; Shockcor, J.; Holmes, E.; Nicholson, J.K. Global metabolic profiling procedures for urine using UPLC-MS. *Nat. Protoc.* **2010**, *5*, 1005–1018. [CrossRef]
58. Dunn, W.B.; Broadhurst, D.; Begley, P.; Zelena, E.; Francis-McIntyre, S.; Anderson, N.; Brown, M.; Knowles, J.D.; Halsall, A.; Haselden, J.N.; et al. Procedures for large-scale metabolic profiling of serum and plasma using gas chromatography and liquid chromatography coupled to mass spectrometry. *Nat. Protoc.* **2011**, *6*, 1060–1083. [CrossRef] [PubMed]
59. Chambers, M.C.; MacLean, B.; Burke, R.; Amodei, D.; Ruderman, D.L.; Neumann, S.; Gatto, L.; Fischer, B.; Pratt, B.; Egertson, J.; et al. A cross-platform toolkit for mass spectrometry and proteomics. *Nat. Biotechnol.* **2012**, *30*, 918–920. [CrossRef] [PubMed]
60. Giacomoni, F.; Le Corguille, G.; Monsoor, M.; Landi, M.; Pericard, P.; Petera, M.; Duperier, C.; Tremblay-Franco, M.; Martin, J.F.; Jacob, D.; et al. Workflow4Metabolomics: A collaborative research infrastructure for computational metabolomics. *Bioinformatics* **2015**, *31*, 1493–1495. [CrossRef] [PubMed]
61. Smith, C.A.; Want, E.J.; O'Maille, G.; Abagyan, R.; Siuzdak, G. XCMS: Processing mass spectrometry data for metabolite profiling using nonlinear peak alignment, matching, and identification. *Anal. Chem.* **2006**, *78*, 779–787. [CrossRef] [PubMed]
62. Gika, H.G.; Theodoridis, G.A.; Wilson, I.D. Liquid chromatography and ultra-performance liquid chromatography-mass spectrometry fingerprinting of human urine: Sample stability under different handling and storage conditions for metabonomics studies. *J. Chromatogr. A* **2008**, *1189*, 314–322. [CrossRef]
63. Winnike, J.H.; Busby, M.G.; Watkins, P.B.; O'Connell, T.M. Effects of a prolonged standardized diet on normalizing the human metabolome. *Am. J. Clin. Nutr.* **2009**, *90*, 1496–1501. [CrossRef] [PubMed]
64. Yin, P.; Peter, A.; Franken, H.; Zhao, X.; Neukamm, S.S.; Rosenbaum, L.; Lucio, M.; Zell, A.; Haring, H.U.; Xu, G.; et al. Preanalytical aspects and sample quality assessment in metabolomics studies of human blood. *Clin. Chem.* **2013**, *59*, 833–845. [CrossRef] [PubMed]
65. Schievano, E.; Morelato, E.; Facchin, C.; Mammi, S. Characterization of markers of botanical origin and other compounds extracted from unifloral honeys. *J. Agric. Food Chem.* **2013**, *61*, 1747–1755. [CrossRef] [PubMed]
66. Pontes, T.A.; Barbosa, A.D.; Silva, R.D.; Melo-Junior, M.R.; Silva, R.O. Osteopenia-osteoporosis discrimination in postmenopausal women by 1H NMR-based metabonomics. *PLoS ONE* **2019**, *14*, e0217348. [CrossRef]
67. Zhao, W.; Wang, G.; Xun, W.; Yu, Y.; Ge, C.; Liao, G. Selection of water-soluble compounds by characteristic flavor in Chahua Chicken muscles based on metabolomics. *Sci. Agric. Sin.* **2020**, *53*, 1627–1642. (In Chinese)
68. Dai, Y.; Lü, C.; He, L.; Yi, C.; Liu, X.; Huang, W.; Chen, J. Metabolic changes in the processing of Yunkang 10 sun-dried green tea based on metabolomics. *Sci. Agric. Sin.* **2020**, *53*, 357–370. (In Chinese)
69. Yang, J.; Chen, T.; Sun, L.; Zhao, Z.; Wan, C. Potential metabolite markers of schizophrenia. *Mol. Psychiatry* **2013**, *18*, 67–78. [CrossRef] [PubMed]
70. Mendez, K.M.; Reinke, S.N.; Broadhurst, D.I. A comparative evaluation of the generalised predictive ability of eight machine learning algorithms across ten clinical metabolomics data sets for binary classification. *Metabolomics* **2019**, *15*, 150. [CrossRef] [PubMed]
71. Mahadevan, S.; Shah, S.L.; Marrie, T.J.; Slupsky, C.M. Analysis of metabolomic data using support vector machines. *Anal. Chem.* **2008**, *80*, 7562–7570. [CrossRef] [PubMed]

separations

MDPI

Communication

Coordination Ion Spray for Analysis of the Growth Hormones Releasing Peptides in Urine—An Application Study

Azamat Temerdashev [1,*], Elina Gashimova [1], Alice Azaryan [1], Yu-Qi Feng [2] and Sanka N. Atapattu [3]

[1] Analytical Chemistry Department, Kuban State University, Krasnodar 350040, Russia; elina.gashimova@yandex.ru (E.G.); alisaazaryan@gmail.com (A.A.)
[2] School of Bioengineering and Health, Wuhan Textile University, Wuhan 430079, China; yqfeng@whu.edu.cn
[3] CanAm Bioresearch Inc., Winnipeg, MB R3T 6C6, Canada; sanka@wayne.edu
* Correspondence: temerdashevaz@gmail.com or analyt@chem.kubsu.ru

Abstract: In this article, a comparison of ionization techniques is provided and discussed. Conventional liquid chromatography with an electrospray ionization source shows higher robustness and repeatability in comparison with liquid chromatography coupled with a coordination ion spray (CIS-MS) source using silver nitrate as the dopant. However, the higher sensitivity and possibility to collect more data in untargeted applications mean CIS-MS is emerging as an instrument used in specific applications. During this research, the limit of detection (LOD) for GHRP-2 and GHRP-6 was established at 0.2 ng/mL, and the lower limit of quantification (LLOQ) was 0.5 ng/mL for CIS-MS. For conventional ESI-MS combined with solid-phase extraction on weak cation exchange columns, the limit of detection was found to be 1 ng/mL, and the lower limit of quantification was 2 ng/mL.

Keywords: doping; GHRP; ESI-MS; CIS-MS; LC-MS; peptides; mass spectrometry

Citation: Temerdashev, A.; Gashimova, E.; Azaryan, A.; Feng, Y.-Q.; Atapattu, S.N. Coordination Ion Spray for Analysis of the Growth Hormones Releasing Peptides in Urine—An Application Study. *Separations* **2024**, *11*, 155. https://doi.org/10.3390/separations11050155

Academic Editors: Ronald Beckett and Chao Kang

Received: 19 April 2024
Revised: 9 May 2024
Accepted: 15 May 2024
Published: 16 May 2024

1. Introduction

Growth-hormone-releasing peptides (GHRPs) are well-known doping agents in professional sports. The abuse of GHRPs is banned by the World Anti-Doping Agency (WADA). Nonetheless, cases of their use, while they are not rife, do occur annually [1]. In the last testing report, WADA revealed that approximately 0.01% of samples were positive in growth-hormone-releasing factor testing [2].

In such testing, even when conducted a few days after GHRPs' administration, there can be issues with their determination, which makes the development of methods for their determination at trace concentration levels extremely important [3–8]. There are various solutions available—for instance, using concentration methods, such as solid-phase extraction, and using high-end equipment. Often, to meet the needs of doping control laboratories, a combination of such approaches is required [9–12]. In this case, the analytes for GHRPs exist in samples in an equilibrium state—where both monocharged and polycharged ions can be observed simultaneously—but this balance can be changed by adding sufficiently strong acids, for example, trifluoroacetic acid.

According to WADA recommendations [13], the minimum required performance limit for NGPRs is 1 ng/mL. Some recently published articles showed the possibility of their determination at up to 50 pg/mL (for dried blood spots); however, most articles showed the possibility of qualitative and quantitative analysis of such compounds at 0.1–1 ng/mL. Usually, the LOD and LOQ described depend on the laboratory equipment used. In the case of well-equipped high-end instruments (especially HRMS instruments), there are no issues in finding concentrations below 1 ng/mL. The main issue with such compounds is their stability and the excretion time in urine [14–16]. The main sample preparation technique for such compounds, like GHRP-2 and GHRP-6, is solid-phase extraction on SCX or WCX sorbents of different volumes. Such sample preparation can be used for significant pre-concentration of the sample and its cleanup [14].

An alternative to this approach is the use of coordination ion spray mass spectrometry. In this case, lithium, sodium, potassium or silver ions are added to the mobile phase [17–19]. An excessive presence of sodium and potassium ions is in most cases unwelcome since the formation of such adducts is not always reproducible and is often associated with a poor quality of the solvents used. However, the occurrence of significant amounts of lithium and silver ions in the mobile phase is unlikely and can easily be used under conditions of sufficient solvent purity.

Technically, the use of a coordination ion spray combined with mass spectrometry (CIS-MS) can solve a number of problems. To give an example, the introduction of singly charged ions can significantly increase the intensity of the single charged ion produced and increase the information content of the mass spectrum during a specific isotopic pattern of adducts with silver. For instance, silver has two isotopes—^{107}Ag and ^{109}Ag with an abundance of 52% and 48%, respectively. Accordingly, a distinct distribution pattern emerges, with the charge being localized on the silver ion that binds to the analytes through coordination involving amino groups. Consequently, the formation of positively charged ions occurs, which facilitates convenient detection using LC-MS.

Previously, the literature described the use of a similar approach for the analysis of sulfur-containing compounds and their mixtures, as well as for the determination of lipids, steroids and phytosterols [20–22].

Another alternative to conventional acidification of the mobile phase is the use of the so-called "wrong-way-round ionization" technique. In this case, the mobile phase will be modified with NH_4OH solution, and protonation of the analytes will be caused by transfer of the ammonia ion in the gaseous phase, with a concentration of ammonia hydroxide of up to 5 mM. However, such a methodology requires chromatographic columns, which are allowed to work with alkaline mobile phases [23,24].

This work presents a comparison study of the efficiency of conventional ESI-MS and CIS-MS and discusses the advantages and disadvantages of this approach in the analysis of releasing peptides.

2. Materials and Methods

Standard samples of GHRP-6, GHRP-2 and Delta sleep-inducing peptide (DSIP, internal standard, IS) were purchased from Canada Peptides (Montreal, QC, Canada). Gradient-grade acetonitrile (Biosolve, Jerusalem, Israel) and deionized water (18.2 MΩ × cm, Milli-Q, Millipore, France) were used for mobile phase preparation. Silver nitrate (>99%) and lithium chloride (>99%) were purchased from Vecton (Vecton, Saint Petersburg, Russia). Strata WCX cartridges (1 mL, 100 mg) for solid-phase extraction were purchased from Phenomenex.

2.1. Preparation of Solutions

We prepared 1 mg/mL stock solutions of analytes in water:methanol solution (50:50, *v:v*) and stored them at −20 °C. For the preparation of calibration solutions, the stock solutions were diluted with acidified water solutions (in 0.1% formic acid) to obtain calibration concentrations of 100, 50, 25, 20, 10, 5, 2.5, 2, 1, 0.5, 0.25, 0.2, 0.1, 0.05, 0.025 and 0.02 ng/mL on the day of analysis, which were stored at 4 °C in the tray of the autosampler. The IS concentration for CIS-MS was 2 ng/mL, it was 10 ng/mL for ESI-MS with SPE and it was 50 ng/mL for "dilute-and-shoot" sample preparation. Quality control solutions were prepared at three levels of concentrations—low (QC low), medium (QC med) and high (QC high) for each sample preparation and analysis technique separately, according to their linear range. For CIS-MS: the QC low solution contained 2 ng/mL of analytes, QC med—20 ng/mL, QC high—50 ng/mL. For ESI-MS combined with SPE: QC low—5 ng/mL, QC med—25 ng/mL, QC high—50 ng/mL. For ESI-MS with "dilute-and-shoot" sample preparation: QC low—25 ng/mL, QC med—50 ng/mL, QC high—100 ng/mL.

2.2. Sample Preparation

2.2.1. "Dilute-and-Shoot" Procedure

We diluted 200 μL of the urine sample with 800 μL of 0.1% formic acid in water:acetonitrile (50:50, *v:v*), which was followed by vortex mixing for 3 min and centrifugation at 10,000 rpm for 10 min. The resulting supernatant, at a volume of 800 μL, was transferred into a glass vial for the following LC-MS/MS analysis.

2.2.2. Solid-Phase Extraction

A 3 mL aliquot of urine samples containing native analytes was loaded onto SPE cartridges. The following protocol was used for WCX cartridges: preconditioning was carried out with a 2.5% aqueous solution of ammonia in water, followed by passing a mixture of acetonitrile:water (10:90, *v:v*). Elution was carried out with a 5% solution of ammonium acetate in methanol.

2.3. Instrumentation

A Thermo TSQ Quantum Access MAX (Thermo Scientific, Waltham, MA, USA), tuned at the full-width at half-maximum (FWHM) resolution of 0.7 Da, equipped with an electrospray ionization source (ESI) and coupled with a Dionex Ultimate-3000 (Thermo Scientific, Waltham, MA, USA) UHPLC system controlled by ThermoXCalibur 2.2 software was used. Separation was carried out on a Phenomenex Kinetex C18 (100 × 2.1 mm, 2.6 μm) analytical column with the respective guard column. The mobile phase consisted of acetonitrile (mobile phase A), 0.1 formic acid (FA) in water (mobile phase B) and 100 μg/mL dopant (potassium, lithium, silver) in water. Elution was performed in the gradient mode with a total run time of 10 min, which included column equilibration before the next analysis. The flow rates were 0.4 mL/min at the main pump and 0.05 mL/min at the additional pump. The column temperature was set at 35 °C, and samples were stored in the autosampler tray at 5 °C to prevent their degradation.

The following detection conditions were used: the heated capillary temperature was maintained at 400 °C, the sheath gas and auxiliary gas (nitrogen) pressure levels were set at 60 and 10 arbitrary units, respectively, the transfer capillary temperature was set at 320 °C, the ionization source voltage was 4 kV and the positive ion detection mode was used.

3. Results

It is known that lithium, silver, potassium and sodium salts form complexes with organic molecules containing hard or soft Lewis basic sites. As a result of such coordination with analytes, positively charged and stable complexes can be detected by mass spectrometry. In the case of coupling mass spectrometry with high-performance liquid chromatography, such a technique can be used to elucidate the structures of mixtures because of the high speed of complex production. The presence of these cations provides the formation of single-charged ions for peptides in the higher masses, free from low-molecular interferences, which also provides a greater signal-to-noise ratio. However, it should be noted that for peptides with masses above 1000 Da, such a methodology can have significant limitations related to the limitations of mass spectrometry and m/z values if triple quadrupole or orbitrap mass analyzers are used.

Previous studies published by the authors of [20] showed promising results that could be utilized to significantly improve the sensitivity of trace and ultra-trace levels of analysis and fully exploit the potential of instruments available in laboratories. Nonetheless, given the difficulties with peptide determination, we decided to test this approach on two of GHRPs that are well-known and in WADA-accredited laboratories: GHRP-2 and GHRP-6.

The principal scheme for our CIS-MS experiments is shown in Figure 1, which presents post-column coordination of the peptides with silver.

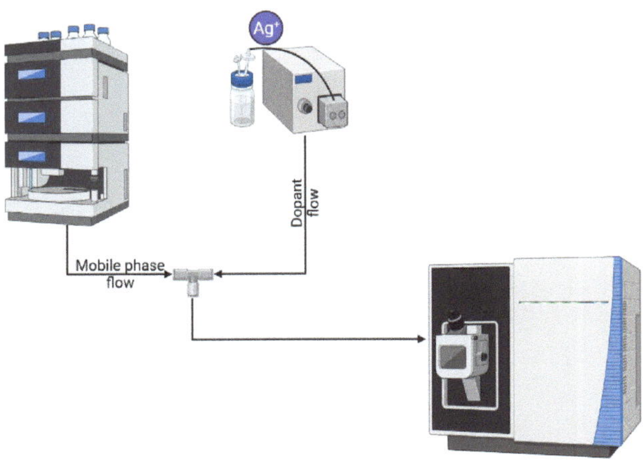

Figure 1. Principal instrumental scheme for CIS-MS experiments.

The introduction of silver nitrate into the mobile phase can lead to chelation of the sorbent, a change in its properties and a loss of reproducibility of results [25–27]. Thus, several factors were taken into account during the post-column process: the flow from the additional pump had to be small and reproducible, and the capillary length had to be relatively short to avoid off-column peak smearing. Since the pressure in the system significantly drops after the column, there was no need to use two pumps with an operating pressure of more than 1000 bar. At the same time, the presence of a T-connector in the system required a sufficiently high-pressure comparable pump to avoid the flow of eluate from the column into the dopant supply line.

Accordingly, a Dionex Ultimate pump with a maximum operating pressure of 600 bar was used. The flow rate was 50 µL/min, and the dopant concentration was optimized experimentally (Figure 2).

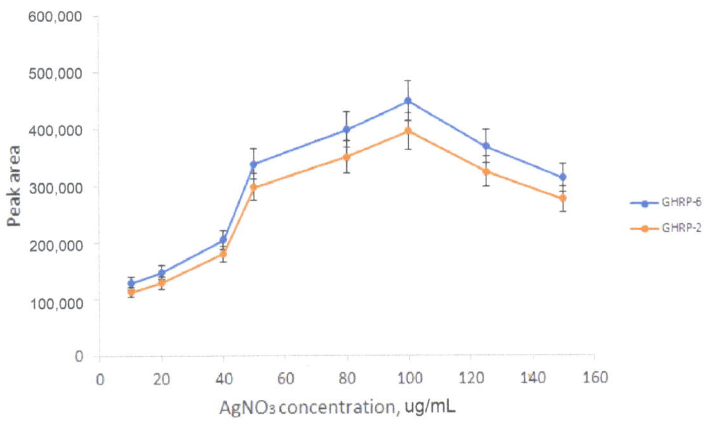

Figure 2. The influence of the dopant concentration on the analytical signal.

An intense $[M+Ag]^+$ ion was obtained, which was the goal of this experiment (Figure 3), and, subsequently, the stability of the ion beam when silver was introduced into the system was assessed. As can be seen from Figure 3, a typical isotopic pattern for silver adducts was observed (^{107}Ag and ^{109}Ag with an abundance of 52% and 48%, respectively), which could be used as an additional confirmation parameter of adduct formation. However, further fragmentation in the collision cell leads to the elimination of silver from the precursor ions, similar to non-derivatized peptide fragmentation.

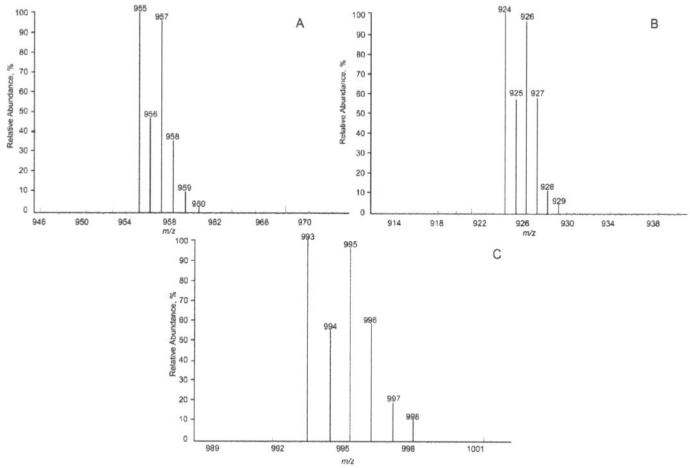

Figure 3. CIS-MS mass spectra of analytes: DSIP (**A**), GHRP-2 (**B**), GHRP-6 (**C**).

To investigate the influence of the dopant concentration on the peak area of the analytes, further experiments were carried out in the absence of a matrix. As can be seen from Figure 4, for the first 20 injections, the analytical signal remained stable, and each subsequent injection led to a loss in the peak area, essentially rendering the system unusable after 35 injections.

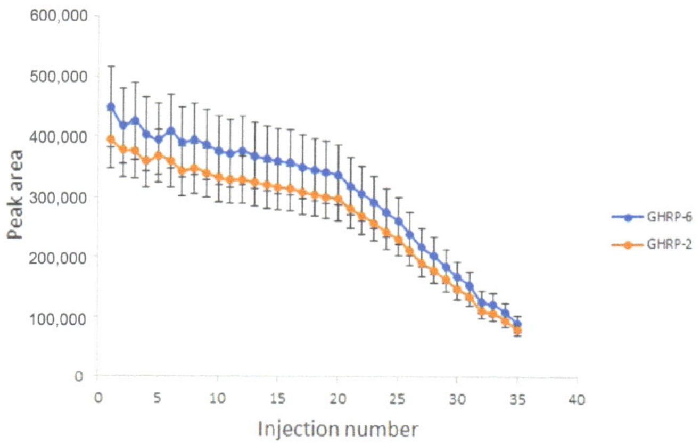

Figure 4. Change in the value of the analytical signal depending on the number of injections ($n = 3$) using silver nitrate as the dopant.

To complete the experiments and prevent false results, an additional ion optics module containing a skimmer, tube lens and transfer capillary was added, because the cleanup procedure of the optics from silver requires more time than the typical cleanup. Instrument tuning was performed after each shutdown. All analysis conducted following the establishment of the metrological parameters of this methodology was completed with ISTD (DSIP). There may have been an additional load on the ion optics added by matrix components since a number of compounds that are not capable of ionization under classical electrospray ionization conditions will be effectively ionized in the presence of dopants, which also accelerate the contamination of the skimmer, transfer capillary and tube lens.

A partial solution to this problem could be the regular use of a divert valve, which sends the mobile phase to the waste line outside the transition scanning time. Alternatively, using lithium or potassium cations as dopants allows one to avoid such difficulties; however, during our research, a lack of effectiveness was noted when using these to form intense ion adducts with these metals (Table 1).

Table 1. Comparison of the different dopants' efficiency at a 100 μg/mL dopant concentration.

Compound	Relative Peak Area of the Analytes with Different Dopants, %		
	CIS-MS (Ag+)	CIS-MS (Li+)	CIS-MS (K+)
GHRP-6	100	62	75
GHRP-2	100	68	88

As can be seen from Table 1, the most abundant signals were obtained when using silver ions as the dopant. In the case of potassium, one of the main issues was its concentration in the mobile phase. To prepare a suitable concentration of dopant in the mobile phase, a preliminary analysis of water and the organic solvent (acetonitrile or methanol) is required to establish the dopant's concentration in the solvent and calculate the necessary concentration and volume of its salt for the mobile-phase components. Even day-to-day concentrations could differ due to its leaching from glass, which makes it unpredictable and, as a result, leads to unstable results. Each 10–15% variability of the dopant concentration leads to significant changes in the molecular ion yield, which cannot be predicted in routine analysis. As such, the use of a naturally occurring dopant is not optimal for post-column derivatization.

Another important point when making the dopant choice is that the efficiency of the analytical signal is increasing. In this case, the most abundant and reproducible ratio between the peak area and area of the internal standard was obtained using silver. In this case, it was possible to achieve less than 15% deviation, while the use of lithium or potassium as the dopant increased that to 20%, which does not meet the FDA criteria for bioanalytical methods. On the other hand, the use of silver as the dopant leads to rapid contamination of the ion optics, while lithium and potassium require much less maintenance of the instrument. Thus, if studies are allowing for extended ranges of determination error and do not require the determination of ultra-trace amounts, then lithium or potassium is advisable as the dopant.

A comparison with the classical approach to determining these compounds using CIS-MS is given in Tables 1–4, and a chromatogram, obtained from real samples, is shown in Figure 5.

Table 2. LOD and LLOQ comparison for CIS-MS and conventional ESI-MS methods ("dilute-and-shoot" and SPE).

Compound	CIS-MS (Ag+)		ESI-MS (SPE)		ESI-MS ("Dilute-and-Shoot")	
	LOD, ng/mL	LLOQ, ng/mL	LOD, ng/mL	LLOQ, ng/mL	LOD, ng/mL	LLOQ, ng/mL
GHRP-6	0.2	0.5	1	2	10	20
GHRP-2	0.2	0.5	1	2	10	20

Table 3. Results of the QC samples' analysis using different techniques.

Compound	ESI-MS (SPE)			ESI-MS ("Dilute-and-Shoot")			CIS-MS (Ag+)		
	QC Low, ng/mL	QC Med, ng/mL	QC High, ng/mL	QC Low, ng/mL	QC Med, ng/mL	QC High, ng/mL	QC Low, ng/mL	QC Med, ng/mL	QC High, ng/mL
GHRP-6	5.6 ± 0.6	27 ± 2	52 ± 5	28 ± 3	51 ± 5	106 ± 8	2.2 ± 0.3	21 ± 2	52 ± 5
GHRP-2	5.3 ± 0.6	24 ± 3	54 ± 5	29 ± 3	53 ± 5	104 ± 8	2.3 ± 0.3	22 ± 2	53 ± 5

Table 4. Observed matrix effects at QC low.

Compound	CIS-MS (Ag+)	ESI-MS (SPE)	ESI-MS ("Dilute-and-Shoot")
	ME, %	ME, %	ME, %
GHRP-6	116 ± 13	112 ± 12	119 ± 22
GHRP-2	114 ± 11	108 ± 9	118 ± 22

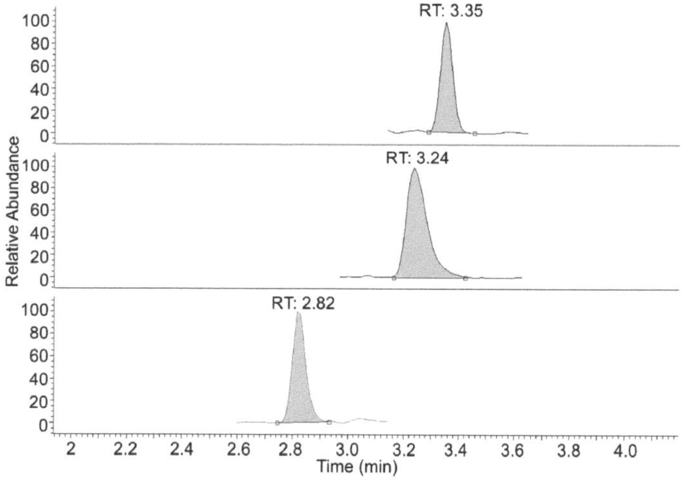

Figure 5. CIS-MS chromatogram of DSIP (t_R 2.82 min), GHRP-2 (t_R 3.24 min) and GHRP-6 (t_R 3.35 min) at 2 ng/mL in the extracted ion current chromatogram mode.

As can be seen from Figure 5, the total ion current collected in the MRM mode does not show any significant matrix peaks. This allows us to assert that the selected transitions have a high specificity and selectivity. It should be noted that the use of isotopic peaks is unsuitable for quantitative and qualitative analyses in this case. It caused the absence of structural information from MRMs, collected from isotopic peaks, and revealed only the elimination of the silver ions from the analyte. More important is a confirmation of the similarity of the derivative fragmentation, which confirms the basic fragments from the structure of the analyte.

According to the FDA guide for bioanalytical methods' validation [28] and the European Medicine Academy (EMA) guidelines [29], the selectivity, linearity, accuracy, precision, limit of detection (LOD), limit of quantification (LOQ), matrix effect, recovery, carryover and stability should be established. In this research, only partial validation was conducted, which caused a significant loss of peak intensity after just 30 injections.

The result reveals that this method is inapplicable for routine analysis and that its applicability is limited to specific applications. In this case, LOD and LLOQ is still mandatory for establishing as well as matrix effects (Tables 2 and 3).

Matrix effects were evaluated through a comparison of the results obtained for real and model solutions spiked with target analytes at the three concentration levels, which were passed through the sample preparation procedure under optimized conditions. The most pronounced matrix effects manifested when we used the dilute-and-inject procedure, which reached 24%, while for CIS-MS, they reached 20%, and the sample preparation scheme least susceptible to matrix effects was the combination with solid-phase extraction (less than 15% at QC low).

At the same time, the possibility of carryover was determined. To do this, after establishing a quality control solution with a high concentration of analytes, a blank sample was analyzed. The absence of peaks of detectable components allows us to state that there was no carryover.

Since the analytes are exogenous, their presence in blank urine samples seems unlikely, which greatly facilitates the task of establishing matrix effects (Table 4) and selectivity during research.

Another important part of the work was the study of the types of ions formed when peptides were coordinated with silver. Under normal conditions, the peptides under study are characterized by the formation of doubly charged ions, while those coordinated around silver form monocharged ions (Table 5) with a characteristic isotopic pattern (Figure 3).

Table 5. MRM transitions for analytes' detection.

Compound	t_R, Min	ESI-MS				CIS-MS			
		Tube Lens, V	Q1, *m/z*	Q3, *m/z*	CE, eV	Tube Lens, V	Q1, *m/z*	Q3, *m/z*	CE, eV
DSIP	2.82	81	$[M+2H]^{2+}$ 389.0	159.0 171.0 255.0	31 29 22	72	$[M+Ag]^+$ 955.0	159.0 171.0 255.0	43 38 34
GHRP-2	3.24	70	$[M+2H]^{2+}$ 409.7	170.1 269.1 550.2	34 15 10	88	$[M+Ag]^+$ 924.0	170.1 269.1 550.2	57 52 61
GHRP-6	3.35	76	$[M+2H]^{2+}$ 437.2	248.0 324.0 129.1	32 28 23	81	$[M+Ag]^+$ 993.0	248.0 324.0 129.1	58 50 62

As can be seen from Table 5, in the optimization of the MRM transition, for the ESI-MS and CIS-MS methods, the same product ions were produced; however, this required a higher collision energy, which may have been caused by the necessity of silver ions' elimination and then analytes' dissociation. Such a fragmentation pathway could possibly lead to early maintenance of the collision cell because of the ion optics contamination with silver. In this case, the combination of ESI and SPE looks to be preferable by a significant margin for routine applications.

4. Discussion

The utilization of silver as the dopant in CIS-MS has advantages for the analysis of GHRP-6 and GHRP-2. Firstly, CIS-MS offers enhanced sensitivity compared to conventional liquid chromatography with an electrospray ionization source. By introducing silver ions into the mobile phase, the generation of positively charged ions is facilitated, leading to improved detectability by LC-MS. This heightened sensitivity enables the determination of GHRPs at trace and ultra-trace concentration levels. Secondly, CIS-MS allows more data to be gathered in untargeted applications. The incorporation of silver ions as dopants permits the formation of intense $[M+Ag]^+$ ions, thereby enriching the information content in the mass spectrum. This proves particularly beneficial in complex analyses, such as the determination of growth-hormone-releasing peptides. Quantitative data from this study indicate that CIS-MS provides an improved LOD and LLOQ for GHRP-6 and GHRP-2 when compared when conventional ESI-MS methods. The LOD and LLOQ values achieved with CIS-MS were 0.2 ng/mL and 0.5 ng/mL, respectively, whereas the values were higher for ESI-MS with solid-phase extraction (SPE) and "dilute-and-shoot" sample preparation. Overall, CIS-MS demonstrates high sensitivity and the capability to generate extensive data, making it a suitable analytical instrument for achieving exceptional sensitivity in the analysis of GHRP-6 and GHRP-2.

It is important, however, to acknowledge the limitations associated with CIS-MS, such as the potential instability of the ion beam when silver is introduced into the system. Beyond that, the main issue of this technique is the lack of resistance of ion optics to contamination by silver ions, which leads to a rapid loss of sensitivity. As a result, even establishing the basic metrological characteristics of the technique requires the device to be repeatedly switched off for maintenance, which is unacceptable for routine applications. Further research and optimization efforts are necessary to address these limitations and fully tap into the potential of CIS-MS in peptide analysis. At the same time, usage of lithium or even potassium salts as the dopant with the preliminary solid-phase extraction technique looks promising in order to achieve a higher sensitivity than conventional techniques for the analysis of such peptides.

Author Contributions: Conceptualization, S.N.A. and A.T.; methodology, A.T.; software, A.A.; validation, E.G. and A.A.; formal analysis, A.A.; investigation, Y.-Q.F.; resources, A.T.; data curation, Y.-Q.F.; writing—original draft preparation, A.T.; writing—review and editing, S.N.A.; visualization, E.G.; supervision, Y.-Q.F.; project administration, A.T.; funding acquisition, Y.-Q.F. and A.T. All authors have read and agreed to the published version of the manuscript.

Funding: This research was funded by the Russian Science Foundation (project no. 24-43-00003), used the scientific equipment of the Center for Environmental Analysis at the Kuban State University and was funded by the National Natural Science Foundation of China (22361132526).

Institutional Review Board Statement: This study was conducted in accordance with the Declaration of Helsinki and approved by the local Ethics Committee of Ochapovskiy Central Clinical Hospital No. 1, protocol No. 122.

Data Availability Statement: Data is contained within the article.

Conflicts of Interest: Sanka N. Atapattu was employed by the CanAm Bioresearch Inc. The remaining authors declare that the research was conducted in the absence of any commercial or financial relationships that could be construed as a potential conflict of interest. The authors declare no conflicts of interest.

References

1. World Anti-Doping Agency, Prohibited List. Available online: https://www.wada-ama.org/sites/default/files/2023-09/2024list_en_final_22_september_2023.pdf (accessed on 19 April 2024).
2. World Anti-Doping Agency, WADA Testing Figures Report. Available online: https://www.wada-ama.org/sites/default/files/2023-01/2021_anti-doping_testing_figures_en.pdf (accessed on 12 March 2024).
3. Gil, J.; Cabrales, A.; Reyes, O.; Morera, V.; Betancourt, L.; Sánchez, A.; García, G.; Moya, G.; Padrón, G.; Besada, V.; et al. Development and validation of a bioanalytical LC-MS method for the quantification of GHRP-6 in human plasma. *J. Pharm. Biomed. Anal.* **2012**, *60*, 19–25. [CrossRef]
4. Coppieters, G.; Deventer, K.; van Eenoo, P.; Judák, P. Combining direct urinary injection with automated filtration and nanoflow LC-MS for the confirmatory analysis of doping-relevant small peptide hormones. *J. Chromatogr. B Anal. Technol. Biomed. Life Sci.* **2021**, *1179*, 122842. [CrossRef] [PubMed]
5. Min, H.; Han, B.; Sung, C.; Park, J.-H.; Lee, K.; Kim, H.; Kim, K.; Son, J.; Kwon, O.-S.; Lee, J. LC-MS/MS method for simultaneous analysis of growth hormone-releasing peptides and secretagogues in human urine. *Mass Spectrom. Lett.* **2016**, *7*, 55–63. [CrossRef]
6. Timms, M.; Hall, N.; Levina, V.; Vine, J.; Steel, R. A high-throughput LC-MS/MS screen for GHRP in equine and human urine, featuring peptide derivatization for improved chromatography. *Drug Test. Anal.* **2014**, *6*, 985–995. [CrossRef] [PubMed]
7. Thevis, M.; Thomas, A.; Schänzer, W. Doping control analysis of selected peptide hormones using LC-MS(/MS). *Forensic Sci. Int.* **2011**, *213*, 35–41. [CrossRef] [PubMed]
8. Cox, H.D.; Hughes, C.M.; Eichner, D. Detection of GHRP-2 and GHRP-6 in urine samples from athletes. *Drug Test. Anal.* **2015**, *7*, 439–444. [CrossRef] [PubMed]

9. Gómez-Guerrero, N.A.; González-López, N.M.; Zapata-Velásquez, J.D.; Martínez-Ramírez, J.A.; Rivera-Monroy, Z.J.; García-Castañeda, J.E. Synthetic Peptides in Doping Control: A Powerful Tool for an Analytical Challenge. *ACS Omega* **2022**, *7*, 38193–38206. [CrossRef] [PubMed]
10. Semenistaya, E.; Zvereva, I.; Thomas, A.; Thevis, M.; Krotov, G.; Rodchenkov, G. Determination of growth hormone releasing peptides metabolites in human urine after nasal administration of GHRP-1, GHRP-2, GHRP-6, Hexarelin, and Ipamorelin. *Drug Test. Anal.* **2015**, *7*, 919–925. [CrossRef]
11. Semenistaya, E.; Zvereva, I.; Krotov, G.; Rodchenkov, G. Solid-phase extraction of small biologically active peptides on cartridges and microelution 96-well plates from human urine. *Drug Test. Anal.* **2016**, *8*, 940–949. [CrossRef]
12. Jing, J.; Tian, T.; Wang, Y.; Xu, X.; Shan, Y. Multi-analyte screening of small peptides by alkaline pre-activated solid phase extraction coupled with liquid chromatography-high resolution mass spectrometry in doping controls. *J. Chromatogr. A* **2022**, *1676*, 463272. [CrossRef]
13. WADA Technical Document-TD2022MRPL Document Number: TD2022MRPL Version Number: 1.0 Minimum Required Performance Levels and Applicable Minimum Reporting Levels for Non-Threshold Substances Analyzed by Chromatographic-mass Spectrometric Analytical Methods. Available online: https://www.wada-ama.org/sites/default/files/resources/files/td2022mrpl_v1.0_final_eng.pdf (accessed on 1 May 2024).
14. Esposito, C.L.; Ac, A.G.; Laszlo, E.; Duy, S.V.; Michaud, C.; Sauvé, S.; Ong, H.; Marleau, S.; Banquy, X.; Brambilla, D. A Quantitative UHPLC-MS/MS Method for the Growth Hormone-Releasing Peptide-6 Determination in Complex Biological Matrices and Transdermal Formulations. *Talanta* **2021**, *233*, 122555. [CrossRef]
15. Reverter-Branchat, G.; Segura, J.; Pozo, O.J. On the Road of Dried Blood Spot Sampling for Antidoping Tests: Detection of GHRP-2 Abuse. *Drug Test. Anal.* **2021**, *13*, 510–522. [CrossRef]
16. Lange, T.; Thomas, A.; Walpurgis, K.; Thevis, M. Fully Automated Dried Blood Spot Sample Preparation Enables the Detection of Lower Molecular Mass Peptide and Non-Peptide Doping Agents by Means of LC-HRMS. *Anal. Bioanal. Chem.* **2020**, *412*, 3765–3777. [CrossRef]
17. Judák, P.; Esposito, S.; Coppieters, G.; van Eenoo, P.; Deventer, K. Doping control analysis of small peptides: A decade of progress. *J. Chromatogr. B Anal. Technol. Biomed. Life Sci.* **2021**, *1173*, 122551. [CrossRef]
18. Zhang, Y.; Huang, C.; Kong, F.; Wang, Y.; Shi, Q.; Zhang, L. Selective molecular characterization of olefins in hydrocarbon mixtures by Ag+ complexation ESI high-resolution mass spectrometry. *Fuel* **2022**, *319*, 123760. [CrossRef]
19. Medvedovici, A.; Lazou, K.; D'Oosterlinck, A.; Zhao, Y.; Sandra, P. Analysis of jojoba oil by LC-coordination ion spray-MS (LC-CIS-MS). *J. Sep. Sci.* **2002**, *25*, 173–178. [CrossRef]
20. Gómez-Guerrero, N.A.; González-López, N.M.; Zapata-Velásquez, J.D.; Martínez-Ramírez, J.A.; Rivera-Monroy, Z.J.; García-Castañeda, J.E. Synthesis of polysulfanes with 20 or more sulfur atoms with characterization by UPLC-(Ag^+)-coordination ion spray-MS. *J. Sulfur. Chem.* **2013**, *34*, 55–66.
21. Cha, E.; Kim, S.; Kim, H.J.; Lee, K.M.; Kim, K.H.; Kwon, O.S.; Lee, J. Sensitivity of GC-EI/MS, GC-EI/MS/MS, LC-ESI/MS/MS, LC-Ag^+ CIS/MS/MS, and GC-ESI/MS/MS for analysis of anabolic steroids in doping control. *Drug Test. Anal.* **2015**, *7*, 1040–1049. [CrossRef]
22. Cha, E.; Lee, K.M.; Park, K.D.; Park, K.S.; Lee, K.W.; Kim, S.M.; Lee, J. Hydroxycholesterol levels in the serum and cerebrospinal fluid of patients with neuromyelitis optica revealed by LC-Ag^+CIS/MS/MS and LC-ESI/MS/MS with picolinic derivatization: Increased levels and association with disability during acute attack. *PLoS ONE* **2016**, *11*, 0167819. [CrossRef]
23. Virus, E.D.; Sobolevsky, T.G.; Rodchenkov, G.M. "Wrong-Way-Round Ionization" and Screening for Doping Substances in Human Urine by High-Performance Liquid Chromatography/Orbitrap Mass Spectrometry. *J. Mass. Spectrom.* **2012**, *47*, 381–391. [CrossRef]
24. Venter, P.; van Onselen, R. Evaluating the "Wrong-Way-Round" Electrospray Ionization of Antiretroviral Drugs for Improved Detection Sensitivity. *Anal. Bioanal. Chem.* **2023**, *415*, 1187–1193. [CrossRef] [PubMed]
25. Seal, J.R.; Havrilla, C.M.; Porter, N.A.; Hachey, D.L. Analysis of unsaturated compounds by Ag+ coordination ionspray mass spectrometry: Studies of the formation of the Ag+/lipid complex. *J. Am. Soc. Mass. Spectrom.* **2003**, *14*, 872–880. [CrossRef] [PubMed]
26. Cecchi, T.; Pucciarelli, F.; Passamonti, P.; Ferraro, S. Influence of metal impurities sorption onto a silica based C18 stationary phase on the HPLC of metal chelating analytes. *J. Liq. Chromatogr. Relat. Technol.* **1999**, *22*, 429–440. [CrossRef]
27. de Pra, M.; Greco, G.; Krajewski, M.P.; Martin, M.M.; George, E.; Bartsch, N.; Steiner, F. Effects of titanium contamination caused by iron-free high-performance liquid chromatography systems on peak shape and retention of drugs with chelating properties. *J. Chromatogr. A* **2020**, *1611*, 460619. [CrossRef]
28. FDA Guidance Document M10 Bioanalytical Method Validation and Study Sample Analysis. Available online: https://www.fda.gov/media/162903/download (accessed on 19 April 2024).
29. European Medicine Academy Guideline on Bioanalytical Method Validation. Available online: https://www.ema.europa.eu/en/documents/scientific-guideline/guideline-bioanalytical-method-validation_en.pdf (accessed on 19 April 2024).

separations

Article

Multiplex Detection of Seven Staphylococcal Enterotoxins Using Liquid Chromatography–Mass Spectrometry Combined with a Novel Capture Molecule

Jing Lv [1], Tingting Liu [1], Xinyu Fang [1], Songyang Han [2], Lina Dong [3], Jiaxin Li [1], Jing Wang [1], Jinglin Wang [1], Shan Gao [1], Lin Kang [1,*] and Wenwen Xin [1,*]

[1] State Key Laboratory of Pathogen and Biosecurity, Beijing Institute of Microbiology and Epidemiology, Academy of Military Medical Sciences (AMMS), Beijing 100071, China; jinglv1205@163.com (J.L.); liutingtingwushi@163.com (T.L.); fxytt0327@163.com (X.F.); ljx1658675586@163.com (J.L.); amms_wj@163.com (J.W.); wilwj10801@sina.com (J.W.); gaoshan845@163.com (S.G.)

[2] College of Life Science and Technology, Beijing University of Chemical Technology, Beijing 100029, China; hansongyang0.0@outlook.com

[3] School of Basic Medicine, Anhui Medical University, Hefei 230032, China; donglina2024@163.com

* Correspondence: kanglin@bmi.ac.cn (L.K.); xinwenwen@bmi.ac.cn (W.X.); Tel.: +86-130-1188-8501 (L.K.); +86-132-6912-8674 (W.X.)

Abstract: Food poisoning caused by Staphylococcal enterotoxins (SEs) is prevalent globally, making efficient detection of these toxins very important. Traditionally, liquid chromatography–mass spectrometry required immunosorbent enrichment by magnetic bead-coupled antibodies obtained by animal-specific immunization. However, this method is time-consuming and costly. In this study, two recombinant protein capture molecules were designed based on the principle of toxins binding to Major Histocompatibility Complex (MHCII) and T cell receptor (TCR) molecules. The two capture molecules are called MHCII and MHCII-D10. The design of the MHCII and TCR-D10 was achieved through searching for the binding site protein sequence of Staphylococcal enterotoxins in the relevant literature, and MHCII-D10 was to link MHCII sequence with TCR-D10 sequence using linker (G4S)3 linking peptide. These capture molecules were shown to effectively bind to seven types of toxins and to capture SEs in various matrices. The digestion time, ratio, and temperature were further optimized, reducing the overall digestion time to just 2 h. The specificity, linearity, sensitivity, precision (RSD%), and recovery of the two methods were verified by liquid chromatography–mass spectrometry. When the MHCII and MHCII-D10 captured the toxins, the limit of quantification (LOD) in the 1 × PBS, plasma, and milk matrices ranged from 1.5625 to 100 fmol/μL, with the recovery rate ranging from 18.4% to 96%. The design of these capture molecules eliminates the need for animal-specific immunization, simplifying the pre-detection process and avoiding ethical concerns. This development holds significant promise for clinical diagnosis and reference.

Keywords: capture molecules; liquid chromatography–mass spectrometry; staphylococcal enterotoxins detection; major histocompatibility complex; T cell receptor

Citation: Lv, J.; Liu, T.; Fang, X.; Han, S.; Dong, L.; Li, J.; Wang, J.; Wang, J.; Gao, S.; Kang, L.; et al. Multiplex Detection of Seven Staphylococcal Enterotoxins Using Liquid Chromatography–Mass Spectrometry Combined with a Novel Capture Molecule. *Separations* **2024**, *11*, 136. https://doi.org/10.3390/separations11050136

Academic Editors: Gavino Sanna and Javier Saurina

Received: 27 March 2024
Revised: 20 April 2024
Accepted: 26 April 2024
Published: 29 April 2024

1. Introduction

Staphylococcus aureus (*S. aureus*) is a common bacteria associated with food poisoning [1]. Indeed, *S. aureus* poses a global food safety problem, as the bacteria can contaminate a range of foods from processed meats and dairy products to salads and cooked food [2]. It typically poisons through ingestion or inhalation, with common symptoms including fever, chills, headache, myalgia, coughing, and difficulty breathing; severe cases may exhibit vomiting, diarrhea, or even multi-organ and systemic damage, as well as immune dysfunction [3].

Staphylococcal enterotoxins (SEs) primarily cause this poisoning effect. These enterotoxins are a type of SEs, a group of monomeric polypeptide chains comprising 238–258 amino acids, with a molecular weight ranging from 26 to 34 kDa [4]. There are approximately 27 distinct serological types of SEs that exhibit structural, functional, and sequential similarities [5]. The most prevalent serotypes associated with food poisoning, due to their superantigen activity, are SEA, SEB, SEC, SED, SEE, SEF, SEG, and SEH [6].

SEs are not only major causes of foodborne illness caused by bacterial toxins [7], but they also risk becoming potent biological warfare agents. With atomization and inhalation toxicity, ready availability, easy preparation, high toxicity, and a lack of medical countermeasures [8], SEs can serve as effective potential biological warfare agents. Therefore, establishing an efficient method for detecting SEs is crucial for clinical diagnosis, disease prevention, and anti-biowarfare applications. Various detection methods are currently employed, including biodetection, immunological molecular biology, biosensor technology, and liquid chromatography–mass spectrometry [5]. Of these, liquid chromatography coupled with mass spectrometry (LC-MS/MS) represents a newly developed analytical method recommended by the designated laboratory of the International Organization for the Prohibition of Chemical Weapons (IOPCW) and commonly employed by the network laboratories of the United Nations Secretary-General's Investigative Mechanism (UNSGM). LC-MS/MS offers several advantages over other techniques, such as a high sensitivity, low sample requirements, a rapid analysis, and a straightforward operation [9,10].

The detection of SEs using LC-MS/MS involves bottom–up methods. These methods rely on immune capture techniques for detecting the toxins in a complex sample, with specific antibodies playing an important role. Conventionally, specific antibodies, in this case polyclonal or monoclonal antibodies, are obtained through animal immunization, resulting in a cumbersome, time-consuming, and costly process. There is some literature on the mechanisms of the MHCII and TCR interactions with SEs and their application to the diagnosis and treatment of SE poisoning. There is also some literature on the use of T cell receptor structural domains for capturing the ELISA detection of SEs [11,12]. In this paper, we designed two capture molecules aiming to replace the traditional antibodies used in the LC-MS/MS detection of SEs, thereby making detection more efficient. These new capture molecules were designed to make use of the superantigen mechanism of SEs binding to MHCII class molecules [13–15]. Our molecules, MHCII and MHCII-D10, proved capable of simultaneously adsorbing seven serotypes of toxins (A, B, C, D, E, G, and H). Compared with the traditional antibodies, the new capture molecules simplified the detection process, eliminating the need for the time-consuming immunization of animals to obtain the antibodies. The capture molecules could be quickly obtained by protein expression and purification alone. Moreover, only one capture molecule was required to capture all seven toxins. Compared with the complicated procedure of obtaining the traditional antibodies, the capture molecules could be obtained quickly, and the detection time was shortened, which is crucial for clinical diagnosis and rapid detection.

As a second aim of our study, we explored the optimal digestive conditions for the LC-MS/MS detection of SEs. The LC-MS/MS detection of protein toxins typically involves the trypsin digestion process, which usually takes between 18 and 24 h [16]. This long digestion time and low yield of the target peptide are challenges for efficient detection. Researchers have attempted to address these issues by developing various digestion protocols, including organic solvent-assisted or surfactant-assisted methods [17–20], microwave-assisted or ultrasound-assisted protocols [21–25], and pressure-assisted methods [26,27]. However, the existing literature suggests that the auxiliary effect of adding organic reagents to *S. aureus* enterotoxin digestion is not optimal [18]. Therefore, this study focused on optimizing the digestion temperature, time, and protein-to-trypsin ratio of SEs. The optimized digestion method can improve the yield of peptides, shorten the digestion time, and improve the detection efficiency [28].

2. Materials and Methods

2.1. Safety Precautions

SEs are highly toxic, so the safety rules for handling toxic contamination and the use of personal protective equipment were strictly observed. The SE-contaminated consumables and solutions were inactivated overnight with 2 M of NaOH.

2.2. Chemicals and Reagents

Acetonitrile (HPLC grade) and formic acid (HPLC grade) (>98%) were purchased from Thermo Fisher Scientific (Waltham, MA, USA). Ammonium bicarbonate (NH_4HCO_3), dithiothreitol (DTT), and iodoacetamide (IAA) were purchased from Sigma (Roedermark, Germany). Sequencing grade modified trypsin was sourced from Promega Corporation (Madison, WI, USA). Ultrapure water was obtained from Wahaha pure water (Hangzhou, China). Dynabeads® M-270 Epoxy magnetic beads were obtained from Thermo Fisher Scientific (USA). The capture molecule MHCII-D10 is a designed sequence, which was synthesized by the GenScript company and purified by the recombinant protein. Synthetic heavy peptides with labeled Lys [13C6; 15N2] (+8 Da) and Arg [13C6; 15N4] (+10 Da) were synthesized by the GenScript company at a high purity grade. Semi-skimmed milk was purchased from local stores. The animal immune polyclonal rabbit antibody was commissioned by Beijing Protein Innovation Company. Sheep plasma was purchased from Solebol. The anti-*Staphylococcus aureus* Enterotoxins A + B + TSST-1 antibody ab190337) were purchased from Abcam Company. Imidazole ($C_3H_4N_2$), disodium hydrogen phosphate (Na_2HPO_4), dihydrogen phosphate (NaH_2PO_4), sodium chloride (NaCl), and sodium hydroxide were obtained from the China National Medicines Corporation Ltd. Urea, reduced glutathione, oxidized glutathione, arginine, and ethylene diamine tetraacetic acid (EDTA) were obtained from the China National Medicines Corporation Ltd. Phosphate-buffered saline ($1 \times$ PBS) and 1 M Tris-HCl were obtained from Solarbio.

2.3. Protein Sequence Information

We obtained the SE serotype sequences from Uniprot (http://www.uniprot.org). These serotypes were SEA, SEB, SEC, SED, SEE, SEH, and SEG. The full length of the seven enterotoxin sequences were 702 bp, 720 bp, 720 bp, 702 bp, 693 bp, 654 bp, and 702 bp, respectively. The protein sequences of the toxins were rearranged to remove the signal peptide. Two restriction sites of NcoI and XhoI, six His tags, and the restriction site R were added to the design. Then, the GenScript company added our designed recombinant protein sequence into the pET-28a(+) carrier plasmid to synthesize the glycerobacteria before carrying out the expression and transformation to produce the recombinant protein [29] The design of the recombinant capture molecules mainly relied on the antigenic peptide binding sites of the SEs and the MHCII and TCR molecules. Upon analysis, we determined that the binding part of the MHCII to the toxin was divided into the α1 chain domain. The α1 domain sequence 1–84 and the amino acid sequence of the MHCII were identified. MHCII-D10 capture molecules link MHCII and TCR-D10 antibodies via a (G4S)3 linker. The designed sequence was sent to the GenScript company, and our recombinant protein sequence was added to the pET-28a(+) carrier plasmid to synthesize glycerobacteria, which were then expressed and transformed into the recombinant protein. The sequence is shown in Figure S1.

2.4. Screening of Specific Marker Peptides for Staphylococcal aureus Enterotoxin

The amino acid sequences of the SEA (P0A0L1), SEB (P01552), SEC (P01553), SED (P20723), SEE (P12993), SEH (P0A0L9), and SEG (P0A0L8) were obtained from the UniProt database. The proteins were digested via an online theoretical digestion method, and the peptides were classified. DNAMAN 7.0 software was used to compare the sequences of the seven toxins and to screen the different peptides. As shown in Figure S2, the ACQUITY UPLC®I-Class-Xevo G2-XS Tof -MS/MS was used in a bottom–up approach to screen the peptides with good signals, and the NCBI non-redundant protein database was used to search for specific peptides.

2.5. Induced Expression and Purification of the Target Protein

In total, 100 μL of the SE-engineered bacteria solution was added to two 5 mL of Luria–Bertani (LB) liquid mediums containing Kanamycin (Kana) and incubated at 180 rpm at 37 °C overnight. Then, 5 mL of the bacterial solution cultured overnight was added to 500 mL of the LB medium containing Kana, one of which was supplemented with Isopropyl β-D-Thiogalactoside (IPTG) inducer at a final concentration of 1 mM. The cultures were shaken at 180 rpm at 37 °C for 4 h until the OD_{600} reached between 0.6 and 0.8. Meanwhile, a blank control group was set up without the inducer.

The bacterial solutions of the induction group and the blank control group were then centrifuged at 8000 rpm for 20 min. After the centrifugation, the supernatant medium was discarded and the precipitation was re-suspended in 80 mL of 1 × PBS. After the suspension, the ultrasonic crushing precipitation was performed, followed by centrifugation at 8000 rpm for 20 min. After the centrifugation, the supernatant was separated from the precipitation, and the precipitation was suspended in 20 mL of 1 × PBS. The blank control group and the induced bacteria were tested with a 4–20% SDS-PAGE gel electrophoresis to verify whether the bacteria induced expression. For the gel electrophoresis, the voltage was set to 180 V, the current to 400 mA, and the run time to 40 min. The expressed strain was then expanded, and the target protein was purified with an affinity chromatography (HisTrap™HP column). After the centrifugation of the sonicated bacterial solution at 180 rpm for 20 min, the bacterial solution was filtered through a 0.45 mm flow-through membrane. The HisTrap™ HP column was first equilibrated with five times the column volume of the equilibration solution. The filtered sample was aspirated onto the column at a flow rate of 2 mL/min and purified according to the principle of a low concentration of imidazole for the binding and a high concentration elution. The equilibrium solution was a solution containing a low 19 nM concentration of imidazole ($C_3H_4N_2$), 39 μM of disodium hydrogen phosphate (Na_2HPO_4), 5.8 μM of sodium dihydrogen phosphate (NaH_2PO_4), and sodium chloride (NaCl), and the eluent was a high concentration equilibrium solution containing a high concentration of 500 μM of imidazole. The proteins were eluted in a gradient of 0–100% at a flow rate of 3 mL/min, and the eluted proteins were divided into 1–7 fractions. The proteins were collected as target proteins in 1–7 tubes. Then, the 4–20% SDS-PAGE gel electrophoresis was performed. The protein concentration was measured with a BCA by mixing the homogeneous proteins with a higher purity.

The recombinant MHCII and MHCII-D10 antibodies were expressed in inclusion bodies. Therefore, the culture medium was added and centrifuged at 8000 rpm for 20 min. After the separation, the precipitate was collected and the bacteria were suspended in 2.5 mL of 8 M urea–imidazole solution at 4 °C overnight. An ultrasonic disruption at 100% was performed by centrifugation at 8000 rpm for 20 min, and the supernatant was retained. The HisTrap™ HP column was first equilibrated with five times the column volume of the equilibration solution. The filtered sample was applied to the column and purified by the principle of a low concentration of imidazole for the binding and a high concentration elution. The equilibration solution was an imidazole-containing eluent, and the eluted proteins were collected as target proteins. After collecting the protein solution, the required volume of a commercial dialysis bag was removed and washed with distilled water. The revenge solution was composed of 1 L of purified water, 2 M of Urea, 0.002 M of reduced glutathione, 0.004 M of oxidized glutathione, 0.2 M of arginine, and 2 mM of ethylene diamine tetraacetic acid (EDTA). The protein solution was dialyzed overnight in 100 mM of Tris-HCl with fluid changes every 4 h. After 48 h, the protein solution was collected. Then, the 4–20% SDS-PAGE gel electrophoresis was performed. The results of the protein purification and gel electrophoresis are shown in Figure S3.

2.6. Optimization of Trypsin Digestion Conditions

The SEC was digested with modified trypsin. In total, 30 μL of SEC and 270 μL of 50 nM of amine bicarbonate (NH_4HCO_3) were added into EP tubes and mixed well. Then, 3 μL of 1 M dithiothreitol (DTT) was added into the mixture and incubated at 56 °C for 1 h. Then, 9 μL of 1 M iodoacetamide (IAA) was added into the mixture and incubated avoiding light for 45 min. The trypsin was added at a trypsin/SEC ratio of 1:100, 1:50, 1:25, and 1:5 and incubated at 37 °C. There was 50 μL of the sample that was taken at 4 h, 8 h, 12 h, 16 h, 24 h, and 28 h, and the digestion was terminated with 10% of formic acid (FA). After termination, the samples were centrifuged at 12,000× *g* for 10 min and analyzed with the LC-MS/MS.

In order to obtain the best digestion time, 50 μL of SEC was added into 300 μL of 50 nM NH_4HCO_3 and mixed well. Then, 3.5 μL of 1 M DTT was added and incubated at 56 °C for 1 h. In total, 10 μL of 1 M IAA was added into the mixture and incubated avoiding light for 45 min. The trypsin was added at a trypsin/SEC ratio of 1:50 and 1:5, and the mixture was incubated at different temperatures (50 °C and 60 °C) for 0.5 h, 1 h, 1.5 h, 2 h, 4 h, 6 h, and 8 h. There was 50 μL of the sample that was taken and terminated with 10% formic acid. After termination, the samples were centrifuged at 12,000× *g* for 10 min and analyzed with the LC-MS/MS.

In order to obtain the optimal digestion temperature, 5 μL of SEC was added to 45 μL each of 50 nM NH_4HCO_3 and mixed well. Then, 0.5 μL of 1 M DTT was added and incubated at 56 °C for 1 h. Then, 1.5 μL of 1 M IAA was added into the mixture and incubated avoiding light for 45 min. The trypsin was added at a trypsin/SEC ratio of 1:50 and 1:5, and the mixture was incubated at different temperatures (50 °C, 60 °C, 65 °C, and 70 °C) for 2 h. After termination with 10% FA, the samples were centrifuged at 12,000× *g* for 10 min and analyzed with the LC-MS/MS.

2.7. The Ability of Recombinant Antibodies to Bind to Toxins

The binding ability of the capture molecules to the toxins was verified by molecular interaction experiments. The capture molecule MHCII was labeled with biotin and then immobilized onto an Streptavidin (SA) sensor at a concentration of 1 μg/mL. Different concentrations of the toxin were then tested to explore and determine the binding capacity of the MHCII capture molecule to the toxin, with concentrations set at 3200 nM, 1600 nM, 800 nM, 400 nM, 200 nM, and 0 nM. Similarly, the recombinant antibody MHCII-D10 was used at a curing concentration of 1 μg/mL, and its binding concentration to the toxin was tested at 200 nM, 100 nM, 50 nM, 25 nM, 12.5 nM, and 0 nM. The concentrations of polyclonal antibodies obtained through animal immunization (pAb), namely pAb-SEA and pAb-SEB, were determined, and the binding concentrations were set at 200 nM, 100 nM, 50 nM, 25 nM, 12.5 nM, and 0 nM. The binding concentrations of the pAb-SEC, pAb-SED, pAb-SEE, pAb-SEH, and pAb-SEG to the toxins were tested at 3200 nM, 1600 nM, 800 nM, 400 nM, 200 nM, and 0 nM. The Octet molecular interaction platform was used for analysis, and the binding ability of the toxin to the capture molecules was determined through data processing.

2.8. SEs Extraction from Complex Matrices

Different concentrations of SEs were added to the 1 × PBS milk and plasma samples at concentration gradients of 100 fmol/μL, 50 fmol/μL, 25 fmol/μL, 12.5 fmol/μL, 6.25 fmol/μL, 3.125 fmol/μL, 1.5625 fmol/μL, and 0.7825 fmol/μL. Each sample was enriched by adding 0.2 mg of magnetic beads before incubation with SEs at room temperature for 1 h.

2.9. Multiplex Immunocapture of SEs

The capture molecules were coupled to the Dynabeads® Capture Molecules Coupling Kit magnetic beads (Thermo Fisher Scientific, USA) at a ratio of 100 μL (10 mg/mL) of beads to 25 μg of capture molecules and incubated for 1 h in complex matrices. We used mass spectrometry to quantify the toxin in the complex matrix after the magnetic bead adsorption, and the results showed that 1 mg of antibody could bind approximately 10 μg of toxin. The capture molecules conjugated to the magnetic beads were incubated with different concentrations of the toxin in the complex matrix. After incubation, the magnetic beads were adsorbed using a magnetic rack. The complex matrix adsorbed by the magnetic beads was removed and washed once with 1 × PBS and twice with 50 mM of NH_4HCO_3. Then, 50 μL of 50 mM NH_4HCO_3. DTT (0.5 μL and 10 mM) was added to the beads, and they were subsequently heated for 1 h at 56 °C to induce the toxin reduction. Next, 1.5 μL of IAA (30 mM) was added, and the mixture was incubated for 45 min in the dark at room temperature. The trypsin was added at a trypsin/SEs ratio of 1:5, and the SEs were digested at 60 °C for 2 h. The digestion was stopped with 2.5 μL of 10% FA, and 5 μL of the labeled peptide mixture was added to the samples. Finally, 10 μL was injected into the UPLC-MS/MS system.

2.10. Liquid Chromatography–Mass Spectrometry Analysis

The UPLC-MS/MS experiment was performed using the ACQUITY UPLC®I-Class-Xevo TQ-S. The peptide was isolated with a C18 reversed-phase chromatography (ACQUITY UPLC@BEH C18 300A 2.1×*50 mm 1.7 μm). The column temperature was 40 °C, and the mobile phases were 0.1% formic acid water (A) and 0.1% formic acid ACN (B). After an initial equilibration step with 95% of water, a gradient from 95% to 10% of the aqueous phase was applied over 10 min at a flow rate of 0.3 mL/min. The ion source used for the mass spectrometry was UniSpray+. The data acquisition was performed in positive ion mode using multi-reaction monitoring (MRM). The capillary voltage and ion source temperature were set to 3.5 kV and 500 °C, respectively. The precursor ions were selected from both unlabeled peptides and internal labeled peptides, and the daughter ions were optimized by adjusting the cone voltage and collision energy.

2.11. Method Evaluation

The performance of the method was evaluated across the milk, 1 × PBS, and sheep plasma matrices. The specificity, linearity, sensitivity, immunocapture recovery, and precision of the proposed method were evaluated.

2.11.1. Specificity

The capture molecules conjugated to the magnetic beads were added to blank matrices (milk, plasma, and 1 × PBS) and the matrices containing Botulinum toxin and Ricin (milk, plasma, and 1 × PBS) before incubation at room temperature. The specificity was verified by mass spectrometry after digestion.

2.11.2. Linearity Range and Sensitivity

An 8-point standard curve was generated for the milk and plasma between 0.7825 fmol/μL and 100 fmol/μL. The internal standard peptide was added after digestion, and the results were analyzed with the UPLC-MS/MS. The calibration curves were generated by linear regression analysis. The lower limit of qualitative (LOD) was defined as three times the signal-to-noise ratio (S/N). The lower limit of quantification (LOQ) was defined as S/N > 10.

2.11.3. Immunocapture Recovery and Precision (RSD)

To evaluate the immunocapture recovery of the SEs from different matrices, we compared the content of the total SES in the eluent with that in the eluate during the immunocapture at three levels (QCL, QCM, and QCH) in triplicate. Three concentration samples were set: 100 fmol/μL, 25 fmol/μL, and 3.125 fmol/μL or 6.25 fmol/μL were added into the samples of PBS, milk, plasma, and other complex substrates, and 0.2 mg of magnetic beads coupled with the MHCII and MHCII- D10 antibodies were added. After incubation at room temperature for 1 h, the digestion was performed. After digestion, the sample concentration was detected on the UPLC-MS/MS. The relative deviation of peptide (RSD%) was measured.

3. Results and Discussion

3.1. Sequence Characterization of Toxin Standards and Selection of Proteotypic Peptides

The SE toxin protein sequences were collected for the theoretical digestion analysis. The Peptide Cutter tool was used to predict the theoretically digested peptides (https://web.expasy.org/peptide_cutter/, accessed on 26 March 2024), and the DNAMAN 7.0 software was used to compare the toxin protein sequences of seven serotypes and to identify different peptide sequences, as shown in Table S1. The ACQUITY UPLC®I-Class-Xevo G2-XS Tof-MS/MS uses a bottom–up approach to match the peptide profiles after trypsin digestion. The peptides with stronger signals were selected by screening. The specific protein sequences were determined by a BLAST search of the National Center for Biotechnology Information (NCBI) non-redundant database. The peptide profile matching was performed using the ACQUITY UPLC®I-Class-Xevo G2-XS Tof-MS/MS after the trypsin digestion. The matching coverage areas were as follows: SEA 74%, SEB 93%, SEC 93%, SED 83%, SEE 99%, SEG 99%, and SEH 96%, as shown in Table S2.

3.2. Optimization of Digestion Conditions

The results showed that the highest peptide yield of peptides was achieved at a ratio of 1:5 at 37 °C. In addition, there was no decrease in the digestion efficiency even after 28 h, indicating that the digestion was continuing. We then increased the temperature and shortened the time to explore the shortest digestion time to obtain the maximum peptide yield. The result is shown in Figure 1a.

We tested the digestion at 50 °C and 60 °C, and the ratios of trypsin to toxin of 1:50 and 1:5. The digestion durations of 0.5 h, 1 h, 1.5 h, 2 h, 4 h, 6 h, and 8 h were tested. After digestion, the peak areas of five peptides were determined with the mass spectrometry. The results indicated that 2 h after digestion, the digestion effect reached its peak at 60 °C, with the highest peptide yield observed at this time. Subsequently, the efficiency of the digestion began to decline. Thus, the shortest digestion time of 2 h resulted in the highest peptide yield. The result is shown in Figure 1b.

Finally, the optimal digestion temperature was determined. By controlling the digestion time, different temperatures were set as 50 °C, 60 °C, 65 °C, and 70 °C, and the digestion ratio of trypsin to protein was set as 1:50 and 1:5. The termination reaction was compared after the digestion for 2 h, and the peak area was determined with the mass spectrometry peptide segment to find out the highest peptide yield at the optimal temperature. The results showed that the optimal digestion temperature was 60 °C at 2 h. The result is shown in Figure 1c.

The recently published literature did not specifically reflect the digestion time and conditions of SEs by modified trypsin [5,6], and only one article explored the kinetic properties of the sequencing grade modified trypsin and the temperature at which the peptide yield was improved [28]. We systematically explored the optimal digestion efficiency and peptide yield from three aspects: the digestion temperature, digestion time, and digestion ratio. The results were significant.

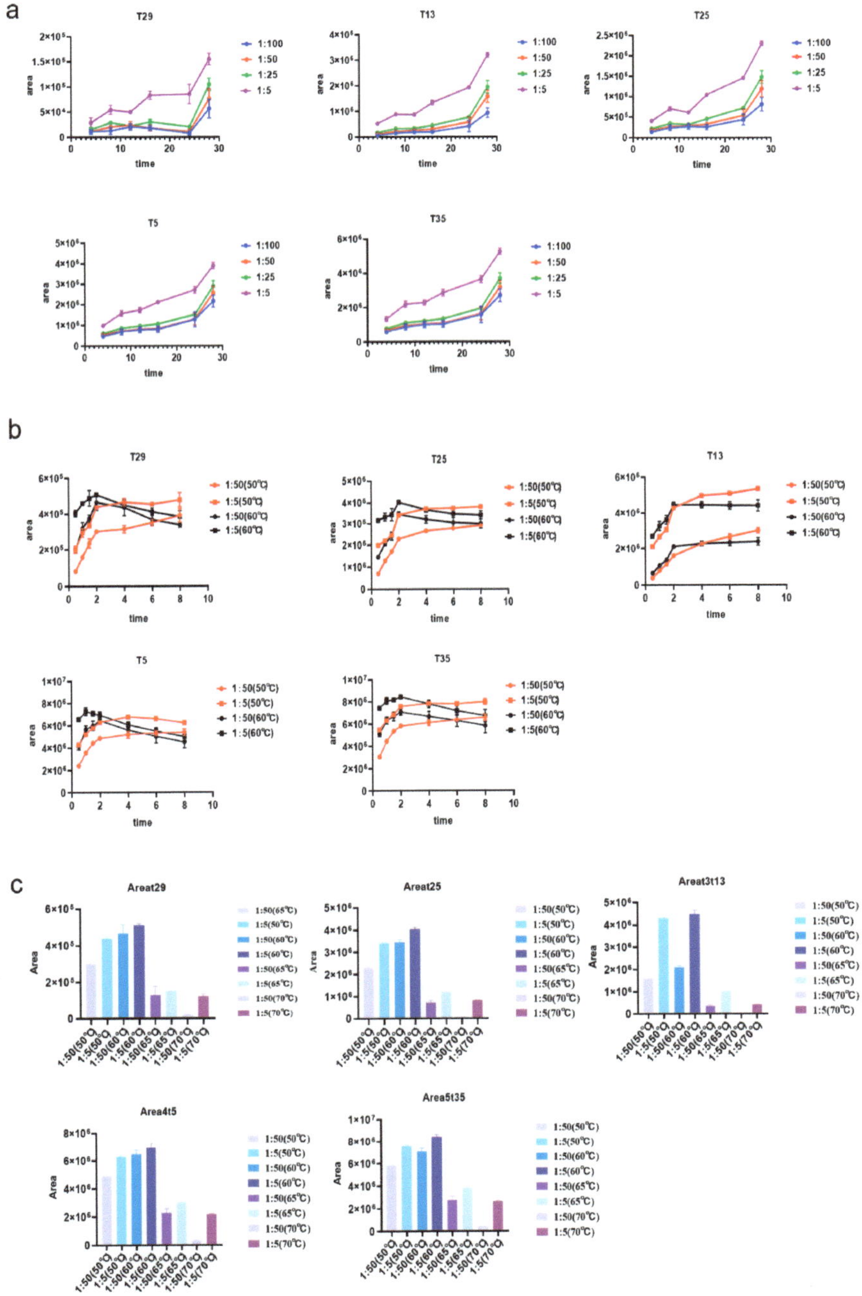

Figure 1. (**a**) The optimal ratio of trypsin to protein was studied at 37 °C by selecting digestion ratios of 1:100, 1:50, 1:25, and 1:5. (**b**) The ratio of the protein and the digestion was set at 1:5 and 1:50, and the digestion temperature was set at 50 °C and 60 °C. Seven time points were selected to explore the optimal digestion time. (**c**) The ratio of protein to trypsin was 1:5 and 1:50. Four digestion temperatures were selected as 50 °C, 60 °C, 65 °C, and 70 °C, to explore the optimal digestion temperature.

3.3. The Ability of Capture Molecules to Bind Toxins

The affinity of the two new capture molecules to various enterotoxins was tested using the Octet molecular interaction platform. Data processing was conducted to obtain the binding ability of the toxins and capture molecules. The Data Analysis 12.0 software was used for the data analysis; the baseline concentration was removed; and the fitting curve analysis was performed according to the dissociation signal data. The value of the correlation affinity constant KD, which equals the dissociation rate constant/binding rate constant (KD = kdis/kon), was obtained, where Kdis reflects the stability of the complex and the percentage of the dissociated complex per second. The faster the dissociation of Kon represents the rate of complex formation (AB). KD represents the size of the binding capacity of the interaction. The results showed that the lowest concentration of the MHCII binding to the seven toxins was 200 nM, with a range of equilibrium dissociation constant (KD) values from 10^{-8} M to 10^{-12} M. Specifically, the KD of the SED was in the range of 10^{-12} M. The lowest binding concentration of the MHCII-D10 with the seven toxins was 12.5 nM. The KD range was 10^{-9}–10^{-11} M, of which the KD of the SED was most optimal at 10^{-11} M. The minimum binding concentration of the specific antibody pAb, obtained by the immunization of animals, to the toxins was 200 nM, and the range KD was 10^{-8}–10^{-12} M. The results showed that two capture molecules, MHCII and MHCII-D10, had the same affinity for toxins as the pAb. The results are presented in Table 1.

Table 1. Protein binding to capture molecules.

Capture Molecules	Protein	KD (M)	Kon (1/Ms)	Kdis (1/s)	R^2	X^2
MHCII	SEA	3.47×10^{-8}	5.60×10^{4}	1.94×10^{-3}	0.9853	0.2332
MHCII	SEB	3.34×10^{-9}	3.39×10^{5}	1.13×10^{-3}	0.9646	1.5755
MHCII	SEC	9.47×10^{-9}	2.53×10^{5}	2.40×10^{-3}	0.9951	1.0342
MHCII	SED	1.0×10^{-12}	3.40×10^{4}	1.0×10^{-7}	0.993	3.4625
MHCII	SEE	1.43×10^{-8}	1.47×10^{5}	2.10×10^{-3}	0.9694	0.7162
MHCII	SEG	6.48×10^{-9}	9.98×10^{4}	6.47×10^{-4}	0.9773	0.7921
MHCII	SEH	4.45×10^{-8}	1.49×10^{4}	6.64×10^{-4}	0.9821	2.1512
MHCII-D10	SEA	3.89×10^{-9}	2.27×10^{5}	8.82×10^{-4}	0.9971	0.0083
MHCII-D10	SEB	2.62×10^{-9}	4.84×10^{5}	1.27×10^{-3}	0.9909	0.1127
MHCII-D10	SEC	3.38×10^{-9}	3.98×10^{5}	1.35×10^{-3}	0.9914	0.0527
MHCII-D10	SED	1.91×10^{-11}	5.00×10^{4}	9.55×10^{-7}	0.9326	3.8586
MHCII-D10	SEE	4.99×10^{-9}	3.18×10^{5}	1.59×10^{-3}	0.9947	0.045
MHCII-D10	SEG	4.42×10^{-9}	1.54×10^{5}	6.80×10^{-4}	0.9979	0.0347
MHCII-D10	SEH	9.19×10^{-9}	1.79×10^{5}	1.65×10^{-3}	0.9849	0.0224
pAb-SEA	SEA	4.84×10^{-9}	1.39×10^{5}	6.72×10^{-4}	0.9984	0.0202
pAb-SEB	SEB	1.76×10^{-9}	2.94×10^{5}	5.18×10^{-4}	0.9949	0.0991
pAb-SEC	SEC	1.36×10^{-8}	8.27×10^{4}	1.12×10^{-3}	0.9600	0.4866
pAb-SED	SED	1.0×10^{-12}	1.56×10^{4}	1.0×10^{-7}	0.9957	2.9424
pAb-SEE	SEE	1.54×10^{-8}	1.17×10^{5}	1.80×10^{-3}	0.9483	0.2682
pAb-SEG	SEG	1.41×10^{-8}	3.13×10^{3}	4.40×10^{-5}	0.9673	2.454
pAb-SEH	SEH	7.00×10^{-8}	1.01×10^{4}	7.05×10^{-4}	0.9986	0.1528

3.4. UPLC-MS/MS (MRM) Method Development

A sensitive UPLC-MS/MS (MRM) method was developed using the MassLynxV4.2 software (Waters) to refine the MRM parameters of SE-specific peptides such as precursor ion, product ion, cone voltage, and collision energy. Table 2 shows the optimization parameters of the precursor ions and product ions of each serotype-labeled peptide and its corresponding isotope-labeled synthetic peptide. For the optimization of the collision energy, Skyline (64 bit) software was used. Using the software algorithms, the product ions with a high response signal and the optimal collision energy were selected.

Table 2. Optimized MRM parameters of specific peptide segment.

Peptide	Amino Acid Sequences	Precursor Ion (*m*/*z*)	Product Ions (*m*/*z*)	Cone (V)	Collision Energy (V)
SEA3	QNTVPLETVK	565.00 (2+)	686.30, 476.10, 380.20	35	15, 25, 20
SEA4	NVTVQELDLQAR	693.70 (2+)	487.30, 396.10, 214.10	35	35, 20, 20
SEA3R*	QNTVPLETV	569.00 (2+)	560.00, 694.90, 894.60	35	20, 15, 20
SEB2	VLYDDNHVSAINVK	529.90 (3+)	744.60, 687.90, 213.20	35	15, 15, 15
SEB4	LGNYDNVR	476.10 (2+)	837.40, 503.25, 86.10	35	15, 15, 25
SEB6	VTAQELDYLTR	654.90 (2+)	909.50, 780.50, 276.20	35	20, 20, 20
SEB6R*	VTAQELDYLT	660.10 (2+)	919.50, 562.40, 201.20	35	20, 30, 20
SEC1	VLYDDHYVSATK	471.20 (3+)	656.60, 600.00, 461.00	35	15, 10, 15
SEC2	TELLNEGLAK	544.30 (2+)	744.42, 631.34, 344.18	35	19, 19, 19
SEC3R*	FLAHDLIYNISD	519.80 (3+)	583.30, 469.60, 356.61	35	15, 15, 20
SED4	NVDVYPIR	488.60 (2+)	762.50, 647.50, 548.30	35	15, 20, 15
SED5	LYNNDTLGGK	548.30 (2+)	818.40, 249.20, 136.10	35	18, 18, 30
SED5R*	LYNNDTLGG	552.10 (2+)	989.50, 826.40, 249.20	35	20, 15, 20
SEE1	NALSNLR	394.40 (2+)	602.30, 489.30, 299.20	35	10, 10, 10
SEE4	QTTVPIDK	451.40 (2+)	472.30, 326.30, 262.10	35	15, 20, 20
SEE4R*	QTTVPID	455.40 (2+)	680.40, 480.50, 270.30	35	15, 15, 20
SEH1	SDEISGEK	432.90 (2+)	662.30, 533.30, 420.40	35	15, 15, 15
SEH3	FATADLAQK	483.00 (2+)	746.40, 473.90, 364.90	35	15, 10, 15
SEH5R*	NVTLQELDI	591.40 (2+)	867.40, 625.40, 214.40	35	20, 20, 15
SEG2	TELENTELANNYK	769.87 (2+)	609.30, 538.26, 344.18	35	27, 27, 27
SEG4	NMVTIQELDYK	677.34 (2+)	1108.59, 345.16, 310.18	35	20, 20, 20
SEG6	FLNIYGDNK	542.80 (2+)	824.50, 596.40, 233.40	35	15, 15, 15
SEG6R*	FLNIYGDN	546.60 (2+)	831.40, 604.30, 233.30	35	15, 15, 15

R* Lys [13C6; 15N2] (+8 Da) and Arg [13C6; 15N4] labeled arginine.

3.5. Method Validation

The performance of the method was evaluated across the milk, 1 × PBS, and sheep plasma matrices. The specificity, linearity, sensitivity, immunocapture recovery, and precision of the proposed method were evaluated.

3.5.1. Specificity

The specificity was evaluated by analyzing analytical matrix blank samples (1 × PBS, milk, and plasma) and samples spiked with Botulinum toxin and Ricin in 1 × PBS, milk, and plasma. In each case, the background of each labeled peptide showed no interference peaks.

3.5.2. Linearity Range and Sensitivity

The SEs were added to 1 × PBS, milk, and plasma to prepare a sample of seven standard calibrations, ranging from 1.5625 fmol/µL to 100 fmol/µL. Each calibrator sample was captured on MHCII molecule beads, digested with trypsin, added with internal standard peptides, and analyzed with the UPLC-MS/MS (MRM). Similarly, seven serotypes of SEs were added to 1 × PBS, milk, and plasma to produce eight standard calibration samples in the range of 0.78125–100 fmol/µL. Each calibration agent sample was captured on MHCII-D10 molecule beads, digested with trypsin, added with internal standard peptide, and analyzed with the UPLC-MS/MS (MRM). Using MHCII as the capture molecule, the qualitative and quantitative detection limit, detection ranges, and linearity of different serotypes of toxins in different substrates are shown in Tables 3–5.

Table 3. Standard curve, linear range, limit of detection, and limit of quantification of different SES isoforms in 1× PBS. (MHCII as the capture molecule).

Peptide	Matrices	Calibration Curve	R^2	LOD (fmol/μL)	LOQ (fmol/μL)	Range (fmol/μL)
SEA4	PBS	Y = 0.005120X − 0.000189	0.997	1.5625	3.125	3.125–100
SEB2	PBS	Y = 0.061606X − 0.049921	0.999	1.5625	3.125	3.125–100
SEC2	PBS	Y = 0.000118X − 0.000156	0.981	1.5625	3.125	3.125–100
SED4	PBS	Y = 0.170828X − 0.403331	0.951	1.5625	3.125	3.125–100
SEE4	PBS	Y = 0.002636X + 0.001088	0.992	1.5625	3.125	3.125–100
SEH3	PBS	Y = 0.009839X − 0.002605	0.990	3.125	6.25	6.25–100
SEG6	PBS	Y = 0.155547X + 0.019418	0.990	3.125	6.25	6.25–100

Table 4. Standard curve, linear range, limit of detection, and limit of quantification of different SES isoforms in plasma. (MHCII as the capture molecule).

Peptide	Matrices	Calibration Curve	R^2	LOD (fmol/μL)	LOQ (fmol/μL)	Range (fmol/μL)
SEA4	Plasma	Y = 0.001202X − 0.000736	0.990	3.125	6.25	6.25–100
SEB2	Plasma	Y = 0.020269X − 0.011634	0.995	1.5625	3.125	3.125–100
SEC2	Plasma	Y = 0.004754X − 0.001350	0.990	3.125	6.25	6.25–100
SED4	Plasma	Y = 0.050879X − 0.069610	0.944	1.5625	3.125	3.125–100
SEE4	Plasma	Y = 0.001211X + 0.002182	0.997	3.125	6.25	6.25–100
SEH3	Plasma	Y = 0.004225X − 0.004713	0.992	6.25	12.5	12.5–100
SEG4	Plasma	Y = 0.157611X − 0.149189	0.987	1.5625	3.125	3.125–100

Table 5. Standard curve, linear range, limit of detection, and limit of quantification of different SES isoforms in milk. (MHCII as the capture molecule).

Peptide	Matrices	Calibration Curve	R^2	LOD (fmol/μL)	LOQ (fmol/μ)	Range (fmol/μL)
SEA4	Milk	Y = 0.001729X − 0.000568	0.998	3.125	6.25	6.25–100
SEB2	Milk	Y = 0.012888X + 0.004582	0.998	1.5625	3.125	3.125–100
SEC2	Milk	Y = 0.003255X − 0.003341	0.965	1.5625	3.125	3.125–100
SED4	Milk	Y = 0.019634X + 0.015828	0.956	1.5625	3.125	3.125–100
SEE4	Milk	Y = 0.002358X − 0.001658	0.993	1.5625	3.125	3.125–100
SEH3	Milk	Y = 0.15241X − 0.280831	0.980	3.125	6.25	6.25–100
SEG6	Milk	Y = 0.001537X − 0.004049	0.956	3.125	6.25	6.25–100

The result shows that one of the specific peptides of the different serotypes was quantified, and the calibration curves were generated by linear regression (Figure S4). Using MHCII as the molecule, the SEA, SEB, SEC, SED, and SEE showed good linearity in the range of 3.125–100 fmol/μL in the PBS substrate, with a coefficient of determination $R^2 > 0.951$. However, the SEE and SEH showed good linearity in the range of 6.25–100 fmol/μL, with $R^2 > 0.990$. In the plasma matrix, the SEB, SED, and SEG showed good linearity in the range of 3.125–100 fmol/μL, with $R^2 > 0.944$. However, the SEA, SEC and SEE showed good linearity in the range of 6.25–100 fmol/μL, with $R^2 > 0.990$. The SEH showed good linearity in the range of 12.5–100 fmol/μL, and $R^2 > 0.992$. In the milk samples, the SEB, SEC, SED, and SEE showed good linearity in the range of 3.125–100 fmol/μL, and $R^2 > 0.956$. The SEA, SEH, and SEG showed good linearity in the range of 6.25–100 fmol/μL, with $R^2 > 0.956$.

Using MHCII-D10 as the capture molecule, the qualitative and quantitative detection lines, detection ranges, and linearity of different serotypes of toxins in different substrates are shown in Tables 6–8. In the PBS matrix samples, the SEA, SEB, SEC, SED, SEE, and SEG showed good linearity in the range of 1.5625–100 fmol/μL, with $R^2 > 0.953$. However, the SEH showed good linearity in the range of 12.5–100 fmol/µL, with $R^2 > 0.964$. In the plasma matrix, the SEB and SED showed good linearity in the range of 1.5625–100 fmol/µL, with $R^2 > 0.935$. The SEG showed good linearity in the range of 3.125–100 fmol/µL, with $R^2 > 0.971$. The SEA, SEC, and SEE showed good linearity in the range of 6.25–100 fmol/µL, with $R^2 > 0.960$. The SEH showed good linearity in the range of 12.5–100 fmol/µL, with $R^2 > 0.997$. In the milk matrix samples, the SED showed a good linearity in the range of 3.125–100 fmol/µL, with $R^2 > 0.986$. The SEA, SEB, SEC, SEE, and SEG had good linearity in the range of 6.25–100 fmol/µL with $R^2 > 0.943$. The SEH had good linearity in the range of 12.5–100 fmol/µL, with $R^2 > 0.967$.

Table 6. Standard curve, linear range, limit of detection, and limit of quantification of different SES isoforms in 1 × PBS. (MHCII-D10 as the capture molecule).

Peptide	Matrices	Calibration Curve	R^2	LOD (fmol/µL)	LOQ (fmol/µL)	Range (fmol/µL)
SEA4	PBS	Y = 0.041789X + 0.003444	0.974	0.78125	1.5625	1.5625–100
SEB2	PBS	Y = 0.038587X + 0.005922	0.998	0.78125	1.5625	1.5625–100
SEC2	PBS	Y = 0.000209X + 0.000238	0.977	0.78125	1.5625	1.5625–100
SED4	PBS	Y = 0.1172X − 0.10204	0.993	0.78125	1.5625	1.5625–100
SEE4	PBS	Y = 0.003741X − 0.000546	0.999	0.78125	1.5625	1.5625–100
SEH3	PBS	Y = 0.033934X + 0.026944	0.964	6.25	12.5	12.5–100
SEG6	PBS	Y = 0.24238X + 0.008165	0.953	0.78125	1.5625	1.5625–100

Table 7. Standard curve, linear range, limit of detection, and limit of quantification of different SES isoforms in plasma. (MHCII-D10 as the capture molecule).

Peptide	Matrices	Calibration Curve	R^2	LOD (fmol/µL)	LOQ (fmol/µL)	Range (fmol/µL)
SEA4	Plasma	Y = 0.003135X + 0.002139	0.960	3.125	6.25	6.25–100
SEB2	Plasma	Y = 0.001340X + 0.000483	0.958	0.7825	1.5625	1.5625–100
SEC2	Plasma	Y = 0.003198X + 0.000597	0.994	3.125	6.25	6.25–100
SED4	Plasma	Y = 0.055135X + 0.004234	0.935	0.7825	1.5625	1.5625–100
SEE4	Plasma	Y = 0.002209X + 0.001913	0.980	3.125	6.25	6.25–100
SEH3	Plasma	Y = 0.120729X − 0.003764	0.997	6.25	12.5	12.5–100
SEG6	Plasma	Y = 0.083461X − 0.040817	0.971	1.5625	3.125	3.125–100

Table 8. Standard curve, linear range, limit of detection, and limit of quantification of different SES isoforms in milk. (MHCII-D10 as the capture molecule).

Peptide	Matrices	Calibration Curve	R^2	LOD (fmol/µL)	LOQ (fmol/µL)	Range (fmol/µL)
SEA4	Milk	Y = 0.002878X + 0.005775	0.973	3.125	6.25	6.25–100
SEB2	Milk	Y = 0.008302X − 0.001616	0.986	3.125	6.25	6.25–100
SEC2	Milk	Y = 0.008098X − 0.000650	0.984	3.125	6.25	6.25–100
SED4	Milk	Y = 0.006779X − 0.005443	0.986	1.5625	3.125	3.125–100
SEE4	Milk	Y = 0.000726X − 0.000004	0.963	3.125	6.25	6.25–100
SEH1	Milk	Y = 0.223869X + 0.588802	0.967	6.25	12.5	12.5–100
SEG6	Milk	Y = 0.027874X − 0.000883	0.943	3.125	6.25	6.25–100

3.5.3. Immunocapture Recovery and Precision (RSD)

The ample recovery was measured with three replicates at three levels (QCL, QCM, and QCH). The results showed that when the MHCII was used for the capture molecule, the immune capture recovery rates were 19.2–70.4% in the PBS matrix, 18.4–80% in the plasma matrix, and 19.2–96% in the milk matrix. The results are given in Tables 9–11. In the PBS substrate, the recovery rates of the SEC immune capture at three levels (QCL, QCM, and QCH) ranged from 30.4% to 70.4%. In the plasma matrix, the recovery rates of the SED at three levels (QCL, QCM, and QCH) ranged from 24% to 80%. The recovery rates of the SEH were 36.7–57.8%. In the milk matrix, the immune capture recovery of the SEA, SEB, SEC, SED, and SEE at three levels (QCL, QCM, and QCH) ranged from 32.4% to 96%. The results are shown in Tables 9–11.

Table 9. Immunocapture extraction recovery and precision (RSD%) of SEs in 1 × PBS (n = 3). (MHCII as the capture molecule).

Peptide	Matrices	Added (fmol/μL)	Found (fmol/μL)	Recovery (%)	RSD%
SEA4	PBS	3.125	2.1 ± 0.04	67.2%	2.7%
		25	6.3 ± 0.8	25.2%	17.1%
		100	41 ± 3	41%	8.1%
SEB2	PBS	3.125	0.6 ± 0.04	19.2%	9.2%
		25	10 ± 0.05	40%	0.7%
		100	30 ± 0.9	30%	3.9%
SEC2	PBS	3.125	2.2 ± 0.01	70.4%	0.7%
		25	7.6 ± 0.6	30.4%	10.3%
		100	43.7 ± 0.05	43.7%	0.2%
SED4	PBS	3.125	0.7 ± 0.01	22.4%	2.2%
		25	6.8 ± 0.04	27.2%	0.7%
		100	63 ± 0.5	63%	0.9%
SEE4	PBS	3.125	0.9 ± 0.1	28.8%	17%
		25	5.7 ± 0.3	22.8%	7.4%
		100	30 ± 0.7	30%	2.9%
SEH3	PBS	6.25	2.9 ± 0.4	46.4%	15.2%
		25	11.5 ± 0.9	46%	9.3%
		100	55.1 ± 0.05	55.1%	0.1%
SEG6	PBS	6.25	2.4 ± 0.01	38.4%	0.7%
		25	7.6 ± 0.4	30.4%	6.3%
		100	44.8 ± 3.2	44.8%	8.8%

Table 10. Immunocapture extraction recovery and precision (RSD%) of SEs in plasma (n = 3). (MHCII as the capture molecule).

Peptide	Matrices	Added (fmol/μL)	Found (fmol/μL)	Recovery (%)	RSD%
SEA4	Plasma	6.25	2.1 ± 0.01	33.6%	0.8%
		25	6.7 ± 0.1	26.8%	2%
		100	45.9 ± 5.0	45.9%	13.6%
SEB2	Plasma	3.125	0.9 ± 0.09	28.8%	12.4%
		25	7.0 ± 0.8	28%	14.5%
		100	32.3 ± 3.4	32.3%	12.9%
SEC2	Plasma	6.25	1.6 ± 0.1	25.6%	9.7%
		25	7.9 ± 0.2	31.6%	2.7%
		100	30.8 ± 1.6	30.8%	6.4%
SED4	Plasma	3.125	0.8 ± 0.01	25.6%	2.2%
		25	6.0 ± 0.1	24%	2.8%
		100	80 ± 0.7	80%	1.1%
SEE4	Plasma	6.25	2.6 ± 0.05	41.6%	2.6%
		25	14.7 ± 1.4	58.8%	11.4%
		100	30 ± 1.1	30%	4.6%
SEH3	Plasma	12.5	5 ± 0.1	40%	2.6%
		25	9.1 ± 0.02	36.4%	0.4%
		100	57.8 ± 0.1	57.8%	0.2%
SEG4	Plasma	6.25	1.4 ± 0.01	22.4%	1.1%
		25	4.6 ± 0.01	18.4%	0.4%
		100	27.4 ± 1.2	27.4%	5.4%

Table 11. Immunocapture extraction recovery and precision (RSD%) of SEs in milk (n = 3). (MHCII as the capture molecule).

Peptide	Matrices	Added (fmol/μL)	Found (fmol/μL)	Recovery (%)	RSD%
SEA4	Milk	6.25	3.6 ± 0.1	57.6%	4.9%
		25	9.2 ± 0.2	36.8%	3.2%
		100	42.4 ± 4.8	42.4%	13.9%
SEB2	Milk	3.125	1.6 ± 0.1	52.1%	12.2%
		25	11.6 ± 1.7	46.5%	17.4%
		100	57.9 ± 6.9	57.9%	14.6%
SEC2	Milk	3.125	1.7 ± 0.02	54.4%	1.9%
		25	8.1 ± 0.5	32.4%	7.5%
		100	61.5 ± 0.08	61.5%	0.1%
SED4	Milk	3.125	3.0 ± 0.02	96%	2%
		25	12.8 ± 0.03	51.2%	3%
		100	89.1 ± 1.7	89.1%	2.4%
SEE4	Milk	3.125	2.3 ± 0.2	73.6%	12%
		25	20 ± 2.2	80%	13.2%
		100	41 ± 2.6	41%	7.8%
SEH3	Milk	6.25	1.6 ± 0.1	25.6%	9.4%
		25	12 ± 1.6	48%	16.4%
		100	55.2 ± 0.09	55.2%	0.2%
SEG6	Milk	6.25	1.2 ± 0.01	19.2%	1.4%
		25	11.2 ± 0.8	44.8%	9.3%
		100	37.1 ± 6.0	37.1%	19.8%

When MHCII-D10 was used as the capture molecule, the immune capture recovery rates were 19.2–75.1% in the PBS matrix, 19.2–85.1% in the plasma matrix, and 22.4–59.2% in the milk matrix. In the plasma matrix, the immune capture recovery of the SED was 51.2–85.1% in the three plasma matrices (QCL, QCM, and QCH). The recovery rates of the SEC immune capture were 30.2–52.7%. In the milk matrix, the immune capture recoveries of the SEA at the three levels (QCL, QCM, and QCH) were 34.4–56%, while those of the SEC were 52.3–58% and those of the SEH were 43.6–59.2%. The results are shown in Tables 12–14.

Table 12. Immunocapture extraction recovery and precision (RSD%) of SEs in 1 × PBS (n = 3). (MHCII-D10 as the capture molecule).

Peptide	Matrices	Added (fmol/μL)	Found (fmol/μL)	Recovery (%)	RSD%
SEA4	PBS	1.5625	0.4 ± 0.05	25.6%	17.6%
		25	7.8 ± 0.9	31.2%	14.6%
		100	66.4 ± 10.2	66.4%	18.8%
SEB2	PBS	1.5625	0.6 ± 0.09	38.4%	18.7%
		25	9.4 ± 1.3	37.6%	18.1%
		100	47.7 ± 1.8	47.7%	4.6%
SEC2	PBS	1.5625	0.5 ± 0.01	32.3%	3.3%
		25	8.2 ± 1.3	32.8%	19.5%
		100	49.6 ± 2.1	49.6%	5.2%
SED4	PBS	1.5625	0.4 ± 0.01	25.6%	4.7%
		25	11.1 ± 1.2	44%	13.6%
		100	44.7 ± 6	44.7%	16.4%
SEE4	PBS	1.5625	0.4 ± 0.06	25.6%	18.4%
		25	4.8 ± 0.5	19.2%	13.8%
		100	28.2 ± 1.1	28.2%	4.6%
SEH3	PBS	12.5	2.5 ± 0.3	20%	14.5%
		25	7 ±0.2	28%	4.1%
		100	75.1 ± 5.1	75.1%	8.4%
SEG6	PBS	1.5625	0.7 ± 0.04	44.8%	6.9%
		25	7.6 ± 0.3	30.4%	5.2%
		100	58.8 ± 2.4	58.8%	5.02%

Table 13. Immunocapture extraction recovery and precision (RSD%) of SEs in plasma (n = 3). (MHCII-D10 as the capture molecule).

Peptide	Matrices	Added (fmol/μL)	Found (fmol/μL)	Recovery (%)	RSD%
SEA4	Plasma	6.25	2.4 ± 0.3	38.4%	0.7%
		25	8.2 ± 0.7	32.8%	10.1%
		100	30.7 ± 0.6	30.7%	2.4%
SEB2	Plasma	1.5625	0.6 ± 0.09	38.4%	18.2%
		25	9.4 ± 1.2	37.6%	15.8%
		100	24.8 ± 1.2	24.8%	6.1%
SEC2	Plasma	6.25	2.8 ± 0.2	44.8%	8.4%
		25	7.5 ± 1	30%	17.1%
		100	52.7 ± 0.1	52.7%	0.3%
SED4	Plasma	1.5625	0.8 ± 0.1	51.2%	19.7%
		25	11.5 ± 0.08	46%	0.9%
		100	85.1 ± 0.1	85.1%	0.1%
SEE4	Plasma	6.25	1.2 ± 0.1	19.2%	5.5%
		25	6.4 ± 0.2	25.6%	2.9%
		100	23.2 ± 0.9	23.2%	4.9%
SEH3	Plasma	12.5	2.7 ± 0.04	21.6%	2.1%
		25	5.2 ± 0.2	20.8%	4.7%
		100	24.1 ± 0.9	24.1%	4.4%
SEG4	Plasma	3.125	2 ± 0.2	64%	11.2%
		25	9.9 ± 0.6	39.6%	7.3%
		100	32.4 ± 1.1	32.4%	4.1%

Table 14. Immunocapture extraction recovery and precision (RSD%) of SEs in milk (n = 3). (MHCII-D10 as the capture molecule).

Peptide	Matrices	Added (fmol/μL)	Found (fmol/μL)	Recovery (%)	RSD%
SEA4	Milk	6.25	3.5 ± 0.3	56%	9.3%
		25	8.6 ± 0.6	34.4%	8.8%
		100	51.6 ± 1.9	51.6%	4.46%
SEB2	Milk	6.25	2.1 ± 0.3	33.6%	15.1%
		25	10.9 ± 1.3	43.6%	14.8%
		100	30 ± 0.1	30.0%	0.6%
SEC2	Milk	6.25	3.6 ± 0.5	57.6%	15.8%
		25	14.8 ± 0.9	58%	7.6%
		100	52.3 ± 0.7	52.3%	1.57%
SED4	Milk	3.125	0.9 ± 0.04	28.8%	5.4%
		25	6.4 ± 0.3	25.6%	4.8%
		100	29.8 ± 0.1	29.8%	0.4%
SEE4	Milk	6.25	3.2 ± 0.2	51.2%	8.5%
		25	10.2 ± 0.7	40.8%	8.1%
		100	27.7 ± 0.9	27.7%	3.9%
SEH1	Milk	12.5	7.4 ± 0.7	59.2%	12.1%
		25	10.9 ± 0.2	43.6%	2.1%
		100	49.1 ± 2	49.1%	4.94%
SEG6	Milk	6.25	1.4 ± 0.09	22.4%	8.2%
		25	6.9 ± 0.3	27.6%	4.6%
		100	23.3 ± 0.7	23.3%	3.9%

From the above results, we can see that in different complex matrix samples, it can be known that the two designed capture molecules have the ability to capture seven toxins.

Liquid chromatography–mass spectrometry using immunosorbent enrichment technology to detect SEs in complex matrices requires antibodies. There are 27 SEs that are enriched for antibodies. Antibodies must be prepared for each toxin, but there is no universal enrichment method. In addition, antibody selection is an issue when testing unknown samples. The capture molecules we have developed can capture at least seven toxins but have the potential to bind all SEs. The capture molecules are versatile and suitable for screening unknown samples [13–15].

4. Discussion

In this study, a new method for the detection of various SEs in complex substrates was established by capture molecules combined with liquid chromatography–mass spectrometry. This study consists of three parts. The first part was the design and acquisition of novel capture molecules and their affinity with SEs. The affinity of the capture molecules was compared with traditional animal-specific immune polyclonal antibodies and SEs. The results showed that the KD values of the new capture molecules binding to the toxin were in the range of 10^{-8}–10^{-12} M, indicating that the binding strength of the capture molecules and the toxin was the same as that of the traditional animal-specific immune polyclonal antibodies. The second part was the detection of toxins in complex substrates with liquid chromatography and mass spectrometry. Detection is usually performed using bottom–up methods, so the pretreatment of the sample is required. In the process of the sample pretreatment, trypsin digestion usually takes a long time and has a low efficiency. Most of the literature studies usually use physical methods, such as microwave-assisted or ultrasonic-assisted methods [22,23], chemical methods, or organic solvent-assisted methods to improve the enzyme digestion efficiency [19], due to the lack of research on modified trypsin digestion of SEs. This paper mainly explored the digestion of Staphylococcus aureus enterotoxin by a modified trypsin enzyme by controlling the temperature, time, and digestion ratio. The optimal conditions for digestion were with a trypsin/SEs ratio of 1:5 and an optimal digestion temperature of 60 °C. A reduction in the digestion time and an improvement of the digestion efficiency followed.

The third part was to verify the specificity, recovery accuracy, and linearity of SEs in complex substrates with liquid chromatography and mass spectrometry to prove whether the novel capture molecules can replace the traditional animal-specific immune antibodies. The results showed that, although the linearity and detection limit of the new trapping molecules were not better than that of the traditional animal-specific immune antibodies, their universality and recovery rate mean that they are potential replacements for the traditional animal-specific immune antibodies.

5. Conclusions

The two capture molecules developed in this paper are different from animal immune antibodies and can be obtained by the purification of recombinant proteins. This new approach not only yields many antibodies but also saves cost and time. Additionally, it eliminates the ethical issues with the use of animals. The two capture antibodies have a good binding effect on the seven serotypes of toxins, though the detection limits of the two capture antibodies appeared higher than that of the specific monoclonal or polyclonal antibodies obtained by animal immunization reported in the literature [6].

Furthermore, this study systematically explored the optimal digestion conditions of modified trypsin for SEs. By controlling the ratio, temperature, and time of the trypsin and toxin digestion, the optimum digestion conditions were identified. This step represents the most time-consuming aspect of the sample processing. Therefore, the endeavor to shorten the digestion time, streamline the processing of the complex matrix samples, and improve the detection efficiency are of great significance for clinical diagnosis and other real-world applications of the detection of SEs.

Supplementary Materials: The following supporting information can be downloaded at https://www.mdpi.com/article/10.3390/separations11050136/s1. Figure S1: sequence diagrams of the capture molecules MHCII and MHCII-D10; Figure S2: sequence alignment map of the SE proteins; Figure S3: Toxin and capture molecules gel electrophoresis; Figure S4: standard curves of the seven serotypes of toxins in three different substrates of the MHCII and MHCII-D10 capture molecule:

1 × PBS, milk, and plasma were used; Table S1: specific peptide sequence; Table S2: peptide profile match rate of the protein sequence; and Table S3: target protein information.

Author Contributions: Conceptualization: W.X., L.K. and S.G.; writing—original draft: J.L. (Jing Lv); data curation/formal analysis/methodology: J.L. (Jing Lv) and T.L.; validation: X.F. and S.H.; contributed resources: L.D., J.W. (Jing Wang) and J.L. (Jiaxin Li); and project administration: J.W. (Jinglin Wang). All authors have read and agreed to the published version of the manuscript.

Funding: This work was supported by a project of the State Key Laboratory of Pathogen and Biosecurity (SKLPBS2102).

Institutional Review Board Statement: Not applicable.

Informed Consent Statement: Not applicable.

Data Availability Statement: The data are contained within this manuscript or the Supplementary Information.

Conflicts of Interest: The authors declare that they have no conflicts of financial and nonfinancial interest.

References

1. Balaban, N.; Rasooly, A. Staphylococcal enterotoxins. *J. Food Microbiol.* **2000**, *61*, 1–10. [CrossRef]
2. Attien, P.; Sina, H.; Moussaoui, W.; Zimmermann-Meisse, G.; Dadié, T.; Keller, D.; Riegel, P.; Edoh, V.; Kotchoni, S.O.; Djè, M.; et al. Mass spectrometry and multiplex antigen assays to assess microbial quality and toxin production of *Staphylococcus aureus* strains isolated from clinical and food samples. *BioMed Res. Int.* **2014**, *2014*, 485620. [CrossRef] [PubMed]
3. Kadariya, J.; Smith, T.C.; Thapaliya, D. *Staphylococcus aureus* and staphylococcal food-borne disease: An ongoing challenge in public health. *BioMed Res. Int.* **2014**, *2014*, 827965. [CrossRef] [PubMed]
4. Bao, K.D.; Letellier, A.; Beaudry, F. Analysis of *Staphylococcus* enterotoxin B using differential isotopic tags and liquid chromatography quadrupole ion trap mass spectrometry. *Biomed. Chromatogr. BMC* **2012**, *26*, 1049–1057. [CrossRef] [PubMed]
5. Lefebvre, D.; Blanco-Valle, K.; Hennekinne, J.A.; Simon, S.; Fenaille, F.; Becher, F.; Nia, Y. Multiplex Detection of 24 Staphylococcal Enterotoxins in Culture Supernatant Using Liquid Chromatography Coupled to High-Resolution Mass Spectrometry. *Toxins* **2022**, *14*, 249. [CrossRef] [PubMed]
6. Lefebvre, D.; Blanco-Valle, K.; Feraudet-Tarisse, C.; Merda, D.; Simon, S.; Fenaille, F.; Hennekinne, J.A.; Nia, Y.; Becher, F. Quantitative Determination of *Staphylococcus aureus* Enterotoxins Types A to I and Variants in Dairy Food Products by Multiplex Immuno-LC-MS/MS. *J. Agric. Food Chem.* **2021**, *69*, 2603–2610. [CrossRef] [PubMed]
7. Tonacini, J.; Stephan, D.; Vogel, G.; Avondet, M.A.; Kalman, F.; Crovadore, J.; Lefort, F.; Schnyder, B. Intact *Staphylococcus* Enterotoxin SEB from Culture Supernatant Detected by MALDI-TOF Mass Spectrometry. *Toxins* **2019**, *11*, 101. [CrossRef] [PubMed]
8. Nia, Y.; Rodriguez, M.; Zeleny, R.; Herbin, S.; Auvray, F.; Fiebig, U.; Avondet, M.A.; Munoz, A.; Hennekinne, J.A. Organization and ELISA-Based Results of the First Proficiency Testing to Evaluate the Ability of European Union Laboratories to Detect Staphylococcal Enterotoxin Type B (SEB) in Buffer and Milk. *Toxins* **2016**, *8*, 268. [CrossRef] [PubMed]
9. Picard, G.; Lebert, D.; Louwagie, M.; Adrait, A.; Huillet, C.; Vandenesch, F.; Bruley, C.; Garin, J.; Jaquinod, M.; Brun, V. PSAQ™ standards for accurate MS-based quantification of proteins: From the concept to biomedical applications. *J. Mass Spectrom. JMS* **2012**, *47*, 1353–1363. [CrossRef]
10. Dupré, M.; Gilquin, B.; Fenaille, F.; Feraudet-Tarisse, C.; Dano, J.; Ferro, M.; Simon, S.; Junot, C.; Brun, V.; Becher, F. Multiplex quantification of protein toxins in human biofluids and food matrices using immunoextraction and high-resolution targeted mass spectrometry. *Anal. Chem.* **2015**, *87*, 8473–8480. [CrossRef]
11. Deacy, A.M.; Gan, S.K.; Derrick, J.P. Superantigen Recognition and Interactions: Functions, Mechanisms and Applications. *Front. Immunol.* **2021**, *12*, 731845. [CrossRef] [PubMed]
12. Sharma, P.; Wang, N.; Kranz, D.M. Soluble T cell receptor Vβ domains engineered for high-affinity binding to staphylococcal or streptococcal superantigens. *Toxins* **2014**, *6*, 556–574. [CrossRef] [PubMed]
13. Rödström, K.E.; Elbing, K.; Lindkvist-Petersson, K. Structure of the Superantigen Staphylococcal Enterotoxin B in Complex with TCR and Peptide–MHC Demonstrates Absence of TCR–Peptide Contacts. *J. Immunol.* **2014**, *193*, 1998–2004. [CrossRef] [PubMed]
14. Buonpane, R.A.; Moza, B.; Sundberg, E.J.; Kranz, D.M. Characterization of T cell receptors engineered for high affinity against toxic shock syndrome toxin-1. *J. Mol. Biol.* **2005**, *353*, 308–321. [CrossRef] [PubMed]
15. Mattis, D.M.; Spaulding, A.R.; Chuang-Smith, O.N.; Sundberg, E.J.; Schlievert, P.M.; Kranz, D.M. Engineering a soluble high-affinity receptor domain that neutralizes staphylococcal enterotoxin C in rabbit models of disease. *Protein Eng. Des. Sel.* **2013**, *26*, 133–142. [CrossRef] [PubMed]
16. Koike, H.; Kanda, M.; Hayashi, H.; Matsushima, Y.; Ohba, Y.; Nakagawa, Y.; Nagano, C.; Sekimura, K.; Hirai, A.; Shindo, T.; et al. Quantification of staphylococcal enterotoxin type A in cow milk by using a stable isotope-labelled peptide via liquid chromatography–tandem mass spectrometry. *Food Addit. Contam. Part A* **2019**, *36*, 1098–1108. [CrossRef] [PubMed]

17. Liang, L.H.; Yang, Y.; Geng, S.; Cheng, X.; Yu, H.L.; Liu, C.C.; Liu, S.L. Rapid Differential Detection of Abrin Isoforms by an Acetonitrile- and Ultrasound-Assisted On-Bead Trypsin Digestion Coupled with LC-MS/MS Analysis. *Toxins* **2021**, *13*, 358. [CrossRef] [PubMed]
18. Oyler, J.M.; Tran, B.Q.; Kilgour, D.P. Rapid Denaturing Organic Digestion Method for Targeted Protein Identification and Characterization. *Anal. Chem.* **2021**, *93*, 5046–5053. [CrossRef] [PubMed]
19. Ostin, A.; Bergström, T.; Fredriksson, S.A.; Nilsson, C. Solvent-assisted trypsin digestion of ricin for forensic identification by LC-ESI MS/MS. *Anal. Chem.* **2007**, *79*, 6271–6278. [CrossRef]
20. Brown, K.A.; Chen, B.; Guardado-Alvarez, T.M.; Lin, Z.; Hwang, L.; Ayaz-Guner, S.; Jin, S.; Ge, Y. A photocleavable surfactant for top-down proteomics. *Nat. Methods* **2019**, *16*, 417–420. [CrossRef]
21. Vesper, H.W.; Mi, L.; Enada, A.; Myers, G.L. Assessment of microwave-assisted enzymatic digestion by measuring glycated hemoglobin A1c by mass spectrometry. *Rapid Commun. Mass Spectrom. RCM* **2005**, *19*, 2865–2870. [CrossRef] [PubMed]
22. Reddy, P.M.; Hsu, W.Y.; Hu, J.F.; Ho, Y.P. Digestion Completeness of Microwave-Assisted and Conventional Trypsin-Catalyzed Reactions. *J. Am. Soc. Mass Spectrom.* **2010**, *21*, 421–424. [CrossRef] [PubMed]
23. Taverna, D.; Norris, J.L.; Caprioli, R.M. Histology-directed microwave assisted enzymatic protein digestion for MALDI MS analysis of mammalian tissue. *Anal. Chem.* **2015**, *87*, 670–676. [CrossRef] [PubMed]
24. Kim, Y.; Lee, D.; Kim, J. Effects of incubation temperature and acetonitrile amount on microwave-assisted tryptic digestion of proteins. *Anal. Biochem.* **2019**, *569*, 31–38. [CrossRef] [PubMed]
25. Hansbauer, E.M.; Worbs, S.; Volland, H.; Simon, S.; Junot, C.; Fenaille, F.; Dorner, B.G.; Becher, F. Rapid Detection of Abrin Toxin and Its Isoforms in Complex Matrices by Immuno-Extraction and Quantitative High Resolution Targeted Mass Spectrometry. *Anal. Chem.* **2017**, *89*, 11719–11727. [CrossRef] [PubMed]
26. Lopez-Ferrer, D.; Petritis, K.; Robinson, E.W.; Hixson, K.K.; Tian, Z.; Lee, J.H.; Lee, S.W.; Tolic, N.; Weitz, K.K.; Belov, M.E.J.M.; et al. Pressurized pepsin digestion in proteomics: An automatable alternative to trypsin for integrated top-down bottom-up proteomics. *Mol. Cell. Proteom.* **2011**, *10*, S1–S11. [CrossRef] [PubMed]
27. LoPez-Ferrer, D.; Petritis, K.; Hixson, K.K.; Heibeck, T.H.; Moore, R.J.; Belov, M.E.; Camp, D.G.; Smith, R.D. Application of Pressurized Solvents for Ultrafast Trypsin Hydrolysis in Proteomics: Proteomics on the Fly. *J. Proteome Res.* **2008**, *7*, 3276–3281. [CrossRef] [PubMed]
28. Finehout, E.J.; Cantor, J.R.; Lee, K.H. Kinetic characterization of sequencing grade modified trypsin. *Proteomics* **2005**, *5*, 2319. [CrossRef]
29. Francis, D.M.; Page, R. Strategies to optimize protein expression in *E. coli*. *Curr. Protoc. Protein Sci.* **2010**, *61*, 5–24. [CrossRef]

foods

Article

Metabolomics on Apple (*Malus domestica*) Cuticle—Search for Authenticity Markers

Kamila Bechynska [1], Jiri Sedlak [2], Leos Uttl [1], Vit Kosek [1], Petra Vackova [1], Vladimir Kocourek [1] and Jana Hajslova [1,*]

1 Department of Food Analysis and Nutrition, University of Chemistry and Technology, Technicka 3, 16628 Prague 6, Czech Republic; bechynsk@vscht.cz (K.B.); uttll@vscht.cz (L.U.); kosekv@vscht.cz (V.K.); vackovap@vscht.cz (P.V.); kocourev@vscht.cz (V.K.)

2 Reserach and Breeding Institute of Pomology Holovousy, Holovousy 129, 50801 Holovousy, Czech Republic; jiri.sedlak@vsuo.cz

* Correspondence: jana.hajslova@vscht.cz; Tel.: +420-6028-33424

Abstract: The profile of secondary metabolites present in the apple cuticular layer is not only characteristic of a particular apple cultivar; it also dynamically reflects various external factors in the growing environment. In this study, the possibility of authenticating apple samples by analyzing their cuticular layer extracts was investigated. Ultra-high-performance liquid chromatography coupled with high-resolution tandem mass spectrometry (UHPLC-HRMS/MS) was employed for obtaining metabolomic fingerprints. A total of 274 authentic apple samples from four cultivars harvested in the Czech Republic and Poland between 2020 and 2022 were analyzed. The complex data generated, processed using univariate and multivariate statistical methods, enabled the building of classification models to distinguish apple cultivars as well as their geographical origin. The models showed very good performance in discriminating Czech and Polish samples for three out of four cultivars: "Gala", "Golden Delicious" and "Idared". Moreover, the validity of the models was tested over several harvest seasons. In addition to metabolites of the triterpene biosynthetic pathway, the diagnostic markers were mainly wax esters. "Jonagold", which is known to be susceptible to mutations, was the only cultivar for which an unambiguous classification of geographical origin was not possible.

Keywords: UHPLC-HRMS/MS; metabolomic fingerprints; classification models; markers; wax esters

Citation: Bechynska, K.; Sedlak, J.; Uttl, L.; Kosek, V.; Vackova, P.; Kocourek, V.; Hajslova, J. Metabolomics on Apple (*Malus domestica*) Cuticle—Search for Authenticity Markers. *Foods* **2024**, *13*, 1308. https://doi.org/10.3390/foods13091308

Academic Editors: Chao Kang and Ronald Beckett

Received: 2 April 2024
Revised: 17 April 2024
Accepted: 21 April 2024
Published: 24 April 2024

1. Introduction

Apple (*Malus* × *domestica* Borkh) is one of the most widely cultivated fruits in the temperate climate zone, with an annual world harvest of around 96 million tonnes [1]. In the Czech Republic, more than 100,000 tonnes of apples were harvested in production orchards in 2023 (based on data from the Central Institute for Supervising and Testing in Agriculture, 2023). Although they are more expensive compared to imported apples, the regional fruits are favored by consumers as they believe that (thanks to the widespread integral farming practices of local producers) fewer pesticides have been used in their production. Unfortunately, under these conditions, dishonest traders tend to falsely declare information when placing this popular fruit on the market. Alongside the intentional false declaration of geographical origin, the other fraudulent practice is the mislabeling of the particular apple cultivar. In detecting such economically motivated frauds, the development of methods to detect them, the aim of which is both to take preventive measures against fraudulent practices and to protect consumers, is a challenging task [2,3].

To date, a variety of instrumental techniques have been published to authenticate the geographical origin and/or cultivar of apples. This includes near-infrared spectroscopy (NIR) [4], fluorescence spectroscopy [5], head-space solid-phase microextraction coupled with gas chromatography and mass spectrometry (SPME-GC-MS) [6,7], isotope ratio mass spectrometry (IR-MS), either separately [2,8,9] or in combination with elemental analysis by inductively coupled plasma mass spectrometry (ICP-MS) [10], and electronic nose and electronic tongue [3]. Further information on these authentication studies is summarized in Table 1. However, as shown here, some of them were performed with a limited number of samples or did not sufficiently cover factors of natural variability (different apple cultivars, growing locations, farming systems, different harvest years, etc.), and in some cases, the description of statistical methods for data processing was insufficient. Under these conditions, generic applicability of the results could be rather difficult.

Table 1. The overview of studies dealing with the authentication of the origin and/or cultivar of apples.

Analytical Method	Description of Apple Samples	Classification Factor	Number of Samples	Number of Classes to Be Distinguished within the Sample Set	Classification Method	Performance of Classification	Reference
NIR	Surface of whole apple fruits	Cultivar	300	3 (Fuji, Red Star, Gala)	NN, SVM, ELM	Calibration set 98% (ELM) Prediction set 97% (ELM)	[4]
		Geographical origin		2 (grown in different Chinese provinces)			
Fluorescent spectroscopy	Apple juice (squeezed with a juice extractor)	Cultivar	89	2 (grown in different Chinese provinces)	PLS	Calibration set 100% Prediction set 96%	[5]
SPME-GC-MS	Apple juice (squeezed with a juicer)	Cultivar	50	6 (Starkrimson, Qinguan, Gala, Jonagold, Golden Delicioius, Fuji)	LDA, SLDA	Predicition set 100% (SLDA)	[6]
		Geographical origin		5 (grown in different counties within Chinese province)		Predicition set 90% (SLDA)	
SPME-GC-MS	Apple juice (squeezed with hand press)	Cultivar	4 (3 kg of apples per sample)	4 (Rijo, Verde, Ribeiro, Azedo)	PLS-DA, HCA	Vague description of model performance	[7]
		Geographical origin		2 (different civil parishes of Madeira)			
IR-MS + conventional methods	Pulp, juice	Cultivar	19	6 (Topaz, Idared, Golden Delicious, Goldrush, Gala, Gloster)	LDA	Insufficient description of models	[2]
		Geographical origin		4 (different regions of Slovenia)			
		Agricultural practice		2 (way of farming organic, conventional)			
IR-MS	Whole apples, peel, pulp, seed	Cultivar	128	4 (Cripps Pink, Gala, Golden Delicious, Granny Smith)	LDA	71% correctly classified samples	[8]
		Geographical origin		4 (grown in different districts of northerm Italy)		99% (LOOCV)	
IR-MS	Peel, petiole, pulp, seed	Geographical origin	48	2 (grown in different districts of northern Italy)	LDA	Limited information on classification models performance	[9]
IR-MS, ICP-MS	Apple juice (concentrated to sugar content 65.0° Brix)	Geographical origin	135	6 (grown in different Chinese provinces)	LDA, PLS-DA	Only description of sample clustering in PLS-DA model without information about model validation	[10]
Electronic nose, electronic tongue	Apple juice (centrifugal juicer)	Cultivar	126	10 (Fuji, Jonagold, Corolla, Gala, Red Delicous, Red Chief Delicious, Cattle Apple, Ralls Janet, Ourin, Tail, Golden Delicous)	LDA, PLS-DA, SVM	100% (prediction ability of PLS-DA) 100% (accuracy testing rate of SVM)	[3]
		Geographical origin		7 (grown in different Chinese provinces)			

ELM—extreme learning machine; HCA—hierarchical cluster analysis; LDA—linear discriminant analysis; LOOCV—leave-one-out cross validation; NN—neural networks; PLS—partial least square; PLS-DA—partial least square discriminant analysis; SLDA—stepwise linear discriminant analysis; SVM—support vector machine.

Various parts and/or processed forms of apples were used within the studies listed in Table 1; nevertheless, only in one of them [4] was the authentication based on the data collected from the apple surface. However, the NIR technique used in that particular study did not allow for the identification of characteristic marker compounds that could be suitable for authentication via targeted analysis. In this context, it is worth noting that the cuticular layer is an interesting matrix in the search for authenticity markers. The cuticular layer contains a number of secondary metabolites whose profile is not only characteristic of the respective cultivar (genotype) but can also be influenced by various external factors in the respective growing location, such as local weather conditions, application of pesticides and growth regulators, diseases and pests. The cuticle, the outer protective layer of the fruit, consists of structural polymers coated with a layer of wax [11]. While the intracuticular waxes are directly incorporated into the cutin, the epicuticular waxes cover a surface of the cutin polymers [12]. The cuticular layer is a complex mixture of secondary metabolites, which include long-chain hydrocarbons and their derivatives, such as carboxylic acids, alcohols, aldehydes and ketones, esters, etc. The other group of typical metabolites are various triterpenoids [13–15].

The composition of cuticular waxes has been analyzed in detail in several studies using different analytical methods. For the wax extraction, several authors have used cuticle membranes enzymatically isolated by pectinase and cellulase [14,16]. On the other hand, simple methods based only on rinsing the apple surface with a solvent (usually chloroform, but also dichloromethane, petroleum ether, hexane, etc.) have also been reported [11,17–19]. The most common method for analyzing the isolated compounds is GC-MS [11,14,16,18], but this method requires derivatization. Reversed phase liquid chromatography coupled with mass spectrometry (RPLC-MS) has been increasingly used as an alternative, using both atmospheric pressure chemical ionization (APCI) and electrospray ionization (ESI) [20–25]. The advantage of LC-ESI-MS is the possibility of the simultaneous detection of different lipid classes, including difficult-to-ionize neutral wax esters when additives are used. The use of high-resolution (HR) MS allows for identification based on molecular or adduct ion exact masse, and the application of collision energy (tandem mass spectrometry, MS/MS) is suitable to obtain information on molecular structure [20].

The aim of the present study was to investigate the potential of UHPL-HRMS-based metabolomic fingerprinting of cuticular layer components, followed by advanced statistics, for the classification of apple cultivars "Gala", "Golden Delicious", "Idared" and "Jonagold", grown either in the Czech Republic or in Poland. To our knowledge, no other published study has used such an approach.

2. Materials and Methods

2.1. Samples

Apple (*Malus domestica*) samples analyzed in this study were collected by the Research and Breeding Institute of Pomology Holovousy Ltd. (Holovousy, Czech Republic). A total of 274 authentic apple samples of known geographical origin and cultivar characterization were provided over a period of 3 harvest years (2020–2022). The cultivars available in this study were "Gala", "Golden Delicious", "Idared" and "Jonagold", originating either from the Czech Republic (CZE) or Poland (POL). A total of 35 samples of the "Gala" cultivar (16 CZE, 19 POL), 37 samples of the "Golden Delicious" cultivar (19 CZE, 18 POL), 32 samples of the "Idared" cultivar (17 CZE, 15 POL) and 33 samples of the "Jonagold" cultivar (13 CZE, 19 POL) were available. After delivery to the laboratory, the samples were stored at 4 °C for a maximum of 3 days before further processing.

2.2. Chemicals

Analytical-grade methanol (MeOH), dichloromethane (DCM), ethyl acetate (EtAC), methyl-tert-butyl-ether (MTBE) and isopropanol (iPrOH) were purchased from Merck (Darmstadt, Germany). Deionized water (dH2O) was obtained from a Milli-Q Integral system (Millipore supplied by Merck (Darmstadt, Germany)). The mobile phase modifiers (ammonium formate, formic acid) were purchased from Sigma-Aldrich (Darmstadt, Germany).

2.3. *Methods*

2.3.1. Sample Preparation

To isolate the metabolites present in the apple cuticular layer, the whole apple was carefully placed into a 1000 mL glass beaker containing 400 mL of solvent mixture. After covering the beaker with parafilm, its content was gently shaken for 10 min. In the experiments undertaken to evaluate the extraction efficiency, the following solvents/mixtures thereof were tested: hexane–EtAc (1:1, *v:v*), DCM, DCM–MeOH (2:1, *v:v*), MTBE–MeOH (10:3, *v:v*) and DCM–MeOH (1:1, *v:v*). The last one was selected as optimal and then used throughout the study.

To obtain maximum yield of apple surface components, this procedure was repeated twice, always washing three representative apples. The combined extracts of each sample were evaporated stepwise to dryness and the residue was stored at −80 °C. Prior to analysis, the residue was reconstituted in a calculated amount of DCM:MeOH (1:1, *v:v*) to obtain a standardized concentration of the extracted material of 33.33 mg/mL. The solution was filtered using syringe filters (pore size 0.22 µm) and a 1.5 mL aliquot of each extract was then transferred to glass vials for LC-MS analysis. The quality control (QC) sample was prepared as a pool of aliquots of all prepared standardized apple extracts.

2.3.2. UHPLC-HRMS/MS Non-Target Screening

For the metabolomic analysis of apple extracts, the UHPLC-HRMS/MS technique was employed. A high-performance liquid chromatograph Dionex UltiMate 3000 RS (Thermo Fisher Scientific, Waltham, MA, USA) coupled with quadrupole-time-of-flight TripleTOF™ 6600 mass spectrometer (Sciex, Concord, ON, Canada) was used for this purpose. For sample components separation, Acquity UPLC BEH C18 column (2.1 × 100 mm, 1.7 µm) (Waters, Milford, MA, USA) was employed. The mobile phase consisted of (A) 5 mM ammonium formate in a mixture of dH_2O:MeOH (95:5, *v:v*) with 0.1% formic acid and (B) 5 mM ammonium formate in a mixture of iPrOH:MeOH:dH_2O (65:30:5, *v:v*) with 0.1% formic acid. The following gradient was used for both positive and negative ionization modes: 0 min (70% A), 2 min (50% A), 7 min (20% A), 13 min (0% A), 20 min (0% A), 20.1 min (70% A) and 22 min (70% A), with a constant flow rate of 0,4 mL/min. The column temperature was kept at 60 °C; the temperature of the autosampler at 5 °C and the sample injection volume was 1 µL.

The mass spectrometer was operated in both positive (ESI+) and negative (ESI-) mode with the following ion source settings: nebulizing gas pressure, 55 psi; drying gas pressure, 55 psi; capillary voltage, +4500 V (for ESI+)/−4000 V (for ESI-); ion temperature, 500 °C. Both MS and MS/MS data were acquired using full-scan and information-dependent acquisition (IDA) methods. The mass range in MS mode was set to 100–1200 m/z, and that in MS/MS mode to 50–1200 m/z. The collision energy was 35 V with the spread of ±15 V. Mass spectrometer calibration was performed regularly after every 10 samples based on APCI calibration solution (Sciex, Concord, Canada).

Samples were injected in a randomized order; QC samples were injected during the entire analytical run (after 10 previous sample injections). Blank samples (extraction solvent mixture) were injected at the beginning of the sequence to capture background features.

2.3.3. Data Processing

The UHPLC-HRMS/MS data obtained were processed using the open-source software MS-Dial (version 4.8) [26]. In the first step, the data were converted into the specific *.ibf format. The peak picking parameters were set as follows for both data acquired in ESI+ and ESI− mode: a minimum signal intensity threshold (peak height) of 10,000; a mass accuracy of 0.01 Da for MS data and 0.025 Da for MS/MS data. For data alignment, retention time tolerance of 0.05 min (ESI+) and 0.3 min (ESI−), along with m/z tolerance of 0.015 Da, were used. In ESI+ mode $[M+H]^+$, $[M+Na]^+$ and $[M+NH4]^+$ adducts were considered, and in ESI- mode, $[M-H]^-$ and $[M+HCOO]^-$ adducts were considered.

The exported data matrices, consisting of all detected features characterized by m/z and retention time, were filtered according to the relative standard deviation (RSD) of the signal intensity (peak area) in the QC samples, with a maximum RSD threshold of 20% for both ESI+ and ESI−. Furthermore, all features with a signal-to-noise ratio (SNR) below 3 were filtered out to obtain the final data matrices.

2.3.4. Statistical Analysis

Prior to statistical analysis, the data were pre-processed to avoid possible misinterpretation of data variability. In this study, the total area sum normalization and logarithmic transformation were performed prior to any univariate statistics, followed by Pareto scaling in the case of multivariate model building.

Within the chemometric processing, the aim was to create models for the classification of apple samples using both univariate and multivariate statistical tools. Principal component analysis (PCA) was used to overview the data. Diagnostics features were selected using *t*-test/analysis of variance (ANOVA), fold change and receiver operating characteristics (ROC) methods. The combination of used methods, where each of them evaluates the feature significance based on a different algorithm, enables the selection of relevant markers. Based on the selected feature subset, both partial least square discriminant analysis (PLS-DA) and orthogonal partial least square discriminant analysis (OPLS-DA) were applied. The developed classification models were validated using 7-fold internal cross validation and characterized by the described variance (R^2X and R^2Y), the predicted variance (Q^2Y), the root mean square error of estimation (RMSEE) and permutation tests for R^2Y and Q^2Y.

All statistical analyses were performed using SIMCA® (Sartorius, Göttingen, Germany), Metaboanalyst (metaboanalyst.ca, accessed on 16 January 2024) and using custom built R scripts.

2.3.5. Marker Identification

All significant features (significance is described by the results of the univariate statistical analysis or variable importance on the projection (VIP) score from PLS-DA/OPLS-DA) used for classification models building were subject of structure identification. For these features, *.mat files (containing both MS and MS/MS spectral information) were exported from MS-Dial and imported into the open-source software SIRIUS 4 [27,28], which also integrates CSI:FingerID [29] and CANOPUS [30,31]. Together, these three tools suggest possible molecular formulae, potential structures and compound classes for a given feature, which is compared with online spectral databases (BioCyc, HMDB, COCONUT) [32–34]. Tentatively identified markers were characterized by elemental formula, mass error and compound name. In addition, a confidence level of markers identification was classified according to the approaches used in previous studies [35–37] for identification of compounds based on LC-MS metabolomic data. The confidence levels range from Level 4 (unknown reproducible signal defined by m/z, retention time and MS spectrum), Level 3 (known compound class with many isomers possibilities), Level 2 (annotated compound based on matched MS/MS spectra and library) and Level 1 (identified compound confirmed with analytical standard) to Level 0 (identified compound including full stereochemistry).

3. Results and Discussion

This study, aimed at the authentication of apple cultivars and their geographical origin, was based on the assumption that characteristic metabolites, authenticity markers, could be identified via the statistical processing of the HPLC-HRMS/MS metabolic fingerprints of cuticular layer extracts. As described in the introduction, several studies have performed well in cultivar or geographical origin classification (Table 1); nevertheless, none of them focused on the analysis of the cuticular layer extracts, which, according to our working hypothesis, have a high application potential for authentication. In the paragraphs below, the steps taken to test this working hypothesis are described.

3.1. Selection of Extraction Solvent/Mixture

The first step was to find an extraction solvent that would enable the reproducible extraction (not necessarily quantitative) of the widest possible range of substances from the apple cuticle. The tested solvents/solvent mixtures, differing in their selectivity, involved hexane–EtAc (1:1, *v:v*), MTBE–MeOH (10:3, *v:v*), DCM, DCM–MeOH (2:1, *v:v*) and DCM–MeOH (1:1, *v:v*). The comparison of the total ion chromatograms of the tested solvents is shown in Supplementary Materials (Figure S1). The suitability of the extraction solvent was assessed by the distribution of chromatographic peaks in terms of their retention times and the total number of features detected via the reversed-phase UHPLC-HRMS method, which is commonly used by the authors of metabolomic studies. Since wax esters, which are known to occur in large amounts in the apple cuticular layer [13], ionize poorly in ESI-mode, only ESI+ was used for the experiments. A similar approach was used in other studies analyzing a similar matrix [20,22]. Based on the above criteria, the best solvent mixture was DCM–MeOH (1:1, *v:v*). A total of 11,581 features corresponding to compounds with a wide range of polarities were detected. Only a slightly lower number of features were detected in DCM and DCM–MeOH (2:1, *v:v*) extracts, while in the chromatograms obtained with the other solvent mixtures, hexane–EtAc (1:1, *v:v*) and MTBE–MeOH (10:3, *v:v*), the more polar metabolites, eluted at lower retention times, were not sufficiently represented.

3.2. UHPLC-HRMS/MS Analysis

A number of analytical strategies have been applied for the investigation of apples' authenticity, either geographically or by variety. As shown in Table 1, in most cases, non-target screening performed by various instrumental techniques, such as NIR, GC-MS or IR-MS, was employed for analyses of various apple parts/forms. In contrast to these approaches, the UHPLC-HRMS/MS technique used in this study allows for not only the acquisition of the metabolomic fingerprints of the respective sample but also the identification of diagnostic markers (without pre-analytical derivatization) that can be used for authentication based on a simpler target screening applicable under routine conditions. The complexity of the apple cuticular layer extract is documented by the total ion chromatograms (TIC) in Figure 1A,B. As expected, wide range of lipophilic compounds was isolated; their main groups are indicated in the figures.

3.3. Chemometric Analysis

Processing of all raw UHPLC-HRMS/MS data with MS-Dial software resulted in the detection of 96,072 features in positive ionization mode and 21,040 features in negative ionization mode. Both of the aligned data matrices obtained were then filtered according to the criteria mentioned in Materials and Methods (Section 2.3.3) based on RSD in the QC sample (20%) and the SNR in the blank sample ($\geq$3). As a result of feature filtering, the final data matrices contained 16,044 features and 2132 features, respectively, in ESI+ and ESI- mode, respectively. These data were used for further processing.

3.3.1. Data Overview

Data visualization via PCA revealed a clustering of the samples based on cultivar (Figure 2A). However, the separation of apples from the Czech Republic and Poland by geographical origin was not pronounced (Figure 2B). The data shown in these figures were obtained for the ESI+ mode; similar trends are documented in the PCA Score plots for the ESI− mode (see Supplementary Materials Figure S2).

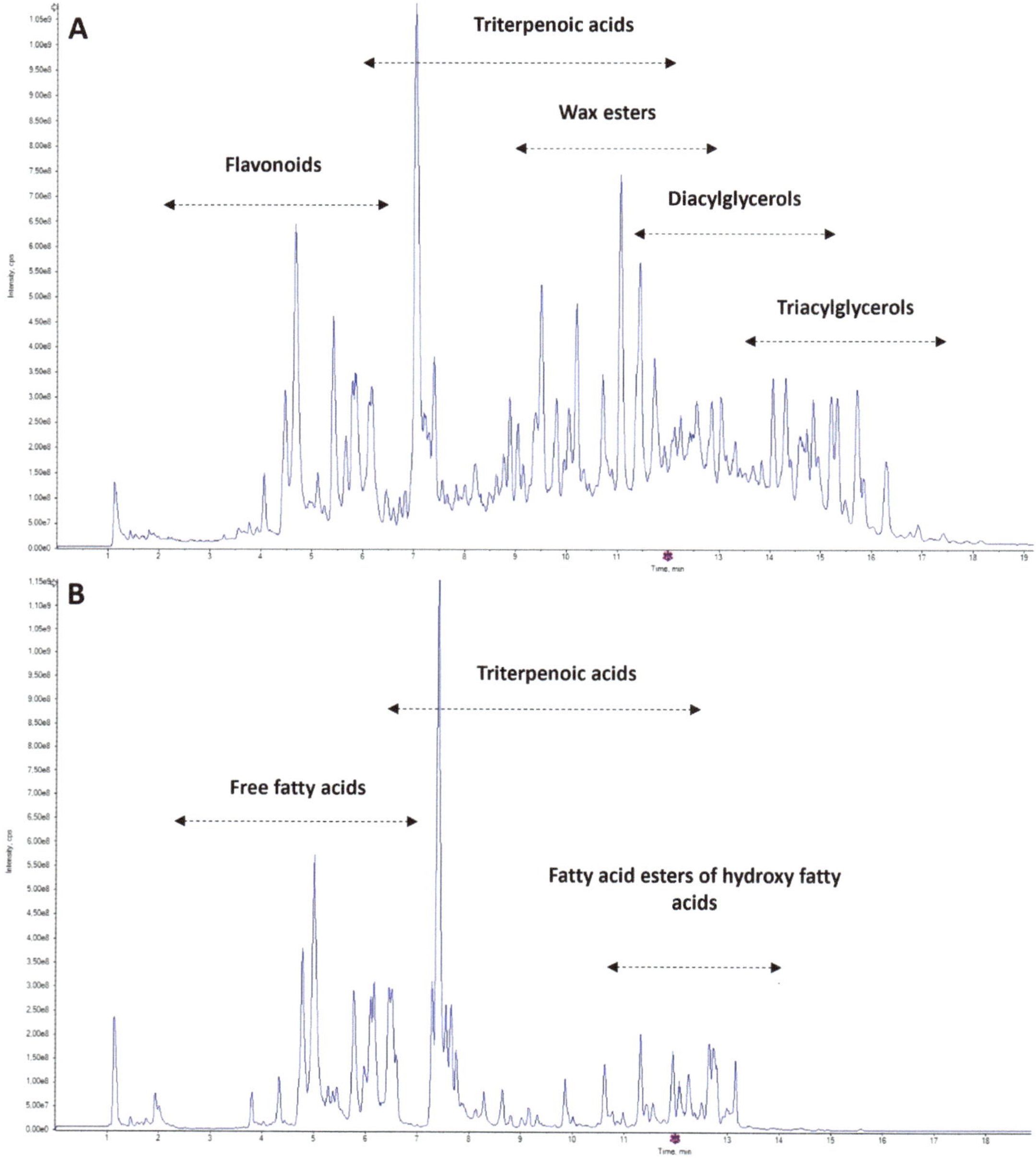

Figure 1. Total ion chromatogram of apple cuticle layer extract, QC sample: (**A**) ESI+ mode and (**B**) ESI− mode.

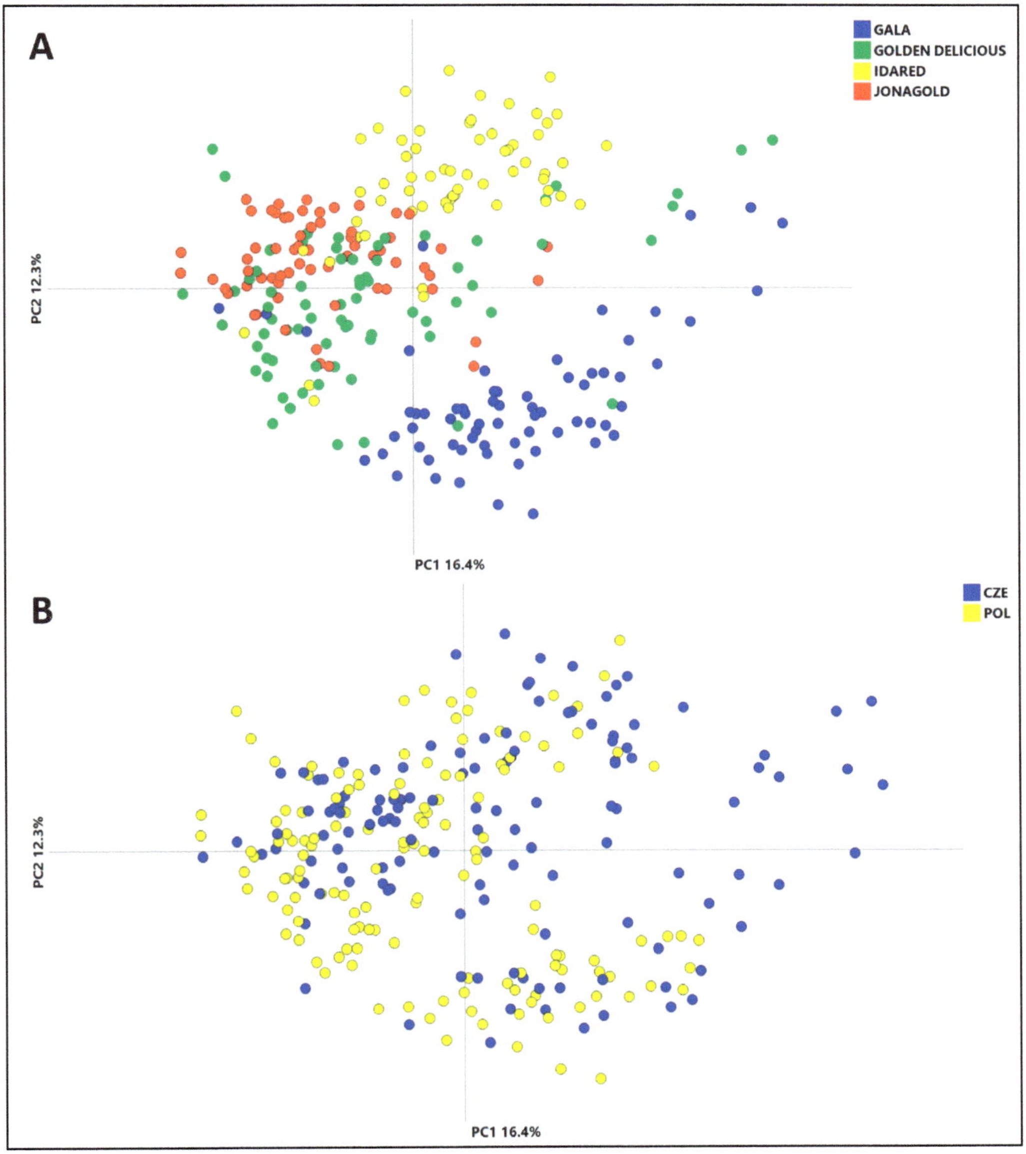

Figure 2. PCA Score plots of the complete dataset of ESI+ features. Samples are colored according to (**A**) their cultivar and (**B**) their geographical origin.

In order to obtain reliable models for the classification of both apple cultivars and geographical origin, two alternative methods of supervised learning were combined. PLS-DA was used for cultivar classification, while a systematic strategy was used for geographical origin classification, as the influence of growing location on the metabolome of the cuticular layer is less pronounced. Individual binary OPLS-DA models were created for each apple cultivar. The use of this specific chemometric method allowed us to reduce the effect of collinearity and model overfitting [38].

3.3.2. Apple Cultivars Classification

Prior to building the PLS-DA classification models, an ANOVA false discovery rate (FDR) p-value threshold of 0.05 was used to filter out features that were unimportant for cultivar differentiation. This process resulted in a subset of 14,551 and 1678 significant features for the ESI+ and ESI- modes, respectively, which were used for building PLS-DA models (all four groups of cultivars included). For the model based on ESI+ data, the validation parameters were $R^2Y = 0.790$ and $Q^2Y = 0.756$; for the ESI- data, similar qualitative results were achieved, with $R^2Y = 0.770$ and $Q^2Y = 0.756$. Permutation tests ($n = 100$) were performed to validate the developed models; for both R^2Y and Q^2Y, the p-value was below 0.01, indicating valid models [39]. In general, a VIP score > 1 is considered as a threshold value for significant features [40]; nevertheless, in the study here presented, this threshold was increased up to 1.5, which resulted in exclusion of less significant metabolites. This way, only the 197 most significant markers (ESI+ and ESI− combined) were selected for further processing. Table 2 provides an overview of 13 metabolites whose molecular structure could be identified using the criteria specified in Materials and Methods (Section 2.3.5).

Table 2. Identified significant metabolites used for apple cultivar classification (compounds sorted in descending order according to the PLS-DA VIP score).

Marker Ion (*m/z*)	Retention Time [min]	Adduct Type	Elemental Formula	Mass Error [ppm]	Tentative Identification	PLSDA VIP Score	Confidence Level
701.7138	14.03	[M+H]+	$C_{48}H_{92}O_2$	−5.4	Wax ester (30:1/18:1)	3.1	2
317.064	2.06	[M+H]+	$C_{16}H_{12}O_7$	−6.7	Isorhamnetine	2.9	3
673.6829	13.69	[M+H]+	$C_{46}H_{88}O_2$	−5	Wax ester (28:1/18:1)	2.9	2
461.1111	2.14	[M-H]−	$C_{22}H_{22}O_{11}$	5.9	Isorhamnetin rhamnoside	2.8	3
671.6652	13.43	[M+H]+	$C_{46}H_{86}O_2$	-8	Wax esters (46:3)	2.8	3
699.691	14.64	[M+Na]+	$C_{46}H_{92}O_2$	−12.1	Wax esters (46:0)	2.7	3
979.8971	14.81	[M+Na]+	$C_{63}H_{120}O_5$	−6.4	TAG (60:2)	2.4	3
509.4234	12.24	[M+H]+	$C_{31}H_{56}O_5$	5.6	DAG (28:2)	2.1	3
533.0917	1.33	[M-H]−	$C_{24}H_{22}O_{14}$	2.6	Luteolin-O-malonyl glucoside	2.1	3
663.3906	5.99	[M+HCOO]−	$C_{39}H_{54}O_6$	1.3	Caffeoylbetulinic acid	2	3
535.4747	12.87	[M-H]−	$C_{34}H_{64}O_4$	3.8	FAHFA (18:1/16:0)	1.9	2
549.3436	3.7	[M+HCOO]−	$C_{30}H_{48}O_6$	1.6	Triterpenic acid	1.6	3
749.6105	13.21	[M-H]−	$C_{49}H_{82}O_5$	2.8	DAG (46:7)	1.6	3

DAG—diacylglycerol; FAHFA—fatty acid ester of hydroxy fatty acid; PLSDA VIP—variable importance in projection of PLSDA model; TAG—triacylglycerol

Flavonoids, wax esters and other lipids together with triterpenoids were the three classes of compounds identified as markers. The first group included isorhamnetin (methylated metabolite of quercetin), isorhamnetin rhamnoside and luteolin malonyl glucoside, secondary metabolites reported to be present at low levels in apple skin [41,42]; all were present at higher levels in "Idared" and "Jonagold" apple cultivars. The second group of significant markers (according to the VIP score) was identified as wax esters, a lipid subclass defined by the LIPID MAPS structure database (LMSD) [43]. Although many studies have indicated the presence of wax esters in the apple cuticular layer [12,44–46], the (tentative) identification of individual representatives relied exclusively on gas chromatography coupled to a flame ionization detector or mass-spectrometry with a simple mass analyzer, where the sample preparation typically involved the hydrolysis of ester bonds and a derivatization step aimed at increasing the volatility of the released fatty acids and alcohols under these conditions; some of the information about the wax structure is lost. Contrary to that approach, in another earlier comprehensive study [47], GC-MS analysis of whole molecules was performed; a database involving electron ionization mass spectra of 154 wax ester standards (various straight-chain and methyl-branched saturated and unsaturated species) was created. As regards LC-MS, several authors have investigated the mass spectra of wax esters obtained via this technique in more detail. In two older studies [48,49], atmospheric pressure ionization (APCI) and two types of mass analyzers (ion trap and Orbitrap) were used for this purpose. The dominant signals in the mass spectra were protonated molecular ions $[M+H]^+$. Also in our study, where ESI was used instead of APCI, the most intensive ions in the wax ester spectra were $[M+H]^+$. On the other hand, some papers report the use of ESI (and ammonium formate was a component of a mobile phase), but the most intensive adducts of wax esters were $[M+NH_4]^+$ [50,51]. In our study employing the Q-ToF mass analyzer, the fragmentation spectra of the $[M+H]^+$ ion were further investigated and compared with those reported by Chen et al. [20], who focused on a systematic investigation of collision-induced dissociation (CID) patterns of different wax ester standards using HPLC-ESI-Q-ToF-MS. Based on this information, the fragmentation spectra of some wax esters present in our samples could be interpreted; an example of one of identified markers as wax ester (28:1/18:1) is shown in Figure 3.

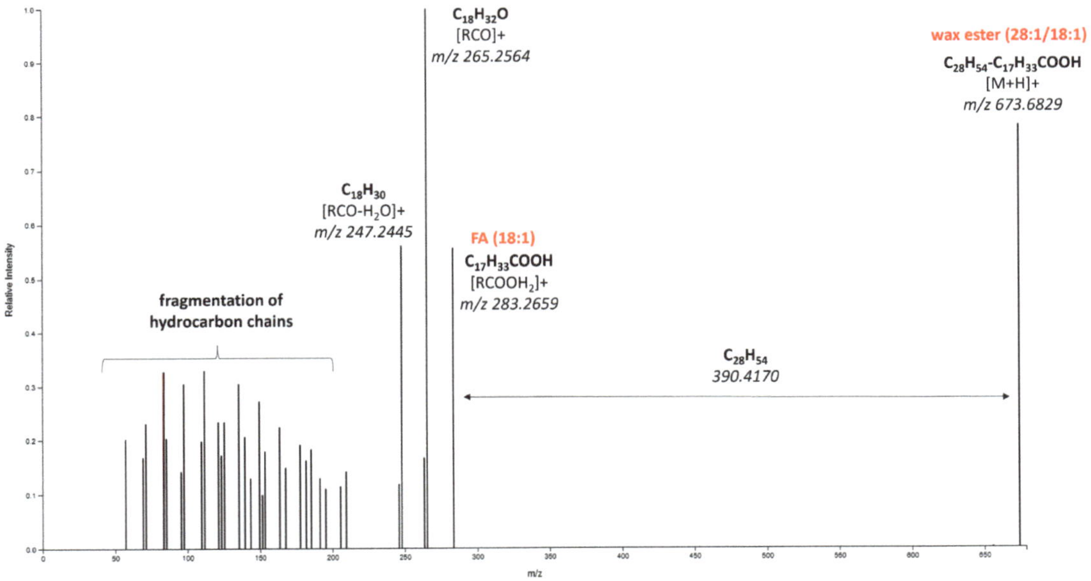

Figure 3. Fragmentation spectrum of marker identified as wax ester (28:1/18:1).

The molecular formula $C_{46}H_{88}O_2$ was calculated for the precursor ion m/z 673.6829 based on its exact mass and isotopic envelope. In the MS/MS spectrum, the base ion m/z 283.2659 was identified as an octadecenoic acid fragment $[RCOOH_2]^+$, which was formed by breaking the ester bond in the wax molecule. The fragments ions m/z 256.2564 and m/z 247.2445 were residues of oleic acid, namely, $[RCO]^+$ and $[RCO-H_2O]^+$, respectively, which is in line with the spectra interpretation introduced by the aforementioned study [20]. The calculated difference between the precursor ion $[M+H]^+$ (m/z 673.6829) and the fatty acyl fragment (m/z 283.2659) 390.4170 corresponded to the elemental composition $C_{28}H_{54}$, i.e., the loss of the octacosenol alkyl moiety. The fragmentation spectra for other wax esters with identified fatty acid and fatty alcohol moieties are shown in File S1. The description of the wax esters' structure (number of carbon atoms in the alcohol; number of double bonds/number of carbons in the fatty acid; number of double bonds) corresponds to the nomenclature used in LMSD for this lipid subclass [52]. Another identified marker belonging to lipids was the fatty acid ester of hydroxy fatty acid (FAHFA, 18:1/16:1). These FAHFAs (previously known as indicators of inflammation in human samples) have only recently been quantified in some foods, including apple skin [53]. The other identified markers representing the class of acylglycerols were DAG (46:7), DAG (28:2) and TAG (60:2).

The last group of identified markers of apple cultivar were triterpenic acids, a group of compounds that was the only one analyzed in the apple cuticle by several authors using the UHPLC-HRMS/MS technique [22,23,25]. Caffeoylbetulinic acid, one of the tentatively identified triterpenic acids, was, together with caffeoyloleanolic acid, recognized earlier as a typical metabolite occurring in the skins of russeted apples [25]. It should be noted that susceptibility to russeting is rather cultivar-specific; some apples, such as Idared, develop this defect only very rarely.

It must be emphasized that despite a relatively small number of markers used for cultivars classification, the performance characteristics of the PSL-DA model were acceptable with $R^2Y = 0.691$, $Q^2Y = 0.666$ and recognition ability of 91% [39]. The boxplots illustrating markers can be found in Supplementary Materials (Figure S3).

3.3.3. Classification of Apple Geographical Origin

The comparison of the PCA cosre plots (Figures 2 and S1) documents a bigger impact of the apple cultivar on the fingerprint of the cuticular-layer metabolome than that of the geographical origin. For this reason, the generated data were investigated more in depth and separately for each variety. The input file was a filtered data matrix with 16,044 and 2132 features, obtained in ESI+ and ESI− mode, respectively. Both t-test and ROC were then applied to the data and the fold change was calculated. An OPLS-DA model was created based on the subset of features that met the t-test FDR p-value < 0.05 and where the area under curve (AUC) value was higher than 0.75 [54]. Apart from the model validation performed, which was performed using seven-round cross-validation [55], the validity of the model was also proved over several harvest seasons. The predictive ability of the model was calculated by inserting samples from the 2020 and 2022 harvest seasons into the model created from samples harvested in 2021. The performance characteristics of all created OPLS-DA models for all cultivars are summarized in Table 3.

As shown in Table 3, all OPLS-DA models, except for the one classifying the geographical origin of "Jonagold" apples (ESI− data), performed satisfactorily [39]. The highest number of features differentiating between apples from Poland and the Czech Republic was found for the "Golden Delicious" cultivar. On the other hand, the worst performance of the developed classification models was found for the "Jonagold" cultivar, which is known to be susceptible to mutations (according to the experts from the Czech Research and Breeding Institute of Pomology Holovousy, there are 23 known mutations in "Jonagold" compared to 4 known mutations in "Golden Delicious", 3 in "Idared" and 8 in "Gala"). Rather high variability of apple metabolite patterns in the case of "Jonagold" cultivar (even within the respective country) was obviously associated with its high susceptibility to mutations. For this reason, it was difficult to identify reliable diagnostic markers. When searching for experiences with the authentication of "Jonagold" in other studies, we found that Chinese authors [6] included this variety in their sample set, together with "Starkrimson", "Qinguan", "Gala", "Golden Delicious" and "Fuji". However, a different approach, the fingerprinting of volatiles in apple juices via the SPME-HS-GC-MS technique, was used by authors. Interestingly, contrary to other cultivars, the authentication of which was 100% successful, in the case of Jonagold, it was only 89%.

Table 3. Parameters of the OPLS-DA models for the classification of apple samples according to the geographical origin (Poland vs. Czech Republic).

OPLS-DA Model Parameters	ESI+				ESI−			
	Gala	Golden Delicious	Idared	Jonagold	Gala	Golden Delicious	Idared	Jonagold
number of features	506	1048	156	11	13	44	24	9
R^2X	0.783	0.596	0.570	0.946	0.667	0.567	0.850	0.921
R^2Y	0.735	0.635	0.886	0.561	0.639	0.738	0.646	0.480
Q^2Y	0.624	0.554	0.809	0.501	0.543	0.686	0.574	0.436
RMSEE	0.265	0.309	0.175	0.335	0.307	0.261	0.308	0.362
p-value of permutation for R^2Y	<0.01	<0.01	<0.01	<0.01	<0.01	<0.01	<0.01	<0.01
p-value of permutation for Q^2Y	<0.01	<0.01	<0.01	<0.01	<0.01	<0.01	<0.01	<0.01
validity of the model over time	82%	65%	85%	88%	78%	77%	78%	88%

R^2X—fraction of X variation described by the model; R^2Y—fraction of Y variation described by the model; Q^2Y—fraction of Y variation predicted by model according to the cross validation; RMSEE—root mean square error of estimation.

All features selected for geographical origin classification (distinguishing between Czech and Polish apple samples), i.e., those with ROC AUC value > 0.75 and *t*-test FDR *p*-value < 0.05, were subjected to the identification process. A total of, 40, 72 and 6 markers were identified for "Gala" (Table S1), "Golden Delicious" (Table S2) and "Idared" 6 (Table S3) cultivars, respectively. With regard to the facts mentioned above, no marker was identified for "Jonagold" cultivar.

The most frequent markers for geographical origin were wax esters, 23 of them for "Gala" and 28 for "Golden Delicious". In addition to fatty acids, hydroxy fatty acids were also bound in wax ester molecules. The example of the fragmentation spectrum of such a marker (*m*/*z* 631.5975; retention time: 13.02 min), which was identified as a hydroxy wax ester (24:0/18:3-O), is shown in Figure 4. The obtained spectra were similar to those of the wax esters discussed in the previous section, but the calculated molecular formula (in this case, $C_{42}H_{78}O_3$) contained three oxygen atoms instead of two. In the fragmentation spectra of these compounds, alike to that shown in Figure 5, there was a visible neutral loss of H_2O molecules from both precursor and fragment ions corresponding to the hydroxy fatty acyl [56]. A total of 27 hydroxy wax esters were identified as markers of geographical origin in the analyzed apples. For those, in which fatty acid composition was tentatively identified, the corresponding fragmentation spectra are summarized in File S1.

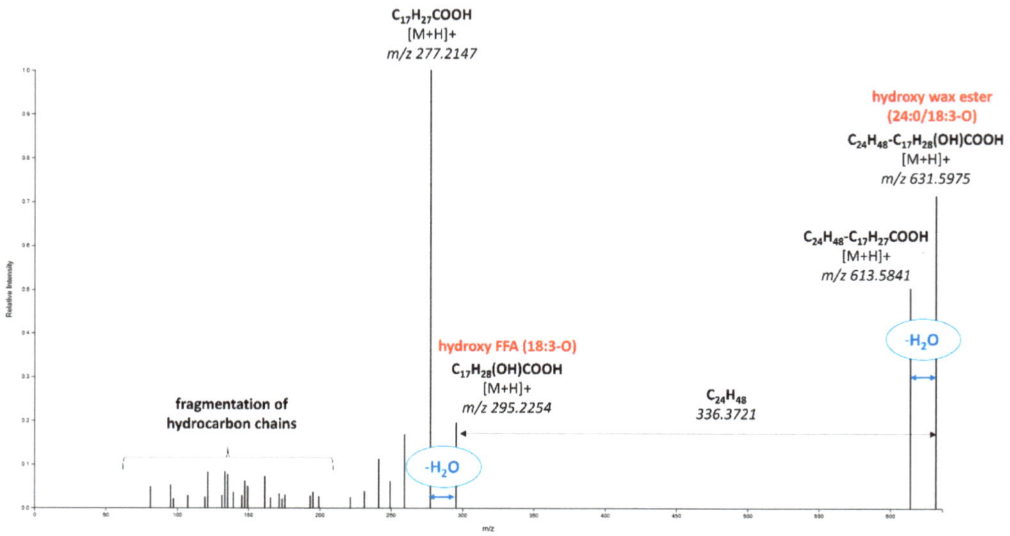

Figure 4. Fragmentation spectra of the marker identified as hydroxy wax ester (24:0/18:3-O).

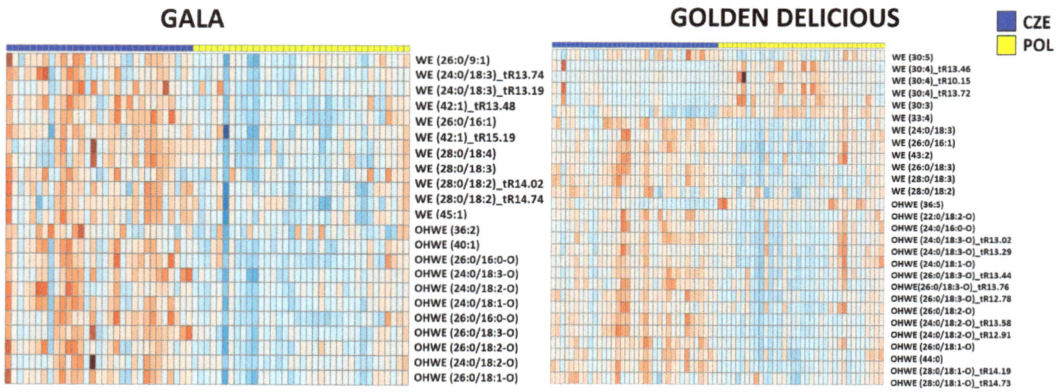

Figure 5. Heatmaps showing fold changes (red and blue colours represent positive and negative fold changes, respectively, darker shadows indicate higher absolute values) of wax ester (WE) and hydroxy wax ester (OHWE) intensities corresponding to the median value of the respective compound in apple samples (on the horizontal axis) harvested either in the Czech Republic (blue) or Poland (yellow).

In general, the most common fatty acids bound in waxes were hexadecanoic acid, hexadecenoic acid (16:0, 16:1) and unsaturated fatty acids containing 18 carbons (18:1, 18:2, 18:3, 18:4) bound to C22-C28 aliphatic alcohol. The heatmaps in Figure 5 show that in both "Gala" and "Golden Delicious" cultivars, the wax esters, as well as hydroxy wax esters, were upregulated in the apples harvested in the Czech Republic (exceptions were shorter chain wax esters, i.e., 30:3, 30:4, 30:5). In "Gala", the differences were more pronounced than in "Golden Delicious"; the fold change of wax esters ranged from 1.21 to 2.37 and from 1.11 to 3,17, respectively. The higher intensities of wax esters signals (i.e., their higher amounts) observed in the samples from the Czech Republic could be due to the higher average altitude (and thus the different climatic conditions) of the orchards where the apples were collected (450 m in the Czech Republic versus 173 m in Poland), which has been previously reported as a parameter associated with thicker cuticles [57].

The next group of identified geographical origin markers were triterpenic acids, specifically ursane-type triterpenic acids. As already mentioned, these compounds, together with their derivatives (oxo, dihydroxy, oxohydroxy), have been identified in the apple cuticle layer by several authors [22,23].

In the case of our data, eight derivatives of ursolic acid were identified among the markers for the cultivar "Gala". Interestingly, all ursenoic acids showed increased intensity in Czech apple samples. The higher intensity was statistically significant based on *t*-test p-value < 0.05 (except for oxohydroxy ursenoic acid), with fold changes ranging from 1.3 to 4. In "Golden Delicious", in addition to the five ursolic acid derivatives, their precursors lupeone, uvaol, amyrin and hydroxybetulin were also found. Among markers for cultivar "Golden Delicious", there was no clear trend. As an example, the boxplots of all ursenoic acids selected as markers are shown in the Supplementary Materials (Figures S4 and S5).

As relatively polar metabolites such as sorbitol, fucose, heptulose and mannose (trace amounts were contained in analyzed extracts), alike some flavonoids, represented by phloretin and chlorogenic acid common apple flavonoids [58,59], were also on the list of geographic origin markers.

4. Conclusions

In this study, the UHPLC-HRMS/MS-based metabolomic fingerprinting of cuticular layer extracts followed by advanced data processing was proved to be an applicable strategy for the authentication of apples cultivars and their geographic origin. The results of this study, within which 274 apple samples of four cultivars harvested either in the Czech Republic or Poland in three seasons were analyzed, can be summarized as follows:

- PCA showed a more pronounced cultivar impact on the metabolites occurring in the apple cuticle compared to that of geographical origin.
- The created PLS-DA models enabled reliable apple cultivar classification; 13 markers encompassing mainly waxes and triterpenoids were identified,
- The created OPLS-DA models enabled the safe classification of geographical origins of "Gala", "Golden Delicious" and "Idared" cultivars; however, for "Jonagold", it was unsuccessful.
- Wax esters, including those with bound hydroxy fatty acids (reported for the first time in apple cuticular wax), represented a significant group of identified markers, the amount of which in "Golden Delicious" and "Gala" cultivars was higher (upregulated) in samples from the Czech Republic compared those from Poland.

Overall, our findings underscore the potential of apple cuticular layer analysis to be used as a robust tool for apple authentication, offering insights into both cultivar and geographical origin distinctions. Moreover, revealing novel compounds that enhance our understanding of apple wax composition is facilitated by this approach.

Supplementary Materials: The following supporting information can be downloaded at https://www.mdpi.com/article/10.3390/foods13091308/s1: Figure S1: Total ion chromatograms (ESI+) showing intensities of compounds extracted by various solvent mixtures. Figure S2: PCA Score plots of the complete dataset of ESI− features. Samples are colored according to their cultivar (A) and geographical origin (B). Figure S3: Boxplots of identified markers for classification of apple cultivar. Figure S4: Boxplots of markers for classification of geographical origin for cultivar "Gala" identified as ursane-type triterpene acids and their derivatives and precursors. Figure S5: Boxplots of markers for classification of geographical origin for cultivar "Golden Delicious" identified as ursane-type triterpene acids and their derivatives and precursors. Table S1: Identification of metabolites used for geographical origin classification of apple cultivar "Gala". Markers are in descending order according to the AUC ROC value. The log2 FC value indicates whether the marker is increased in Czech samples (log2 FC > 0) or in Polish samples (log2 FC < 0). Table S2: Identification of metabolites used for geographical origin classification of apple cultivar "Golden Delicious". Markers are in descending order according to the AUC ROC value. The log2 FC value indicates whether the marker is increased in Czech samples (log2 FC > 0) or in Polish samples (log2 FC < 0). Table S3: Identification of metabolites used for geographical origin classification of apple cultivar "Idared". Markers are in descending order according to the AUC ROC value. The log2 FC value indicates whether the marker is increased in Czech samples (log2 FC > 0) or in Polish samples (log2 FC < 0). File S1: Fragmentation spectra of all wax esters with identified fatty acid and fatty alcohol moiety.

Author Contributions: Conceptualization, K.B., J.S., V.K. (Vladimir Kocourek) and J.H.; methodology, K.B., L.U. and P.V.; software, K.B.; formal analysis, K.B.; investigation, K.B., L.U. and P.V.; resources, J.S.; data curation, K.B.; writing—original draft preparation, K.B. and J.H.; writing—review and editing, J.H., V.K. (Vladimir Kocourek) and V.K. (Vit Kosek); supervision, J.H., V.K. (Vit Kosek), V.K. (Vladimir Kocourek) and J.H.; project administration, J.S., V.K. (Vladimir Kocourek) and J.H. All authors have read and agreed to the published version of the manuscript.

Funding: This project was funded by the Ministry of Agriculture of the Czech Republic, grant NAZV QK1910104 "Research of metabolomic methods for laboratory authentication of apples geographicity". Regarding the instrumental research facilities, this study was supported by the METROFOOD-CZ research infrastructure project (MEYS grant no. LM2023064).

Institutional Review Board Statement: Not applicable.

Informed Consent Statement: Not applicable.

Data Availability Statement: The original contributions presented in the study are included in the article and Supplementary Materials, further inquiries can be directed to the corresponding author.

Acknowledgments: The authors acknowledge Michaela Rektorisova and Klara Navratilova for their support of this research.

Conflicts of Interest: The authors declare no conflicts of interest.

References

1. Shanbandeh, M. Apple Production Worldwide in 2022. Available online: https://www.statista.com/statistics/237605/production-of-apples-worldwide-by-region-2007/ (accessed on 29 February 2024).
2. Bat, K.; Vidrih, R.; Nečemer, M.; Mozetič Vodopivec, B.; Mulič, I.; Kump, P.; Ogrinc, N. Characterization of Slovenian apples with respect to their botanical and geographical origin and agricultural production practice. *Food Technol. Biotechnol.* **2012**, *50*, 107–116.
3. Wu, H.; Yue, T.; Yuan, Y. Authenticity Tracing of Apples According to Variety and Geographical Origin Based on Electronic Nose and Electronic Tongue. *Food Anal. Methods* **2018**, *11*, 522–532. [CrossRef]
4. Li, C.; Li, L.; Wu, Y.; Lu, M.; Yang, Y.; Li, L. Apple Variety Identification Using Near-Infrared Spectroscopy. *J. Spectrosc.* **2018**, *2018*, e6935197. [CrossRef]
5. Bian, H.; Sheng, L.; Yao, H.; Ji, R.; Yu, Y.; Chen, R.; Wei, D.; Han, Y. Application of fluorescence spectroscopy in classifying apple juice according to the variety. *Optik* **2021**, *231*, 166361. [CrossRef]
6. Guo, J.; Yue, T.; Yuan, Y. Feature selection and recognition from nonspecific volatile profiles for discrimination of apple juices according to variety and geographical origin. *J. Food Sci.* **2012**, *77*, C1090–C1096. [CrossRef] [PubMed]
7. Medina, S.; Perestrelo, R.; Santos, R.; Pereira, R.; Câmara, J.S. Differential volatile organic compounds signatures of apple juices from Madeira Island according to variety and geographical origin. *Microchem. J.* **2019**, *150*, 104094. [CrossRef]
8. Mimmo, T.; Camin, F.; Bontempo, L.; Capici, C.; Tagliavini, M.; Cesco, S.; Scampicchio, M. Traceability of different apple varieties by multivariate analysis of isotope ratio mass spectrometry data. *Rapid Commun. Mass Spectrom.* **2015**, *29*, 1984–1990. [CrossRef] [PubMed]
9. Brombin, V.; Mistri, E.; Bianchini, G. Multi stable isotope ratio analysis for the traceability of northern Italian apples. *Food Chem. X* **2022**, *16*, 100514. [CrossRef] [PubMed]
10. Liu, X.; Zhao, Y.; Mu, J.; Zhang, J.; Zhang, A. Determination of geographical origin of concentrated apple juice through analysis of stable isotopic and mineral elemental fingerprints: Preliminary results. *J. Sci. Food Agric.* **2021**, *101*, 3795–3803. [CrossRef]
11. Belding, R.D.; Blankenship, S.M.; Young, E.; Leidy, R.B. Composition and Variability of Epicuticular Waxes in Apple Cultivars. *J. Am. Soc. Hort. Sci.* **1998**, *123*, 348–356. [CrossRef]
12. Yang, Y.; Zhou, B.; Zhang, J.; Wang, C.; Liu, C.; Liu, Y.; Zhu, X.; Ren, X. Relationships between cuticular waxes and skin greasiness of apples during storage. *Postharv. Biol. Technol.* **2017**, *131*, 55–67. [CrossRef]
13. Wu, W.; Jiang, B.; Liu, R.; Han, Y.; Fang, X.; Mu, H.; Farag, M.; Simal-Gandara, J.; Prieto, M.A.; Chen, H.; et al. Structures and functions of cuticular wax in postharvest fruit and its regulation: A comprehensive review with future perspectives. *Engineering* **2023**, *23*, 118–129. [CrossRef]
14. Leide, J.; Xavier de Souza, A.; Papp, I.; Riederer, M. Specific characteristics of the apple fruit cuticle: Investigation of early and late season cultivars 'Prima' and 'Florina' (*Malus domestica* Borkh.). *Sci. Hortic.* **2018**, *229*, 137–147. [CrossRef]
15. Shepherd, T.; Wynne Griffiths, D. The effects of stress on plant cuticular waxes. *New Phytol.* **2006**, *171*, 469–499. [CrossRef]

16. Fernandez-Moreno, J.-P.; Malitsky, S.; Lashbrooke, J.; Biswal, A.K.; Racovita, R.C.; Mellerowicz, E.J.; Jetter, R.; Orzaez, D.; Aharoni, A.; Granell, A. An efficient method for medium throughput screening of cuticular wax composition in different plant species. *Metabolomics* **2016**, *12*, 73. [CrossRef]
17. Lara, I.; Belge, B.; Goulao, L.F. A Focus on the Biosynthesis and Composition of Cuticle in Fruits. *J. Agric. Food Chem.* **2015**, *63*, 4005–4019. [CrossRef] [PubMed]
18. Li, F.; Min, D.; Song, B.; Shao, S.; Zhang, X. Ethylene effects on apple fruit cuticular wax composition and content during cold storage. *Postharv. Biol. Technol.* **2017**, *134*, 98–105. [CrossRef]
19. Li, F.; Min, D.; Ren, C.; Dong, L.; Shu, P.; Cui, X.; Zhang, X. Ethylene altered fruit cuticular wax, the expression of cuticular wax synthesis-related genes and fruit quality during cold storage of apple (*Malus domestica* Borkh. C.v. Starkrimson) fruit. *Postharv. Biol. Technol.* **2019**, *149*, 58–65. [CrossRef]
20. Chen, J.; Green, K.B.; Nichols, K.K. Characterization of Wax Esters by Electrospray Ionization Tandem Mass Spectrometry: Double Bond Effect and Unusual Product Ions. *Lipids* **2015**, *50*, 821–836. [CrossRef]
21. Tada, A.; Jin, Z.-L.; Sugimoto, N.; Sato, K.; Yamazaki, T.; Tanamoto, K. Analysis of the constituents in jojoba wax used as a food additive by LC/MS/MS. Shokuhin Eiseigaku Zasshi. *J. Food Hyg. Soc. Jpn* **2005**, *46*, 198–204. [CrossRef]
22. McGhie, T.K.; Hudault, S.; Lunken, R.C.M.; Christeller, J.T. Apple peels, from seven cultivars, have lipase-inhibitory activity and contain numerous ursenoic acids as identified by LC-ESI-QTOF-HRMS. *J. Agric. Food Chem.* **2012**, *60*, 482–491. [CrossRef] [PubMed]
23. Poirier, B.C.; Buchanan, D.A.; Rudell, D.R.; Mattheis, J.P. Differential Partitioning of Triterpenes and Triterpene Esters in Apple Peel. *J. Agric. Food Chem.* **2018**, *66*, 1800–1806. [CrossRef] [PubMed]
24. Andre, C.M.; Greenwood, J.M.; Walker, E.G.; Rassam, M.; Sullivan, M.; Evers, D.; Perry, N.B.; Laing, W.A. Anti-inflammatory procyanidins and triterpenes in 109 apple varieties. *J. Agric. Food Chem.* **2012**, *60*, 10546–10554. [CrossRef] [PubMed]
25. Andre, C.M.; Larsen, L.; Burgess, E.J.; Jensen, D.J.; Cooney, J.M.; Evers, D.; Zhang, J.; Perry, N.B.; Laing, W.A. Unusual immuno-modulatory triterpene-caffeates in the skins of russeted varieties of apples and pears. *J. Agric. Food Chem.* **2013**, *61*, 2773–2779. [CrossRef] [PubMed]
26. Tsugawa, H.; Kind, T.; Nakabayashi, R.; Yukihira, D.; Tanaka, W.; Cajka, T.; Saito, K.; Fiehn, O.; Arita, M. Hydrogen Rearrangement Rules: Computational MS/MS Fragmentation and Structure Elucidation Using MS-FINDER Software. *Anal. Chem.* **2016**, *88*, 7946–7958. [CrossRef] [PubMed]
27. Dührkop, K.; Fleischauer, M.; Ludwig, M.; Aksenov, A.A.; Melnik, A.V.; Meusel, M.; Dorrestein, P.C.; Rousu, J.; Böcker, S. SIRIUS 4: A rapid tool for turning tandem mass spectra into metabolite structure information. *Nat. Methods* **2019**, *16*, 4. [CrossRef] [PubMed]
28. Dührkop, K.; Nothias, L.-F.; Fleischauer, M.; Reher, R.; Ludwig, M.; Hoffmann, M.A.; Petras, D.; Gerwick, W.H.; Rousu, J.; Dorrestein, P.C.; et al. Systematic classification of unknown metabolites using high-resolution fragmentation mass spectra. *Nat. Biotechnol.* **2021**, *39*, 4. [CrossRef] [PubMed]
29. Dührkop, K.; Shen, H.; Meusel, M.; Rousu, J.; Böcker, S. Searching molecular structure databases with tandem mass spectra using CSI:FingerID. *Proc. Natl. Acad. Sci. USA* **2015**, *112*, 12580–12585. [CrossRef] [PubMed]
30. Kim, H.W.; Wang, M.; Leber, C.A.; Nothias, L.-F.; Reher, R.; Kang, K.B.; van der Hooft, J.J.J.; Dorrestein, P.C.; Gerwick, W.H.; Cottrell, G.W. NPClassifier: A Deep Neural Network-Based Structural Classification Tool for Natural Products. *J. Nat. Prod.* **2021**, *84*, 2795–2807. [CrossRef]
31. Djoumbou Feunang, Y.; Eisner, R.; Knox, C.; Chepelev, L.; Hastings, J.; Owen, G.; Fahy, E.; Steinbeck, C.; Subramanian, S.; Bolton, E.; et al. ClassyFire: Automated chemical classification with a comprehensive, computable taxonomy. *J. Cheminform.* **2016**, *8*, 61. [CrossRef]
32. Karp, P.D.; Billington, R.; Caspi, R.; Fulcher, C.A.; Latendresse, M.; Kothari, A.; Keseler, I.M.; Krummenacker, M.; Midford, P.E.; Ong, Q.; et al. The BioCyc collection of microbial genomes and metabolic pathways. *Brief. Bioinform.* **2019**, *20*, 1085–1093. [CrossRef] [PubMed]
33. Sorokina, M.; Merseburger, P.; Rajan, K.; Yirik, M.A.; Steinbeck, C. COCONUT online: Collection of open natural products database. *J. Cheminform.* **2021**, *13*, 2. [CrossRef] [PubMed]
34. Wishart, D.S.; Guo, A.C.; Oler, E.; Wang, F.; Anjum, A.; Peters, H.; Dizon, R.; Sayeeda, Z.; Tian, S.; Lee, B.L.; et al. HMDB 5.0: The Human Metabolome Database for 2022. *Nucleic Acids Res.* **2022**, *50*, D622–D631. [CrossRef] [PubMed]
35. Reisdorph, N.A.; Walmsley, S.; Reisdorph, R. A Perspective and Framework for Developing Sample Type Specific Databases for LC/MS-Based Clinical Metabolomics. *Metabolites* **2019**, *10*, 8. [CrossRef] [PubMed]
36. Blaženović, I.; Kind, T.; Ji, J.; Fiehn, O. Software Tools and Approaches for Compound Identification of LC-MS/MS Data in Metabolomics. *Metabolites* **2018**, *8*, 31. [CrossRef]
37. Sumner, L.W.; Amberg, A.; Barrett, D.; Beale, M.H.; Beger, R.; Daykin, C.A.; Fan, T.W.-M.; Fiehn, O.; Goodacre, R.; Griffin, J.L.; et al. Proposed minimum reporting standards for chemical analysis Chemical Analysis Working Group (CAWG) Metabolomics Standards Initiative (MSI). *Metabolomics* **2007**, *3*, 211–221. [CrossRef] [PubMed]
38. Bylesjö, M.; Rantalainen, M.; Cloarec, O.; Nicholson, J.K.; Holmes, E.; Trygg, J. OPLS discriminant analysis: Combining the strengths of PLS-DA and SIMCA classification. *J. Chemom.* **2006**, *20*, 341–351. [CrossRef]

39. Triba, M.N.; Le Moyec, L.; Amathieu, R.; Goossens, C.; Bouchemal, N.; Nahon, P.; Rutledge, D.N.; Savarin, P. PLS/OPLS models in metabolomics: The impact of permutation of dataset rows on the K-fold cross-validation quality parameters. *Mol. BioSyst.* **2015**, *11*, 13–19. [CrossRef]
40. Cocchi, M.; Biancolillo, A.; Marini, F. Chapter Ten—Chemometric Methods for Classification and Feature Selection. In *Comprehensive Analytical Chemistry*; Jaumot, J., Bedia, C., Tauler, R., Eds.; Data Analysis for Omic Sciences: Methods and Applications; Elsevier: Amsterdam, The Netherlands, 2018; Volume 82, pp. 265–299. [CrossRef]
41. Pennington, J.A.T. Food Composition Databases for Bioactive Food Components. *J. Food Comp. Anal.* **2002**, *15*, 419–434. [CrossRef]
42. Ramírez-Ambrosi, M.; López-Márquez, D.M.; Abad-García, B.; Dapena, E.; Berrueta, L.Á.; Gallo, B. Comparative study of phenolic profile of fruit and juice samples of a progeny of 'Meana' × 'Florina' from an Asturian cider apple breeding program. *Eur. Food Res. Technol.* **2015**, *241*, 769–784. [CrossRef]
43. Sud, M.; Fahy, E.; Cotter, D.; Brown, A.; Dennis, E.A.; Glass, C.K.; Merrill Jr, A.H.; Murphy, R.C.; Raetz, C.R.H.; Russell, D.W.; et al. LMSD: LIPID MAPS structure database. *Nucleic Acids Res.* **2007**, *35*, D527–D532. [CrossRef] [PubMed]
44. Veraverbeke, E.A.; Lammertyn, J.; Saevels, S.; Nicolaï, B.M. Changes in chemical wax composition of three different apple (*Malus domestica* Borkh.) cultivars during storage. *Postharv. Biol. Technol.* **2001**, *23*, 197–208. [CrossRef]
45. Klein, B.; Falk, R.B.; Thewes, F.R.; Anese, R.d.O.; dos Santos, I.D.; Ribeiro, S.R.; Donadel, J.Z.; Brackmann, A.; Barin, J.S.; Cichoski, A.J.; et al. Dynamic controlled atmosphere: Effects on the chemical composition of cuticular wax of 'Cripps Pink' apples after long-term storage. *Postharv. Biol. Technol.* **2020**, *164*, 111170. [CrossRef]
46. Morice, I.M.; Shorland, F.B. Composition of the surface waxes of apple fruits and changes during storage. *J. Sci. Food Agric.* **1973**, *24*, 1331–1339. [CrossRef] [PubMed]
47. Urbanova, K.; Vrkoslav, V.; Valterova, I.; Haková, M.; Cvacka, J. Structural characterization of wax esters by electron ionization mass spectrometry. *J. Lipid Res.* **2011**, *53*, 204–213. [CrossRef] [PubMed]
48. Vrkoslav, V.; Urbanová, K.; Cvacka, J. Analysis of wax ester molecular species by high performance liquid chromatography/atmospheric pressure chemical ionisation mass spectrometry. *J. Chromatogr. A* **2010**, *1217*, 4184–4194. [CrossRef] [PubMed]
49. Vrkoslav, V.; Urbanova, K.; Haková, M.; Cvacka, J. Analysis of wax esters by silver-ion high-performance liquid chromatography–tandem mass spectrometry. *J. Chromatogr. A* **2013**, *1302*, 105–110. [CrossRef] [PubMed]
50. Sekosan, G. LC-MS Identification of Wax Esters in Cloudy Canola Oil. *LCGC Suppl.* **2014**, *12*, 16–21.
51. Iven, T.; Herrfurth, C.; Hornung, E.; Heilmann, M.; Hofvander, P.; Stymne, S.; Feussner, I. Wax ester profiling of seed oil by nano-electrospray ionization tandem mass spectrometry. *Plant Methods* **2013**, *9*, 24. [CrossRef]
52. Liebisch, G.; Fahy, E.; Aoki, J.; Dennis, E.A.; Durand, T.; Ejsing, C.S.; Fedorova, M.; Feussner, I.; Griffiths, W.J.; Köfeler, H.; et al. Update on LIPID MAPS classification, nomenclature, and shorthand notation for MS-derived lipid structures. *J. Lipid Res.* **2020**, *61*, 1539–1555. [CrossRef]
53. Liberati-Čizmek, A.-M.; Biluš, M.; Brkić, A.L.; Barić, I.C.; Bakula, M.; Hozić, A.; Cindrić, M. Analysis of Fatty Acid Esters of Hydroxyl Fatty Acid in Selected Plant Food. *Plant Foods Hum. Nutr.* **2019**, *74*, 235–240. [CrossRef] [PubMed]
54. Mandrekar, J.N. Receiver Operating Characteristic Curve in Diagnostic Test Assessment. *J. Thorac. Oncol.* **2010**, *5*, 1315–1316. [CrossRef] [PubMed]
55. Westerhuis, J.A.; Hoefsloot, H.C.J.; Smit, S.; Vis, D.J.; Smilde, A.K.; van Velzen, E.J.J.; van Duijnhoven, J.P.M.; van Dorsten, F.A. Assessment of PLSDA cross validation. *Metabolomics* **2008**, *4*, 81–89. [CrossRef]
56. Spickett, C.M.; Pitt, A.R. Oxidative Lipidomics Coming of Age: Advances in Analysis of Oxidized Phospholipids in Physiology and Pathology. *Antioxid. Redox Signal.* **2015**, *22*, 1646–1666. [CrossRef] [PubMed]
57. Tessmer, M.A.; Antoniolli, L.R.; Appezzato-da-Glória, B. Cuticle of "Gala" and "Galaxy" apples cultivars under different environmental conditions. *Braz. Arch. Biol. Technol.* **2012**, *55*, 709–714. [CrossRef]
58. Lommen, A.; Godejohann, M.; Venema, D.P.; Hollman, P.C.; Spraul, M. Application of directly coupled HPLC-NMR-MS to the identification and confirmation of quercetin glycosides and phloretin glycosides in apple peel. *Anal. Chem.* **2000**, *72*, 1793–1797. [CrossRef]
59. Awad, M.A.; de Jager, A.; van Westing, L.M. Flavonoid and chlorogenic acid levels in apple fruit: Characterisation of variation. *Sci. Hort.* **2000**, *83*, 249–263. [CrossRef]

separations

MDPI

Article

Isolation and Identification of Bitter Compounds in Ginseng (*Panax ginseng* C. A. Mey.) Based on Preparative High Performance Liquid Chromatography, UPLC-Q-TOF/MS and Electronic Tongue

Yang Chen [1], Ziwei Liao [1,2], Zhe Wang [1], Wanyin Shi [1] and Jian Xu [1,*]

1 Hubei Provincial Key Laboratory of Quality and Safety of Traditional Chinese Medicine Health Food, Jing Brand Research Institute, Jing Brand Co., Ltd., Daye 435100, China; yangchen886@hotmail.com (Y.C.); ziwei10liao@hotmail.com (Z.L.); wang1985zhe02@hotmail.com (Z.W.); shiwan0501yin@outlook.com (W.S.)

2 Hubei Traditional Chinese Medicine Formula Granule Engineering & Technology Research Center, Huangshi 435000, China

* Correspondence: xujian@jingpai.com

Abstract: As a traditional Chinese medicinal herb, ginseng (*Panax ginseng* C. A. Mey.) is commonly used to treat common diseases, for example, esophageal cancer and myasthenia gravis. Furthermore, ginseng is also processed into a functional food additive that is utilized to improve the freshness of chicken soup and make health wine. Unfortunately, ginseng (*Panax ginseng* C. A. Mey.) has already shown a noticeable bitterness during its application process. In this research, the bitter substances in ginseng (*Panax ginseng* C. A. Mey.) after two common preparation processes (water extraction and ethanol extraction) were separated, purified and identified by preparative high performance liquid chromatography (prep-HPLC), high performance liquid chromatography with diode array detector (HPLC-DAD), ultra-performance liquid chromatography coupled with high-resolution quadrupole time-of-flight mass spectrometry (UPLC-Q-TOF/MS) and an electronic tongue. The results indicated that compared with the other four bitter compounds, the ginsenoside Rb1 had the highest bitterness value, followed by 20(S)-ginsenoside Rg2, ginsenoside Rg1, ginsenoside Rf and ginsenoside Rb3. Upon the evaluation of results to reduce the bitterness of ginseng extract, we found that the composite embedding system of chitosan adsorption in the ginseng carrageenan gel microsphere ($K/M_C/M_{CG}$) could effectively reduce the bitterness.

Keywords: ginseng (*Panax ginseng* C. A. Mey.) extract; bitter substances; UPLC-Q-TOF/MS; electronic tongue; reduce the bitterness

Citation: Chen, Y.; Liao, Z.; Wang, Z.; Shi, W.; Xu, J. Isolation and Identification of Bitter Compounds in Ginseng (*Panax ginseng* C. A. Mey.) Based on Preparative High Performance Liquid Chromatography, UPLC-Q-TOF/MS and Electronic Tongue. *Separations* **2024**, *11*, 114. https://doi.org/10.3390/separations11040114

Academic Editor: Wojciech Piekoszewski

Received: 28 February 2024
Revised: 3 April 2024
Accepted: 6 April 2024
Published: 7 April 2024

1. Introduction

Ginseng (*Panax ginseng* C. A. Mey.), considered "The Nine of Immortality", is a famous perennial herb that belongs to the Araliaceae Panax and has been utilized as a traditional medicine for more than 200 years in China, North Korea and Russia [1]. According to their cultivation methods, ginseng can be divided into three categories: wild ginseng (growth in a forest of mixed coniferous and broad-leaved trees, shrubs and weeds); garden ginseng (artificial planting and growth in an artificial pool bed); understory ginseng (artificial sowing and growth of primitive ginseng in deep mountains and dense forests) [2]. Among these, garden ginseng has the highest yield, accounting for 90% of China's total yield, and is widely used for processing health products and food [3].

In China, the active substances in ginseng have been extracted and used as functional raw materials to process into traditional medicines for treating chronic diseases, for example, diabetes and tumors [4,5]. Ginseng has also been developed as a cosmetic skincare material and anti-aging agent [6]. In the processing of many foods and beverages, ginseng has been used as an additive to improve the aroma and taste of the product [7,8]. In Southeast Asia, ginseng extract has been dissolved in beverages, being used as a food additive to enhance the body's immune system [9]. In ginseng chicken soup, a well-known Chinese dish, ginseng was a main condiment with a health care function. Unfortunately, the application of ginseng in food products and health care products has been restricted due to the obvious bitterness [10]. To date, the specific source of this bitterness and the contribution of bitter substances to the bitterness of ginseng were unclear.

Bitter taste is considered as one of the basic tastes, which widely exists in a variety of drugs and food. Compared with other tastes, although the preference and acceptability of a bitter taste is slightly lower, it is more characteristic taste of some food or health products such as citrus [11], green tea [12] and faba bean [13]. Studies have shown that naturally occurring bitter compounds mainly include polyphenolic compounds, such as ester catechins, which are important bitter components in green tea [14]; terpene compounds, for which previous research reports have shown that the bitterness of bitter melon is regulated through the terpenes [15]; flavonoid glycoside compounds, such as the bitter taste of the *Forsythia suspense* (Thunb.) Vahl herbal which was attributed to phillyrin [16]; and alkaloid compounds, like theobromine in chocolate which is one of the reasons for its bitter taste [17].

In the stems, leaves and roots of ginseng (*Panax ginseng* C. A. Mey.), saponins are considered as the key functional components. Thus far, researchers have found and identified various ginsenosides in ginseng. Lee et al. [18] identified 58 ginsenosides in ginseng by UPLC-Q-TOF/MS technology, of which 39 ginsenosides were found by quantitative analysis. In total, 803 ginsenosides were identified by off-line three-dimensional liquid chromatography/Q-Orbitrap mass spectrometry [19]. Ginseng has excellent pharmacological effects, and its ginsenosides have been reported to have anti-aging effects on skin through enhancing immune function, resisting melanin formation, inhibiting oxidation, and elevating the concentration of collagen and hyaluronic acid [20]. In addition, saponins in ginseng have antioxidant effects and anti-cancer effects and ameliorate intestinal mucosal damage by enhancing HUR and c-Myc levels [21,22]. Others have shown that saponins in ginseng, such as the ginsenoside Rg1, might have the effect of reducing neuronal apoptosis via ERK/CREB/BDNF signaling to improve learning and memory in epileptic rats [23]. However, there is currently limited literature on the study of ginseng flavor. Thus far, the literature has indicated that total ginsenosides have a bitter taste, but it is not clear which specific type of ginsenoside cause the bitterness.

In this research, the effects of different extraction methods on the bitter value of ginseng extract were studied. Two common methods (water extraction and alcohol extraction) were utilized to obtain experimental samples. The bitterness compounds were analyzed and identified by instrumental analysis. In addition, the method of reducing bitterness in the application process of the condiment was also displayed.

2. Materials and Methods

2.1. Chemicals and Materials

Ginseng was sourced from the local market. The product was harvested in Tieli city, Heilongjiang province, in mid-to-late September 2022. Ginsenoside Rg1, ginsenoside Rf, ginsenoside Rb1, 20(S)-ginsenoside Rg2 and ginsenoside Rb3 of 98.0% purity (HPLC graded) were purchased from National Institutes for Food and Drug Control (Beijing, China). D941 macroporous weak base ion-exchange resin column was purchased from Xi'an Lanxiao Technology Co., Ltd. (Xi'an, China).

2.2. Sample Preparation

Ginseng was crushed through using a Yunbang YB-1000A grinder (Yongkang, China). Based on the use of ginseng in the food industry, two production processes, namely ethanol extraction (for alcoholic beverages) and pure water extraction (for non-alcoholic beverages), were conducted to obtain a bitter taste sample. The ginseng samples extracted with pure water consisted of 200 g of ginseng powder added to 2000 g of distilled water and extracted for 2 h at 80 °C. After standing for 4 h, ginseng samples extracted with pure water were centrifuged for 20 min at 8000 rpm through using a Hettich ROTO SILENTA 630RS centrifuge (Munich, Germany). Afterwards, the supernatant was obtained and fat-soluble pigment impurities were removed by petroleum ether. Finally, the sample was freeze-dried and prepared with pure water at a concentration of 2.5 mg/mL, which was filtered through 0.22 μm pores and stored at −30 °C until analysis.

The ginseng samples extracted with ethanol consisted of 200 g of ginseng powder added to 2000 g of 55% ethanol solution and extracted for 2.5 h at 65 °C. After cooling, the ginseng samples extracted with 55% ethanol solution were then centrifuged for 20 min at 8000 rpm using a Hettich ROTO SILENTA 630RS centrifuge (Munich, Germany) and the supernatant was collected. The fat-soluble pigment in the supernatant was removed by using the D941 macroporous weak base ion-exchange resin column while the permeate was collected. Finally, the sample was freeze-dried and prepared with 55% ethanol solution at a concentration of 2.5 mg/mL, which was filtered through 0.22 μm pores and stored at −30 °C until analysis.

2.3. HPLC-DAD Analysis

The bitter substances in ginseng samples were analyzed by HPLC-DAD (Waters 2695 series, USA). First, the mobile phase and the elution gradient were adjusted and optimized. The HPLC technique was used to detect the chemical composition of ginseng samples after water extraction and ethanol extraction on an Agilent ZORBAX SB-Aq C18 column (4.6 mm × 250 mm, 5 μm). The solvents were acetonitrile (A) and ultra-pure water (B). A total of 10 μL of the sample was injected into the HPLC, and the column temperature was 30 °C. The flow rate was set at 1.0 mL/min, and the detection wavelength was 203 nm. The program of elution was used as follows for pure water extraction and ethanol extraction of the ginseng samples: 0–30 min, 19% Solvent A; 30–40 min, 19–24% Solvent A; 40–43 min, 24–29% Solvent A; 43–50 min, 29–28% Solvent A; 50–60 min, 28% Solvent A; 60–85 min, 28–36% Solvent A; 85–88 min, 36–45% Solvent A; 88–100 min, 45% Solvent A.

2.4. Preparation and Collection of the Fractions

According to the analysis results of HPLC-DAD, each of the fractions in both ginseng samples was obtained by peak-based information through using prep-HPLC (Agilent 1260 Infinity Series, Santa Clara, California, USA) on a Daisogel C18-10 um-120A (20 mm × 250 mm, 7 μm, Agilent Technologies, Santa Clara, California, USA). The solvents were acetonitrile (A) and ultra-pure water (B). 15 mL of the sample were injected into the prep-HPLC, and the column temperature was 30 °C at a flow rate of 20.0 mL/min. The detection wavelength was 203 nm. For pure water extraction and ethanol extraction of ginseng samples: 0–30 min, 19% Solvent A; 30–40 min, 19–24% Solvent A; 40–43 min, 24–29% Solvent A; 43–50 min, 29–28% Solvent A; 50–60 min, 28% Solvent A; 60–85 min, 28–36% Solvent A; 85–88 min, 36–45% Solvent A; 88–100 min, 45% Solvent A. Each of the samples was tested three times.

2.5. Distillate Collection and Treatment

The samples were repeatedly prepared and collected by using prep-HPLC, and the fractions with the same peak time were combined. The collected ingredients were then concentrated by using a N-1300D-W rotary evaporator (Tokyo, Japan) at a temperature of 40 °C. The organic solvent was removed by vacuum treatment combined with rotary evaporation, and the final sample was freeze-dried twice and kept in a dry environment.

2.6. Recognition of the Bitter Fraction

The freeze-dried sample was dissolved in ultra-pure water for recognition of bitter fractions through an electronic tongue system with an initial concentration of 50 μg/mL. The TS-5000Z system (Intelligent Sensor Technology Co., Ltd., Tokyo, Japan) with two reference electrodes and five lipid/membrane electrodes was used to measure taste values, including sourness (CA0), saltiness (CT0), bitterness (C00), astringency (AE1) and umami (AAE), and this was used to verify the taste of different samples [24]. In order to determine whether the sample has a bitter taste, each fraction of the two ginseng samples (water extraction and ethanol extraction) collected using prep-HPLC was determined through an electronic tongue system.

2.7. Qualitative Analysis of Bitter Compounds

Bitter substances were identified via an ultra-performance liquid chromatography-quadrupole-time-of-flight high-resolution mass spectrometry system (Waters Xevo G2 QTof series, Milford, MA, USA) connected to a UPLC system (Waters ACQUITY series, Milford, MA, USA). Before the formal experiment, the mobile phase and the elution gradient were adjusted and optimized by UPLC-Q-TOF/MS. The mobile phase consisted of ultra-pure water containing 0.1% formic acid (A) and acetonitrile containing 0.1% formic acid (B). In total, 0.5 μL of the sample was injected into the UPLC-Q-TOF/MS, and the column temperature was 40 °C. The flow rate was set at 0.4 mL/min. For pure water extraction and ethanol extraction of ginseng samples, the parameters were as follows: 0 min, 100% Solvent A; 5 min, 95% Solvent A; 20 min, 60% Solvent A; 28 min, 10% Solvent A; 30 min, 100% Solvent A; 32 min, 100% Solvent A. Each of the samples was tested three times. The characterization of bitter fractions was carried out on the electro spray ionization source (ESI) in the negative ion mode. The mass spectrometry working parameters were set as follows: source temperature, 40 °C; desolvation gas temperature, 500 °C; flow rates of cone and desolvation gas, 60 L/h and 800 L/h, respectively; capillary voltage, 3.0 kV. The data collection range was m/z 50~1200.

2.8. Quantitation Analysis of Bitter Compounds

According to the results of qualitative analysis, the standard compounds corresponding to each identified bitter compound were dissolved in methanol to prepare a standard compound solution with an initial concentration of 0.5 mg/mL. Then, the initial concentration of the standard solution was diluted in multiples of 2 to obtain a set of standard compound diluted solutions. The regression equations of each standard compound were obtained by evaluating the linear relationship between peak area and concentration under the same chromatographic working parameters used for ginseng samples (water extraction and ethanol extraction). The concentration of bitter substances in ginseng was calculated by a regression equation. Each of the samples was tested three times.

2.9. Bitter Taste Evaluation of Samples Using Electronic Tongue

In order to evaluate the taste characteristics of different samples, the TS-5000Z system (Intelligent Sensor Technology Co., Ltd., Tokyo, Japan) was used for electronic tongue measurement. All electrode sensors were preconditioned in a standard activation solution for more than 24 h, and the sensor check was calibrated before each sample measurement. The output parameters included "umami", "saltiness", "sourness", "bitterness" and "astringency" for the first taste, as well as "richness", "aftertaste-B (aftertaste of bitterness)" and "aftertaste-A (aftertaste of astringency)" as described by Sora [25]. According to the quantitative results of bitter samples, the original bitter standard solution was obtained by ultrasonic dissolution of the corresponding standard solution of each bitter sample in water. Each bitterness sample was tested by the TS-5000Z system.

2.10. Preparation of Debittered Samples

The composite embedding system of chitosan adsorption in the ginseng (*Panax ginseng* C. A. Mey.) carrageenan gel microsphere ($K/M_C/M_{CG}$) was built in the following ways. Firstly, 0.5 g of carrageenan powder was dispersed in 100 mL of ultra-pure water and stirred for 4 h at 65 °C to ensure complete dissolution. When the system was cooled to 30 °C, 0.8 g of ginseng extract sample (pure water extraction or ethanol extraction) was added to a carrageenan solution and stirred at 400 rpm/min for 3 h at 30 °C to form the final K/M_C transport system, which contained 0.5% (w/v) carrageenan and 0.8% ginseng extract. At room temperature, the ginseng extract/carrageenan mixture was injected into 10% (w/v) calcium chloride solution to form carrageenan gel microspheres. Then, these gel microspheres were filtered with filter paper. During the filtration process, ultra-pure water with pH 4 was used to remove the excess calcium chloride solution on the surface of gel microspheres. The smaller gel microspheres were added to 0.5% chitosan solution. After 30 min mixing at room temperature, 10% w/v calcium chloride solution was injected into the system, and gel microspheres were filtered and collected, which were finally washed with ultra-pure water with pH 7 to remove excess calcium ions on the surface. The washed gel microspheres were freeze-dried and collected, which was the sample after debittering.

2.11. Statistical Analysis

The data collected by UPLC-Q-TOF/MS were processed and analyzed by Agilent Mass Hunter software (Qualification Analysis 10.0). The data were processed using SPSS 26.0 statistical software ($p < 0.05$), and the experimental data were represented as the mean ± standard deviation.

3. Results and Discussion

3.1. Separation of the Bitter Compounds from Ginseng

The analysis method of various components in ginseng samples (water extraction and ethanol extraction) was established by using HPLC-DAD. The components in the two samples were separated and obtained using prep-HPLC, and the above operation was repeated 15 times. Then, the same fractions were merged, concentrated, freeze-dried and stored. As shown in Figure 1, 19 fractions from the ethanol-extracted samples and 6 fractions from the pure-water-extracted samples were collected. Each of the components in the both ginseng samples was then evaluated through an electronic tongue system. It was evident that Components 2 and 7 present in the pure-water-extracted ginseng sample had a bitter taste (Table 1 and Figure 1B). In addition, Fractions 2, 5, 7, 8 and 12 present in the ethanol-extracted samples also had high bitterness values (Table 1 and Figure 1A). Compared with pure water extraction of ginseng samples, ethanol extraction of ginseng contained more soluble bitter compounds, indicating that different preparation methods had a significant impact on the basic properties of bitter compounds in ginseng extracts.

Table 1. Recognition of bitterness fractions in ginseng samples [a] through an electronic tongue system.

Tasteless	Code [b]	Bitterness	Code [c]	Bitterness
0	1	−0.21 ± 0.02		
0	2	1.62 ± 0.03	2	1.60 ± 0.03
0	3	−0.27 ± 0.01	3	−0.26 ± 0.02
0	4	−0.31 ± 0.03		
0	5	13.72 ± 0.37		
0	6	−0.34 ± 0.04		
0	7	11.21 ± 0.32	7	11.24 ± 0.21
0	8	6.69 ± 0.51		
0	9	−0.32 ± 0.02		
0	10	−0.36 ± 0.03	10	−0.38 ± 0.04
0	11	−0.33 ± 0.04	11	−0.35 ± 0.04
0	12	7.21 ± 0.45		
0	13	−0.41 ± 0.03	13	−0.39 ± 0.04
0	14	−0.29 ± 0.04		
0	15	−0.31 ± 0.02		
0	16	−0.36 ± 0.04		
0	17	−0.35 ± 0.02		
0	18	−0.29 ± 0.02		
0	19	−0.34 ± 0.03		

[a]: the bitter taste samples of ginseng were obtained through pure water extraction and ethanol extraction, respectively; [b]: a total of 19 fractions from the ethanol-extracted ginseng samples were detected through an electronic tongue system; [c]: a total of 6 fractions from the pure-water-extracted samples were detected through an electronic tongue system.

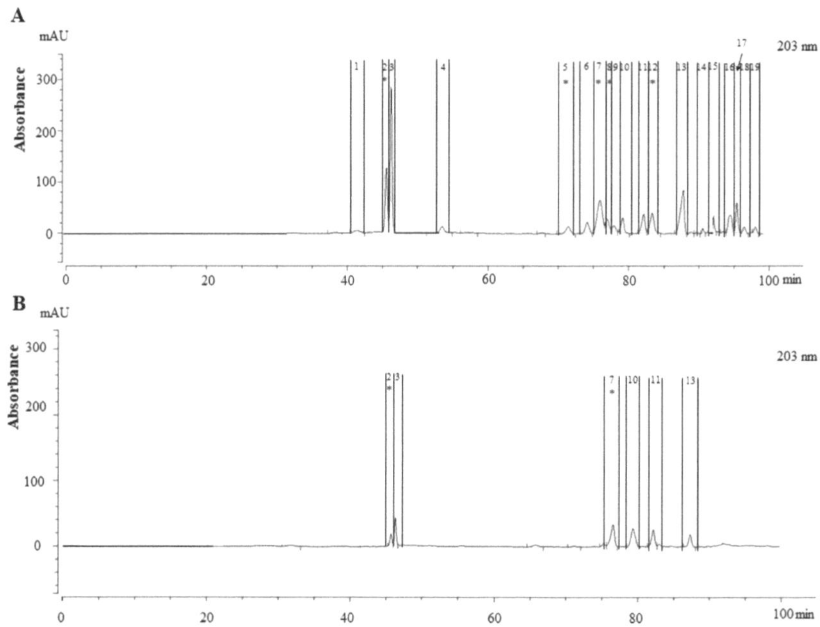

Figure 1. (**A**) Prep-HPLC chromatogram of ginseng extracted with ethanol solution; (**B**) prep-HPLC chromatogram of ginseng extracted with pure water; *: fraction with a bitter taste.

3.2. *Qualitative Analysis of Bitter Fractions*

The characteristic UV absorption peaks of bitter compounds at 203 nm in the five fractions were detected by HPLC-DAD (Table 2), indicating that all of the bitter compounds were saponin compounds [26]. Saponins were the main active substances in ginseng, which belonged to the secondary metabolites of ginseng cells. This indicated that the bitter substances in ginseng were inherent, rather than generated during the extraction process. The identity of bitter substances in ginseng needs to be further analyzed. UPLC-Q-TOF/MS technology was utilized to analyze the structure of the five components. The precursor ion fragments and major product ions-of each bitter compound are shown in Table 2, respectively. The possible cleavage pathways of bitter compounds in ginseng are shown in Figure 2.

Fraction 2: As shown in Table 2 and Figure 2A, in the negative ion mode of Fraction 2, the $[M\text{-}H]^-$ ion with m/z 799.48575 can be detected. After removing one or two glucose residues (Glc, 162 u), high-abundance ion fragments with m/z 637.43224 ($[M\text{-}H\text{-}Glc]^-$) and m/z 475.37895 ($[M\text{-}H\text{-}2Glc]^-$) were obtained. Therefore, Fraction 2 was identified as the ginsenoside Rg1.

Fraction 5: As shown in Table 2 and Figure 2B, the quasi-molecular ion peak of Fraction 5 was detected at m/z 799.48416 ($[M\text{-}H]^-$) through UPLC-Q-TOF/MS testing. The high-abundance ion fragment m/z 637.43105 ($[M\text{-}H\text{-}Glc]^-$) was believed to be obtained by the loss of a molecule of glucose residue (Glc, 162 u). When a molecule of glucose residue was lost at m/z 637.43105, the fragment ion ($[M\text{-}H\text{-}2Glc]^-$) was obtained at m/z 475.37831. Based on the above analysis, Fraction 5 was considered as the ginsenoside Rf.

Fraction 7: The mass spectrometry of the m/z 1107.59562 ion showed major product ions with m/z 945.54233 and 553.29492 in Table 2 and Figure 2C. The fragment ion m/z 945.54233 ($[M\text{-}H\text{-}Glc]^-$) was obtained by the loss of a glucose residue (Glc, 162 u) from the excimer ion peak m/z 1107.59562 ($[M\text{-}H]^-$). It was speculated that the $[M\text{-}H\text{-}3Glc\text{-}68u]^-$ ion with m/z 553.29492 resulted from the loss of three glucose residues (3Glc, 324 u), followed by the loss of a branch chain (68 u) observed at the position of C-5 after rearrangement. To sum up, Fraction 7 was considered as the ginsenoside Rb1.

Table 2. Identification of bitterness fractions in ginseng samples [a].

Code [b]	RT (min) [c]	TMUAW (nm) [d]	[M-H]$^-$	Major Product Ions	Identification
2	15.83	202.31	799.48575	637.43224; 475.37895	ginsenoside Rg1
5	18.93	203.79; 234.68	799.48416	637.43105; 475.37831	ginsenoside Rf
7	20.01	201.76; 276.93	1107.59562	945.54233; 553.29492	ginsenoside Rb1
8	19.79	205.53; 263.95; 385.40	784.49303	637.43184; 475.37893	20(S)-ginsenoside Rg2
12	21.03	198.10; 275.71; 365.45	1077.58591	945.54273; 783.48965	ginsenoside Rb3

[a]: the bitter taste samples of ginseng were obtained through pure water extraction and ethanol extraction, respectively; [b]: corresponding to the serial numbers of bitter substances in Figure 1A,B; [c]: retention time; [d]: the maximum ultraviolet absorption wavelengths.

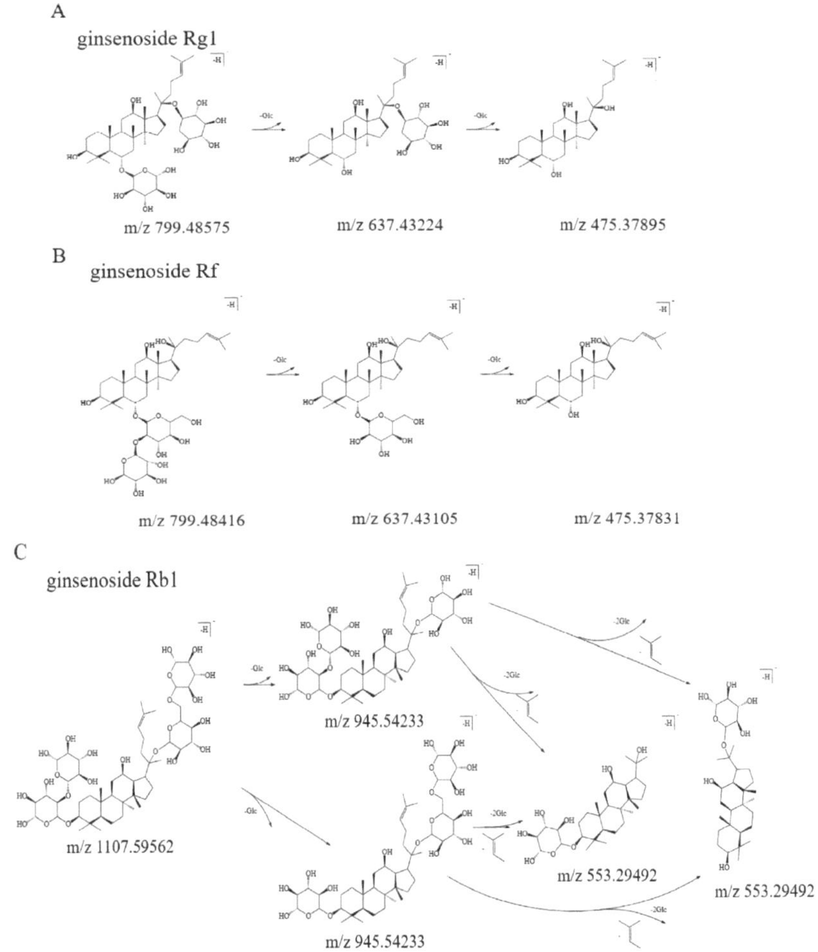

Figure 2. *Cont.*

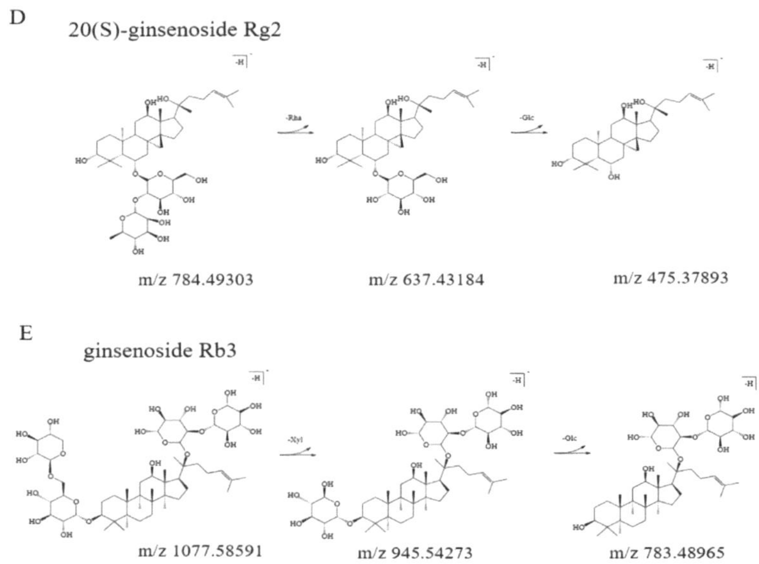

Figure 2. The possible cleavage pathways of bitter compounds in ginseng.

Fraction 8: The mass spectrometry of the m/z 784.49303 ion showed major product ions with m/z 637.43184 and 475.37893 in Table 2 and Figure 2D. The $[M\text{-}H\text{-}Rha]^-$ ion with m/z 637.43184 resulted from the loss of a rhamnose residue (Rha, 147 u). The $[M\text{-}H\text{-}Rha\text{-}Glc]^-$ ion with m/z 475.37893 was obtained when the m/z 637.43184 lost a glucose residue (Glc, 162 u). The above results indicated that Fraction 8 was the 20(S)-ginsenoside Rg2.

Fraction 12: According to the negative ion mode of mass spectrometry data in Table 2 and Figure 2E, the quasi-molecular ion peak of Fraction 12 was observed at m/z 1077.58591 ($[M\text{-}H]^-$), and the ion peak at m/z 945.54273 was a molecular ion peak that lost a xylose residue (Xyl, 132 u). The ion ($[M\text{-}H\text{-}Xyl\text{-}Glc]^-$) of m/z 783.48965 was obtained after losing a glucose residue (Glc, 162 u). From this, Fraction 12 was inferred as the ginsenoside Rb3.

To ensure the accuracy of the identification results, standard compounds were subsequently used for revalidation. The standard compounds were analyzed using HPLC-DAD and UPLC-Q-TOF/MS, respectively, under the same HPLC and MS working parameters. The qualitative analysis results showed that the bitter compounds in the ginseng sample had the same retention time and mass spectrometry data as the speculated standard compounds. Therefore, the accuracy of the identification results was further verified.

3.3. *Quantitation of the Bitter Fractions*

The quantitative detection of bitter fractions in ginseng was conducted by using an external standard quantification method (in Table 3). The bitter compounds in the sample (extracted with pure water and ethanol) are shown in Figure 3, respectively. In ethanol-extracted ginseng samples, the levels of the ginsenoside Rg1 were the highest (196.16 $\pm$ 2.17 μg/mL). It was clear that the content of the 20(S)-ginsenoside Rg2 were the lowest (15.78 $\pm$ 0.92 μg/mL). Interestingly, two bitter compounds were detected in ginseng samples extracted with pure water, where the levels of the ginsenoside Rb1 (23.67 $\pm$ 0.17 μg/mL) were higher than that of the ginsenoside Rg1 (19.42 $\pm$ 0.15) μg/mL. Compared with ethanol extraction, the level of the ginsenoside Rg1 in ginseng samples extracted with pure water was lower, while the level of the ginsenoside Rb1 was (23.67 $\pm$ 0.17) μg/mL. A possible explanation was that the bitter compounds were easily soluble in polar solvents, especially ethanol solutions.

Table 3. Parameters of analytical methods of bitter compounds [a].

Bitter Compound	Linear Regression Equation	R^2	Linearity Range (mg/L)	LOD (μg/mL)	LOQ (μg/mL)
ginsenoside Rg1	Y = 0.0525X − 0.1264	1.000	6.72~537.57	1.0723	3.4532
ginsenoside Rf	Y = 0.0516X + 0.0105	1.000	5.39~431.54	0.4565	1.6671
ginsenoside Rb1	Y = 0.0374X + 0.0089	1.000	7.32~585.61	1.0931	3.2864
20(S)-ginsenoside Rg2	Y = 0.0544X + 0.4095	0.999	4.20~671.52	0.3504	1.0857
ginsenoside Rb3	Y = 0.0370X − 0.0051	1.000	12.97~518.79	2.7841	7.9123

[a]: the bitter taste samples of ginseng were obtained through pure water extraction and ethanol extraction, respectively.

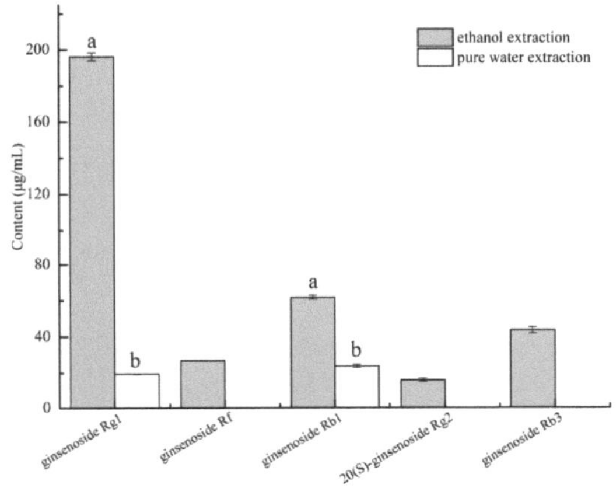

Figure 3. The content of bitter compounds in ginseng (pure water extraction and ethanol extraction). (The letters a and b above the peak indicate significant differences between different extraction methods ($p < 0.05$)).

3.4. Bitterness Analysis of Ginsenoside Rg1, Ginsenoside Rf, Ginsenoside Rb1, 20(S)-Ginsenoside Rg2 and Ginsenoside Rb3 by Electronic Tongue

The experimental reference solution was composed of potassium chloride and tartaric acid. Tasteless represented the output of the reference solution (odorless point), which was −13 for a sour taste and −6 for a salty taste. Based on this, when the taste value of the sample was lower than tasteless, it indicated that the sample had no taste, and vice versa. Based on the quantitative results of the five bitter compounds in ginseng samples, the corresponding standard samples of each bitter compound were dissolved in ultra-pure water and diluted to the corresponding concentration. Then, five bitter compounds were detected through an electronic tongue system at this concentration, and the results indicated that the five bitter substances exhibited different bitterness values. The content level of the bitter compounds in ginseng cannot represent their contribution to the total bitterness of ginseng samples. Electronic tongue analysis was an excellent method for comparing their contribution to the bitterness. The bitterness values of the ginsenoside Rg1, ginsenoside Rf, ginsenoside Rb1, 20(S)-ginsenoside Rg2 and ginsenoside Rb3 were 5.91 ± 0.23, 7.26 ± 0.31, 14.29 ± 0.25, 12.63 ± 0.12 and 9.88 ± 0.17, respectively (in Figure 4A). This indicated that among all bitter compounds, the ginsenoside Rb1 had the strongest bitterness, followed by ginsenoside Rb3, 20(S)-ginsenoside Rg2, ginsenoside Rf and ginsenoside Rg1. The astringency values were 4.09 ± 0.17, 4.52 ± 0.15, 9.45 ± 0.13, 8.61 ± 0.11 and 6.38 ± 0.12, respectively. The aftertaste-B values were 0.13 ± 0.08, 0.11 ± 0.11, 0.14 ± 0.01, 0.09 ± 0.02 and 0.08 ± 0.01, respectively, and aftertaste-B was the bitter taste that was perceived after the food or drug had been swallowed or spit out. As shown in Figure 4A, the five ginsenosides also exhibited salty and umami tastes, which could be related to the fresh amino acids or inorganic salts extracted during the extraction process [27–29]. Although ginsenosides contain umami and salty tastes, the bitterness was not affected. Thus, the ginsenoside Rg1, ginsenoside Rf, ginsenoside Rb1, 20(S)-ginsenoside Rg2, and ginsenoside Rb3 could be considered as bitter substances in ginseng.

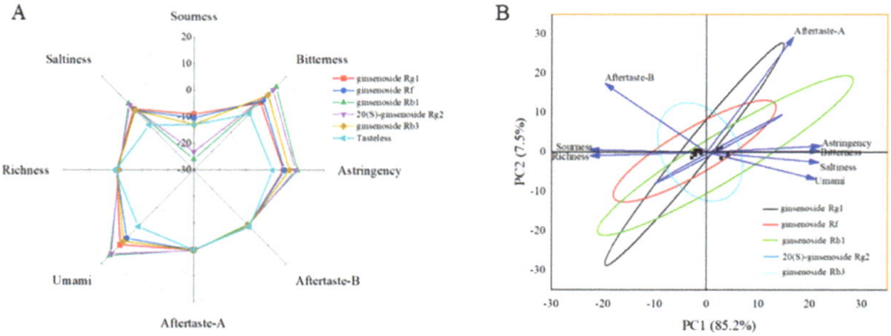

Figure 4. (**A**) Electronic tongue taste properties of five samples; (**B**) the biplot of PCA for the electronic tongue.

Principal component analysis (PCA) is the most widely used data dimensionality reduction algorithm, which could be used to analyze the main taste indicators detected by the electronic tongue [30]. The PCA score plots indicated the relationship between variables using the variable information based on PC1 and PC2. As shown in Figure 4B, the main taste indicators belonging to PC1 and PC2 had a clear separation trend in spatial layout. The bitterness, astringency, umami and saltiness indicators were distributed on the positive half axis of the X-axis, while sour indicators were distributed on the negative half axis of the X-axis, indicating that the main taste indicators of the five samples were bitterness, astringency, umami and saltiness. The variance contribution rates of PC1 and PC2 were 85.2% and 7.5%, respectively, with a cumulative variance contribution rate of 92.7%, which could effectively reflect the overall taste information of the different bitter taste samples. Additionally, the five samples were distributed in four quadrants, and they were close to each other with overlapping areas, indicating that there was no significant difference in the taste characteristics of the five substances and they all presented typical bitterness.

3.5. Analysis of Bitterness Elimination Effect

Thus far, many studies have shown that embedding is an effective way to remove bitterness. For example, in order to improve the sensory quality of a product, the mixture of gelatin and soy protein isolate can be used as a carrier to weaken or mask the bitterness of casein hydrolysate by spray drying [31]. Similarly, studies have shown that a two-phase gel lotion embedding system could significantly reduce the bitterness value of bitter peptides. The principle of this process is to use two edible materials to gradually embed bitter peptides, so as to increase the stability of the product and reduce the bitterness [32]. Embedding is also a commonly used method for removing bitterness in industry. Therefore, we adopt this method to reduce or eliminate the bitterness of ginseng. The bitterness value of four ginseng extracts (pure water extraction, ethanol extraction, removing bitterness from ginseng samples (extracted with pure water) (RBGPW) and removing bitterness from ginseng samples (extracted with ethanol solution) (RBGES)) were determined through an electronic tongue system in order to effectively demonstrate the bitterness removal effect of the embedding process in this experiment. It was worth mentioning that compared with ginseng samples extracted with pure water and ethanol, RBGPW and RBGES had higher concentrations of bitter compounds. Then, UPLC-Q-TOF/MS and HPLC methods were used to analyze and quantify the content of bitter compounds in the four ginseng extracts mentioned above. After embedding treatment, the bitterness value of the ginseng samples (pure water extraction and ethanol extraction) significantly decreased (Figure 5A). For instance, for ginseng samples of pure water extraction, the bitterness value decreased from 21.13 to 2.24, while ethanol-extracted ginseng samples showed a higher degree of reduction. At this point, as shown in Figure 5A,B, in the electronic tongue experiment, the concentration of bitter compounds in the ginseng sample after embedding treatment was higher than that of the ginseng sample without embedding, but the ginseng sample after embedding treatment had a lower bitterness value. This indicated that embedding treatment could significantly reduce the bitterness of ginseng and improve the taste of the product. Interestingly, as shown in Figure 5A, the bitterness values of ginseng samples obtained by the two extraction methods were significantly reduced after embedding, indicating that the effect of embedding on reducing bitterness was not affected by the sample extraction method.

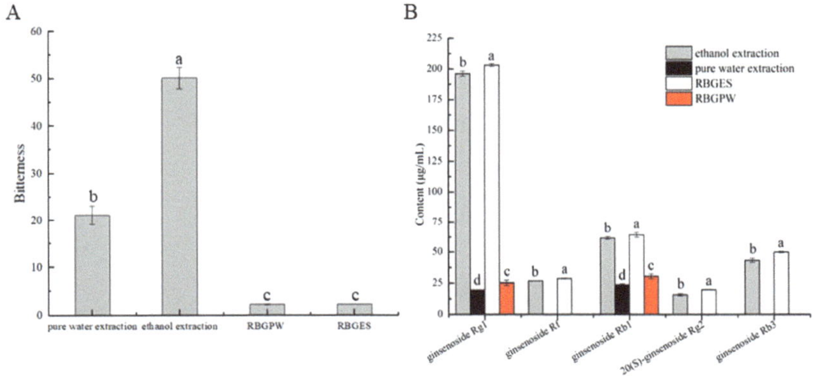

Figure 5. (**A**) The bitterness value of four kinds of ginseng samples; (**B**) the concentration of bitter compounds in four kinds of ginseng samples; RBGPW: removing bitterness from ginseng samples (extracted with pure water); RBGES: removing bitterness from ginseng samples (extracted with ethanol solution). The different lowercase letters (a, b, c and d) represent significant differences between different extraction and deodorization methods ($p < 0.05$).

4. Conclusions

In this research, the bitter compounds ginsenoside Rg1, ginsenoside Rf, ginsenoside Rb1, 20(S)-ginsenoside Rg2 and ginsenoside Rb3 in the ginseng samples (after water extraction and ethanol extraction) were characterized, separated and purified by HPLC-DAD, prep-HPLC, subjected to bitterness value determination through using an electronic tongue and identified by UPLC-Q-TOF/MS. The electronic tongue analysis results indicated that all five ginsenosides had a strong bitterness. Additionally, the research results indicated that the bitter substances in ginseng samples are actually produced during their own secondary metabolism process, rather than appearing during the extraction process, as their content was influenced by the extraction method. In addition, electronic tongue analysis indicated that compared with the other four bitter compounds, the ginsenoside Rb1 had the highest bitterness value, followed by 20(S)-ginsenoside Rg2, ginsenoside Rg1, ginsenoside Rf and ginsenoside Rb3. Simultaneously, the bitterness of ginseng samples was significantly reduced through the embedding method, and the effect was not affected by the extraction method. This research achievement is beneficial for improving the taste of ginseng in food and expanding its application pathways.

Author Contributions: Conceptualization, Y.C., Z.L. and J.X.; methodology, Y.C. and Z.W.; validation, Y.C. and W.S.; formal analysis, W.S.; investigation, Z.L.; resources, Z.W. and J.X.; data curation, Y.C.; writing—original draft preparation, Y.C. and J.X.; writing—review and editing, Y.C. and W.S.; visualization, Y.C. and J.X.; supervision, Y.C. and J.X.; project administration, Y.C. and Z.W.; funding acquisition, J.X. and Z.W. All authors have read and agreed to the published version of the manuscript.

Funding: This work was supported by the National Key Research and Development Program of China (2020ACA007-03).

Data Availability Statement: Data are contained within the article.

Acknowledgments: The authors express their gratitude to Kaijin Pan and Hongjun Li from Hubei Key Laboratory of Quality and Safety of Traditional Chinese Medicine Health Food (Jing Brand Co., Ltd.) for investigation and validation of the ginseng samples.

Conflicts of Interest: All authors were employed by the company Jing Brand Co., Ltd. The authors declare that the research was conducted in the absence of any commercial or financial relationships that could be construed as a potential conflict of interest.

References

1. Wang, Z.; Zhang, Z.; Liu, J.; Guo, M.; Li, H. Panax Ginseng in the treatment of Alzheimer's disease and vascular dementia. *J. Ginseng Res.* **2023**, *47*, 506–514. [CrossRef] [PubMed]
2. Zhang, H.; Abid, S.; Ahn, J.C.; Mathiyalagan, R.; Kim, Y.-J.; Yang, D.-C.; Wang, Y.; Cho, I.-H. Characteristics of Panax ginseng Cultivars in Korea and China. *Molecules* **2020**, *25*, 2635. [CrossRef] [PubMed]
3. Cui, L.; Xu, L.; Yu, H.; Feng, Q.; Yan, H.; Shahid, A.; Chen, Q.; Jibo, R.; Gao, X.; Wang, D. An efficient isolation and purification of broad partition coefficient range ginsenosides from roots of *Panax quinquefolium* L. by linear gradient counter-current chromatography coupled with preparative HPLC. *J. Sep. Sci.* **2023**, *46*, 2300046. [CrossRef] [PubMed]
4. Liu, P.; Zhang, Z.; Cai, Y.; Yang, Y.; Yuan, J.; Chen, Q. Inhibition of the pyroptosis-associated inflammasome pathway: The important potential mechanism of ginsenosides in ameliorating diabetes and its complications. *Eur. J. Med. Chem.* **2023**, *253*, 115336. [CrossRef]
5. Xie, N.; Wu, C.; Ge, Q.; Zhou, J.; Long, F.; Mao, Q.; Li, S.; Shen, H. Structure-specific antitumor effects and potential gut microbiota-involved mechanisms of ginseng polysaccharides on B16F10 melanoma-bearing mice. *Food Funct.* **2023**, *14*, 796–809. [CrossRef]
6. Park, N.J.; Bong, S.K.; Lee, S.; Jung, Y.; Jegal, H.; Kim, J.; Kim, S.K.; Kim, Y.K.; Kim, S.N. Compound K improves skin barrier function by increasing SPINK5 expression. *J. Ginseng Res.* **2020**, *44*, 799–807. [CrossRef]
7. Li, K.; Wang, L.; Zhou, R.; Fan, H.; Sui, J. Amelioration of alcohol-induced liver injury in mice by ginsenosides in ginseng wine. *J. Funct. Foods.* **2019**, *54*, 281–288. [CrossRef]
8. Cui, S.; Wang, J.; Yang, L.; Wu, J.; Wang, X. Qualitative and quantitative analysis on aroma characteristics of ginseng at different ages using E-nose and GC–MS combined with chemometrics. *J. Pharm. Biomed. Anal.* **2015**, *102*, 64–77. [CrossRef]
9. Xu, X.; Sanjeevram, D.; Mi, X.; Hye-Ryung, P.; Yeon-Ju, K. Immune-enhancing efficacy of Curtobacterium proimmune K3 lysates isolated from Panax ginseng beverages in cyclophosphamide-induced immunosuppressed mice. *J. Funct. Foods.* **2022**, *92*, 105020. [CrossRef]
10. Hee, S.C.; Soo-Yeun, L. Modification of ginseng flavors by bitter compounds found in chocolate and coffee. *J. Food Sci.* **2012**, *77*, S202–S210.
11. Gupta, A.K.; Dhua, S.; Kumar, V.; Naik, B.; Magwaza, L.S.; Ncama, K.; Opara, U.L.; McClements, D.J.; Mishra, P. Current and emerging applications in detection and removal of bitter compounds in citrus fruit juice: A critical review. *Food Biosci.* **2023**, *55*, 102995. [CrossRef]
12. Yu, Y.; Zhu, X.; Ouyang, W.; Chen, M.; Jiang, Y.; Wang, J.; Hua, J.; Yuan, H. Effects of electromagnetic roller-hot-air–steam triple-coupled fixation on reducing the bitterness and astringency and improving the flavor quality of green tea. *Food Chem. X* **2023**, *19*, 100844. [CrossRef]
13. Karolkowski, A.; Meudec, E.; Bruguière, A.; Offer, A.C.M.; Bouzidi, E.; Levavasseur, L.; Sommerer, N.; Briand, L.; Salles, C. Faba Bean (*Vicia faba* L. minor) Bitterness: An Untargeted Metabolomic Approach to Highlight the Impact of the Non-Volatile Fraction. *Metabolites.* **2023**, *13*, 964. [CrossRef] [PubMed]
14. Deng, S.; Zhang, G.; Oluwaseun, O.A.; Mo, Z.; Mao, J.; Zhang, H.; Liu, X.; Ma, M.; Wang, Q.; Liu, H. Bitter and astringent substances in green tea: Composition, human perception mechanisms, evaluation methods and factors influencing their formation. *Food Res. Int.* **2022**, *157*, 111262. [CrossRef] [PubMed]
15. Deng, Y.; Ma, Y.; Liu, H.; Zhang, Y.; Wei, Z.; Liu, G.; Tang, X.; Jia, X. Structure determination, bitterness evaluation and hepatic gluconeogenesis inhibitory activity of triterpenoids from the *Momordica charantia* fruit. *Food Chem.* **2022**, *372*, 131224. [CrossRef] [PubMed]
16. Xu, Y.; Sun, Q.; Chen, W.; Han, Y.; Gao, Y.; Ye, J.; Wang, H.; Gao, L.; Liu, Y.; Yang, Y. The Taste-Masking Mechanism of Chitosan at the Molecular Level on Bitter Drugs of Alkaloids and Flavonoid Glycosides from Traditional Chinese Medicine. *Molecules* **2022**, *27*, 7455. [CrossRef] [PubMed]
17. Drewnowski, A. The science and complexity of bitter taste. *Nutr. Rev.* **2001**, *59*, 163–169. [CrossRef] [PubMed]
18. Lee, J.; Choi, B.-R.; Kim, Y.-C.; Choi, D.; Lee, Y.-S.; Kim, G.-S.; Baek, N.-I.; Kim, S.-Y.; Lee, D. Comprehensive Profiling and Quantification of Ginsenosides in the Root, Stem, Leaf, and Berry of Panax ginseng by UPLC-QTOF/MS. *Molecules* **2017**, *22*, 2147. [CrossRef] [PubMed]

19. Li, J.; Wang, H.; Xu, X.; Wang, H.; Li, X.; Hu, Y.; Chen, B.; Liu, M.; Gao, X.; LI, H.; et al. An off-line three-dimensional liquid chromatography/Q-Orbitrap mass spectrometry approach enabling the discovery of 1561 potentially unknown ginsenosides from the flower buds of *Panax ginseng*, *Panax quinquefolius* and *Panax notoginseng*. *J. Chromatogr. A* **2022**, *1675*, 463177.
20. Cong, L.; Ma, J.; Zhang, Y.; Zhou, Y.; Cong, X.; Hao, M. Effect of anti-skin disorders of ginsenosides—A Systematic Review. *J. Ginseng Res.* **2023**, *47*, 605–614. [CrossRef]
21. Li, X.; Cao, D.; Sun, S.; Wang, Y. Anticancer therapeutic effect of ginsenosides through mediating reactive oxygen species. *Front. Pharmacol.* **2023**, *14*, 1215020. [CrossRef] [PubMed]
22. Wang, A.; Liang, X.; Chen, W.; Liu, Y.; Zhu, Y.; Zhang, D.; Hu, L.; Li, R. Total ginsenosides promotes intestinal epithelial proliferation via affecting polyamine-mediated HuR on post-transcriptional control. *Food Biosci.* **2023**, *55*, 102970. [CrossRef]
23. Zhang, J.; Zhu, C.; Jin, Y.; Shen, W.; Pan, Y.; Shen, Y. Ginsenoside Rg1 improved learning and memory ability and reduces neuronal apoptosis in epileptic rats through ERK/CREB/BDNF signal pathway. *Biochem. Biophys. Res. Commun.* **2023**, *675*, 26–32. [CrossRef] [PubMed]
24. Zhao, W.; He, J.; Yu, Z.; Wu, S.; Li, J.; Liu, J.; Liao, X. In silico identification of novel small molecule umami peptide from ovotransferrin. *Int. J. Food Sci Tech.* **2022**, *57*, 2628–2635. [CrossRef]
25. Sora, H.; Tatsuro, H.; Miho, K.; Kenlchi, K.; Hideyuki, O.; Satoru, T.; Satoshi, S.; Hideyuki, Y.; Kaoru, S.; Takayuki, M. Comparison of taste characteristics between koji mold-ripened cheese and Camembert cheese using an electronic tongue system. *J. Dairy Sci.* **2023**, *106*, 6701–6709.
26. Hsu, B.-Y.; Jen, C.-T.; Inbaraj, B.S.; Chen, B.-H. A comparative study on analysis of ginsenosides in American ginseng root residue by HPLC-DAD-ESI-MS and UPLC-HRMS-MS/MS. *Molecules* **2022**, *27*, 3071. [CrossRef] [PubMed]
27. Zhu, L.; Xu, L.; Huang, Y.; Xie, C.; Dou, D.; Xu, J. Correlations between ecological factors and the chemical compositions of mountainous forest cultivated ginseng. *J. Food Compos. Anal.* **2022**, *114*, 104867. [CrossRef]
28. Zhang, Z.; Chen, W.; Tao, L.; Wei, X.; Gao, L.; Gao, Y.; Suo, J.; Yu, W.; Hu, Y.; Yang, B. Ethylene treatment promotes umami taste-active amino acids accumulation of *Torreya grandis* nuts post-harvest by comparative chemical and transcript analyses. *Food Chem.* **2023**, *408*, 135214. [CrossRef] [PubMed]
29. Iiyama, S.; Ezaki, S.; Toko, K. Analysis of saltiness and bitterness of inorganic salts using taste sensors. *Sens. Mater.* **2001**, *13*, 137–144.
30. Fraga, C.G.; Clowers, B.H.; Moore, R.J.; Zink, E.M. Signature-discovery approach for sample matching of a nerve-agent precursor using liquid chromatography−mass spectrometry, XCMS, and chemometrics. *Anal. Chem.* **2010**, *82*, 4165–4173. [CrossRef]
31. Fávaro-Trindade, C.S.; Santana, A.d.S.; Monterrey-Quintero, E.S.; Trindade, M.A.; Netto, F.M. The use of spray drying technology to reduce bitter taste of casein hydrolysate. *Food Hydrocolloids* **2010**, *24*, 336–340. [CrossRef]
32. Gao, Y.; Li, X.; Xie, Y.; Huang, X.; Cheng, C.; McClements, D.J.; Zhang, L.; Chen, X.; Zou, L.; Wei, L. Encapsulation of bitter peptides in diphasic gel double emulsions: Bitterness masking, sustained release and digestion stability. *Food Res. Int.* **2022**, *162*, 112205. [CrossRef] [PubMed]

foods

MDPI

Article

High-Resolution Mass Spectrometry-Based Metabolomics for Increased Grape Juice Metabolite Coverage

Sébastien Nicolas [1], Benjamin Bois [2], Kevin Billet [1,†], Rémy Romanet [1,3], Florian Bahut [1], Jenny Uhl [4], Philippe Schmitt-Kopplin [4,5] and Régis D. Gougeon [1,*]

1 Procédés Alimentaires et Microbiologiques, PAM UMR A 02.102, Université de Bourgogne-Institut Agro, Institut Universitaire de la Vigne et du Vin-Jules Guyot, F-21000 Dijon, France; sebastien.nicolas@u-bourgogne.fr (S.N.); kevin.billet@inrae.fr (K.B.); remy.romanet@sayens.fr (R.R.); florian.bahut@u-bourgogne.fr (F.B.)

2 Centre de Recherches de Climatologie, Biogéosciences UMR 6282, CNRS-Université de Bourgogne, Institut Universitaire de la Vigne et du Vin-Jules Guyot, F-21000 Dijon, France; benjamin.bois@u-bourgogne.fr

3 DIVVA Platform, PAM UMR A 02.102, Institut Universitaire de la Vigne et du Vin-Jules Guyot, F-21000 Dijon, France

4 Research Unit Analytical Biogeochemistry, Helmholtz Zentrum München, Ingolstaedter Landstrasse 1, 85764 Neuherberg, Germany; philippe.schmittkopplin@helmholtz-munich.de (P.S.-K.)

5 Analytische Lebensmittel Chemie, Technische Universität München, Maximus-von-Imhof-Forum 2, 85354 Freising, Germany

* Correspondence: regis.gougeon@u-bourgogne.fr

† Present address: Unité BIA—Equipe Polyphénols Réactivités, Centre Bretagne-Normandie, F-35653 Le Rheu CEDEX, France.

Abstract: The composition of the juice from grape berries is at the basis of the definition of technological ripeness before harvest, historically evaluated from global sugar and acid contents. If many studies have contributed to the identification of other primary and secondary metabolites in whole berries, deepening knowledge about the chemical composition of the sole flesh of grape berries (i.e., without considering skins and seeds) at harvest is of primary interest when studying the enological potential of widespread grape varieties producing high-added-value wines. Here, we used non-targeted DI-FT-ICR-MS and RP-UHPLC-Q-ToF-MS analyses to explore the extent of metabolite coverage of up to 290 grape juices from four *Vitis vinifera* grape varieties, namely Chardonnay, Pinot noir, Meunier, and Aligoté, sampled at harvest from 91 vineyards in Europe and Argentina, over three successive vintages. SPE pretreatment of samples led to the identification of more than 4500 detected C,H,O,N,S-containing elemental compositions, likely associated with tens of thousands of distinct metabolites. We further revealed that a major part of this chemical diversity appears to be common to the different juices, as exemplified by Pinot noir and Chardonnay samples. However, it was possible to build significant models for the discrimination of Chardonnay from Pinot noir grape juices, and of Chardonnay from Aligoté grape juices, regardless of the geographical origin or the vintage. Therefore, this metabolomic approach opens access to a remarkable holistic molecular description of the instantaneous composition of such a biological matrix, which is the result of complex interplays among environmental, biochemical, and vine growing practices.

Keywords: Chardonnay; Pinot noir; Meunier; Aligoté; molecular fingerprints; mesocarp

Citation: Nicolas, S.; Bois, B.; Billet, K.; Romanet, R.; Bahut, F.; Uhl, J.; Schmitt-Kopplin, P.; Gougeon, R.D. High-Resolution Mass Spectrometry-Based Metabolomics for Increased Grape Juice Metabolite Coverage. *Foods* **2024**, *13*, 54. https://doi.org/10.3390/foods13010054

Academic Editor: Pedro Vitoriano de Oliveira

Received: 7 November 2023
Revised: 17 December 2023
Accepted: 19 December 2023
Published: 22 December 2023

1. Introduction

Within the frame of winemaking, the composition and the analysis of grape juice—here considered as the sole flesh of berries—are of great importance throughout ripening up to harvest, as the juice contains primary metabolites, including sugars (notably glucose and fructose) and organic acids (notably L-tartaric and L-malic acids) [1], whose concentration evolution is at the basis of the decision to harvest [2]. Many studies have shown that their concentration at harvest is controlled by several factors, among which grape variety, vine growing practices, soil types, and environmental conditions with heat and water stress being at the center of the current concerns about climate change [2,3]. Amino acids also constitute major primary metabolites and fundamental nutrients for yeasts during alcoholic fermentation [4].

Grape juices also contain several known secondary metabolites, whose occurrence has also been the object of many studies. The major families include aroma compounds with C13-norisoprenoids, terpenes, benzene derivatives, alcohols, esters, and thiols, often present as glycosidic precursors [5–12]. Also included are phenolics with acid phenols in particular, peptides, carotenoids, and some flavonols, among which quercetin and hormones with abscisic acid (ABA) are a well know example, and vitamins [13–17]. A few thousand secondary metabolites have thus been observed and quantified so far in grape and wine matrices, thanks to various targeted analytical methodologies.

Furthermore, the last two decades have seen the advent of targeted and non-targeted metabolomic approaches, combined with advanced statistical tools, aiming at providing more comprehensive information about the actual chemical diversity, including low-concentrated metabolites, of various biological samples and body fluids [18,19]. In the case of grape and wine matrices, and as far as non-volatile compounds are concerned, Mass Spectrometry coupled to Liquid Chromatography (LC-MS) has certainly been central to the highest number of metabolomic analyses. LC-MS-based metabolomic analyses of grape berries have thus been applied to compare varieties, or to evaluate various impacts of vineyard practices or environmental parameters on the grape metabolome [20–24]. However, matrix effects are also central to LC-MS-based metabolomics, and particular attention has already been paid to sample preparation and to solvent extraction to get the best metabolite coverage in grape analysis [25,26]. These authors showed that a solvent containing approximately equal amounts of methanol and chloroform and up to 20% water allowed the detection of up to 4500 features by ElectroSpray Ionization Reverse Phase Ultra Performance Liquide Chromatography coupled with Time of Flight-Mass Spectrometry (ESI RP-UPLC-ToF-MS) of grape berries. However, as in several other studies, the different compartments of the berry (skin, flesh and seeds) were not analyzed separately.

Alternatively, the development of ultra-high resolution mass spectrometry with Direct Infusion Fourier Transform-Ion Cyclotron Resonance Mass Spectrometry (DI-FT-ICR-MS) has offered new perspectives for exploring the chemical complexity of bio/geo/chemical matrices [27]. Thanks to extraordinarily high sensitivity and mass resolution, DI-FT-ICR-MS non-targeted analyses of musts, wines, and spirits could, for example, provide unprecedented fingerprints [28], discriminate six varieties of grapes [29], identify forest-related specific fingerprints in barrel-aged wines [30], or simultaneously identify hundreds of volatile and non-volatile compounds in gin [31]. These analyses highlighted both the small amounts and the very minimal preparation steps required for remarkably efficient metabolite coverage in wine and spirit analysis. Although DI-FT-ICR-MS could be applied directly on methanol-diluted grape juice samples [32] thanks to the high dynamic range of the signals, grape juice fractionation by SPE is supposed to provide the highest metabolite coverage through the removal of highly abundant primary sugar metabolites, thus enhancing the detection of lower concentration secondary metabolites. In this context, combining DI-FT-ICR-MS with the highest resolution and thus compositional selectivity with the possibility to identify isobars by RP/HILIC-UHPLC-Q-ToF-MS^2 has proven remarkably efficient for increasing the range of unknown detectable metabolites in life sciences and foods [33].

In this study, an extensive set of grape juices from two white grape varieties, *V. vinifera* Chardonnay and Aligoté, and two red grape varieties, Pinot noir and Meunier, collected at harvest in 13 different regions representing 91 vineyards in Europe and Argentina, were systematically analyzed in negative ionization mode by DI-FT-ICR-MS and RP-UHPLC-Q-TOF-MS in order to evaluate and compare the impact of two minimal sample preparations on the performance of metabolite coverage. These sample preparations were either diluted (non SPE), or SPE fractionated before analysis. We further used this dataset to probe the capacity of the combined methodologies to characterize the metabolite coverage and discriminate grape juices according to different parameters such as variety or geographical origin of the grape.

2. Materials and Methods

2.1. Grape Berry Collection

Grape juices were obtained throughout a random collection of berries from 91 distinct vineyards (Table 1), covering five countries (Argentina, France, Germany, Italy, and Portugal), 13 wine producing regions (Adige Valley, Alsace, Beaujolais, Bordeaux, Burgundy, Champagne, Douro, Languedoc, Piemont, Rheingau, Tarn, Uco Valley, and Württemberg), three vintages (2019, 2020, 2021) and four *Vitis vinifera* varieties (Aligoté, Chardonnay, Meunier, and Pinot noir). For each considered grape variety within a given plot and for a given vintage, grape bunches were randomly picked from incoming harvest cases throughout the harvest duration. To reflect the genuine natural conditions of the vineyards, we complied with the winegrowers' decision to harvest (Table S1). Then, pseudo-biological replicates were made by randomly picking two pools of 100 berries within these grape bunches. In some cases (not enough berries), only one pool of 100 berries was used for a vineyard/vintage. The berries were then frozen to prevent any further development. Grape juices were finally obtained by manually pressing the thawed berries in plastic bags, and the collected juice was then frozen before further sample preparation (−20 °C).

Table 1. Details of the number of samples collected for each variety in function of regions and vintages.

Region	Aligoté			Chardonnay			Meunier			Pinot Noir		
	2019	2020	2021	2019	2020	2021	2019	2020	2021	2019	2020	2021
Adige valley										10		
Alsace										2	4	4
Beaujolais				1								
Bordeaux	1		2	1	2	2	1		2	1	2	2
Burgundy	4	4	2	20	35	8				29	26	10
Champagne				2		1	1		1	2		1
Douro				4								
Gaillac	1	2	2	1	2	2	1	2	2	1	2	2
Languedoc	2	2		6	6		2	2		6	6	
Piedmont				2	2					2	3	
Rheingau										14		
Uco Valley				5	6					8	6	
Württemberg				2			2					

2.2. Sample Preparation

Samples were defrosted at ambient temperature prior to vortexing and centrifugation (10 min, 10,000 rpm, 4 °C). The further preparation steps were according to Roullier-Gall, et al. [34]. To perform solid phase extraction (SPE), the samples' supernatant was first acidified to pH2 with pure formic acid (LC-MS grade). Bond Elut C18 cartridges (100 mg, 1 mL, 120 µm, Agilent, Les Ulis, France) were then conditioned with 2 mL methanol (LC-MS grade), followed by 1 mL of acidified (pH2, formic acid) ultra-pure water (18.2 MΩ, Millipore, Merck, Darmstadt, Germany). Samples were then filtered through cartridges, which were washed with 1 mL of pH2 water prior to elution with 1 mL of methanol. All filtrations were performed using a CHROMABOND SPE vacuum manifold at −0.2 Bar. Before DI-FT-ICR-MS analysis, SPE extracts were 1/20 diluted in methanol. For RP-UHPLC-Q-TOF-MS, SPE extracts were 1/2 diluted in 5% acetonitrile. Non SPE preparation (direct dilution of the juices) was performed by diluting to 1/500 in methanol and to 1/5 in 5% acetonitrile for DI-FT-ICR-MS and RP-UHPLC-Q-ToF-MS, respectively.

2.3. DI-FT-ICR-MS

Fourier Transform-Ion Cyclotron Resonance Mass Spectrometry (FT-ICR-MS) was used with direct infusion of samples to an Apollo II electrospray ionization (ESI) source, working in negative mode ESI (−), coupled to a 12T FT-ICR-MS (SolariX, Bruker Daltonics, Bremen, Germany). Mass spectra were acquired in negative ionization mode with a flow rate of 120 μL/h within a mass range of 92–1000 Da. A total of 400 scans were accumulated for each sample. Raw spectra were calibrated using Compass DataAnalysis 4.2 (Bruker Daltonics, Bremen, Germany), and peaks with a signal-to-noise ratio (S/N) above 3 were considered. The two matrices (SPE; non SPE) were then obtained by aligning all spectra of each type within a 0.5-ppm alignment error (defined as the ratio of the difference between two aligned masses ($m/z_1 - m/z_2$) to one of these masses (m/z_1) $\times 10^6$). Molecular formulae were then assigned using an in-house developed software tool (NetCalc v2.0) [35].

2.4. RP-UHPLC-Q-ToF-MS

Ultra-High-Pressure Liquid Chromatography (UHPLC, Dionex Ultimate 3000, Thermo Fischer Scientific, Waltham, MA USA) was used for separation, coupled to a MaXis plus MQ ESI-Q-ToF mass spectrometer (Bruker, Bremen, Germany). Reverse Phase-Liquid Chromatography (RP-LC) was performed with an Acquity BEH C18 1.7 μm, 100 × 2.1 mm column (Waters, Guyancourt, France). Elution was performed using a gradient of water (A) and acetonitrile (B), both acidified with 0.1% (v/v) formic acid. For elution (40 °C), 5% (v/v) of eluent B was used from 0 to 1.1 min followed by a gradual increase of eluant B up to 99.5%, reached at 6.40 min. Detection was performed in negative ionization mode, within 100 to 1500 Da range, using an electrospray at two bars of pressure for the nebulizer and 10 L/min for nitrogen dry gas flow, an end plate offset of 500 V, and capillary voltage of 4500 V. To recalibrate spectra, four times diluted calibrant ESI-L Low Concentration Tuning Mix (Agilent, Les Ulis, France) was injected at the beginning of each run. Before batch analysis, the mass spectrometer was calibrated using undiluted Tuning Mix in enhanced quadratic mode, which allowed for the alignment of peaks with errors < 0.5 ppm. The two matrices (SPE; non SPE) were then obtained using the Metaboscape software (Bruker, Bremen, Germany, V 8.0.1), with the T-Rex 3D algorithm, using an intensity threshold of 4000 and considering $[M-H]^-$, $[M-H-H_20]^-$, $[M+Cl-]^-$ ions.

2.5. Data Analysis

All statistical analyses and plot generation were performed within the R environment [36] (v 4.3.0). First, a data sanity check was performed over all datasets [37], leading to two samples being removed. Two other samples missing in one dataset (DI-FT-ICR-MS non SPE) were also removed from all datasets, leading to a total of 291 samples to be analyzed by each methodology. For data description, only features present in at least four samples were considered. As the four datasets (SPE, non SPE, for DI-FT-ICR-MS, and RP-UHPLC-Q-ToF-MS) were obtained separately, alignment among the datasets was performed as follows. The two DI-FT-ICR-MS datasets were aligned by matching the assigned formulas. For alignment using RP-UHPLC-Q-ToF-MS (UHPLC-Q-ToF-MS (SPE) vs. UHPLC-Q-ToF-MS (non SPE) or UHPLC-Q-ToF-MS vs. FT-ICR-MS), features identified in RP-UHPLC-Q-ToF-MS where first grouped within a 2-ppm alignment error window to identify potential isobars. The grouped m/z mean was then used to perform alignment with DI-FT-ICR-MS data within a 5-ppm alignment error mass range. For the RP-UHPLC-Q-ToF-MS vs. RP-UHPLC-Q-ToF-MS data alignment, a retention time (RT) tolerance of 10 s was used. In case of multiple matches, the lower mass differences and/or RT delta were selected. Prior to statistical analysis, batch effect was checked and corrected using the DBnorm package with the ber method [38,39]. Multivariate analysis was performed on features present in at least 33% of the samples, and any zero value was replaced with 2/3 of the minimum value for each feature [40]. Principal Component Analysis (PCA) and Orthogonal Partial Least Squares–Discriminant Analysis (OPLS–DA, n permutation = 500) were performed using ropls [37], after Log10 transformation and pareto scaling for all datasets. Redundancy Analysis (RDA) and Variance Partition were used to compare the impact of vintage, region, and cultivar [41,42] using the vegan package [43]. The significance of RDA models was tested with 1000 permutations. Univariate statistics were generated with the rstatix package [44]. Comparisons across treatment groups were performed using the Kruskal–Wallis test followed by Dunn's *post-hoc* pairwise comparison. Intensities Fold Change (FC) between varieties were pairwise tested using either Student *t*-test or Mann–Whitney U-test. *p*-values were corrected (p.adjust) using the False Discovery Rate method (FDR) [45], and adjusted *p*-values < 0.05 were considered significant. To aid biological interpretation, the MetaCyc (v 26.0), Plant Metabolic Network (PMN, v 15.0), and GrapeCyc (v 9.0.1) databases were used to perform annotation on the identified ions of interest using the MetaboAnnotation package [46].

3. Results and Discussion

3.1. Global Metabolome of Grape Juices

The main objective of this study was to explore the extent of grape juice metabolite coverage, which can possibly be reached through integrated mass spectrometry-based metabolomics. In contrast with most LC-MS-based molecular analyses of grape berries [20–26], this study focused on the sole flesh of grape berries (mesocarp), i.e., without considering skins and seeds. To that purpose, up to 290 grape juice samples from four grape varieties (all characterized by uncolored flesh), were collected at harvest from 91 vineyards in Europe and Argentina (Table 1). The diversity in geographical origins was aimed at introducing the largest possible variability in compositions. Figure 1 gathers characteristic features resulting from the different analytical strategies considered here, when applied to all the samples, regardless of the geographical origin, the vintage, or the variety. The ESI (−) ionization mode was selected as a good compromise for the detection of the rather polar hydrophilic compounds consistently expected from acidic grape juices. In the case of DI-FT-ICR-MS, this mode has been shown to favor a larger variety in composition and abundance of compounds and a smaller number of adducts in wines, as well as a better resolution than positive ionization [47].

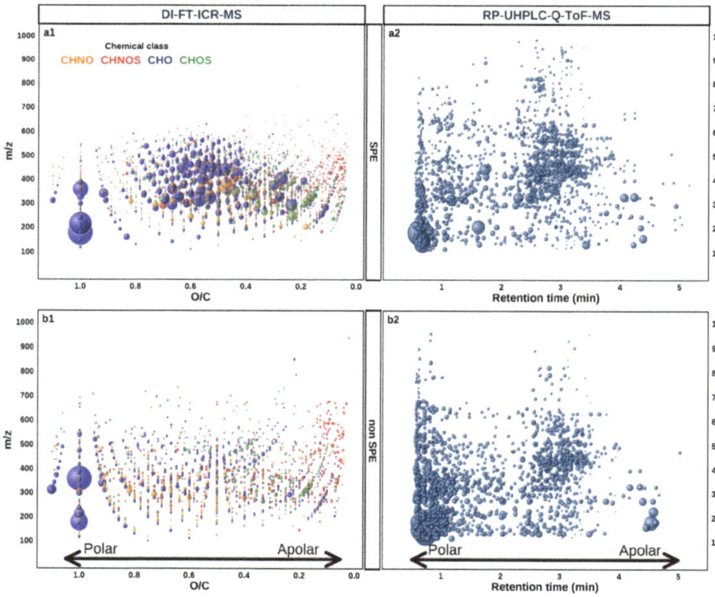

Figure 1. Display of the global grape juice metabolome detected by the different analytical strategies for SPE pretreated samples (**top**) and non SPE pretreated samples (**bottom**); (**a1**,**b1**) mass vs. O/C van

Krevelen diagram representation of all detected DI-FT-ICR-MS mass peaks transformed into assigned elemental composition with an assignment error below 0.5 ppm, and a O/C ratio below 1.1. Dot sizes are proportional to the relative intensity of corresponding mass peaks, and with the following color codes: CHO (blue), CHON (orange), CHOS (green), CHONS (red); (**a2**,**b2**) all detected RP-UHPLC-Q-ToF-MS features represented as 2D maps projecting mass vs. RT (retention time).

The ultra-high-resolution power of DI-FT-ICR-MS readily allowed the detection of up to 9000 mass signals with a signal-to-noise ratio >3 after SPE pretreatment of grape juices. Upon pre-processing these data, including mass difference network analysis, 4629 observed mass signals present in at least four samples could be converted into unambiguous elemental compositions based on the main isotopic elements (^{12}C, ^{14}N, ^{16}O, ^{32}S, ^{1}H) within an assignment error window of 0.5-ppm (defined as the ratio of the difference between the experimental mass and the exact mass ($m/z_{exp} - m/z_{exact}$) to the experimental mass (m/z_{exp}) $\times 10^6$), thus providing an as yet unprecedented representation of the chemical diversity of the sole flesh of ripe grape berries (Figures 1(a1,b1), 2a,b, S1 and S2). Whether samples had been SPE pretreated or not, H/C vs. O/C van Krevelen diagrams (Figures 2a,b, S1 and S2) revealed a remarkable metabolome coverage with likely tens of thousands of compounds (see below, Section 3.2). This metabolome, which comprises peptides and amino acids, fatty and organic acids [30,48], and extensive homologous series of conjugated compounds, i.e., alkylated compounds [49], and/or glycosylated compounds [50–52] consistently with such carbohydrate-rich matrices, provided a snapshot of the numerous metabolic pathways involved in the ripening of grape berries. Our results further revealed that common metabolites to SPE and non SPE pretreated samples could be found throughout the van Krevelen diagram regions, and covered a high range of polarity, suggesting that the hydrophilic character of FT-ICR-MS detected masses were weakly influenced by SPE adsorption behaviors, with likely high (−) ionization efficiency (Figure 1(a1,b1)). However, if both pretreatments led to the detection of metabolites over similar mass ranges (200–750 Da), the SPE pretreatment allowed for the detection of more abundant mass peaks between ca. 300 and 550 Da (Figure 1(a1)), characterized by medium polarity as expressed by their narrower range of O/C ratios, and mostly found in the peptide/nucleic acids and related conjugated compound region (Figures 2a,b and S1). Such mass peaks would exhibit lower ESI (−) ionization efficiency and be suppressed otherwise without SPE pretreatment.

Because of a significantly lower resolving power of RP-UHPLC-Q-ToF-MS, the 1648 and 1599 (Figure 3, Table S2) detected features for SPE and non SPE pretreatments, respectively, could not be straightforwardly and unambiguously transformed into elemental compositions, and the corresponding results were more rigorously presented as 2D maps projecting mass values as a function of retention times (Figure 1(a2,b2)). As confirmed by the density distributions of the detected mass peak intensity (Figure S2), the SPE pretreatment considerably reduced the number of hydrophilic low molecular weight metabolites (retention times < 1.5 min), thus favoring the detection of higher frequencies of less hydrophilic and heavier metabolites (mass > 250 Da, RT > 2.5 min), up to 1000 Da. Interestingly, some common buckets (RT; mass) to SPE and non SPE pretreated samples could be found at low retention times, indicating that SPE pretreatment did not remove all the highly hydrophilic metabolites initially present in the juice. However, it must be noted that under our experimental conditions (RP, C18 column), for simple grape juice dilution (non SPE) and to a lesser extent for SPE pretreated grape juices, the LC step did not prevent high abundance of detected mass peaks for retention times < 1 min (Figures 1(a2,b2) and S2), which indicates that for low RTs, chromatographic separation could be considered inefficient, and which points to the need for HILIC-UHPLC-Q-ToF-MS for a further increase of the metabolite coverage [34].

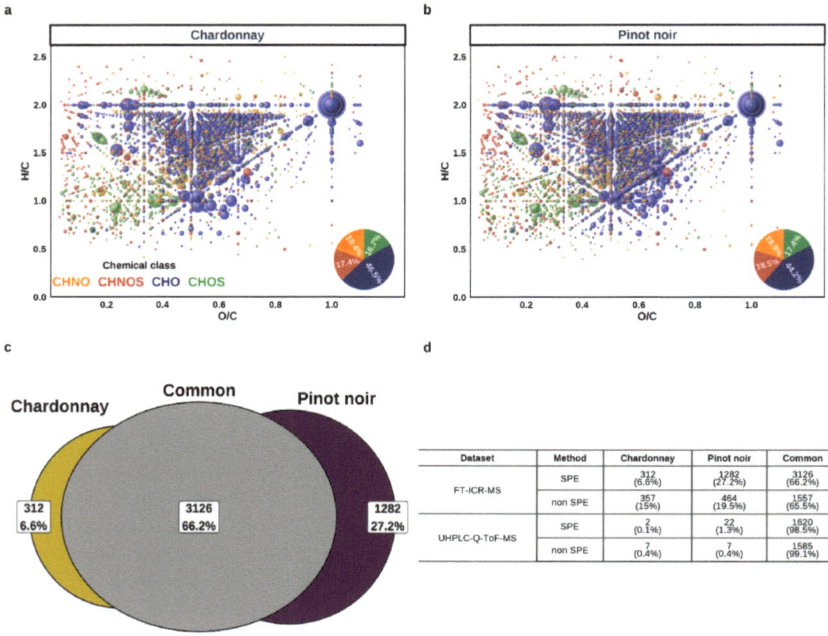

d

Dataset	Method	Chardonnay	Pinot noir	Common
FT-ICR-MS	SPE	312 (6.6%)	1282 (27.2%)	3126 (66.2%)
	non SPE	357 (15%)	464 (19.5%)	1557 (65.5%)
UHPLC-Q-ToF-MS	SPE	2 (0.1%)	22 (1.3%)	1620 (98.5%)
	non SPE	7 (0.4%)	7 (0.4%)	1585 (99.1%)

Figure 2. Comparison of the global Chardonnay and Pinot noir metabolomes obtained by DI-FT-ICR-MS, considering all geographical regions and vintages. H/C vs. O/C van Krevelen diagram representation for Chardonnay (**a**) and Pinot noir (**b**) with the following color codes for elemental compositions: CHO (blue), CHON (orange), CHOS (green), CHONS (red). These diagrams represent mass peaks annotated into assigned elemental composition with an assignment error below 0.5 ppm, and an O/C ratio below 1.1, a H/C ratio below 1.5, and present in at least four samples for each variety. Dot sizes are proportional to the relative intensity of corresponding mass peaks. Comparison between annotated mass peaks for the two cultivars is reported in a Venn Diagram (**c**). Summary of comparisons of numbers of detected features between the two cultivars for the four analytical methods of this study (**d**).

Figure 2, which focuses on all the detected features in Chardonnay and Pinot noir samples only, provides a striking illustration of the high compositional similarity between grape juices of a red and a white wine grape variety. Considering the FT-ICR-MS results of SPE pretreated samples, 66.2% of the annotated mass peaks detected in at least four samples of Chardonnay and four samples of Pinot noir were common to the two varieties, whereas only 6.6% and 27.2% were only detected in Chardonnay and Pinot noir juices, respectively (Figure 2c). It must be noted that, although care had been taken to prevent juice/skin contact while Pinot noir berries were pressed, some skin extraction could possibly have occurred. This would contribute to the higher relative proportion of annotated masses specific to Pinot noir in the polyphenolic region of the van Krevelen diagram (H/C between 0.8 and 1.2; O/C around 0.5 and 0.7; Figure 2b). Figure 2d further gathers similar consideration for the other analytical methods, which shows that, under our experimental conditions, RP-UHPLC-Q-ToF-MS analyses of these grape juices could hardly detect features (present in at least four samples within each variety) specific to each of the varieties, with at best 22 out of a total of 1644, for SPE pretreated Pinot noir samples.

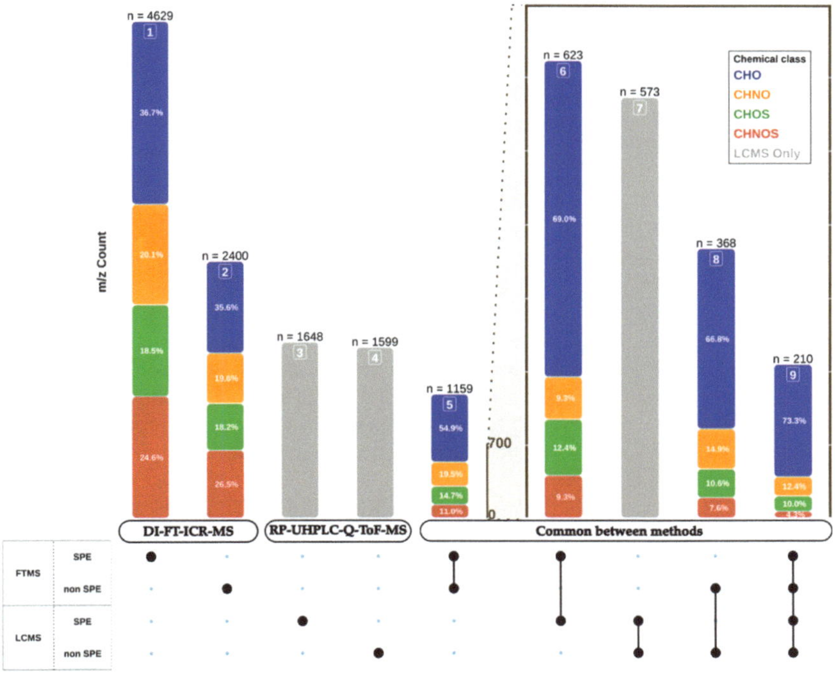

Figure 3. Histogram (upper part) of the number of unambiguously assigned elemental compositions based on DI-FT-ICR-MS (column 1–2) and RP-UHPLC-Q-ToF-MS (column 3–4) detected *m*/*z* values with the following color codes for elemental compositions: CHO (blue), CHNO (orange), CHOS (green), CHNOS (red). Single dots below associate elemental compositions to the pretreatment used for the two MS analyses, and connected dots indicate that these compositions are common to two or more pretreatments and analyses, after peak alignment (column 5–9).

3.2. Aligning the Ultra-High-Resolution Power of DI-FT-ICR-MS with the UHPLC-Controlled High Resolution of RP-UHPLC-Q-ToF-MS

As shown in Figure 3, and in agreement with the literature [26,53], SPE pretreatment enabled the best metabolite coverage by DI-FT-ICR-MS with up to 4629 assigned elemental formulas, whereas for non SPE, only 2400 mass peaks could be assigned an elemental formula. This is a direct representation of the filtering impact of SPE pretreatment on ESI (−) DI-FT-ICR-MS leading to the significant reduction of adduct-formation salts, and of carbohydrates-in particular, sugars with concentrations at harvest being classically higher than 200 g/L-which contribute to a competition for ionization in the electrospray [54]. Figure 3 also indicates that whatever the pretreatment used (SPE or non SPE), up to 1159 mass peaks (and corresponding elemental compositions) were systematically detected in at least four samples, by DI-FT-ICR-MS. The same consideration for RP-UHPLC-Q-ToF-MS analyses led to 573 buckets (RT; mass) systematically found whatever the pretreatment used. This represented about a third of the number of buckets detected either by SPE (1648) or non SPE (1599) pretreated samples analyzed by RP-UHPLC-Q-ToF-MS. However, without alignment with DI-FT-ICR-MS data, it was unrealistic to assign elemental compositions to mass values related to these buckets, because of the lower resolving power of RP-UHPLC-Q-ToF-MS.

In previous papers [34], we have shown that the alignment of highly resolved mass peaks detected by DI-FT-ICR-MS with isobaric LC-separated mass peaks detected by LC-MS could clearly increase the scope of detectable unknown metabolites in wines. Despite a lower resolution, RP-UHPLC-Q-ToF-MS mass peaks with hits in DI-FT-ICR-MS peaks could then be assigned an unambiguous elemental formula (further validated by

MS[2]), whereas multiple RP-UHPLC-Q-ToF-MS-detected retention times for a common mass peak would count the possible DI-FT-ICR-MS detected isobars. All results from all possible alignments were gathered in Table S2. A first striking feature was that after alignment within a 2-ppm alignment error window, distinct numbers of elemental compositions appeared to be common/unique to DI-FT-ICR-MS and RP-UHPLC-Q-ToF-MS analyses, depending on the aligned sets of data, illustrating the complementarity of the two platforms for grape juice analyses [34,47]. A maximum of 210 elemental compositions appeared to be common to the four analytical strategies (Figure 3, Table S2). Looking more closely at common compositions to DI-FT-ICR-MS and RP-UHPLC-Q-ToF-MS for non SPE pretreatment, Table 2 and Figure 3 show that 368 DI-FT-ICR-MS compositions (representing 12.2% of assigned compositions) found hits in RP-UHPLC-Q-ToF-MS spectra. Consistent with the filtering effect and the reduction of ion suppression of SPE, up to 623 hits (11.4%) were observed between the two MS methods, after SPE pretreatment. Since a given mass value observed in RP-UHPLC-Q-ToF-MS could be associated with multiple retention times, the count of RP-UHPLC-Q-ToF-MS hits could be higher than for DI-FT-ICR-MS, as shown both for SPE and non SPE pretreated samples, with up to 1035 (62.8%) and 825 (51.6%) aligned peaks, respectively.

Table 2. Focus on the number of features related to DI-FT-ICR-MS and RP-UHPLC-Q-ToF-MS mass peak alignments for SPE and non SPE pretreatments.

Method	DI-FT-ICR-MS			RP-UHPLC-Q-ToF-MS			
	n (Total)	Chemical Class	n Chemical Class	n Total [1]	n Chemical Class [1]	n Isobars Mean [2]	n Isobars Range [2]
SPE	623 (11.4%)	CHO	430 (69%)	1035 (62.8%)	794 (76.7%)	1.5	1–6
		CHNO	58 (9.3%)		71 (6.9%)	1.1	1–3
		CHNOS	58 (9.3%)		77 (7.4%)	1.2	1–6
		CHOS	77 (12.4%)		93 (9%)	1.1	1–4
non SPE	368 (12.2%)	CHO	246 (66.8%)	825 (51.6%)	623 (75.5%)	1.8	1–13
		CHNO	55 (14.9%)		99 (12%)	1.4	1–4
		CHNOS	28 (7.6%)		47 (5.7%)	1.3	1–4
		CHOS	39 (10.6%)		56 (6.8%)	1.2	1–3

[1] "n" corresponds to the number of common elemental compositions between the two analytical methods within a 2-ppm mass peak alignment error. [2] "n isobaric mean" is the average number of retention times associated with a given elemental composition and "n isobaric range" indicates the range of retention times associated with a given elemental composition (RP-UHPLC-Q-ToF-MS analyses).

If the average number of observed retention times was between one and two for a given mass value, Table 2 further showed that some DI-FT-ICR-MS mass peaks could frequently be associated with up to four retention times, and in some case up to 13 distinct retention times. Assuming a restrictive 2-ppm alignment error window for the alignment procedure within RP-UHPLC-Q-ToF-MS spectra, and considering that only about 12% of the DI-FT-ICR-MS peaks found hits in RP-UHPLC-Q-ToF-MS peaks, one could hypothesize that the actual chemical diversity probed by DI-FT-ICR-MS in SPE treated grape juice samples is certainly of a few tens of thousands of metabolites. Furthermore, upon breaking down common mass peaks into distinct compositions, Table 2 showed that most of the mass peaks aligned between the two MS methods were assigned CHO-based elemental formulas. This result highlighted the complementary power of DI-FT-ICR-MS to more evenly probe the diversity of S/N-containing metabolites (Figure 3). It must be noted that the case of 13 distinct retention times was observed for non SPE samples (Table 2), for a mass peak at 215.03279 Da easily assigned the $C_6H_{12}O_6$ $[M+Cl]^-$ ion formula, corresponding to a chlorine adduct of glucose [55]. The observation of up to 13 retention times over the 1–5.74 min range in RP-UHPLC-Q-ToF-MS spectra raised the question of competition for ionization in non SPE pretreated samples, with the possible need to better adjust the dilution. An explanation would be that highly abundant glucose and fructose molecules, considered here as "pollutants", are eluted throughout the chromatographic time frame.

3.3. Grape Juice Discrimination by ESI (−) DI-FT-ICR-MS Applied to SPE Pretreated Samples

With such a high dimensionality of the data set, both in terms of individuals (grape juice samples, Table 1) and variables (assigned elemental compositions, Figure 3), it is possible to explore the ability of metabolomics to discriminate specific molecular fingerprints among subsets of samples. Of the four grape varieties considered in this study, Chardonnay and Pinot noir were the most widely represented in terms of vintages and geographical origins and therefore terroirs, thus providing an unprecedented array of likely subtle variations in metabolite expressions. However, if the discrimination between red grape (Pinot noir) and white grape (Chardonnay) cultivars can be straightforward when considering skin extracts because of the higher polyphenol concentrations for the former, discrimination based on the sole berry juices, regardless of both the geographical origin and the vintage, remains challenging, as demonstrated by Figure 2. As shown by a variance partition study (Figure S3) of a data subset corresponding to DI-FT-ICR-MS analyses (SPE pretreatment) of Chardonnay and Pinot noir grape juices for vintages 2019 and 2020, the variety and the geographical origin could significantly explain only 15.9% and 10.4% of the variance, respectively, whereas nearly 75% of the variance appeared unexplained by these parameters or the vintage. Furthermore, the Redundancy Analysis (RDA) of mass peaks, which significantly contribute to the geographical origin-related variance, showed that the Languedoc region would be more different from Burgundy than the Southern Hemisphere region of the Uco Valley in Argentina. These results (Figure S3) emphasized that, for such a diversity of grape juice origins, environmental conditions along with vineyard practices can indeed contribute to significantly modulating metabolite expressions even within a single grape variety, thus introducing possibly significant noise into targeted analyses. In contrast, non-targeted analyses can provide comprehensive transient chemical snapshots of highly subtle metabolite expressions, which are considered to integrate contributions from every plant–environment interaction associated with the multiple vineyard conditions, and provided that the sampling is large enough, it is possible to apply robust multivariate statistical analyses to such a dataset.

Figure 4a,b show that OPLS-DA models could significantly discriminate Chardonnay from Pinot noir grape juices, and Chardonnay from Aligoté grape juices (two white cultivars), respectively. In both cases, the robustness of the model was guaranteed by Q2-values > 0.9, for the quality of prevision, and R2Y-values > 0.97, for the goodness of the fit. Furthermore, similar significant discriminations could also be obtained with non SPE pretreated samples, and again with whatever the sample pretreatment used (SPE/non SPE) with RP-UHPLC-Q-ToF-MS (Table S3). In contrast, attempts to discriminate between the two red cultivars (Pinot noir and Meunier) failed through cross-validation steps, and the non-supervised PCA statistical analysis appeared to be controlled by the geographical origins of the vineyards regardless of the grape variety, with the first two components explaining more than 48% of the variance (Figure 4c). A much smaller set of samples (with only four geographical origins and two vintages, Table 1) could possibly contribute to this model failure for the two red grape cultivars, but interestingly, when reduced to the same geographical origins as those for Meunier, data subsets for Pinot noir and Chardonnay could still lead to a significantly discriminant model (Table S3). This result suggests that Pinot noir and Meunier grape juices could be too similar, and their discrimination would be primarily driven by vineyard characteristics, with Champagne (emblematic land of Meunier) being best discriminated (Figure 4c).

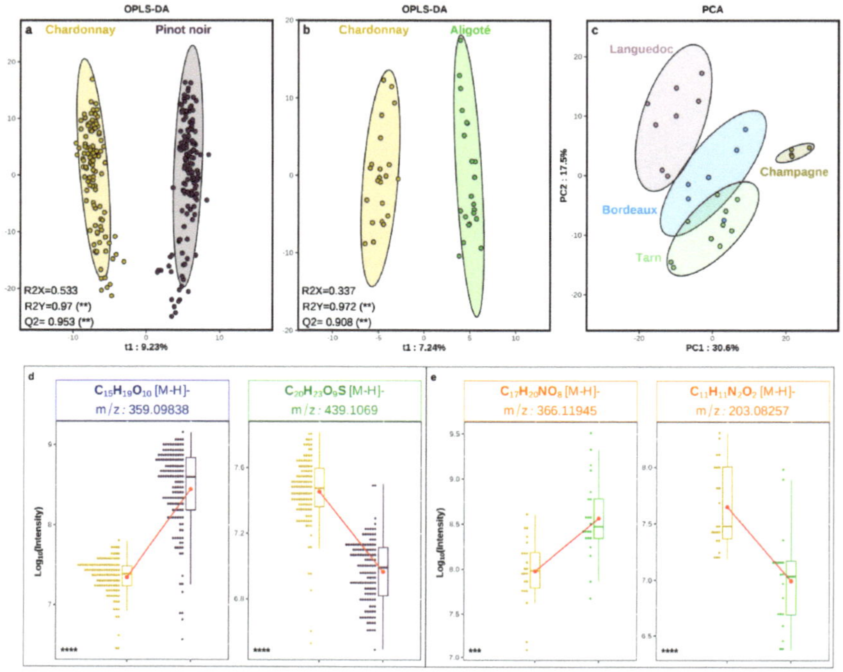

Figure 4. Multivariate statistical analyses of DI-FT-ICR-MS metabolomic analyses of SPE pretreated samples, regardless of the geographical origin or the vintage; supervised OPLS-DA discrimination of Chardonnay/Pinot noir grape juices (**a**) and Chardonnay/Aligoté grape juices (**b**); non supervised PCA analysis of Pinot noir and Meunier grape juices (**c**); each dot in (**a–c**) refers to a grape juice sample; jitter plots and boxplots showing examples of VIP m/z peaks for Chardonnay and Pinot noir (**d**) and for Chardonnay and Aligoté (**e**). Each jitter plot provides a density distribution of the corresponding mass peak intensity (expressed as the Log_{10} value) among Chardonnay (yellow) and Pinot noir (purple) samples (**d**), and among Chardonnay (yellow) and Aligoté (green) samples (**e**); **, *** and **** indicate significance with p value at 1E−2, 1E−3, and 1E−4, respectively.

The molecular fingerprints for both Pinot noir and Chardonnay were dominated by CHO compositions with up to 80.7% (142 assigned elemental formulas) and 50.9% (31 assigned elemental formulas) of Variable Importance in Projection (VIPs), respectively (Figure S4). SPE pretreated Pinot noir grape juices clearly appeared much richer in glycosylated homologous series of polyphenolic structures (Figure S2). Although care had been taken to prevent juice/skin contact while Pinot noir berries were pressed, some skin extraction could not be completely ruled out, and these homologous series would possibly be partly attributed to skin metabolites. Interestingly, when considering VIPs which discriminated among non SPE treated samples (Figure S5), CHO compositions were still dominating fingerprints, but with significantly fewer mass values (13 CHO elemental compositions representing 59.1% for Pinot noir, and 26 CHO compositions representing 43.3% for Chardonnay). However, up to 35% of Chardonnay VIPs were S-containing compositions, including 15 CHNOS (25%) and 6 CHOS (10%), which confirmed the relatively high importance of N/S-containing metabolites for this cultivar.

Finally, Figure 3 reports examples of VIP elemental compositions among tens of others and likely hundreds of related compounds (Table S4), with the 359.09838 m/z peak, to which the $[C_{19}H_{19}O_{10}]^-$ ion composition could be assigned for Pinot noir, and the 439.1068 m/z peak, to which the $[C_{20}H_{23}O_9S]^-$ ion composition could be assigned for Chardonnay. The latter S-containing formula was better observed as part of a three-membered homologous series with an O/C ratio of around 0.4 in non SPE pretreated samples (Figure S4) but no pertinent structural assignment could be found in accessible databases. In contrast, the CHO marker for Pinot noir, could likely correspond to a glycosidic adduct of syringic acid or isomers, which have been identified in other *Vitis vinifera* red grapes [56], and whose concentration in berries could be modulated by vine growing management. As witnessed by the four associated RT from RP-UHPLC-Q-ToF-MS results, up to four glycosidic isomers could possibly be more abundant in Pinot noir grape juices, depending for instance on the O-glycosylation position. As to examples of VIPs discriminant for Chardonnay and Aligoté, Figure 3 shows that the 203.08257 m/z peak, to which the $[C_{11}H_{11}N_2O_2]^-$ ion composition can be assigned, was significantly more abundant in Chardonnay grape juices, whereas the 366.11945 m/z peak, to which the $[C_{17}H_{20}NO_8]^-$ ion composition can be assigned, was significantly more abundant in Aligoté. A consistent structural assignment to the Chardonnay VIP could be Tryptophan, as this grape variety has been shown to be among the most concentrated in this precursor of indoleacetic acid [57,58]. Interestingly, the Aligoté marker shown in Figure 3 could correspond to indolelactic acid glucoside, another glucoside already identified in white grapes as a contributor to the "phenolic taste" of white wines [59]. This would be the first identification of this metabolite in the rarely studied Aligoté grape variety.

4. Conclusions

In this study, a large series of juices of ripe grape berries from four different varieties, sampled in various vineyards internationally and over three successive vintages, were analyzed by DI-FT-ICR-MS and RP-UHPLC-Q-ToF-MS, to explore the possible extent of metabolome coverage. Samples were either SPE pretreated or not, before analysis. DI-FT-ICR-MS analyses of SPE pretreated samples clearly provided higher metabolite coverage, with only about 13% of the 4629 assigned elemental compositions being common to RP-UHPLC-Q-ToF-MS detected masses. Our results revealed that the sole flesh of Chardonnay, Pinot noir, Meunier, and Aligoté berries could likely contain tens of thousands of compounds transiently present throughout the ripening period up to harvest. This result is even more remarkable given that such chemical diversity does not include the many seed- and skin-related metabolites such as polyphenols. Additionally, when considering Chardonnay and Pinot noir, the two most represented grape varieties in our sampling, we have shown that up to 75% of this chemical diversity is common to all juices, thus emphasizing the fact that many similar metabolic pathways must be involved in the ripening of these two varieties. Only 15.9% of the variance of Pinot noir and Chardonnay metabolomes appeared to be explained by the variety. However, thanks to the high dimensionality of our sampling and of the detected metabolomes, it was possible to build significant models for the discrimination of Chardonnay from Pinot noir grape juices, and of Chardonnay from Aligoté grape juices, regardless of the geographical origin or the vintage. Several tens of related chemical markers could be identified, including, for example, glycosides for Pinot noir. In summary, an excellent complementarity between the two analytical methods was shown, with FT-ICR-MS being a rapid, reproducible, and highly precise tool for non-targeted sample screening, and RP-UHPLC-Q-ToF-MS complementing ideally with the possible resolution of isomers through the chromatographic dimension, and the MS/MS fragmentation tools for structural identifications. Such results are of primary importance when studying the enological potential of emblematic grape varieties such as Pinot noir and Chardonnay, which can produce high-added-value wines from many different vineyards around the world.

Supplementary Materials: The following supporting information can be downloaded at: https://www.mdpi.com/article/10.3390/foods13010054/s1, Figure S1: H/C vs. O/C van Krevelen diagrams representing all detected DI-FT-ICR-MS mass peaks transformed into assigned elemental compositions, Figure S2: Display of the global grape juice metabolome detected by the different analytical strategies, Figure S3: Multivariate statistical analysis of Chardonnay and Pinot noir grape juice metabolomes, detected by DI-FT-ICR-MS after SPE pretreatment, Figure S4: S-plot and van Krevelen diagrams of SPE prepared samples (Chardonnay and Pinot noir grape juices) analyzed by DI-FT-ICR-MS, Figure S5: S-plot and van Krevelen diagrams of non SPE prepared samples (Chardonnay and Pinot noir grape juices) analyzed by DI-FT-ICR-MS, Table S1: Summary of the harvest date ranges for the different vintages, varieties and hemispheres, Table S2: Summary of all DI-FT-ICR-MS and RP-UHPLC-Q-ToF-MS mass peaks alignments broken down by common elemental compositions, Table S3: Statistical analyses parameters for the different grape variety discriminations shown in Figure 4, Table S4: Summary of all DI-FT-ICR-MS and RP-UHPLC-Q-ToF-MS mass peak alignments that significatively discriminate Chardonnay and Pinot noir.

Author Contributions: Conceptualization, B.B., P.S.-K. and R.D.G.; methodology, S.N., K.B., R.R., F.B., J.U., P.S.-K. and R.D.G.; software, S.N.; validation, S.N., P.S.-K. and R.D.G.; formal analysis, S.N.; investigation, S.N., K.B., R.R., F.B. and J.U.; resources, S.N., B.B., P.S.-K. and R.D.G.; data curation, S.N., K.B., R.R., F.B. and J.U.; writing—original draft preparation, S.N., P.S.-K. and R.D.G.; writing—review and editing, S.N., B.B., K.B., R.R., F.B., J.U., P.S.-K. and R.D.G.; visualization, S.N.; supervision, B.B., P.S-K. and R.D.G.; project administration, B.B. and R.D.G.; funding acquisition, B.B. and R.D.G. All authors have read and agreed to the published version of the manuscript.

Funding: This research was funded by the Bureau Interprofessionnel des Vins de Bourgogne (BIVB), the Comité Champagne, and the Catena Institute, through the CLIMCHANGE project, and the European Funds for Regional Development (FEDER) through the FEDER-FSE Bourgogne 2014–2020 METABOLOM program, grant no. BG0022832.

Data Availability Statement: Data is contained within the article or Supplementary Materials.

Acknowledgments: Fernando Buscema and Roy Urvieta (Catena Institute, Mendoza, Argentina) are warmly thanked for providing grape juice samples, and for fruitful discussions; Héloise Mahé and Christine Monamy (BIVB), Sébastien Debuisson (Comité Champagne), Fernando Alvès (Symington, Portugal), Danielle Dellavalle (Agricola Dellavalle), Agnès Destrac (Bordeaux Sciences Agro), Olivier Yobregat, Laurent Audeguin and Jean-Yves Cahurel (IFV), Florian Haas (Laimburg Research Center), Marianne Henner (Chambre d'Agriculture d'Alsace), and Manfred Stoll and Magali Blank (Forschungsanstalt Geisenheim), are warmly thanked for providing grape juice samples; Marianna Lucio and Mourad Harir (Helmholtz Zentrum München) are also warmly thanked for fruitful discussions. Laurence Noret is thanked for her technical support.

Conflicts of Interest: The authors declare no conflict of interest.

References

1. Conde, C.; Silva, P.; Fontes, N.; Dias, A.C.P.; Tavares, R.M.; Sousa, M.J.; Agasse, A.; Delrot, S.; Gerós, H. Biochemical Changes throughout Grape Berry Development and Fruit and Wine Quality. *Food* **2007**, *1*, 1–22.
2. van Leeuwen, C.; Barbe, J.-C.; Darriet, P.; Destrac-Irvine, A.; Gowdy, M.; Lytra, G.; Marchal, A.; Marchand, S.; Plantevin, M.; Poitou, X.; et al. Aromatic Maturity Is a Cornerstone of Terroir Expression in Red Wine: This Article Is Published in Cooperation with Terclim 2022 (XIVth International Terroir Congress and 2nd ClimWine Symposium), 3–8 July 2022, Bordeaux, France. *OENO One* **2022**, *56*, 335–351. [CrossRef]
3. Allamy, L.; van Leeuwen, C.; Pons, A. Impact of Harvest Date on Aroma Compound Composition of Merlot and Cabernet-Sauvignon Must and Wine in a Context of Climate Change: A Focus on Cooked Fruit Molecular Markers. *OENO One* **2023**, *57*, 99–112. [CrossRef]
4. Gobert, A.; Tourdot-Maréchal, R.; Sparrow, C.; Morge, C.; Alexandre, H. Influence of Nitrogen Status in Wine Alcoholic Fermentation. *Food Microbiol.* **2019**, *83*, 71–85. [CrossRef] [PubMed]
5. Sefton, M.A.; Francis, I.L.; Williams, P.J. The Volatile Composition of Chardonnay Juices: A Study by Flavor Precursor Analysis. *Am. J. Enol. Vitic.* **1993**, *44*, 359–370. [CrossRef]
6. Pinu, F.R.; Edwards, P.J.B.; Jouanneau, S.; Kilmartin, P.A.; Gardner, R.C.; Villas-Boas, S.G. Sauvignon Blanc Metabolomics: Grape Juice Metabolites Affecting the Development of Varietal Thiols and Other Aroma Compounds in Wines. *Metabolomics* **2014**, *10*, 556–573. [CrossRef]
7. Hjelmeland, A.K.; Zweigenbaum, J.; Ebeler, S.E. Profiling Monoterpenol Glycoconjugation in *Vitis vinifera* L. Cv. Muscat of Alexandria Using a Novel Putative Compound Database Approach, High Resolution Mass Spectrometry and Collision Induced Dissociation Fragmentation Analysis. *Anal. Chim. Acta* **2015**, *887*, 138–147. [CrossRef]
8. Cebrián-Tarancón, C.; Oliva, J.; Cámara, M.Á.; Alonso, G.L.; Rosario Salinas, M. Analysis of Intact Glycosidic Aroma Precursors in Grapes by High-Performance Liquid Chromatography with a Diode Array Detector. *Foods* **2021**, *10*, 191. [CrossRef]
9. De Rosso, M.; Lonzarich, V.; Navarini, L.; Flamini, R. Identification of New Glycosidic Terpenols and Norisoprenoids (Aroma Precursors) in *C. arabica* L. Green Coffee by Using a High-Resolution Mass Spectrometry Database Developed in Grape Metabolomics. *Curr. Res. Food Sci.* **2022**, *5*, 336–344. [CrossRef]

10. Wei, Y.; Chen, Z.; Zhang, X.K.; Duan, C.Q.; Pan, Q.H. Comparative Analysis of Glycosidic Aroma Compound Profiling in Three *Vitis vinifera* Varieties by Using Ultra-High-Performance Liquid Chromatography Quadrupole-Time-of-Flight Mass Spectrometry. *Front. Plant Sci.* **2021**, *12*, 694979. [CrossRef]
11. Caffrey, A.J.; Lerno, L.A.; Zweigenbaum, J.; Ebeler, S.E. Direct Analysis of Glycosidic Aroma Precursors Containing Multiple Aglycone Classes in *Vitis vinifera* Berries. *J. Agric. Food Chem.* **2020**, *68*, 3817–3833. [CrossRef]
12. Rienth, M.; Vigneron, N.; Darriet, P.; Sweetman, C.; Burbidge, C.; Bonghi, C.; Walker, R.P.; Famiani, F.; Castellarin, S.D. Grape Berry Secondary Metabolites and Their Modulation by Abiotic Factors in a Climate Change Scenario—A Review. *Front. Plant Sci.* **2021**, *12*, 262. [CrossRef] [PubMed]
13. Evers, M.S.; Roullier-Gall, C.; Morge, C.; Sparrow, C.; Gobert, A.; Alexandre, H. Vitamins in Wine: Which, What for, and How Much? *Compr. Rev. Food Sci. Food Saf.* **2021**, *20*, 2991–3035. [CrossRef] [PubMed]
14. Mattivi, F.; Guzzon, R.; Vrhovsek, U.; Stefanini, M.; Velasco, R. Metabolite Profiling of Grape: Flavonols and Anthocyanins. *J. Agric. Food Chem.* **2006**, *54*, 7692–7702. [CrossRef] [PubMed]
15. Martínez-Gil, A.M.; Gutiérrez-Gamboa, G.; Garde-Cerdán, T.; Pérez-Álvarez, E.P.; Moreno-Simunovic, Y. Characterization of Phenolic Composition in Carignan Noir Grapes (*Vitis vinifera* L.) from Six Wine-Growing Sites in Maule Valley, Chile. *J. Sci. Food Agric.* **2018**, *98*, 274–282. [CrossRef] [PubMed]
16. Deluc, L.G.; Quilici, D.R.; Decendit, A.; Grimplet, J.; Wheatley, M.D.; Schlauch, K.A.; Mérillon, J.M.; Cushman, J.C.; Cramer, G.R. Water Deficit Alters Differentially Metabolic Pathways Affecting Important Flavor and Quality Traits in Grape Berries of Cabernet Sauvignon and Chardonnay. *BMC Genom.* **2009**, *10*, 212. [CrossRef]
17. Ferrandino, A.; Lovisolo, C. Abiotic Stress Effects on Grapevine (*Vitis vinifera* L.): Focus on Abscisic Acid-Mediated Consequences on Secondary Metabolism and Berry Quality. *Environ. Exp. Bot.* **2014**, *103*, 138–147. [CrossRef]
18. Nicholson, J.K.; Lindon, J.C.; Holmes, E. "Metabonomics": Understanding the Metabolic Responses of Living Systems to Pathophysiological Stimuli via Multivariate Statistical Analysis of Biological NMR Spectroscopic Data. *Xenobiotica* **1999**, *29*, 1181–1189. [CrossRef]
19. Pinu, F.R. Grape and Wine Metabolomics to Develop New Insights Using Untargeted and Targeted Approaches. *Fermentation* **2018**, *4*, 92. [CrossRef]
20. Toffali, K.; Zamboni, A.; Anesi, A.; Stocchero, M.; Pezzotti, M.; Levi, M.; Guzzo, F. Novel Aspects of Grape Berry Ripening and Post-Harvest Withering Revealed by Untargeted LC-ESI-MS Metabolomics Analysis. *Metabolomics* **2011**, *7*, 424–436. [CrossRef]
21. Savoi, S.; Wong, D.C.J.; Degu, A.; Herrera, J.C.; Bucchetti, B.; Peterlunger, E.; Fait, A.; Mattivi, F.; Castellarin, S.D. Multi-Omics and Integrated Network Analyses Reveal New Insights into the Systems Relationships between Metabolites, Structural Genes, and Transcriptional Regulators in Developing Grape Berries (*Vitis vinifera* L.) Exposed to Water Deficit. *Front. Plant Sci.* **2017**, *8*, 1124. [CrossRef]
22. Savoi, S.; Herrera, J.C.; Carlin, S.; Lotti, C.; Bucchetti, B.; Peterlunger, E.; Castellarin, S.D.; Mattivi, F. From Grape Berries to Wines: Drought Impacts on Key Secondary Metabolites. *OENO One* **2020**, *54*, 569–582. [CrossRef]
23. Moro, L.; da Mota, R.V.; Purgatto, E.; Mattivi, F.; Arapitsas, P. Investigation of Brazilian Grape Juice Metabolomic Profile Changes Caused by Methyl Jasmonate Pre-Harvest Treatment. *Int. J. Food Sci. Technol.* **2023**, *58*, 3224–3233. [CrossRef]
24. Narduzzi, L.; Stanstrup, J.; Mattivi, F. Comparing Wild American Grapes with *Vitis vinifera*: A Metabolomics Study of Grape Composition. *J. Agric. Food Chem.* **2015**, *63*, 6823–6834. [CrossRef]
25. Theodoridis, G.; Gika, H.; Franceschi, P.; Caputi, L.; Arapitsas, P.; Scholz, M.; Masuero, D.; Wehrens, R.; Vrhovsek, U.; Mattivi, F. LC-MS Based Global Metabolite Profiling of Grapes: Solvent Extraction Protocol Optimisation. *Metabolomics* **2011**, *8*, 175–185. [CrossRef]
26. Nelson, A.R.; Toyoda, J.; Chu, R.K.; Tolić, N.; Garayburu-Caruso, V.A.; Saup, C.M.; Renteria, L.; Wells, J.R.; Stegen, J.C.; Wilkins, M.J.; et al. Implications of Sample Treatment on Characterization of Riverine Dissolved Organic Matter. *Environ. Sci. Process Impacts* **2022**, *24*, 773–782. [CrossRef]
27. Schmitt-Kopplin, P.; Hemmler, D.; Moritz, F.; Gougeon, R.D.; Lucio, M.; Meringer, M.; Müller, C.; Harir, M.; Hertkorn, N. Systems Chemical Analytics: Introduction to the Challenges of Chemical Complexity Analysis. *Faraday Discuss.* **2019**, *218*, 9–28. [CrossRef]
28. Cooper, H.J.; Marshall, A.G. Electrospray Ionization Fourier Transform Mass Spectrometric Analysis of Wine. *J. Agric. Food Chem.* **2001**, *49*, 5710–5718. [CrossRef] [PubMed]
29. Catharine, R.R.; Cunha, I.B.S.; Fogaça, A.O.; Facco, E.M.P.; Godoy, H.T.; Daudt, C.E.; Eberlin, M.N.; Sawaya, A.C.H.F. Characterization of Must and Wine of Six Varieties of Grapes by Direct Infusion Electrospray Ionization Mass Spectrometry. *J. Mass. Spectrom.* **2006**, *41*, 185–190. [CrossRef] [PubMed]
30. Gougeon, R.D.; Lucio, M.; Frommberger, M.; Peyron, D.; Chassagne, D.; Alexandre, H.; Feuillat, F.; Voilley, A.; Cayot, P.; Gebefügi, I.; et al. The Chemodiversity of Wines Can Reveal a Metabologeography Expression of Cooperage Oak Wood. *Proc. Natl. Acad. Sci. USA* **2009**, *106*, 9174–9179. [CrossRef] [PubMed]
31. Dou, Y.; Mäkinen, M.; Jänis, J. Analysis of Volatile and Nonvolatile Constituents in Gin by Direct-Infusion Ultrahigh-Resolution ESI/APPI FT-ICR Mass Spectrometry. *J. Agric. Food Chem.* **2023**, *71*, 7082–7089. [CrossRef] [PubMed]
32. Oliveira, B.G.; Tosato, F.; Folli, G.S.; de Leite, J.A.; Ventura, J.A.; Endringer, D.C.; Filgueiras, P.R.; Romão, W. Controlling the Quality of Grape Juice Adulterated by Apple Juice Using ESI(-)FT-ICR Mass Spectrometry. *Microchem. J.* **2019**, *149*, 104033. [CrossRef]

33. Pieczonka, S.A.; Lucio, M.; Rychlik, M.; Schmitt-Kopplin, P. Decomposing the Molecular Complexity of Brewing. *NPJ Sci. Food* **2020**, *4*, 11. [CrossRef] [PubMed]
34. Roullier-Gall, C.; Witting, M.; Tziotis, D.; Ruf, A.; Gougeon, R.D.; Schmitt-Kopplin, P. Integrating Analytical Resolutions in Non-Targeted Wine Metabolomics. *Tetrahedron* **2015**, *71*, 2983–2990. [CrossRef]
35. Tziotis, D.; Hertkorn, N.; Schmitt-Kopplin, P. Kendrick-Analogous Network Visualisation of Ion Cyclotron Resonance Fourier Transform Mass Spectra: Improved Options for the Assignment of Elemental Compositions and the Classification of Organic Molecular Complexity. *Eur. J. Mass. Spectrom.* **2011**, *17*, 415–421. [CrossRef]
36. R Core Team. *R: A Language and Environment for Statistical Computing*; R Foundation for Statistical Computing: Vienna, Austria, 2023. Available online: https://www.R-project.org (accessed on 6 November 2023).
37. Thévenot, E.A.; Roux, A.; Xu, Y.; Ezan, E.; Junot, C. Analysis of the Human Adult Urinary Metabolome Variations with Age, Body Mass Index, and Gender by Implementing a Comprehensive Workflow for Univariate and OPLS Statistical Analyses. *J. Proteome Res.* **2015**, *14*, 3322–3335. [CrossRef]
38. Bararpour, N.; Gilardi, F.; Carmeli, C.; Sidibe, J.; Ivanisevic, J.; Caputo, T.; Augsburger, M.; Grabherr, S.; Desvergne, B.; Guex, N.; et al. DBnorm as an R Package for the Comparison and Selection of Appropriate Statistical Methods for Batch Effect Correction in Metabolomic Studies. *Sci. Rep.* **2021**, *11*, 5657. [CrossRef]
39. Giordan, M. A Two-Stage Procedure for the Removal of Batch Effects in Microarray Studies. *Stat. Biosci.* **2014**, *6*, 73–84. [CrossRef]
40. Walach, J.; Filzmoser, P.; Hron, K. Chapter Seven—Data Normalization and Scaling: Consequences for the Analysis in Omics Sciences. In *Data Analysis for Omic Sciences: Methods and Applications*; Jaumot, J., Bedia, C., Tauler, R.B.T.-C.A.C., Eds.; Elsevier: Amsterdam, The Netherlands, 2018; Volume 82, pp. 165–196. ISBN 0166-526X.
41. Uthe, H.; van Dam, N.M.; Hervé, M.R.; Sorokina, M.; Peters, K.; Weinhold, A. A Practical Guide to Implementing Metabolomics in Plant Ecology and Biodiversity Research. *Adv. Bot. Res.* **2021**, *98*, 163–203. [CrossRef]
42. ter Braak, C.J.F. Redundancy Analysis Includes Analysis of Variance-Simultaneous Component Analysis (ASCA) and Outperforms Its Extensions. *Chemom. Intell. Lab. Syst.* **2023**, *240*, 104898. [CrossRef]
43. Oksanen, J.; Simpson, G.L.; Blanchet, F.G.; Kindt, R.; Legendre, P.; Minchin, P.R.; O'Hara, R.B.; Solymos, P.; Stevens, M.H.H.; Szoecs, E.; et al. Vegan: Community Ecology Package. R Package version 2.6-4. 2022. Available online: https://CRAN.R-project.org/package=vegan (accessed on 6 November 2023).
44. Kassambara, A. Rstatix: Pipe-Friendly Framework for Basic Statistical Tests. R Package version 0.7.2. 2023. Available online: https://cran.r-project.org/package=rstatix (accessed on 6 November 2023).
45. Benjamini, Y.; Hochberg, Y. Controlling the False Discovery Rate: A Practical and Powerful Approach to Multiple Testing. *J. R. Stat. Soc. Ser. B (Methodol.)* **1995**, *57*, 289–300. [CrossRef]
46. Rainer, J.; Vicini, A.; Salzer, L.; Stanstrup, J.; Badia, J.M.; Neumann, S.; Stravs, M.A.; Verri Hernandes, V.; Gatto, L.; Gibb, S.; et al. A Modular and Expandable Ecosystem for Metabolomics Data Annotation in R. *Metabolites* **2022**, *12*, 173. [CrossRef] [PubMed]
47. Roullier-Gall, C.; Witting, M.; Gougeon, R.D.; Schmitt-Kopplin, P. High Precision Mass Measurements for Wine Metabolomics. *Front. Chem.* **2014**, *2*, 102. [CrossRef] [PubMed]
48. Roullier-Gall, C.; Boutegrabet, L.; Gougeon, R.D.; Schmitt-Kopplin, P. A Grape and Wine Chemodiversity Comparison of Different Appellations in Burgundy: Vintage vs. Terroir Effects. *Food Chem.* **2014**, *152*, 100–107. [CrossRef] [PubMed]
49. Wu, Z.; Rodgers, R.P.; Marshall, A.G. Two- and Three-Dimensional van Krevelen Diagrams: A Graphical Analysis Complementary to the Kendrick Mass Plot for Sorting Elemental Compositions of Complex Organic Mixtures Based on Ultrahigh-Resolution Broadband Fourier Transform Ion Cyclotron Resonance. *Anal. Chem.* **2004**, *76*, 2511–2516. [CrossRef]
50. Fougère, L.; Zubrzycki, S.; Elfakir, C.; Destandau, E. Characterization of Corn Silk Extract Using HPLC/HRMS/MS Analyses and Bioinformatic Data Processing. *Plants* **2023**, *12*, 721. [CrossRef]
51. Perdue, E.M.; Green, N.W. Isobaric Molecular Formulae of C, H, and O: A View from the Negative Quadrants of van Krevelen Space. *Anal. Chem.* **2015**, *87*, 5079–5085. [CrossRef]
52. Palacio Lozano, D.C.; Jones, H.E.; Ramirez Reina, T.; Volpe, R.; Barrow, M.P. Unlocking the Potential of Biofuels: Via Reaction Pathways in van Krevelen Diagrams. *Green. Chem.* **2021**, *23*, 8949–8963. [CrossRef]
53. Forcisi, S.; Moritz, F.; Lucio, M.; Lehmann, R.; Stefan, N.; Schmitt-Kopplin, P. Solutions for Low and High Accuracy Mass Spectrometric Data Matching: A Data-Driven Annotation Strategy in Nontargeted Metabolomics. *Anal. Chem.* **2015**, *87*, 8917–8924. [CrossRef]
54. King, R.; Bonfiglio, R.; Fernandez-Metzler, C.; Miller-Stein, C.; Olah, T. Mechanistic Investigation of Ionization Suppression in Electrospray Ionization. *J. Am. Soc. Mass. Spectrom.* **2000**, *11*, 942–950. [CrossRef]
55. Cai, Y.; Concha, M.C.; Murray, J.S.; Cole, R.B. Evaluation of the Role of Multiple Hydrogen Adducts in Electrospray Mass Spectrometry. *J. Am. Soc. Mass. Spectrom.* **2002**, *13*, 1360–1369. [CrossRef]
56. Gardiman, M.; De Rosso, M.; De Marchi, F.; Flamini, R. Metabolomic Profiling of Different Clones of *Vitis vinifera* L. Cv."Glera" and "Glera Lunga" Grapes by High-Resolution Mass Spectrometry. *Metabolomics* **2023**, *19*, 25. [CrossRef]
57. Meng, N.; Ren, Z.-Y.; Yang, X.-F.; Pan, Q.-H. Effects of Simple Rain-Shelter Cultivation on Fatty Acid and Amino Acid Accumulation in 'Chardonnay' Grape Berries. *J. Sci. Food Agric.* **2018**, *98*, 1222–1231. [CrossRef]

58. Fernández-Cruz, E.; Cerezo, A.B.; Cantos-Villar, E.; Troncoso, A.M.; García-Parrilla, M.C. Time Course of L-Tryptophan Metabolites When Fermenting Natural Grape Musts: Effect of Inoculation Treatments and Cultivar on the Occurrence of Melatonin and Related Indolic Compounds. *Aust. J. Grape Wine Res.* **2019**, *25*, 92–100. [CrossRef]
59. Gawel, R.; Day, M.; Van Sluyter, S.C.; Holt, H.; Waters, E.J.; Smith, P.A. White Wine Taste and Mouthfeel as Affected by Juice Extraction and Processing. *J. Agric. Food Chem.* **2014**, *62*, 10008–10014. [CrossRef]

foods

Article

Effect of Air Drying on the Metabolic Profile of Fresh Wild and Artificial *Cordyceps sinensis*

Tao Wang [1], Chuyu Tang [1], Mengjun Xiao [1], Zhengfei Cao [1], Min He [1], Jianzhao Qi [2], Yuling Li [3,*] and Xiuzhang Li [1,*]

[1] State Key Laboratory of Plateau Ecology and Agriculture, Qinghai University, Xining 810016, China; 13085500761@163.com (T.W.); chuyutang0410@163.com (C.T.); 15574237597@163.com (M.X.); c1474477969@163.com (Z.C.); hi-mi1228@163.com (M.H.)
[2] College of Chemistry and Pharmacy, Northwest A&F University, Xianyang 712100, China; qjz@nwafu.edu.cn
[3] Qinghai Academy of Animal and Veterinary Science, Xining 810016, China
* Correspondence: 1991990033@qhu.edu.cn (Y.L.); xiuzhang@qhu.edu.cn (X.L.)

Abstract: Fresh and dried *Cordyceps sinensis* are widely used by the public for medicinal and health purposes. However, the differences between them have not been examined. In this study, fresh wild and artificial *C. sinensis* (WFC and AFC) were dried to obtain dried wild and artificial *C. sinensis* (WDC and ADC). Non-targeted GC-MS was used to analyze the metabolic profile characteristics of the four groups of samples. The results showed that air drying significantly altered the composition and content of *C. sinensis*, mainly in the form of higher abundance of organic acids and derivatives and lower abundance of lipids and lipid-like molecules in fresh *C. sinensis*. Hierarchical cluster analysis (HCA) and quantitative analyses showed that air drying increased the abundance of Valine, Zinniol, Urocanate, Vulpinic acid, and Uridine 5′-diphosphate, and decreased Xanthotoxol, Vitexin-4-o-glucoside, Val-trp, and Wogonin. These differentially accumulated metabolites (DAMs) were also shown to be potential biomarkers for *C. sinensis*. KEGG enrichment analysis identified lysine biosynthesis as the most significantly enriched pathway. Annotation of these DAMs to lysine biosynthesis revealed that citrate cycle and pyruvate metabolism entered lysine biosynthesis via 2-oxohlutarate and Homocitrate, respectively, resulting in significant enrichment of L-saccharopine and L-lysine content was significantly higher. Alanine, aspartate, and Glutamate metabolism synthesized more L-aspartate to promote L-lysine synthesis. Thus, high levels of L-lysine result in lysine degradation and pymolysine, which are the most active metabolic pathways during the drying of fresh *C. sinensis* and indirectly lead to differences in metabolic profiles.

Keywords: *Cordyceps sinensis*; metabonomics; air drying; lysine biosynthesis

Citation: Wang, T.; Tang, C.; Xiao, M.; Cao, Z.; He, M.; Qi, J.; Li, Y.; Li, X. Effect of Air Drying on the Metabolic Profile of Fresh Wild and Artificial *Cordyceps sinensis*. *Foods* **2024**, *13*, 48. https://doi.org/10.3390/foods13010048

Academic Editor: Gianfranco Picone

Received: 13 November 2023
Revised: 12 December 2023
Accepted: 20 December 2023
Published: 21 December 2023

1. Introduction

C. sinensis is a rare traditional Chinese medicinal material. It has extremely high nutritional value and medicinal activity and is regarded as "medicine and food homology" by the National Health Commission of the People's Republic of China [1]. It is mainly distributed in Tibet, Sichuan, Qinghai, Yunnan, and Gansu provinces in China, and grows in alpine meadow areas at 3, 500–5, 200 m. The extremely high altitude leads to the environment presenting low temperatures, low oxygen, variable temperatures, and high radiation and other climatic characteristics; in response to a variety of environmental factors, the metabolic profile of wild *C. sinensis* undergoes drastic changes. This also means that wild *C. sinensis* and artificial *C. sinensis* have different pharmacological functions [2,3]. The "Yuewang Yaozhen" and "Compendium of Materia Medica" recorded its sweet taste, which has the effects of tonifying kidneys and lungs, hemostasis, and resolving phlegm. Dry *C. sinensis* can be used to treat kidney deficiency and essence deficiency, impotence and spermatorrhea, lumbar and knee pain, chronic cough and asthma, and spleen and stomach diseases [4,5]. Modern pharmacological studies have shown that *C. sinensis* has many pharmacological effects, such as immune regulation, anti-tumor, anti-apoptosis, anti-oxidation, relieving renal failure, and treating cardiovascular diseases [6–9]. Its pharmacological effects are closely related to polysaccharides, nucleosides, sterols, flavonoids, cyclic peptides, phenols, anthracenes, polyketides, and alkaloids. A recent study showed that the water extract of *C. sinensis* had a significant anti-inflammatory effect on the infection of the compound influenza virus [10,11]. *C. sinensis* contains a water-insoluble polysaccharide, β-(1,3) glucan. As a component of the fungal cell wall, it is also the core structure of immunoactivity polysaccharides, which stimulates immune receptors such as Dectin-1 to trigger innate immune response and enhance individual immunity [12,13]. Cordycepin, a kind of adenosine, has been shown to inhibit the differentiation of T cells into regulatory T cells (Treg, a suppressive phenotype of T cells) and delay tumor growth in tumor-bearing mice [14,15]. According to the theory of traditional Chinese medicine, the medicinal function of *C. sinensis* lies in the combination of multiple metabolites. A single metabolite may also play an important role, but it cannot fully reflect the medicinal value of *C. sinensis*. Similarly, modern pharmacology and traditional Chinese medicine theory believe that the medicinal function of *C. sinensis* comes from a large number of special metabolites [16]. The difference is that traditional Chinese medicine tends to use dry *C. sinensis* and modern pharmacology chooses fresh *C. sinensis*. This results in confusion for ordinary consumers when choosing fresh or dried *C. sinensis* for physical health care.

The most common processing methods for *C. sinensis* are low-temperature freeze drying, high-temperature drying, and air drying. Studies have shown that preservation at 4 °C can keep *C. militaris* fresh for at least 12 months [17]. Unfortunately, they only analyzed the activity of fungi and the ability to produce fruiting bodies and did not point out the changes in metabolites. In recent years, the emerging vacuum freeze-drying method, in a vacuum low-temperature environment, has freeze dried *C. sinensis* quickly while maintaining the appearance, shape, and color of fresh products, and can well retain the protein type and content of *C. sinensis* and antioxidant activity, such as antioxidant dismutase [18,19]. However, it has high requirements for the operation of instruments and methods and cannot be widely used by consumers. On the contrary, hot air drying is not conducive to the retention of proteins, especially proteins with relative molecular weight of 25–100 kDa, which are prone to structural changes during drying and which directly leads to the loss of their functions [20]. Fortunately, this method retains higher sterols, such as ergosterol, cholesterol, and stigmasterol. The authors conjectured that the change in sterol content in *C. sinensis* under high-temperature treatments enhances cell membrane stability and helps to resist environmental changes [21,22]. The study also pointed out that dried *C. sinensis* products are resistant to storage, not easy to corrupt, have increased flavor, and are more suitable for cooking [23]. At present, hot air drying and steam dehydration drying are the most widely used methods. The advantage of this method is that it is not affected by external climate factors and is more suitable for large-scale continuous factory operations. The disadvantages are that the drying time is long, the loss of nutrients is significant, and the production energy consumption is large and thus does not meet the requirements of energy conservation and emission reduction.

What is encouraging is that the drying method of *C. sinensis* is also stipulated in the "Pharmacopoeia of the People's Republic of China 2020" as air drying and drying at low temperatures [24]. What needs to be considered is whether and how the metabolites of *C. sinensis* will change after air drying and whether these metabolite changes will give fresh and air-dried *C. sinensis* different medicinal values. This is one of the urgent problems in the medicinal field of *C. sinensis*. Based on non-targeted metabolomics, wild and artificial fresh *C. sinensis* were air dried in this study, and the DAMs of *C. sinensis* before and after air drying were measured. Finally, the response of *C. sinensis* to air drying was determined to provide a theoretical basis for its medicinal research and guide consumers to choose the *C. sinensis* corresponding to their needs.

2. Materials and Methods

2.1. Sample Collection

AFC (6 repeats) samples were purchased from Shenzhen Dongyangguang Industrial Development Co., Ltd. in Shenzhen China (https://www.dyg.cn/, accessed on 12 November 2023), and WFC (6 repeats) samples were purchased from Qinghai Qingqitang Trading Co., Ltd. (Xining, China), and were collected from Zaduo county, Yushu city, Qinghai province, China (95°38′12″ E, 33°8′15″ N, elevation: 4436 m). On 20 May 2023, six roots of each of the AFC (6 repeats) and WFC (6 repeats) were taken to be air dried at 22 °C while protected from light and without the application of additional treatments; they were weighed once every 6 h until the mass no longer changed significantly on two consecutive occasions. The moisture content was determined to be 13.23% in accordance with the standard of dried *C. sinensis*, and it was determined that the drying was completed to obtain 6 roots of ADC and WDC, respectively (Figure 1).

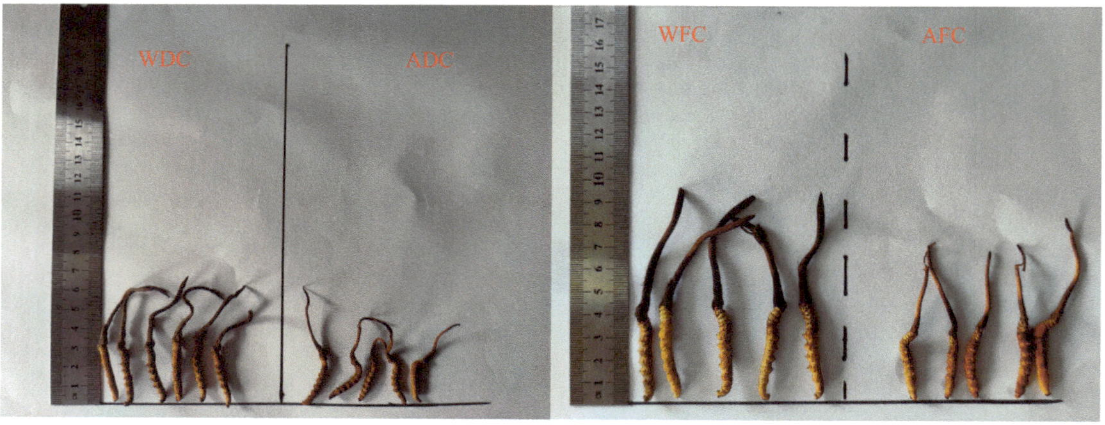

Figure 1. Four groups' samples of *C. sinensis*. (WFC) wild fresh *C. sinensis*; (AFC) artificially fresh *C. sinensis*; (WDC) wildly dried *C. sinensis*; (ADC) artificially dried *C. sinensis*.

2.2. Sample Pretreatment

Four groups of *C. sinensis* samples were rinsed three times using sterile distilled water and then stroma and sclerotia mixture was ground in liquid nitrogen. A 50 mg sample was taken in a 2 mL EP tube and 250 mL of liquid extractant (methanol:water = 4:1) was added and used to present the metabolites. The sample was vortexed and shaken at −10 °C for 5 min (50 Hz) to allow the metabolites to be fully extracted and was then centrifuged at 4 °C, 15,000 rpm, for 20 min. The supernatant was collected through a 5 μm filter membrane to remove the larger particulate matter and microorganisms and transferred to the injection vials for LC-MS analysis, and 6 replicates each [25]. Aliquots of all samples were mixed to make quality control (QC) samples, processed as above, for testing the stability of the system and to ensure reliable data for the assay.

2.3. UPLC-MS/MS Analysis

The chromatographic column of Agilent 1290 Infinity LC (100 nm × 2.1 mm, 1.7 μm) was adopted. The UPLC conditions were column temperature 40 °C, flow rate 0.2 mL·min^{-1}, mobile phase A (H_2O:2 mM ammonium acetate:2 mM ammonia water = 5:47.5), and mobile phase B was acetonitrile. The chromatographic elution procedure in ESI positive mode was: 0–0.5 min, A:B = 5:95, 0.5–7 min, A:B = 35:65, 7–8 min, A:B = 60:40, 8–9 min, A:B = 5:95, 9–12 min, and B was maintained at 95%. The ESI negative mode is: 0–0.5 min, A:B = 5:95, 0.5–7 min, A:B = 10:90, 7–8 min, A:B = 35:65, 8–9 min, A:B = 5:95, 9–12 min, and B was maintained at 95%. Both A and B are linear changes [26].

The samples were kept at 4 °C throughout the injection process, and the stability of the system was tested and evaluated by inserting QC samples into the sample queue using continuous analysis with randomized sequential injection. The measured data were subjected to a simple screening of parameters such as retention time, mass-to-charge ratio, and so on. Then, peak extraction and peak area quantification were performed with a mass deviation of 5 ppm, a signal intensity deviation of 30%, a signal-to-noise ratio of 3, a minimum signal intensity of 100,000, summed ions, etc. The background ions were removed with blank samples, and the molecular ion peaks and fragment ions were compared with the mzCloud (https://www.mzcloud.org/, accessed on 12 November 2023), mzVault, and Masslist databases to predict the molecular formula, and the quantitative results were normalized to obtain the final identification and quantitative results of the data.

The ESI source conditions were set as follows: Ion Source Gas1 (Gas1) as 60, Ion Source Gas2 (Gas2) as 60, curtain gas (CUR) as 30, source temperature: 600 °C, IonSpray Voltage Floating (ISVF) ±5500 V. In MS only acquisition, the instrument was set to acquire over the m/z range 60–1000 Da, and the accumulation time for TOF MS scan was set at 0.20 s/spectra. In auto MS/MS acquisition, the instrument was set to acquire over the m/z range 25–1000 Da, and the accumulation time for product ion scan was set at 0.05 s/spectra. The product ion scan is acquired using information dependent acquisition (IDA) with high-sensitivity mode selected. The parameters were set as follows: the collision energy (CE) was fixed at 35 V with ±15 eV; declustering potential (DP), 60 V (+) and −60 V (−); exclude isotopes within 4 Da, candidate ions to monitor per cycle: 10.

2.4. Data Processing and Metabolite Identification

Data were collected and processed using MassLynx 4.1 for total peak area normalization. Multivariate statistical analyses of the data were performed using the R language ropls package, including principal component analysis (PCA), orthogonal partial least squares-discriminant analysis (OPLS-DA), and S-plot analysis of multiplicity of differences. Differential metabolite data were obtained by screening the minimum criteria for differential metabolite screening by simultaneously satisfying the following conditions: $p \leq 0.05$, variable importance in projection (VIP) > 1, S-Plot > $|0.8|$. The relative quantitative hierarchical cluster plots of metabolites and differential metabolites were performed by the pheatmap program package in Rv 3.3.2. based on the data obtained by MS/MS in KEGG (https://www.genome.jp/kegg/pathway, accessed on 12 November 2023), Metlin (http://metlin.scripps.edu, accessed on 12 November 2023), and MoNA (https://mona.fiehnlab.ucdavis.edu, accessed on 12 November 2023) databases to further match the annotations and obtain accurate information regarding metabolites. Based on the MetPA analysis, the relative response values of metabolic pathways were obtained according to the relative response values of identified metabolites in metabolic pathways and the dimensionality reduction algorithm, from which the correlation coefficients between metabolic pathways were calculated and the metabolic pathway association network diagrams were drawn.

3. Results

3.1. Overview of Metabolite Profiling

In this study, 24 samples were studied, including AFC, WFC, ADC, and WDC. The GC-MS total ion current (TIC) chromatograms of four groups *C. sinensis* samples were obtained by non-targeted metabolomics detection (Figure 2). There were significant differences in the peak, position, and height between the four groups' samples, focusing on the time of 0.5–7 min, and the differential metabolites of the four groups samples were observed. The metabolic profile showed that a total of 2638 metabolites (Table S1) were identified and annotated, which were divided into 18 superclasses and 38 classes (Table S2). The most abundant superclasses were organic acids and derivatives (300 metabolites, 24.4101%), followed by lipids and lipid-like molecules (218 metabolites, 17.7380%), organoheterocyclic compounds (168 metabolites, 13.6697%), benzenoids (151 metabolites, 12.2864%), phenylpropanoids and polyketides (140 metabolites, 11.3914%), and organic oxygen compounds (139 metabolites, 11.3100%) (Figure 3A). The most abundant classes were carboxylic acids and derivatives (249 metabolites, 20.2640%), followed by organooxygen compounds (138 metabolites, 11.2286%), fatty acyls (105 metabolites, 8.5435%), benzene and substituted derivatives (97 metabolites, 7.8926%), prenol lipids (67 metabolites, 5.4516%), flavonoids (56 metabolites, 4.5566%), steroids and steroid derivatives (43 metabolites, 3.4988%), phenols (31 metabolites, 2.5224%), indoles and derivatives (25 metabolites, 2.0342%), and pyrimidine nucleosides (24 metabolites, 1.9528%). This is consistent with the main metabolite types of fungi.

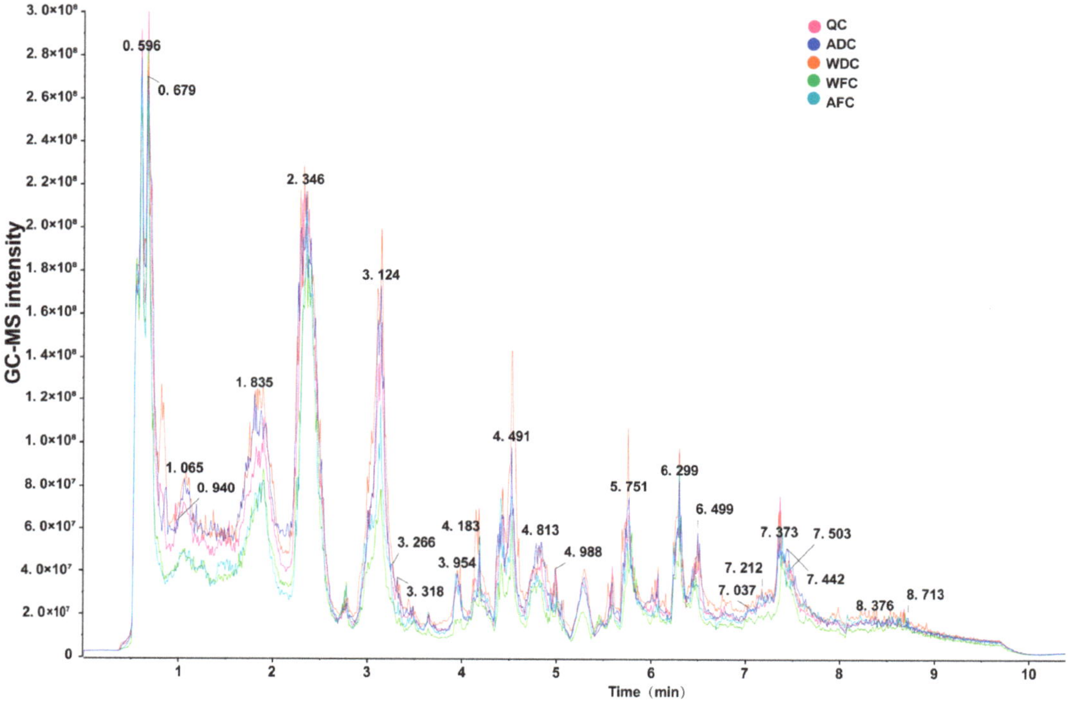

Figure 2. Total ion chromatograms of QC (quality control), ADC, WDC, WFC, and AFC non-targeted metabolomics methods. The numbers in the figure are retention times.

For the expression of the top six superclasses, differential expression analysis was performed in four groups samples (Figure 3B). The results showed that fresh *C. sinensis* contained more organic acids and derivatives and lower abundance of lipids and lipid-like molecules ($p < 0.05$) than dried *C. sinensis* (WFC vs. WDC, AFC vs. ADC). It is worth noting that WFC lacks organoheterocyclic compounds and organic oxygen compounds.

3.2. Multivariate Statistical Analysis

To determine the differences in metabolites among the four groups samples, a multivariate statistical analysis was performed using MS-normalized peak intensity as an indicator of metabolite abundance. Based on the obtained metabolite data set, the dimensionality reduction results of PCA (principal component analysis) showed that ADC, AFC, WDC, and WFC were significantly divided into four groups samples with a relatively long distance (Figure 4A), indicating that the metabolites of *C. sinensis* from different sources were significantly different. Air drying had a great effect on the metabolites of *C. sinensis*. Cluster analysis and UMAP (uniform manifold approximation and projection) found that ADC and WDC cluster together, and AFC and WFC cluster together (Figure 4B,C).

This showed that the DAMs between fresh (WFC and AFC) and dried (WDC and ADC) *C. sinensis* were significant.

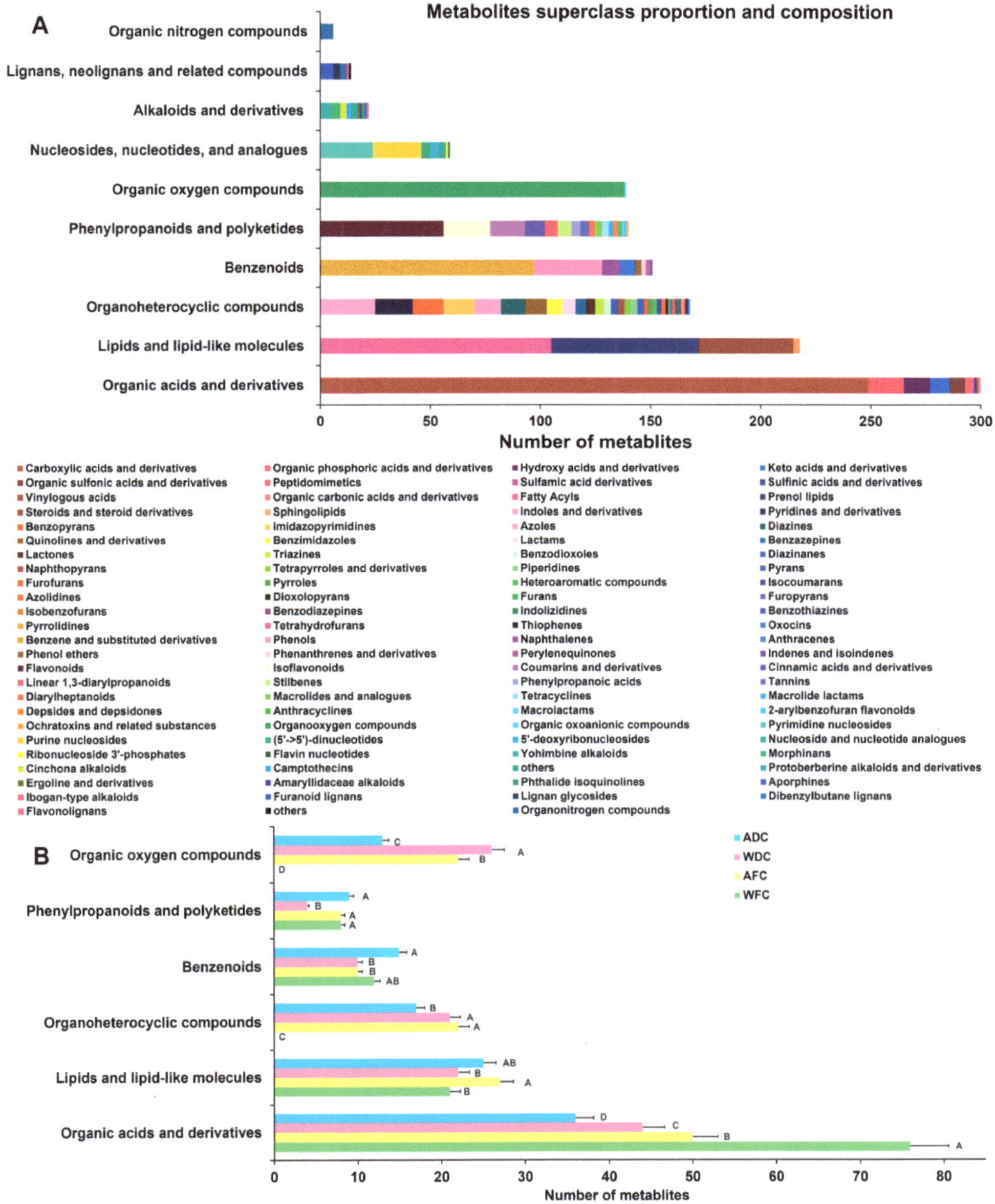

Figure 3. (**A**) Classification histogram of metabolites identified in the top 10 superclasses. (**B**) The expression of the top six superclasses in the four groups' sample. Different capital letters in the same group showed significant differences at the level of $p < 0.05$.

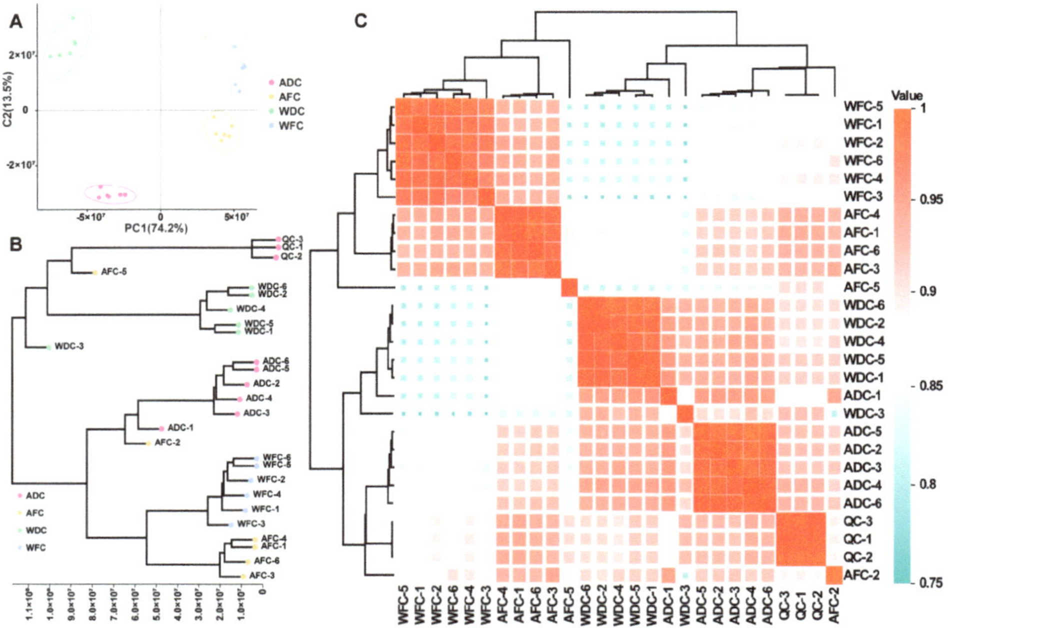

Figure 4. Multivariate statistical analysis of four groups samples. (**A**) PCA analysis of four groups samples. (**B**) Cluster analysis of four groups' samples. (**C**) UMAP of four groups samples.

In PCA analysis, the first two PCs accounted for 87.7% of the total variance (PC1 = 74.2%, PC2 = 13.5%), indicating that the separation of metabolites from samples was clean, and the effect of principal component analysis was better. However, PCA analysis cannot remove intra-group errors and random errors, cannot overcome the integrity of the data, and cannot accurately distinguish between-group differences and differential metabolites between samples. OPLS-DA is a supervised discriminant analysis statistical method. By removing the data changes unrelated to the categorical variable Y in the independent variable X, the classification information is mainly concentrated on one principal component. In this study, OPLS-DA can better obtain the difference information between groups and predict the grouping of samples by establishing a model between metabolite expression and grouping relationship. In the OPLS-DA score plot, different species of *C. sinensis* were significantly distinguished (Figure S1A,B), which was consistent with the results of PCA. It was verified that the metabolites of the four groups samples were significantly different.

In the OPLS-DA score plot, fresh and dried *C. sinensis* were significantly distinguished, indicating that there were significant differences in the metabolic profiles of fresh and dried *C. sinensis*. The OPLS-DA model had high interpretability ($R^2_{X(\text{WFC vs. WDC})} = 0.832$, $R^2_{Y(\text{WFC vs. WDC})} = 1$, $R^2_{X(\text{AFC vs. ADC})} = 0.811$, and $R^2_{Y(\text{AFC vs. ADC})} = 0.997$) and predictability ($Q^2_{\text{WFC vs. WDC}} = 0.997$, and $Q^2_{\text{AFC vs. ADC}} = 0.997$). After 200 permutation tests (Figure S1C,D), it was found that all green Q^2 values on the left side were less than the yellow R^2 values, and the validation intercepts of R^2 and Q^2 were $R^2_{\text{WFC vs. AFC}} = 0.86$, $Q^2_{\text{WFC vs. AFC}} = -0.15$, $R^2_{\text{WDC vs. ADC}} = 0.82$, and $Q^2_{\text{WDC vs. ADC}} = -0.08$, respectively, indicating that the model was not overfitted and the results were reliable. Compared with WDC, WFC has more abundant Lpc 18:2, Acetyl coenzyme a, Glycerophosphocholine, Medermycin, N-n′-diphenyl-p-phenylenediamine, Ng, ng-dimethyl-l-arginine, and Chlormadinone acetate. In WDC, Oleamide, L-arginine, Tyramine, Adenosine, Ser-leu-ile-gly-lys-val-amide, Dehydrophtosphingosine, and S-methyl-5′-thioadensine were more abundant (Figure S1E). AFC is rich in betaine, Lpc 18:2, Medermycin, L-carnitine, Fluvoxamine, Acetyl coenzyme a, and N-N′-diphenyl-p-phenylenediamine. In ADC, Chlormadinone acetate, Fenpropidin, Glycerophosphocholine, S-methyl-5′-thioadensine, Oleamide, His-arg, Adenosine, Terfenadine, and Acetylcarnitine were detected more (Figure S1F).

The results showed that the abundance of Lpc 18:2, Acetyl coenzyme a, and N-N′-diphenyl-p-phenylenediamine in fresh *C. sinensis* was higher. Oleamide and Adenosine abundance was higher in dried *C. sinensis*. These DAMs may be involved in the air-drying process.

3.3. Screening of DAMs and Hierarchical Cluster Analysis (HCA)

The screening conditions for significant DAMs were VIP ≥ 1 and *T*-test $p < 0.5$. To visualize the similarities and differences in metabolites in four groups samples, scatter plots and Venn diagrams of DAMs were used to describe them in detail (Figure 5A,B). By comparing DAMs in fresh and air-dried *C. sinensis* samples, a total of 427 DAMs (Tables S3 and S4) were detected, 237 were upregulated and 190 were downregulated. Among them, organic acids and derivatives, and lipids and lipid-like molecules are the main superclasses. The contents of fatty acids and conjugates, linolenic acids and derivatives, and amino acids, peptides, and analogues in fresh *C. sinensis* samples were higher than those in dried *C. sinensis* samples, such as Oleic acid, Linoleic acid, Phenylalanine, etc. On the other hand, levels of carbohydrates and carbohydrate conjugates, pyrimidine nucleosides, and cinchona alkaloids decreased. For example, Hydroquinidine, Uridine, Uracild, etc. (Tables S5 and S6).

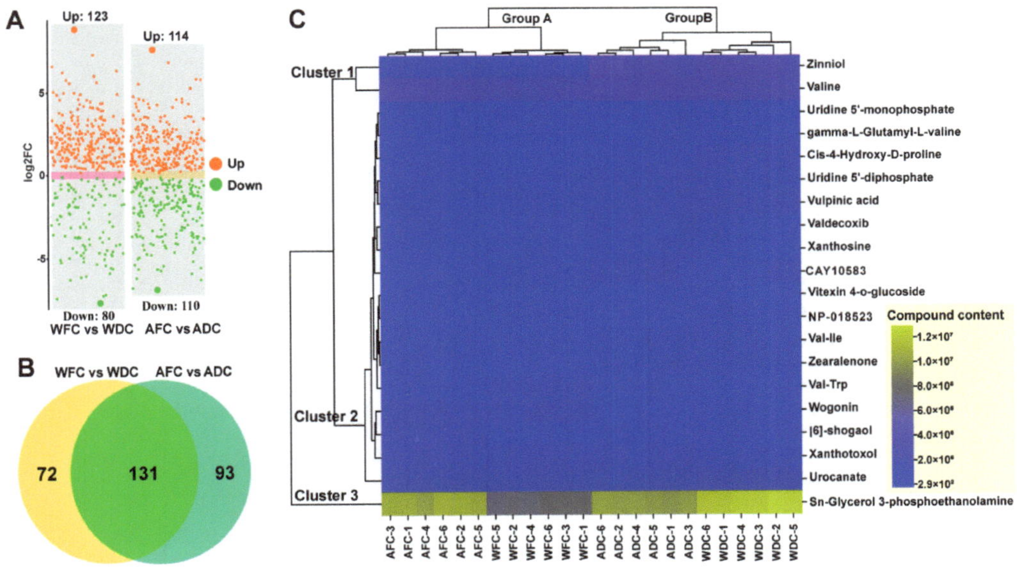

Figure 5. DAMs analysis of the four groups' samples. (**A**) scatterplot of DAMs. (**B**) Numbers of upregulated and downregulated DAMs in four groups samples. (**C**) Dendrogram and heatmap of four groups samples (*C. sinensis*) from different origins based on top 20 differential metabolites content.

This indicated that the air-drying process led to dramatic changes in the metabolites of fresh *C. sinensis*. Such changes may lead to changes in the pharmacological effects of fresh and dried *C. sinensis*. This study reported the types and quantities of metabolic variations, which will contribute to the further study of its pharmacological functions.

Based on the top 20 metabolites, the HCA of four groups sample metabolites was carried out. Different metabolites were clustered into one class, which showed that the expression patterns of different samples were closer. The clustering results of the differences showed that the expression of different samples was different among the samples. The four groups' samples were clustered into two groups. Group A is AFC and WFC, and group B is ADC and WDC, which is consistent with the results of previous sample clustering and UMAP analysis (Figure 4B,C). This indicated that there were significant differences in metabolites between fresh samples and air-dried samples, but the expression of metabolites in each group was similar. At the same time, it can be divided into three clusters according to the content of metabolites in the sample (Figure 5C). Cluster 1 had higher metabolite content, such as Zinniol and Valine, cluster 2 had lower overall content, and cluster 3 had only Sn-glycerol-3-phosphethanolamine, but its content was the highest. To further compare the differences in the content of these metabolites, it is necessary to analyze them quantitatively.

The results showed that the abundances of Valine, Zinniol, Lysine, Cay10583, Vulpinic acid, Uridine 5′-diphosphate, Uridine 5′-monophosphate, and Val-ile were significantly higher in dried *C. sinensis* than in fresh *C. sinensis* ($p < 0.05$) (Table 1). This suggests that drying increases the levels of these metabolites. In contrast, the contents of Xanthotoxol, Vitexin-4-o-glucoside, Val-trp, and Wogonin were higher in fresh *C. sinensis*. This shows that these metabolites are destroyed by air drying.

Table 1. The relative abundance of metabolite top 20 was differentially expressed in four groups samples of *C. sinensis*. Capital letters indicate a significant difference at the $p < 0.05$ level.

Compound	ADC (Relative Abundance)	AFC (Relative Abundance)	WDC (Relative Abundance)	WFC (Relative Abundance)
Sn-Glycerol 3-phosphoethanolamine	$1.06 \times 10^7 \pm 9.48 \times 10^4$ C	$1.11 \times 10^7 \pm 9.17 \times 10^4$ B	$1.28 \times 10^7 \pm 1.43 \times 10^5$ A	$7.73 \times 10^6 \pm 9.88 \times 10^4$ D
Valine	$3.61 \times 10^6 \pm 2.86 \times 10^4$ A	$3.34 \times 10^6 \pm 1.22 \times 10^5$ B	$3.26 \times 10^6 \pm 3.88 \times 10^4$ B	$2.34 \times 10^6 \pm 3.32 \times 10^4$ C
Zinniol	$3.72 \times 10^6 \pm 3.04 \times 10^5$ A	$7.05 \times 10^5 \pm 2.70 \times 10^4$ B	$3.59 \times 10^6 \pm 2.18 \times 10^5$ A	$8.88 \times 10^5 \pm 7.58 \times 10^4$ B
Xanthotoxol	$6.65 \times 10^5 \pm 3.63 \times 10^4$ B	$6.70 \times 10^5 \pm 8.54 \times 10^4$ B	$5.22 \times 10^5 \pm 1.65 \times 10^4$ C	$7.80 \times 10^5 \pm 3.40 \times 10^4$ A
NP-018523	$1.20 \times 10^4 \pm 7.87 \times 10^2$ C	$8.15 \times 10^2 \pm 1.26 \times 10^2$ D	$1.52 \times 10^5 \pm 6.86 \times 10^3$ A	$7.60 \times 10^4 \pm 2.56 \times 10^3$ B
Urocanate	$1.21 \times 10^6 \pm 2018 \times 10^4$ B	$9.68 \times 10^5 \pm 1.20 \times 10^4$ C	$1.45 \times 10^6 \pm 1.40 \times 10^4$ A	$6.61 \times 10^5 \pm 9.16 \times 10^3$ D
CAY-10583	$6.01 \times 10^5 \pm 1.20 \times 10^5$ A	$2.00 \times 10^4 \pm 1.47 \times 10^3$ C	$3.23 \times 10^5 \pm 2.01 \times 10^5$ AB	$7.28 \times 10^4 \pm 4.19 \times 10^3$ B
Cis-4-Hydroxy-D-proline	$2.37 \times 10^5 \pm 2.32 \times 10^3$ C	$2.96 \times 10^5 \pm 2.72 \times 10^3$ B	$6.56 \times 10^5 \pm 1.52 \times 10^4$ A	$7.11 \times 10^4 \pm 8.93 \times 10^2$ D
Vulpinic acid	$5.63 \times 10^5 \pm 3.23 \times 10^4$ B	$1.41 \times 10^5 \pm 1.41 \times 10^3$ C	$7.13 \times 10^5 \pm 4.46 \times 10^4$ A	$1.77 \times 10^5 \pm 4.91 \times 10^3$ C
Uridine 5′-diphosphate	$5.09 \times 10^5 \pm 8.44 \times 10^3$ A	$2.78 \times 10^5 \pm 8.31 \times 10^3$ B	$4.93 \times 10^5 \pm 1.72 \times 10^4$ A	$1.52 \times 10^5 \pm 5.17 \times 10^3$ C
[6]-shogaol	$3.85 \times 10^5 \pm 3.28 \times 10^4$ B	$8.80 \times 10^5 \pm 1.48 \times 10^4$ A	$2.35 \times 10^5 \pm 1.07 \times 10^4$ C	$9.03 \times 10^4 \pm 6.88 \times 10^3$ D
γ-L-Glutamyl-L-valine	$2.82 \times 10^5 \pm 8.87 \times 10^3$ C	$3.96 \times 10^5 \pm 1.77 \times 10^4$ A	$3.38 \times 10^5 \pm 1.62 \times 10^4$ B	$1.72 \times 10^5 \pm 7.52 \times 10^3$ D
Uridine 5′-monophosphate	$4.39 \times 10^5 \pm 5.62 \times 10^3$ A	$3.69 \times 10^5 \pm 2.87 \times 10^3$ B	$3.67 \times 10^5 \pm 5.16 \times 10^3$ B	$2.04 \times 10^5 \pm 1.84 \times 10^3$ C
Xanthosine	$4.60 \times 10^5 \pm 1.82 \times 10^4$ A	$8.98 \times 10^4 \pm 1.27 \times 10^4$ B	$5.87 \times 10^4 \pm 3.18 \times 10^3$ C	$6.85 \times 10^4 \pm 5.10 \times 10^3$ BC
Zearalenone	$1.09 \times 10^5 \pm 1.39 \times 10^4$ B	$1.68 \times 10^5 \pm 1.60 \times 10^4$ A	$1.71 \times 10^5 \pm 2.20 \times 10^3$ A	$7.05 \times 10^4 \pm 3.76 \times 10^3$ C
Val-Ile	$1.29 \times 10^5 \pm 3.11 \times 10^3$ B	$6.93 \times 10^4 \pm 2.17 \times 10^3$ D	$2.75 \times 10^5 \pm 6.30 \times 10^3$ A	$1.06 \times 10^5 \pm 2.81 \times 10^3$ C
Vitexin 4-o-glucoside	$1.89 \times 10^4 \pm 9.78 \times 10^3$ B	$9.92 \times 10^4 \pm 9.71 \times 10^2$ A	$1.33 \times 10^4 \pm 5.41 \times 10^3$ B	$1.38 \times 10^4 \pm 3.66 \times 10^2$ B
Val-Trp	$9.89 \times 10^3 \pm 1.83 \times 10^3$ C	$1.54 \times 10^5 \pm 9.06 \times 10^3$ B	$3.16 \times 10^4 \pm 3.49 \times 10^3$ C	$5.17 \times 10^5 \pm 1.01 \times 10^4$ A
Wogonin	$7.53 \times 10^4 \pm 6.10 \times 10^3$ B	$6.88 \times 10^5 \pm 5.49 \times 10^4$ A	$6.62 \times 10^4 \pm 8.28 \times 10^3$ B	$6.12 \times 10^5 \pm 2.39 \times 10^4$ A
Valdecoxib	$3.45 \times 10^5 \pm 1.09 \times 10^5$ A	$2.00 \times 10^5 \pm 6.26 \times 10^4$ AB	$1.97 \times 10^5 \pm 8.74 \times 10^4$ AB	$1.15 \times 10^4 \pm 3.63 \times 10^2$ B

3.4. KEGG Annotation and Metabolite Enrichment Analysis

KEGG is a database that contains rich information about metabolite pathways and interactions. The pathway information may explain the biochemical function of compounds that can be activated by different metabolites. The first 20 enriched KEGG results showed that the DAMs contained in metabolic pathways were the most abundant in WFC vs. WDC and AFC vs. ADC (Figure 6). In addition, biosynthesis of plant secondary metabolites, lysine biosynthesis, glycerolipid metabolism, cysteine and methionine metabolism, fatty acid biosynthesis, starch and sucrose metabolism, and taurine and hypotaurine metabolism are common enrichment pathways. This suggests that these pathways are highly responsive metabolic pathways in air-dried samples and that lysine biosynthesis is one of the most significant pathways. The author constructed Figure 7 with reference to the pathway of lysine biosynthesis in KEGG (Figure S2). Among them, blue and red highlight the metabolites annotated by KEGG. The results showed that after air drying, the citrate cycle and pyruvate metabolism entered lysine biosynthesis through 2-oxoglutarate and Homocitrate, respectively, resulting in a significant increase in L-saccharopine and L-lysine content. Alanine, aspartate, and glutamate metabolism synthesize more L-aspartate to promote the synthesis of L-lysine. The high content of L-lysine entered lysine degradation and pyrrolysine to resist air drying.

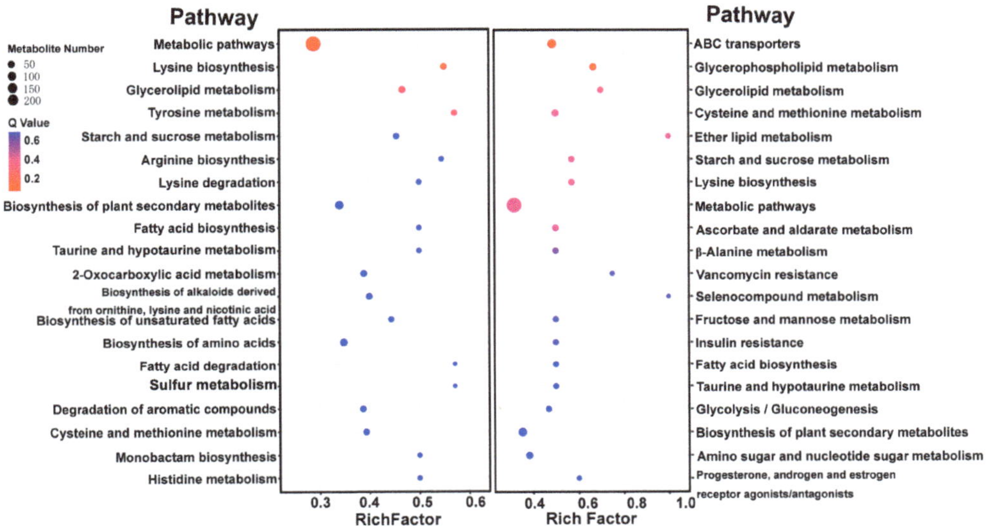

Figure 6. Differentially expressed metabolic pathway (top 20) in the four groups' samples of *C. sinensis*. On the left is WFC vs. ADC and on the right is AFC vs. ADC.

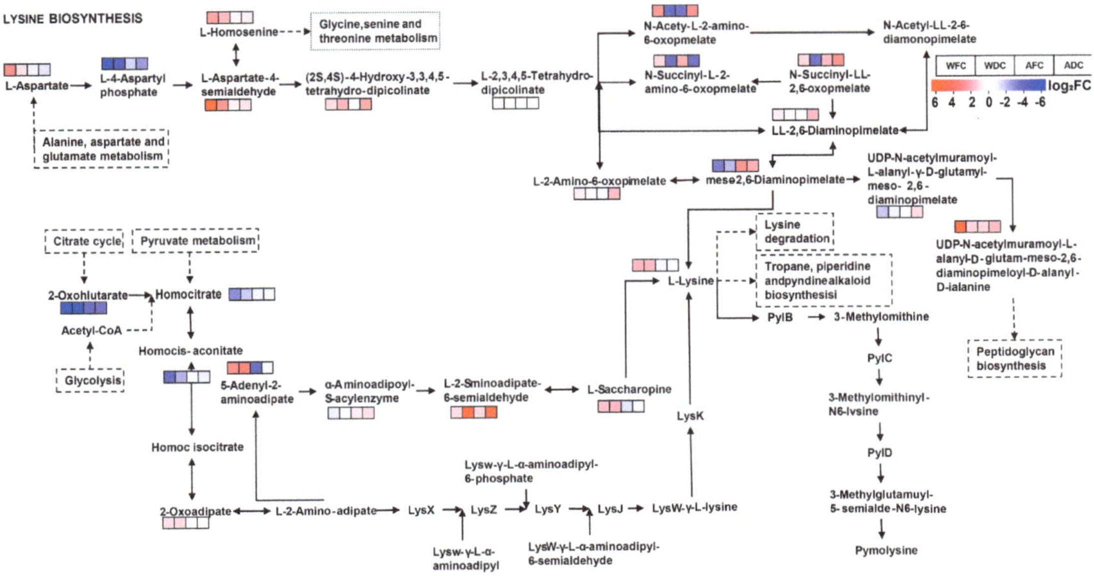

Figure 7. KEGG maps of key DAMs in the four groups' samples of *C. sinensis*. The colored box in front of each metabolite indicates the corresponding $\log_2$ FC value. The dashed boxes indicate different metabolic pathways.

4. Discussion

C. sinensis is widely loved by the public and is of interest to pharmacologists because of its diversity of metabolites and high medicinal value. However, scholars have focused most of their attention on fresh *C. sinensis* and have not performed much research on dried *C. sinensis*. The original source of Chinese medicinal materials is dried *C. sinensis*. Therefore, a comparative analysis of the metabolites of fresh and dried *C. sinensis* was conducted to clarify their metabolic characteristics and to speculate on their possible medicinal value and to ultimately meet the needs of the public. The results showed that fresh *C. sinensis* was richer in organic acids and derivatives while the abundance of lipids and lipid-like molecules was higher in dried *C. sinensis*. There was no significant difference in benzenoid, phenylpropanoid, or polyketide contents. It should be noted that dried *C. sinensis* may either lack or contain only a small amount of organic heterocyclic compounds, organic oxygen compounds, or other metabolites.

It has been reported that the loss of volatile compounds, such as organic acids and their derivatives, organic heterocyclic compounds and organic oxygen compounds (which are prone to decomposition reactions), phenols, acids, and alkanes, exacerbates the decrease in their abundance in dry environments [27,28]. In contrast, lipids and lipid-like molecules, due to their stable chemical structure, undergo only slow oxidation during the drying process to form short-chain acids, short-chain aldehydes, and short-chain ketones, maintaining a high abundance while promoting amino acid metabolism and carbohydrate metabolism [29–31]. Finally, there was a decrease in the abundance of amino acids, peptides, and analogues and an increase in the abundance of fatty acids and conjugates as well as lineolic acids and their derivatives [32,33]. Finally, Valine, Zinniol, Lysine, Cay10583, Vulpinic acid, Uridine 5′-diphosphate, Uridine 5′-monophosphate, and Valile were enriched in dried *C. sinensis*. The abundance of Xanthotoxol, Vitexin-4-o-glucoside, Val-trp, and Wogonin decreased. Strong lipid oxidation occurs in frozen (−20 °C) fresh *C. sinensis* [34], which promotes the synthesis of various amino acids, aldehydes, and ketones and releases energy to resist low-temperature environments. The drying process may not activate this process, and the abundance is relatively stable [35]. In addition, it has also been reported that amino acids and polysaccharides are stress-responsive substances, such as proline, arginine, citrulline, and tryptophan [36–40], which can significantly improve the individual's ability to resist stress [41,42].

These studies have indicated that amino acid metabolism is active in the process of adversity, including but not limited to citrulline metabolism, arginine metabolism, histidine metabolism, tryptophan metabolism, glycine metabolism, and proline metabolism [43–47], resulting in dramatic changes in metabolites. In addition, this study also found that the content of valine and lysine in dried *C. sinensis* was higher, and more importantly, lysine biosynthesis was one of the significant enrichment pathways for drying.

It has been shown that lysine can improve the oxygen free-radical scavenging ability of individuals, inhibit the increase in reactive oxygen species and its possible subsequent damage, and can serve as an essential antioxidant [48,49]. It can also fight viral pathogens by destroying cell membrane structure and can improve T-cell lymphocytes to maintain non-innate immune activity, with strong antibacterial activity and the ability to regulate individual immunity [50]. In vitro mouse models have also demonstrated anti-apoptotic effects in lysine, reducing apoptosis of primary cardiomyocytes by eliminating the harmful effects of H_2O_2 [51]. All the above studies have demonstrated that lysine has the pharmacological effect of improving individual immune capacity and treating heart failure, while dried *C. sinensis* has higher lysine abundance and lysine biosynthesis intensity. Therefore, when the above symptoms occur, drying *C. sinensis* is recommended first.

Whether it is species or content, there are considerable differences in metabolites between fresh and dried *C. sinensis*, resulting in differences in their medicinal functions. Adenosine, cordycepin, cordycepin acid, and polysaccharides are generally considered to be the main medicinal components of *C. sinensis* [52]. Among them, adenosine content was used as a biomarker to evaluate the quality of *C. sinensis*. Adenosine can effectively inhibit the excessive excitation of neurons in the central nervous system and play an anticonvulsant role by inhibiting the release of synaptic front-end neurotransmitters [53]. The molecular structure of 3′-deoxyadenosine is CsHyON, which is essentially one of the derivatives of adenosine and belongs to nucleosides, nucleotides, and analogues [54]. Studies have confirmed that nucleosides, nucleotides, and analogues have strong bactericidal, antiviral, and anticancer potential [55]. This study found that there was no significant difference between fresh and dried *C. sinensis* (Tables S5 and S6). Therefore, both fresh and wild *C. sinensis* may be selected for the of viral diseases or cancer. For the sake of economic cost, fresh *C. sinensis* is recommended.

In summary, air drying significantly changed the metabolic profile of *C. sinensis* by enhancing lysine biosynthesis, resulting in different pharmacological effects and medicinal values of fresh and air-dried *C. sinensis*. This study analyzed these metabolites and possible medicinal differences in detail, providing scientific guidance for the public's choice of medicine and health care.

5. Conclusions

In this study, we compared the metabolic profiles of dried and fresh samples of wild and artificial *C. sinensis* and found differences in the metabolites of fresh and air-dried *C. sinensis*. Air drying significantly altered the composition and content of *C. sinensis*, which was mainly characterized by higher abundance of organic acids and derivative and lower abundance of lipids and lipid-like molecules in fresh *C. sinensis*. A total of 427 DAMs were identified, of which 237 were upregulated in expression, mainly in the following areas: carbohydrates and carbohydrate conjugates, pyrimidine nucleosides, and cinchona alkaloids. The expression of 190 DAMs was downregulated, mainly in fatty acids and conjugates, lineolic acids and derivatives, amino acids, peptides, and analogues. HCA and quantitative analyses showed that air drying increased levels of Valine, Zinniol, Urocanate, Vulpinic acid, and Uridine 5′-diphosphate and decreased levels of Xanthotoxol, Vitexin-4-o-glucoside, Val-trp, and Wogonin. These DAMs were also shown to be potential biomarkers for *C. sinensis*. KEGG enrichment analysis revealed that lysine biosynthesis was the most significantly enriched pathway. Annotation of these DAMs to lysine biosynthesis revealed that citrate cycle and pyruvate metabolism enter lysine biosynthesis via 2-oxohlutarate and Homocitrate, respectively, resulting in significantly higher L-saccharopine and L-lysine content. Alanine, Aspartate, and Glutamate metabolism synthesized more L-aspartate to promote L-lysine synthesis. Thus, high levels of L-lysine go into lysine degradation, and pymolysine is thus the most active metabolic pathway during the drying of fresh *C. sinensis* and indirectly leads to differences in metabolic profiles.

Supplementary Materials: The following supporting information can be downloaded at: https://www.mdpi.com/article/10.3390/foods13010048/s1, Figure S1: (A,B) OPLS-DA score plot and ellipses represent the 95% confidence region for Hoteling's T^2. (C,D) Statistical validation of the OPLS-DA model using 200 permutation analysis. (E,F) Corresponding S-plot of OPLS-DA. The green value represents the VIP $\geq$ 1. (A,C,E): WFC vs. WDC; (B,D,F): AFC vs. ADC; Figure S2: Kegg pathway of lysine biosynthesis; Table S1: Detailed information of metabolites detected in ADCs, AFCs, WDCs, and WFCs; Table S2: Detailed classification information of metabolites detected in ADCs, AFCs, WDCs, and WFCs; Table S3: Details of the differential metabolites detected between WFCs and WDCs; Table S4: Details of the differential metabolites detected between AFCs and ADCs; Table S5: Differential metabolite classification information between WFCs and WDCs; Table S6: Differential metabolite classification information between AFCs and ADCs.

Author Contributions: T.W. designed the study, collected, and analyzed the original data, and wrote the manuscript; J.Q., Y.L. and X.L. participated in data analysis and manuscript revision and directed the study; and C.T., M.X., Z.C. and M.H. participated in the production of the graphs and provided the analysis methods for TB tools-II software. All authors have read and agreed to the published version of the manuscript.

Funding: This research was funded by the major science and technology projects of Qinghai Province [2021-SF-A4]; and the Chinese Academy of Sciences—People's Government of Qinghai Province on Sanjiangyuan National Park (LHZX-2022-01).

Institutional Review Board Statement: Not applicable.

Informed Consent Statement: Not applicable.

Data Availability Statement: Data are contained within the article and Supplementary Materials.

Acknowledgments: The authors thank Qi's team from Northwest A&F University for their suggestions on the experiment. And thanks to K.D. (Kejia De) of Qinghai University for the experimental equipment.

Conflicts of Interest: The authors declare no conflict of interest.

References

1. Hou, Y.; Jiang, J.G. Origin and concept of medicine food homology and its application in modern functional foods. *Food Funct.* **2013**, *12*, 1727–1741. [CrossRef] [PubMed]
2. Shalik, R.S.; Maan, B.R.; Zuzana, M.; Eryuan, L. Habitat Ecology of *Ophiocordyceps sinensis* in Western Nepal. *Mt. Res. Dev.* **2017**, *37*, 216–223.
3. Liu, L.; Guan, L.L.; Zhao, H.X.; Huang, Y.; Mou, Q.Y.; Liu, K.; Chen, T.Y.; Wang, X.Y.; Zhang, Y.; Wei, B.; et al. Modeling habitat suitability of Houttuynia cordata Thunb (Ceercao) using MaxEnt under climate change in China. *Ecol. Inform.* **2021**, *63*, 101324. [CrossRef]
4. Zheng, Y.L.; Mei, Q.X.; Li, W.J.; Tang, Z.F.; Qian, Z.M.; Tao, S.C.; Dong, P.P. Discussion on the medicinal history and modern administration method of *Cordyceps sinensis*. *J. Chin. Med. Mater.* **2017**, *40*, 2722–2725.
5. Li, G.R.; Guo, Y.F.; Yu, J.F.; Liang, G.H.; Fang, Q.M.; Li, W.J. Effects of different drying processes on quality of Chinese Cordyceps. *J. Fungal Res.* **2020**, *18*, 132–138.
6. Wang, S.Y.; Shiao, M.S. Pharmacological functions of Chinese medicinal fungus *Cordyceps sinensis* and related species. *J. Food Drug Anal.* **2020**, *8*, e15. [CrossRef]
7. Xu, J.; Huang, Y.; Chen, X.X.; Zheng, S.C.; Chen, P.; Mo, M.H. The mechanisms of pharmacological activities of *Ophiocordyceps sinensis* fungi. *Phytother. Res.* **2016**, *30*, 1572–1583. [CrossRef]
8. Opeyemi, J.O.; Tang, J.; Tola, A.; Auberon, F.; Omolara, O.; Zhen, O.Y. The genus Cordyceps: An extensive review of its traditional uses, phytochemistry and pharmacology. *Fitoterapia* **2018**, *129*, 293–316.
9. Ng, T.B.; Wang, H.X. Pharmacological actions of *Cordyceps*, a prized folk medicine. *J. Pharm. Pharmacol.* **2005**, *57*, 1509–1519. [CrossRef]
10. Li, S.P.; Su, Z.R.; Dong, T.X.; Tsim, K.W.K. The fruiting body and its caterpillar host of *Cordyceps sinensis* show close resemblance in main constituents and anti-oxidation activity. *Phytomedicine* **2002**, *9*, 319–324. [CrossRef]
11. Wang, J.Q.; Nie, S.P.; Kan, L.J.; Chen, H.H.; Cui, S.W.; Phillips, A.O.; Phillips, G.O.; Xie, M.Y. Comparison of structural features and antioxidant activity of polysaccharides from natural and cultured *Cordyceps sinensis*. *Food Sci. Biotechnol.* **2017**, *26*, 55–62. [CrossRef] [PubMed]
12. Zhu, Y.L.; Yu, X.F.; Ge, Q.; Li, J.; Wang, D.J.; Wei, Y.; Zhen, O.Y. Antioxidant and anti-aging activities of polysaccharides from *Cordyceps cicadae*. *Int. J. Biol. Macromol.* **2020**, *157*, 394–400. [CrossRef] [PubMed]
13. Chen, S.P.; Wang, J.Q.; Fang, Q.Y.; Dong, N.; Nie, S.P. Polysaccharide from natural *Cordyceps sinensis* ameliorated intestinal injury and enhanced antioxidant activity in immunosuppressed mice. *Food Hydrocoll.* **2019**, *89*, 661–667. [CrossRef]
14. Jin, Y.; Meng, X.; Qiu, Z.D.; Su, Y.P.; Yu, P.; Qu, P. Anti-tumor and anti-metastatic roles of cordycepin, one bioactive compound of *Cordyceps militaris*. *Saudi J. Biol. Sci.* **2018**, *25*, 991–995. [CrossRef]
15. Qian, G.M.; Pan, G.F.; Guo, J.Y. Anti-inflammatory and antinociceptive effects of cordymin, a peptide purified from the medicinal mushroom *Cordyceps sinensis*. *Nat. Prod. Res.* **2012**, *26*, 2358–2362. [CrossRef] [PubMed]
16. Cui, H.R.; Zhang, J.Y.; Cheng, X.H.; Zheng, J.X.; Zhang, Q.; Zheng, R.; You, L.Z.; Han, D.R.; Shang, H.C. Immunometabolism at the service of traditional Chinese medicine. *Pharmacol. Res.* **2022**, *176*, e106081. [CrossRef] [PubMed]
17. Sun, H.; Hu, T.; Guo, Y.B.; Liang, Y. Preservation affects the vegetative growth and fruiting body production of *Cordyceps militaris*. *World J. Microbiol. Biotechnol.* **2018**, *34*, e166. [CrossRef]
18. Shi, H.; Zhang, M.; Devahastin, S. New development of efficient processing techniques on typical medicinal fungi: A Review. *Food Rev. Int.* **2019**, *36*, 39–57. [CrossRef]
19. Tian, Y.; Li, D.M.; Luo, W.H.; Zhu, Z.W.; Li, W.J.; Qian, Z.M.; Li, G.R.; Sun, D.W. Rapid freezing using atomized liquid nitrogen spray followed by frozen storage below glass transition temperature for *Cordyceps sinensis* preservation: Quality attributes and storage stability. *LWT* **2020**, *123*, 109066. [CrossRef]
20. Qian, Z.M.; Fan, J.J.; Li, C.H.; Yang, F.Q.; Li, W.J.; Zhong, Z.G.; Yao, Z.Z. Effects of different drying conditions on protein in *Cordyceps*. *China J. Chin. Mater. Med.* **2019**, *44*, 1983–1988.

21. Bok, J.W.; Lermer, L.; Chilton, J.; Klingeman, H.G.; Towers, G.N. Antitumor sterols from the mycelia of *Cordyceps sinensis*. *Phytochemistry* **1999**, *51*, 891–898. [CrossRef] [PubMed]
22. Yuan, J.P.; Wang, J.H.; Liu, X.; Kuang, H.C.; Zhao, S.Y. Simultaneous determination of free ergosterol and ergosteryl esters in *Cordyceps sinensis* by HPLC. *Food Chem.* **2007**, *105*, 1755–1759. [CrossRef]
23. Barido, F.H.; Jang, A.; Pak, J.I.; Kim, D.Y.; Lee, S.K. Investigation of taste-related compounds and antioxidative profiles of retorted samgyetang made from fresh and dried *Cordyceps militaris* mushrooms. *Food Sci. Anim. Resour.* **2020**, *40*, 772–784. [CrossRef] [PubMed]
24. Chinese Pharmacopoeia Commission. *Pharmacopoeia of the People's Republic of China 2020*; China Medical Science Press: Beijing, China, 2020; pp. 115–117.
25. Want, E.J.; Masson, P.; Michopoulos, F.; Wilson, I.D.; Theodoridis, G.; Plumb, R.S.; Shockcor, J.; Loftus, N.; Holmes, E.; Nicholson, J.K. Global metabolic profiling of animal and human tissues via UPLC-MS. *Nat. Protoc.* **2012**, *8*, 17–32. [CrossRef]
26. Barri, T.; Dragsted, L.O. UPLC-ESI-QTOF/MS and multivariate data analysis for blood plasma and serum metabolomics: Effect of experimental artifacts and anticoagulant. *Anal. Chim. Acta* **2013**, *768*, 118–128. [CrossRef]
27. Yang, F.Q.; Feng, K.; Zhao, J.; Li, S.P. Analysis of sterols and fatty acids in natural and cultured *Cordyceps* by one-step derivatization followed with gas chromatography-mass spectrometry. *J. Pharm. Biomed. Anal.* **2009**, *49*, 1172–1178. [CrossRef]
28. Qiu, X.H.; Li, C.; Han, R.C. Analysis of volatile components in different *Ophiocordyceps sinensis* and insect host products. *Molecules* **2020**, *25*, 1603. [CrossRef]
29. Hsu, T.H.; Shiao, L.H.; Hsieh, C.; Chang, D.M. A comparison of the chemical composition and bioactive ingredients of the Chinese medicinal mushroom DongChongXiaCao, its counterfeit and mimic, and fermented mycelium of *Cordyceps sinensis*. *Food Chem.* **2002**, *78*, 463–469. [CrossRef]
30. Gomez, L.L.; Cobas, N.; Franco, I.; Martínez, S.S. Fatty acid profiles and lipid quality indices in canned European eels: Effects of processing steps, filling medium and storage. *Food Res. Int.* **2020**, *136*, 109601. [CrossRef]
31. Wan, W.L.; Xu, J.; Shi, J.Y.; Zhang, X.; Wang, A.; Dong, X.J.; Miao, S.Y. The interaction of dietary niacin and lipid affects the growth, antioxidant capacity and lipid metabolism of male *Eriocheir sinensis*. *Aquac. Rep.* **2023**, *30*, e101541. [CrossRef]
32. Li, C.; Li, Z.; Fan, M.; Cheng, W.; Long, Y.; Ding, T.; Ming, L. The composition of *Hirsutella sinensis*, anamorph of *Cordyceps sinensis*. *J. Food Compos. Anal.* **2006**, *19*, 800–805. [CrossRef]
33. Li, R.; Jiang, X.L.; Guan, H.S. Optimization of mycelium biomass and exopolysaccharides production by *Hirsutella* sp in submerged fermentation and evaluation of exopolysaccharides antibacterial activity. *Afr. J. Biotechnol.* **2010**, *9*, 196–203.
34. Fan, L.C.; Xian, C.N.; Tang, S.J.; Ding, W.; Xu, C.H.; Wang, X.C. Effect of frozen storage temperature on lipid stability of hepatopancreas of *Eriocheir sinensis*. *LWT* **2022**, *154*, e112513. [CrossRef]
35. Hama, M.C.; Abdelsamed, I.; Elshamy, T.Y.; Yusuke, K.; Hirohumi, Y.; Kana, T.; Ayari, O.; Masaaki, N.J.; Ban, S.; Hiroshi, I.; et al. New alkaloidal metabolites from cultures of entomopathogenic fungus *Cordyceps* takaomontana NBRC 101754. *Fitoterapia* **2019**, *139*, 104364. [CrossRef] [PubMed]
36. Joshi, V.; Fernie, A.R. Citrulline metabolism in plants. *Amino Acids* **2017**, *49*, 1543–1559. [CrossRef] [PubMed]
37. Seo, S.K.; Kwon, B. Immune regulation through tryptophan metabolism. *Exp. Mol. Med.* **2023**, *55*, 1371–1379. [CrossRef] [PubMed]
38. Sidney, M.M. Arginine Metabolism Revisited12. *J. Nutr.* **2016**, *146*, 2579S–2586S.
39. María, E.; Alvarez, A.S.; László, S. Proline metabolism as regulatory hub. *Trends Plant Sci.* **2022**, *27*, 39–55.
40. Opeyemi, J.; Olatunji, Y.F.; Oyenike, O.O.; Tang, J.; Wei, Y.; Zhen, O.Y.; Su, Z.L. Polysaccharides purified from *Cordyceps cicadae* protects PC12 cells against glutamate-induced oxidative damage. *Carbohydr. Polym.* **2016**, *153*, 187–195.
41. Zeng, Z.; Mou, D.; Luo, L.; Zhong, W.; Duan, L.; Zou, X. Different cultivation environments affect the yield, bacterial community and metabolites of *Cordyceps cicadae*. *Front. Microbiol.* **2021**, *12*, 669785. [CrossRef] [PubMed]
42. Zhang, H.; Yue, P.; Tong, X.; Gao, T.; Peng, T.; Guo, J. Comparative analysis of fatty acid metabolism based on transcriptome sequencing of wild and cultivated *Ophiocordyceps sinensis*. *Peer J.* **2021**, *9*, e11681. [CrossRef] [PubMed]
43. Marcos, J.; Renau, N.; Valverde, O.; Aznar-Lain, G.; Gracia-Rubio, I.; Gonzalez-Sepulveda, M.; Pérez-Jurado, L.A.; Ventura, R.; Segura, J.; Pozo, O.J. Targeting tryptophan and tyrosine metabolism by liquid chromatography tandem mass spectrometry. *J. Chromatogr. A* **2016**, *1434*, 91–101. [CrossRef] [PubMed]
44. Pan, S.J.; Fan, M.; Liu, Z.G.; Li, X.; Wang, H.J. Serine, glycine and one-carbon metabolism in cancer. *Int. J. Oncol.* **2020**, *11*, 158–170. [CrossRef] [PubMed]
45. Neila, H.M.; Emerson, S.L.; Pritesh, L. Dysregulation of tryptophan catabolism in metabolic syndrome. *Metab. Syndr. Relat. Disord.* **2018**, *16*, 135–142.
46. Brosnan, M.E.; Brosnan, J.T. Histidine metabolism and function. *J. Nutr.* **2020**, *150*, 2570S–2575S. [CrossRef] [PubMed]
47. Shauni, L.G.; Elien, H.; Kim, D.K.; Kim, R.K. The ins and outs of serine and glycine metabolism in cancer. *Nat. Metab.* **2021**, *3*, 131–141.
48. Zhang, X.; Liu, Q.; Zhou, W.; Li, P.; Alolga, R.N.; Qi, L.W.; Yin, X.J. A comparative proteomic characterization and nutritional assessment of naturally and artificially cultivated *Cordyceps sinensis*. *J. Proteom.* **2018**, *181*, 24–35. [CrossRef] [PubMed]
49. Hyun, S.H.; Lee, S.Y.; Sung, G.H.; Kim, S.H.; Choi, H.K. Metabolic profiles and free radical scavenging activity of *Cordyceps* bassiana fruiting bodies according to developmental stage. *PLoS ONE* **2013**, *8*, e73065. [CrossRef]

50. Venkatesh, M.; Xi, Q.X.; Goh, E.T.L.; Koh, S.K.; Toh, T.Y.J.; Barathi, V.A.; Mobashar, H.U.T.F.; Madhavi, L.S.C.; Jayasudha, V.; Darren, S.J.T.; et al. Rational substitution of ε-lysine for α-lysine enhances the cell and membrane selectivity of pore-forming melittin. *J. Med. Chem.* **2020**, *63*, 3522–3537.
51. Wang, W.; Meng, X.; Wang, J.; Li, Y. Improved heart failure by rhein lysinate is associated with p38MAPK pathway. *Exp. Ther. Med.* **2018**, *16*, 2046–2051. [CrossRef]
52. Wei, J.; Zhou, X.; Dong, M.; Yang, L.F.; Zhao, C.; Lu, R.L.; Bao, G.H.; Hu, F.L. Metabolites and novel compounds with anti-microbial or antiaging activities from *Cordyceps fumosorosea*. *AMB Expr.* **2022**, *12*, 40. [CrossRef] [PubMed]
53. Zhang, L.G.; Wu, T.; Olatunji, O.J.; Tang, J.; Wei, Y.; Zhen, O.Y. N6-(2-hydroxyethyl)-adenosine from *Cordyceps cicadae* attenuates hydrogen peroxide induced oxidative toxicity in PC12 cells. *Metab. Brain Dis.* **2019**, *34*, 1325–1334. [CrossRef] [PubMed]
54. Ahmad, S.; Tatsuya, K.; Hideo, D.; Enoch, Y.P. Insight into cordycepin biosynthesis of *Cordyceps militaris*: Comparison between a liquid surface culture and a submerged culture through transcriptomic analysis. *PLoS ONE* **2017**, *12*, e0187052.
55. Liu, Y.; Guo, Z.J.; Zhou, X.W. Chinese Cordyceps: Bioactive components, antitumor effects and underlying mechanism—A review. *Molecules* **2022**, *27*, 6576. [CrossRef]

separations

MDPI

Article

Optimization of QuEChERS Method for Antibiotic Residue Analysis in Animal Foods via Response Surface Methodology

Xiaoqiong Wu, Yun Lin, Xiang Zhang, Nan Ouyang and Ying Zhou *

Jiaxing Center for Disease Control and Prevention, Jiaxing 314050, China; dearwuxiaoqiong@163.com (X.W.)
* Correspondence: zhouyingand@sina.cn; Tel.: +86-0573-8368-3808

Abstract: The present study employed a modified QuEChERS method to systematically analyze the presence of fifteen quinolone and seven tetracycline antibiotic residues in local animal food. Additionally, a multi-level four-factor Box–Behnken design (BBD) within the framework of response surface methodology (RSM) was utilized to evaluate the factors impacting the detection efficiency of the sample pretreatment procedure. Optimization was performed via Design Expert® 10.0.3, and the factors, including the volume of the acetonitrile, the addition of formic acid, the duration of the extraction, and the addition of EDTA, were combined with experimental design until an optimal solution was reached. Finally, the sample was tested via ultra-high performance liquid chromatography-quadrupole-linear ion trap mass spectrometry (UPLC/MS/MS) in both multiple reaction monitoring (MRM) and enhanced product ion (EPI) scan modes on a QTRAP® 5500 instrument (AB SCIEX instruments, Framingham, MA, USA). The overall average recoveries from actual samples fortified with 22 antibiotics at three levels ranged from 73.8% to 98.5% based on the use of matrix-fortified calibration, with variations ranging from 5.80 to 12.4% (n = 6). The limits of detection and quantification were 0.3 μg kg^{-1} and 1.0 μg kg^{-1}, respectively. Lastly, the modified method was applied to practical sample analysis in the daily risk monitoring and assessment of food safety with satisfactory stability and robustness.

Keywords: antibiotics residue; response surface methodology (RSM); quinolones; tetracyclines; multiple reaction monitoring (MRM)

Citation: Wu, X.; Lin, Y.; Zhang, X.; Ouyang, N.; Zhou, Y. Optimization of QuEChERS Method for Antibiotic Residue Analysis in Animal Foods via Response Surface Methodology. *Separations* **2023**, *10*, 459. https://doi.org/10.3390/separations10080459

Academic Editors: Ronald Beckett and Chao Kang

Received: 30 June 2023
Revised: 21 July 2023
Accepted: 24 July 2023
Published: 21 August 2023

1. Introduction

Antibiotic resistance, the ability of bacteria to withstand antibiotics, is now recognized as one of the most serious global threats to human health [1–3]. Naturally occurring resistance that can ultimately lead to incurable bacterial infections could be accelerated by the improper use of antibiotics in human beings and animals [4]. Except for the misuse of antibiotics in human medicine, antibiotics' misuse in livestock is also a major contributor to the emergence of antibiotic resistance [5,6]. To satisfy the growing global demand for animal protein, antibiotics have been massively and increasingly used in farmed animal industries for different purposes, including overdoses for disease prevention and subtherapeutic doses for growth stimulation [7–13]. In September 2017, a report from the World Health Organization (WHO) corroborated that the world is running out of antibiotics [14–17]. Antibiotic misuse, if left unchecked, can drag human beings into a post-antibiotic era whereby minor injuries or common infections become fatal diseases again [18].

To protect the public from health risks, nations and related organizations had to establish broader maximum-residue limits (MRLs) for further surveillance of antibiotic residues in animal food [19,20]. Therefore, more efficient and robust detection methods were promptly developed in the past few years to satisfy increasingly rigorous regulatory requirements [21–27]. Triple quadrupole mass spectrometer with multiple reaction monitoring (MRM) scan modes that follow the requirement of ECD 2002/657/EC should be the preferred method of detection of antibiotic residues in animal food [28–30]. The

Qtrap system (AB Sciex) with the scan mode of multiple reaction monitoring—information-dependent acquisition—enhanced product ion scan mode (MRM—IDA—EPI) was found to efficiently gather comprehensive information from samples in a single run. Consequently, the identification of antibiotic residues in locally sourced animal-derived food was reaffirmed through the successful comparison of antibiotic spectra from the samples with reference spectra. Quinolones and tetracyclines, being the most extensively employed veterinary antibiotics, have contributed to the development of antibiotic resistance, thereby adversely impacting the treatment of severe bacterial infections [31–33]. In this study, a diverse range of samples, fortified with fifteen quinolone and seven tetracycline antibiotics at the recommended concentration (RC) of 1 $\mu g\ L^{-1}$, were subjected to analysis using liquid chromatography-tandem mass spectrometry with a QTRAP 5500 instrument.

Sample pretreatment for the separation and concentration of antibiotic residues is also a critical step in the whole analysis process of quinolone and tetracycline detection. There are, alternatively, two purification methods, including QuEChERS and solid-phase extraction. The latter, using commercial cartridges, has been widely used in daily work based on former research (Figure A1) [34]. It can be performed in an automated SPE system (e.g., Reeko, Fotector plus, USA) during non-working hours and minimize human involvement (Table A1), but high-fat or high-protein samples should be excluded, given that they frequently cause blockage in SPE cartridges and heavily prolong the pretreatment process. Since its development in 2003, QuEChERS has gained widespread acceptance for various sample preparation techniques [35–40]. It was initially introduced as a cost-effective and time-efficient method for analyzing multi-residue samples containing relatively polar compounds. During the extraction process, the efficiency of the QuEChERS sample preparation method is known to be influenced by several factors. Thus, it is imperative to comprehensively optimize this method for the detection of quinolone and tetracycline residues in this study. As a collection of statistical and mathematical techniques, response surface methodology (RSM) has vital applications in the design, development, and improvement of novel or existing product designs [41,42], especially in multi-variable analysis [43,44]. In the present study, response surface methodology with a multi-level four-factor Box–Behnken design (BBD) was applied to simultaneously evaluate the recovery rate of quinolone and tetracycline residues in sample pretreatment.

2. Materials and Methods

2.1. Standards and Stock Solutions

Standards of the following twenty-two antibiotics were all purchased from Dr. Ehrenstorfer GmbH (Augsburg, Germany): Oxytetracycline (Oxytetracycline hydrochloride, 96.5%), Tetracycline (Tetracycline hydrochloride, 98.0%), Doxycycline (Doxycycline hyclate, 98.7%), Demeclocycline (98.0%), Methacycline (99.0%), Minocycline (99.0%), Chlortetracycline (Chlortetracycline hydrochloride, 94.6%), Enrofloxacin (99.0%), Norfloxacin (99.1%), Pefloxacin (Pefloxacin methanesulfonate dehydrate, 99.0%), Ciprofloxacin (Ciprofloxacin hydrochloride, 94.0%), Ofloxacin (99.0%), Sarafloxacin (Sarafloxacin hydrochloride, 97.0%), Enoxacin (95.8%), Lomefloxacin (Lomefloxacin hydrochloride, 99.5%), Pipemidic acid (99.1%), Nalidixic acid (99.0%), Oxolinic acid (98.0%), Flumequine (98.5%), Cinoxacin (99.0%), Danofloxacin (Danofloxacin mesylate, 94.0%), and Difloxacin (Difloxacin hydrochloride, (99.2%)). Individual standards were weighed using an electronic balance (Metter Toledo, MS 205DU), dissolved in methanol at a concentration of 1.0 mg mL^{-1} or ethanol solution supplemented with potassium hydroxide (for antibiotics that were practically insoluble in methanol), and provisionally stored at -28 °C.

2.2. Reagents and Chemicals

Methanol and acetonitrile of HPLC grade used in this study were procured from Merck (Darmstadt, Germany). Ethanol of HPLC grade and formic acid ($\geq$98%) were acquired from Aladdin (Shanghai, China). Potassium hydroxide (G.R.) was purchased from Macklin (Shanghai, China). Cleanert® C18 for QuEChERS was purchased from

Agela Technologies (Beijing, China). Ultra-pure water was obtained from a Milli-Q water purification system from Millipore (Bedford, MA, USA). Oasis® HLB SPE cartridges (6 cc, 200 mg) were purchased from Waters (Milford, MA, USA). Millipore filters (0.22 μm, polytetrafluoroethylene) were obtained from ANPEL Lab (Shanghai, China). Sodium chloride (A.R.) and sodium sulfate (A.R.) were calcined in a muffle furnace prior to use. Citric acid (A.R.), disodium hydrogen phosphate (A.R.), and disodium ethylenediamine tetraacetic acid (EDTA, A.R.) for sample preparation were purchased from Sinopharm Chemical Reagent (Shanghai, China).

2.3. Instrumentation and Software

HPLC analysis was conducted on a Shimadzu LC-30AD system with a Waters BEH C18 column (1.7 μm 2.1 mm × 100 mm, Waters, Milford, MA, USA). Regarding mass spectrometric detection, all experiments were carried out in the MRM-IDA-EPI scan mode using AB SCIEX QTRAP® 5500 (AB SCIEX instruments, Framingham, MA, USA). The compounds were ionized in a Turbo V™ Ion Source (ESI) interface in the positive ionization mode. An Analyst® software v. 1.6.2 (AB SCIEX instruments, Foster City, Canada) was utilized to remotely control the chromatograph and mass spectrometer. A capillary voltage of 5.50 kV and desolvation temperature of 500 °C were applied to the ESI source. Nitrogen produced by the generator (Claind Nitro35, Tremezzina, Italy) was used as the cone gas (50 psi), desolvation gas (50 psi), and collision gas. Quantitative analysis of the experiments was conducted using MultiQuant® 3.0.1. The optimization of QuEChERS using the response surface method was performed on Design Expert® 10.0.3.

2.4. Sample Collection and Processing

The samples were collected from local markets or supermarkets distributed randomly across neighborhoods of the whole city, including swine, poultry, eggs, milk, and eight cultured aquatic products (*Parabramis pekinensis, Carassius auratus, Ctenopharyngodon idella, Ophiocephalus argus Cantor, Macrobranchium nipponense, Macrobrachium rosenbergii, Penaeus chinensis, Procambarus clarkia, Eriocheir sinensis*, and *Larimichthys crocea*). The edible portion of the aforementioned samples was ground and homogenized using a Mixer (BÜCHI, B400, Eastern Switzerland) and stored in polypropylene bottles at −28 °C for the ensuing analyses.

3. Results

In the sample pretreatment process using the SPE extraction method, 0.1 mol L^{-1} EDTA-Mcllvaine buffer solution (pH = 4.0), a non-volatile, non-poisonous, low-cost, and eco-friendly solution, was used as the extraction solution for quinolone and tetracycline residues [34,45–47]. Furthermore, when handling a large amount of sample, the SPE method could enhance stability by increasing the level of automation without increasing human operational time (Table A1). However, during the flowing sample purification procedure, the aqueous solution extracted from high-fat samples, particularly egg-containing samples, may potentially increase the risk of blockage in SPE cartridges and significantly prolong the detection time.

3.1. Experimental Design

The QuEChERS sample preparation method could be applied as an alternative for a quick analysis of quinolone and tetracycline residues in high-fat samples. As documented in former research, the efficiency of the QuEChERS sample preparation method for quinolone and tetracycline residues could be affected by multifarious factors, including the volume of acetonitrile, pH value of the extracted solvent, and duration of the extraction process, among others [35–40].

Attributed to the unique chemical structure of quinolones and tetracyclines, agent–agent interactions [48] between antibiotics and metal ions from the experimental environment could influence the recovery rate. As in the EDTA-Mcllvaine buffer solution of the

SPE method, EDTA acts as a screening agent for metal ions that could also be introduced in the extraction process of the QuEChERS method. As illustrated in Figure 1, when the amount of added EDTA ranged from 0 to 0.2 g in 2 g of the sample using the QuEChERS preparation, the medium recovery rate of 22 antibiotics progressively plateaued at approximately 100 mg. According to the result of this experiment, 0.1 g of EDTA was selected as the auxiliary reagent in the extraction process.

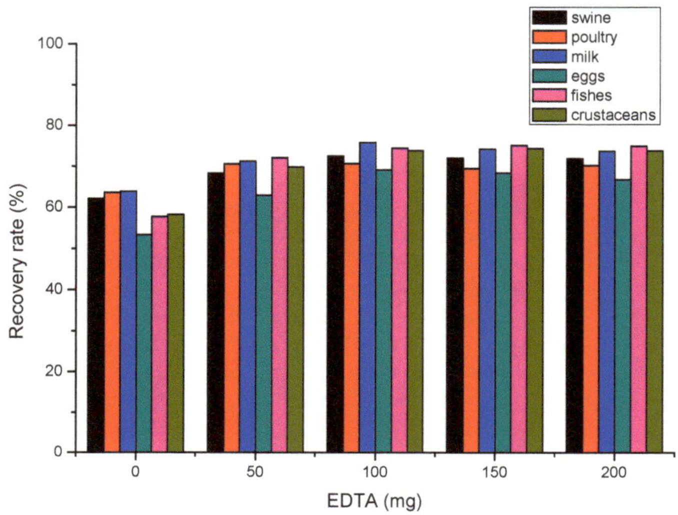

Figure 1. Variation in recovery rate with EDTA (Enrofloxacin).

3.2. Optimization of QuEChERS Method Using the Response Surface Method (RSM)

The QuEChERS sample preparation method is multivariable for the optimization of the extraction process. Its efficiency could be affected by multifarious factors, including the volume of the acetonitrile (A), the additive amount of formic acid (B), the additive amount of EDTA (C), and the extraction time (D). Nevertheless, the univariate experiment has limited ability to evaluate the interactions of the extraction conditions in sample preparation. To avoid interactions among extraction conditions while examining the optimal extraction process, possible factors were comprehensively optimized via the response surface method (RSM), including the volume of the acetonitrile (A), the additive amount of formic acid (B), the additive amount of EDTA (C) and the extraction time (D). The median recovery rate of the 22 antibiotics from standard samples (2.0 μg kg^{-1}, $n = 3$) was chosen as the response. RSM analysis was able to model the relationship between the response (recovery rate) and the four factors. Based on a previous study [48], the respective low and high levels for factors were coded.

The model's F-value of 15.79 implied that the model was significant. This model can be used to navigate the design space. The final equation in terms of actual factors is as follows:

$$\begin{aligned}\text{Recovery} &= 84 + 21.48 \times A + 7.44 \times B + 7.58 \times C + 8.39 \times D + 5.07 \times AB + 1.95 \times AC \\ &+ 4.48 \times AD + 1.32 \times BC + 4.23 \times BD + 4.67 \times CD - 22.11 \times A^2 - 9.80 \times B^2 - 7.41 \times C^2 \\ &- 14.45 \times D^2\end{aligned} \quad (1)$$

The p-value is generally employed to assess the significance of variables and can also reflect interactions among independent variables [49]. A smaller p-value indicates that the corresponding variable is more significant [50]. The ANOVA for the response surface quadratic model is summarized in Table 1; in this case, A, B, C, D, A^2, B^2, C^2, and D^2 were all significant model terms, and the variable volume of the acetonitrile (A) and extraction time (D) were more significant for the recovery rate.

Table 1. ANOVA for the recovery rate according to the response surface quadratic model.

Source	Sum of Squares	*df*	Mean Square	*F*-Value	*p*-Value
Model	12,099.48	14	864.25	15.79	<0.0001
A-MeCN	5538.40	1	5538.40	101.21	<0.0001
B-HCOOH	664.54	1	664.54	12.14	0.0036
C-EDTA	690.08	1	690.08	12.61	0.0032
D-Time	845.04	1	845.04	15.44	0.0015
A^2	3171.17	1	3171.17	57.95	<0.0001
B^2	622.75	1	622.75	11.38	0.0045
C^2	356.24	1	356.24	6.51	0.0231
D^2	1354.08	1	1354.08	24.74	0.0002
Cor total	12,865.58	28			

R^2 (Pred) = 0.7 R^2 (Adj) = 0.8809. Significant at a 95% confidence degree ($p < 0.05$).

As displayed in Figure 2, the recovery rate of the 22 antibiotics was superior among the solutions with the following settings: acetonitrile (A) = 8 mL, formic acid (B) = 150 μL, EDTA (C) = 0.1 g, and time for extraction (D) = 8 min. Verification tests were carried out six times under the above-mentioned optimized conditions. The median recovery rates of the 22 antibiotics from six parallel tests were 75.4%, 81.6%, 85.9%, 73.6%, 77.9%, and 82.5%, with errors ranging from 7.8% to 15.1%. The sample preparation of the QuEChERS method was finally optimized as the best solution from RSM, and the method is outlined in Figure 3.

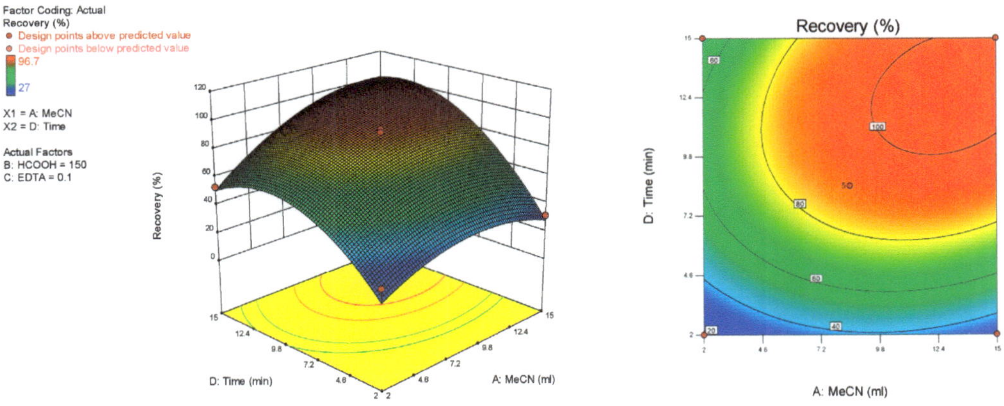

(**a**) (**b**)

Figure 2. (**a**) Response surface plot; (**b**) response surface contour plots of optimization results.

Concerning purification, the addition of a cleaning agent, namely C18 powder, was evaluated in fish to identify the highest detection rate of antibiotic residue. The crude extraction of the spiked sample was adequately purified when 150 mg of C18 powder was added (see Figure 4).

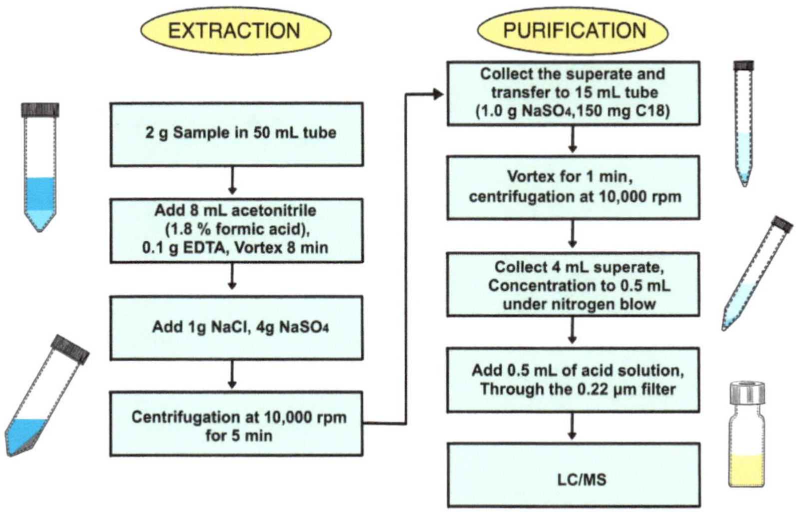

Figure 3. Scheme of QuEChERS sample preparation method.

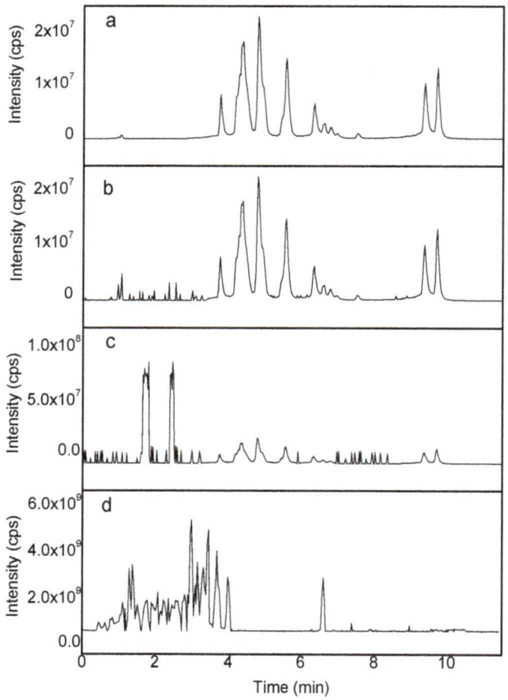

Figure 4. Variations in total ion chromatogram with C18 addition (spiked in fishes): (**a**) 150 mg; (**b**) 100 mg; (**c**) 50 mg; (**d**) no addition.

3.3. Optimization of Chromatographic Conditions and Mass Spectrometry

Quinolines and tetracyclines both contain several O and N atoms; consequently, it is easy to obtain protons and a high response in the positive ion mode. To obtain the two ion pairs for quantification, a mixed standard solution of 22 antibiotics at a concentration of

100 μg L^{-1} was infused into the QTRAP mass spectrometer at a flow rate of 7.0 μL min^{-1} to achieve automatic analyte optimization via the ESI in the positive mode. Under the optimal mass spectrometry conditions, including declustering potential and collision energy, every antibiotic was assigned two pairs of abundant ions for qualitative and quantitative analysis with high sensitivity. To achieve minimum retention time and symmetric shape of ionic peaks, the elution type, flow rate, and gradient were optimized in this study using the C18 chromatographic column. Therefore, several classical compositions of the mobile phase were performed, including acetonitrile, methanol, water, as well as water with ammonium acetates or formic acid. Finally, water (A) and acetonitrile (B), which were both supplemented with 0.1% formic acid, were chosen as the optimal mobile phase. The final gradient elution at a total flow rate of 0.3 mL min^{-1} was as follows: 0–0.5 min, 5–20% B; 0.5–2.0 min, 20–25% B; 2.0–7.0 min, 25–45% B; 7.0–10.0 min, 45–90% B; 10.0–12.0 min, 90% B; and 12.1–13.0 min, 95–5%. The column oven was maintained at a temperature of 40 °C, and the injection volume was 10.0 μL. The representative total ion chromatogram (TIC) was merged in Figure A2. The retention time (RT) and MS information for each antibiotic, including precursor and product ions, DP, and CE, are presented in Table 2.

Table 2. Retention time and MS parameters of the 22 antibiotics.

No.	Compound	Retention Time (min)	CAS No.	Precursor Ion (*m*/*z*)	Product Ion (*m*/*z*)	Declustering Potential (V)	Collision Energy (eV)
1	Pipemidic acid	2.88	51940-44-4	304.3	217.1 * 189.0	70 70	18 27
2	Enoxacin	3.32	74011-58-8	321.4	303.3 * 233.9	80 80	22 33
3	Minocycline	3.59	10118-90-8	458.5	441.4 * 352.4	80 80	20 30
4	Norfloxacin	3.64	70458-96-7	320.3	302.3 * 276.3	80 80	26 35
5	Ofloxacin	3.66	82419-36-1	362.2	318.3 * 261.2	80 80	26 38
6	Pefloxacin	3.70	70458-92-3	334.3	290.3 * 233.2	80 80	27 25
7	Tetracycline	3.74	60-54-8	445.4	410.4 * 427.7	80 80	24 19
8	Ciprofloxacin	3.75	85721-33-1	332.2	314.3 * 288.3	80 80	25 33
9	Methacycline	3.77	914-00-1	443.3	426.4 * 201.2	60 60	18 10
10	Oxytetracycline	3.78	79-57-2	461.4	426.4 * 443.6	80 80	25 17
11	Danofloxacin	3.82	112398-08-0	358.3	340.3 * 82.0	80 80	12 35
12	Lomefloxacin	3.87	98079-51-7	352.3	265.2 * 308.3	80 80	33 28
13	Enrofloxacin	3.96	93106-60-6	360.3	316.4 * 342.3	80 80	25 35
14	Doxycycline	4.00	564-25-0	445.5	428.5 * 154.0	80 80	24 35

Table 2. *Cont.*

No.	Compound	Retention Time (min)	CAS No.	Precursor Ion (*m/z*)	Product Ion (*m/z*)	Declustering Potential (V)	Collision Energy (eV)
15	Demeclocycline	4.07	64-73-3	465.3	430.4 * 448.4	75 75	23 28
16	Sarafloxacin	4.38	98105-99-8	386.3	342.3 * 299.3	80 80	25 38
17	Difloxacin	4.51	98106-17-3	400.1	356.1 * 299.1	80 80	28 41
18	Chlortetracycline	5.13	57-62-5	479.3	444.4 * 462.3	80 80	24 28
19	Cinoxacin	5.29	28657-80-9	263.1	244.1 * 188.8	80 80	25 35
20	Oxolinic acid	5.53	14698-29-4	262.1	244.1 * 155.9	70 70	26 40
21	Nalidixic acid	7.09	389-08-2	233.1	187.0 * 244.1	68 68	18 34
22	Flumequine	7.43	42835-25-6	262.2	244.1 * 202.1	70 70	19 32

CAS: chemical abstracts service; *: quantitative ion.

The EPI scan mode could be activated in the IDA experiment when the ionic intensity exceeded the threshold of 1000 cps. The scan time (including pauses) was 1.57 s for all MRM transitions. EPI mass spectra were acquired over a mass range of *m/z* 50–500 at a scan rate of 10,000 Da s^{-1}.

3.4. Fragmentation Approach for Quinolones and Tetracyclines

In the positive mode of electrospray ionization (ESI) mass spectrometry, the proton first binds to the protonation site, usually at the N atom or O atom, and then triggers cleavage by migrating to the reactive center. Although the most basic site in tetracyclines is the dimethylamino group, protonated tetracyclines initially dissociate via the loss of H_2O or NH_3 from the acylamino group. Regarding tetracycline, demeclocycline, and chlortetracycline, there is no OH at C–6 sites, and, consequently, all presented with successive losses of H_2O and NH_3. As depicted in Figure 5, tetracyclines without the tertiary OH at C–6 initially lose only NH_3 [51].

According to the spectra acquired from the EPI mode (Figure 6), the reactive center of quinolones was located in the carboxylic acid group. The abundant fragment ion $[M+H–H_2O]^+$ was formed due to the dehydration of –COOH, while another abundant fragment ion was characterized by the decarboxylation of this group. The neutral loss of *m/z* 20 Da and *m/z* 30 Da was most probably formed due to the dissociation of –HF or $–CH_2CH_3$. Moreover, another characteristic neutral loss of $–CH=CH–NH_2$ (*m/z* 20 Da) was produced from the cracking of the azine ring [52].

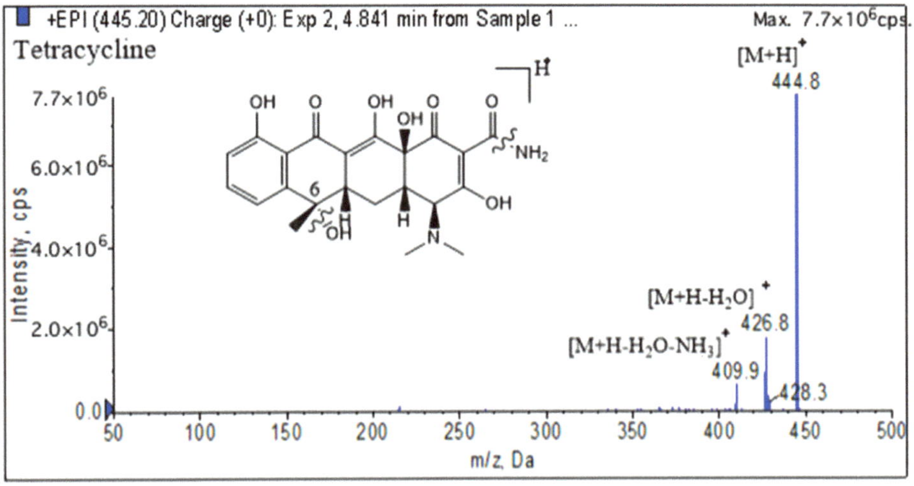

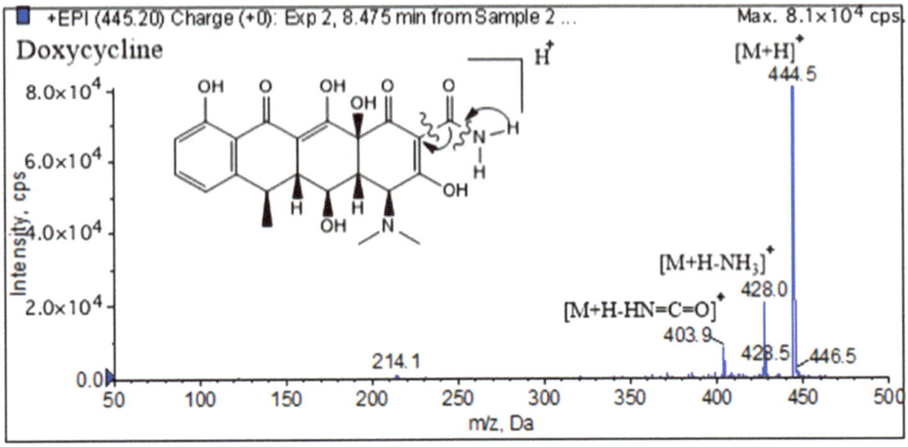

Figure 5. Production spectra and proposed fragmentation pathway of tetracycline and doxycycline.

3.5. Method Validation

3.5.1. Matrix Effect

To evaluate the matrix effects in LC-MS detection, six distinct types of antibiotic-free samples were used as matrix-matched blanks on the ionization of 22 antibiotic residues. The equation is as follows [53]:

$$ME = \frac{A_{Matrix}}{A_S} \times 100\% \tag{2}$$

where A_{Matrix} represents the peak area of the standard solution with the matrix-matched blank, and A_S stands for the peak area of the standard solution in the initial mobile phase. The percentages of the matrix effects of the 22 antibiotics at three different concentrations (2, 20, 200 ng mL^{-1}) ranged from 84.7% to 119.3%. When ion suppression and ion enhancement at the chosen levels were considered, the blank matrix-matching standard curve was adopted to eliminate the effect of the matrix.

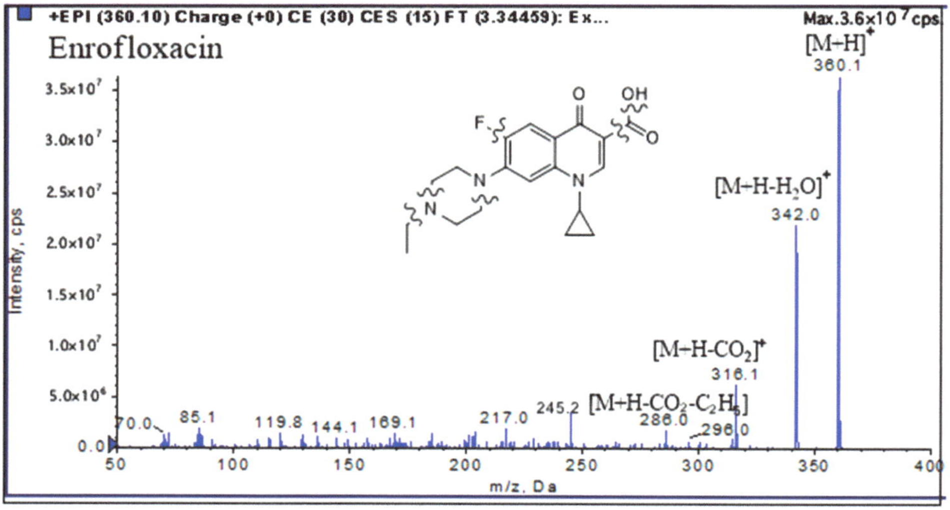

Figure 6. Production spectra and proposed fragmentation pathway of enrofloxacin.

3.5.2. Linearity and Sensitivity

Satisfactory linearities (R > 0.99) were obtained for 22 antibiotics in blank matrix-matched curves over concentrations ranging from 0.5 ng mL^{-1} to 200.0 ng mL^{-1}. The sensitivity of the proposed method was measured according to the limit of detection (LOD) and the limit of quantification (LOQ) values. LOD and the LOQ were calculated using the following equations [53]:

$$\mathrm{LOD} = C_S \frac{3}{S/N} \tag{3}$$

$$\mathrm{LOQ} = C_S \frac{10}{S/N} \tag{4}$$

where S/N denotes the average signal-to-noise ratio, and C_S represents the concentration of the specific antibiotic. The estimated values were tested using suitable spiked samples containing the 22 antibiotics at the corresponding concentrations. When the concentration ranged between 0.5 and 200.0 ng mL^{-1} (0.5, 1.0, 2.0, 5.0, 10.0, 20.0, 50.0, 100.0, 200.0, with R.S.D. under 10%, n = 6), the LOD and LOQ values were 0.3 µg kg^{-1} and 1.0 µg kg^{-1}, respectively, demonstrating the sensitivity of the method for antibiotic residues.

3.5.3. Accuracy and Precision

The accuracy and precision of the method were measured using the intra- and inter-day recoveries and relative standard deviation (R.S.D.). Therefore, the standard mixed solutions of 22 antibiotics were spiked into distinct types of samples, including swine, poultry, eggs, milk, fish, and crustacea, and 18 spiked samples (six types at three concentrations of 2, 20, 200 µg kg^{-1}, shown in Table A2) were obtained. Notably, these spiked samples were detected three times intra-day and three times inter-day. As anticipated, the recovery of 22 antibiotic residues (73.8–98.5%) fell within the recommended guidelines of 60–120% (GB/T 27404-2008, China) [54]. The precision of the analysis measured as the relative standard deviation (R.S.D.) of the recovery, which ranged from 5.80% to 12.4%, was well under the criteria of 30% (GB/T 27404-2008, China) [54].

3.6. Sample Analyses

After validation of the analytical methodology through the above experimentation, it was applied for detection using various real food samples, including swine, poultry, eggs, milk, and nine cultured aquatic products (*Parabramis pekinensis, Carassius auratus, Ctenopharyngodon idella, Ophiocephalus argus Cantor, Macrobranchium nipponense, Macrobrachium rosenbergii, Penaeus chinensis, Procambarus clarkia, Eriocheir sinensis*, and *Larimichthys crocea*). In the last six years of detection (2017–2022, total of 781 samples), quinoline or tetracycline residues from swine, eggs, milk, and *Eriocheir sinensis* were essentially not detected. Compared with the research led by Prof. Treiber [55], tetracycline residues, including tetracycline, oxytetracycline, and chlortetracycline, were also occasionally detected but never exceeded the MRL (200 ug kg^{-1}, GB31650-2019, China [56]) in poultry, *Macrobranchium nipponense, Macrobrachium rosenbergii, Penaeus chinensis* and *Procambarus clarkia*. Quinoline residues were generally detected in cultured aquatic products, with the exception of *Eriocheir sinensis*. The detection rate of quinoline residues was highest in fishes (*Parabramis pekinensis, Carassius auratus, Ctenopharyngodon idella, and Ophiocephalus argus Cantor*), ranging from 11.36% to 37.51%, with the over-limit rate (MRL, 100 μg kg^{-1}, GB 31650-2019, China [56]) ranging from 1.85% to 9.07%. Furthermore, enrofloxacin and ciprofloxacin were the dominant detected residues among the 22 antibiotics. Meanwhile, among the 12 types of samples, *Parabramis pekinensis*, with a medium detected concentration of enrofloxacin and ciprofloxacin of 179.9 μg kg^{-1} and 21.4 μg kg^{-1}, respectively, contributed the maximum detected frequency and value.

A simplified risk assessment of enrofloxacin and ciprofloxacin from fish could be calculated using the following equations:

$$\text{EXP} = \frac{\text{Average consumption (g)} \times \text{Average detection value } (\mu\text{g/kg}) \times \text{p}}{\text{BW}} \quad (5)$$

$$\text{MOS} = \frac{\text{ADI}}{\text{Daily dietary exposure}} \quad (6)$$

where EXP represents the daily dietary exposure of enrofloxacin and ciprofloxacin, and MOS denotes the margin of safety. The average consumption of fish is 24.3 g/d, according to the Scientific Research Report on Dietary Guidelines for Chinese Residents of 2021. p represents the effect of food processing and was excluded from this simplified assessment; BW stands for the average body weight (60 kg and 30 kg for adults and children, respectively). According to the National food safety standard GB 31650-2019 [56], the sum of enrofloxacin and ciprofloxacin residues in fish should not exceed 100 μg kg^{-1}, and the related acceptable daily intake (ADI) is 2.0 μg/(kg·d). The results revealed that consuming fish with high levels of quinolone residues may increase the risk of adverse events in children.

4. Conclusions

The analytical approach for antibiotic residue using LC-QTRAP-MS developed in this study is reliable and effective in daily risk monitoring and assessment for food safety. The mass spectrum of each antibiotic obtained from the EPI mode could be used as a corroboration of positive samples. Furthermore, the optimization of the sample pretreatment using the response surface method (RSM) enhanced work efficiency. The analysis of real food origin samples validated the robustness and applicability of the modified QuEChERS method. Over the past six years, massive antibiotic residues have been detected in food from animal origin. Detection rates exceeding the maximum residue limits (MRL, GB 31650-2019, China) have been decreasing year by year from 2020 owing to the strict legal requirements imposed by the National Food Safety Standard—Maximum residue limits for veterinary drugs in foods (GB 31650-2019, China).

Author Contributions: Conceptualization, Y.Z.; methodology, Y.Z.; software, Y.Z.; validation, X.W. and N.O.; formal analysis, Y.Z., X.W. and X.Z.; investigation, Y.L.; resources, Y.L. and X.W.; data curation, Y.Z.; writing, Y.Z. All authors have read and agreed to the published version of the manuscript.

Funding: This research was funded by Jiaxing Science and Technology Bureau, grant number 2019AD32050, 2023AY11055 and 2023AD11048.

Data Availability Statement: Not applicable.

Acknowledgments: The authors would like to thank each other for the support in daily work and thank the help from the Zhejiang Center for Disease Control and Prevention.

Conflicts of Interest: The authors declare no conflict of interest.

Appendix A

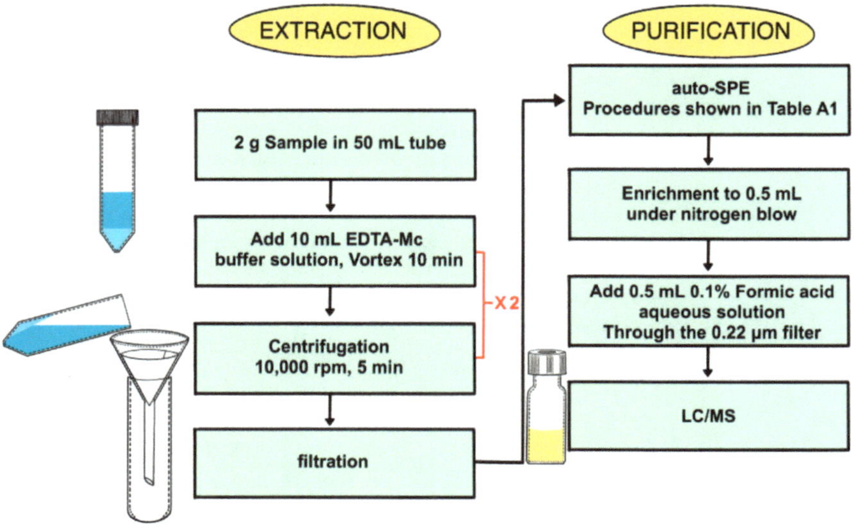

Figure A1. Scheme of solid-phase extraction.

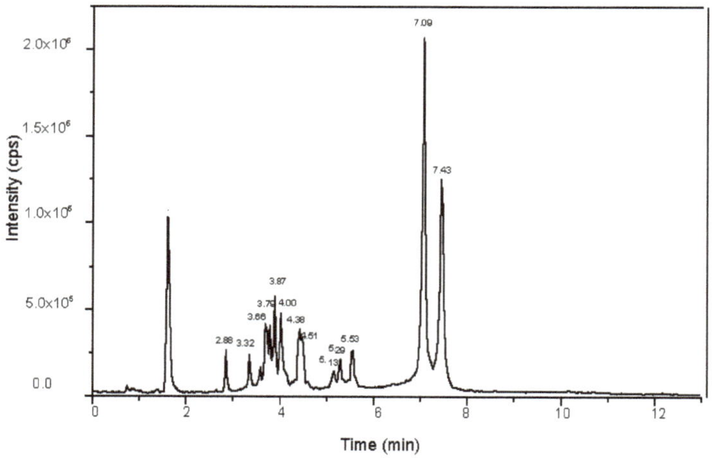

Figure A2. LC-MS/MS chromatograms of 22 antibiotics (50 ng mL^{-1}).

Table A1. Procedures for automated solid-phase extraction.

NO.	Step	Source	Output	Flow Rate (mL/min)	Volume (mL)	Time (min)
1	Rinse sample path	CH_3OH				2.8
2	Rinse sample path	H_2O				2.8
3	Rinse plunger	CH_3OH	Solvent	10	6	1.1
4	Rinse plunger	H_2O	Solvent	10	6	1.1
5	Load sample		Waste	2	20	22.3
6	Rinse	5% CH_3OH	Solvent	3	3	4.3
7	Rinse syringe	CH_3OH		10	10	1.6
8	Elute	CH_3OH	Collect	10	5	0.9
9	Air push		Collect	10	5	1.1
10	End					

Total time: 41.8 min.

Table A2. Recoveries and R.S.D. of 22 antibiotics spiked into fish at three levels.

Compound	Background µg/kg	Fortification µg/kg	Average Recovery Rate %	RSD $n = 3$ %	Compound	Background µg/kg	Fortification µg/kg	Average Recovery Rate %	RSD $n = 3$ %
Pipemidic acid	ND	2.0 20.0 200.0	74.2 78.7 77.8	8.85 6.95 5.86	Lomefloxacin	ND	2.0 20.0 200.0	82.9 89.2 86.3	9.85 9.07 8.41
Enoxacin	ND	2.0 20.0 200.0	73.9 79.1 80.4	9.12 7.63 7.45	Enrofloxacin	ND	2.0 20.0 200.0	90.2 98.5 95.4	5.87 6.15 5.98
Minocycline	ND	2.0 20.0 200.0	73.8 82.6 75.9	9.92 8.43 8.31	Doxycycline	ND	2.0 20.0 200.0	86.2 88.4 84.9	12.4 10.7 11.3
Norfloxacin	ND	2.0 20.0 200.0	80.5 78.3 85.4	9.01 7.65 6.14	Demeclocycline	ND	2.0 20.0 200.0	79.3 88.4 89.0	11.9 9.90 10.2
Ofloxacin	ND	2.0 20.0 200.0	75.2 82.7 86.4	10.2 8.72 6.96	Sarafloxacin	ND	2.0 20.0 200.0	78.9 88.9 78.5	8.68 6.12 5.99
Pefloxacin	ND	2.0 20.0 200.0	89.1 92.4 88.5	9.56 9.17 7.88	Difloxacin	ND	2.0 20.0 200.0	83.7 84.0 85.9	8.69 6.81 7.04
Tetracycline	ND	2.0 20.0 200.0	78.6 83.0 82.7	12.1 10.7 9.57	Chlortetracycline	ND	2.0 20.0 200.0	78.4 88.1 82.7	11.7 8.94 9.10
Ciprofloxacin	ND	2.0 20.0 200.0	82.2 78.6 89.5	9.07 8.19 7.54	Cinoxacin	ND	2.0 20.0 200.0	75.5 80.2 83.4	9.94 8.71 8.07
Methacycline	ND	2.0 20.0 200.0	77.3 85.1 86.4	10.8 9.76 8.33	Oxolinic acid	ND	2.0 20.0 200.0	77.3 83.9 79.8	10.4 9.29 9.38
Oxytetracycline	ND	2.0 20.0 200.0	74.8 76.1 75.5	11.8 10.1 9.78	Nalidixic acid	ND	2.0 20.0 200.0	79.2 84.6 81.7	8.93 7.10 5.81
Danofloxacin	ND	2.0 20.0 200.0	75.7 82.3 77.1	9.64 8.78 9.15	Flumequine	ND	2.0 20.0 200.0	87.4 92.1 88.3	8.94 5.89 6.37

References

1. Qiao, M.; Ying, G.G.; Singer, A.C.; Zhu, Y.G. Review of antibiotic resistance in China and its environment. *Environ. Int.* **2018**, *110*, 160–172. [CrossRef] [PubMed]
2. Liu, Y.-Y.; Wang, Y.; Walsh, T.R.; Yi, L.-X.; Zhang, R.; Spencer, J.; Doi, Y.; Tian, G.; Dong, B.; Huang, X.; et al. Emergence of plasmid-mediated colistin resistance mechanism MCR-1 in animals and human beings in China: A microbiological and molecular biological study. *Lancet Infect. Dis.* **2016**, *16*, 161–168. [CrossRef] [PubMed]
3. Zhang, Q.Q.; Ying, G.G.; Pan, C.G.; Liu, Y.S.; Zhao, J.L. Comprehensive Evaluation of Antibiotics Emission and Fate in the River Basins of China: Source Analysis, Multimedia Modeling, and Linkage to Bacterial Resistance. *Environ. Sci. Technol.* **2015**, *49*, 6772–6782. [CrossRef] [PubMed]

4. Da, V. Origins and Evolution of Antibiotic Resistance. *Microbiologia* **1996**, *12*, 9.
5. Do, K.-H.; Byun, J.-W.; Lee, W.-K. Antimicrobial Resistance, Adhesin and Toxin Genes of Porcine Pathogenic *Escherichia coli* Following the Ban on Antibiotics as the Growth Promoters in Feed. *Pak. Vet. J.* **2021**, *41*, 519–523.
6. Hao, R.-Z.; Zhao, R.-T.; Qiu, S.-F.; Wang, L.-G.; Song, H.-B. Antibiotics crisis in China. *Science* **2015**, *348*, 1100–1101. [CrossRef] [PubMed]
7. Ullah, M.; Rasool, F.; Khan, N.; Ali, S.; Sheikh, A.A. Antibiotic Resistance and its Gene Profile in *Escherichia coli* Isolated from Diseased FarmRaised Carps in Punjab, Pakistan. *Pak. Vet. J.* **2023**, *41*, 1–7.
8. Ishaq, K. Occurrence and Antimicrobial Susceptibility of Proteus Mirabilis from Chicken Carcass. *Pak. Vet. J.* **2022**, *42*, 576–579. [CrossRef]
9. Nawaz, Z.; Aslam, B.; Zahoor, M.A.; Siddique, A.B.; Rafique, A.; Aslam, R.; Qamar, M.U.; Ali, S.; Mubeen, M. Frequency of Extended Spectrum Beta Lactamase Producing Escherichia Coli in Fresh and Frozen Meat. *Pak. Vet. J.* **2021**, *41*, 102–106. [CrossRef]
10. Anwar, M.A.; Aziz, S.; Ashfaq, K.; Aqib, A.I.; Shoaib, M.; Naseer, M.A.; Alvi, M.A.; Muzammil, I.; Bhutta, Z.A.; Sattar, H.; et al. Trends in Frequency, Potential Risks and Antibiogram of E. coli Isolated from Semi-Intensive Dairy Systems. *Pak. Vet. J.* **2022**, *4*, 167–172.
11. Saeed, E.; Amer, A.A.; Keshta, H.G.; Khalifa, E.A. Prevalence and Antibiogram of Salmonella Enterica Isolated from Raw Dromedary Camel Milk in Matrouh Governorate, Egypt. *Int. J. Vet. Sci.* **2022**, *11*, 168–174.
12. Abouelhag, H.A.; Khairy, E.A.; Marie, H.S.; Khalaf, D.D. Prevalence, Antibiogram Pattern and Virulence Genes Profile of Bacillus cereus Isolated from Buffalo Milk. *Int. J. Vet. Sci.* **2021**, *10*, 234–239.
13. Bakry, N.M.; Awad, W.S.; Ahmed, S.A. Virulence Genes, Antibiotic Resistance and Phylotyping of Escherichia coli O157 Recovered from Diarrheic Calves. *Int. J. Vet. Sci.* **2021**, *10*, 1–7.
14. Nelson, F.E.; Jensen, L.S.; Mcginns, J. Studies on the Stimulation of Growth by Dietary Antibiotics: 1. Changes in Growth Response of Chicks to Antibiotics over a three year Period. *Poult. Sci.* **1963**, *42*, 906–909. [CrossRef]
15. Carlet, J.; Rambaud, C.; Pulcini, C. Save Antibiotics: A call for action of the World Alliance Against Antibiotic Resistance (WAAAR). *BMC Infect. Dis.* **2014**, *14*, 436–440. [CrossRef] [PubMed]
16. The World Is Running out of Antibiotics, WHO Report Confirms. Available online: https://www.who.int/news/item/20-09-2017-the-world-is-running-out-of-antibiotics-who-report-confirms (accessed on 20 September 2017).
17. Merkatoris, P.; Schleining, J.; Kruil, A.; Borts, D.; Fajt, V. In vitro Elution of Penicillin, Ampicillin, Tetracycline, Tulathromycin, and Florfenicol From Plaster of Paris Beads. *Front. Vet. Sci.* **2020**, *7*, 585423. [CrossRef] [PubMed]
18. Woman Killed by a Superbug Resistant to Every Available Antibiotic—Scientific American. Available online: https://www.scientificamerican.com/article/woman-killed-by-a-superbug-resistant-to-every-available-antibiotic/ (accessed on 13 January 2017).
19. Commission Regulation. Commission Decision (EU) No 37/2010 of December 2009 on pharmacologically active substances and their classification regarding maximum residue limits in foodstuffs of animal origin. *Off. J. Eur. Union* **2010**, *15*, 1–72.
20. Codex Alimentatius Commission; Food and Agriculture Organization of the United Nations; World Health Organization. Maximum Residue Limits (MRLs) and Risk Management Recommendations (RMRs) for Residues of Veterinary Drugs in Foods. (CX/MRL 2-2021). Available online: https://www.fao.org/fao-who-codexalimentarius/codex-texts/maximum-residue-limits/en/ (accessed on 27 July 2023).
21. Getahun, M.; Abebe, R.B.; Sendekie, A.K.; Woldeyohanis, A.E.; Kasahun, A.E.; Regmi, B. Evaluation of Antibiotics Residues in Milk and Meat Using Different Analytical Methods. *Int. J. Anal. Chem.* **2023**, *2023*, 1–13. [CrossRef]
22. Chung, H.-H.; Lee, J.-B.; Chung, Y.-H.; Lee, K.-G. Analysis of sulfonamide and quinolone antibiotic residues in Korean milk using microbial assays and high performance liquid chromatography. *Food Chem.* **2009**, *113*, 297–301. [CrossRef]
23. Hernandez-Arteseros, J.A.; Barbosa, J.; Compaó, R.; Prat, M. Analysis of quinolone residues in edible animal products. *J. Chromatogr. A* **2002**, *945*, 1–24. [CrossRef]
24. Muscarella, M.; Magro, S.L.; Palermo, C.; Centonze, D. Validation according to European Commission Decision 2002/657/EC of a confirmatory method for aflatoxin M 1 in milk based on immunoaffinity columns and high performance liquid chromatography with fluorescence detection. *Anal. Chim. Acta* **2007**, *594*, 257–264. [CrossRef]
25. Önal, A. Overview on liquid chromatographic analysis of tetracycline residues in food matrice. *Food Chem.* **2011**, *127*, 197–203. [CrossRef]
26. Sinem, R.; Sezer, A.; Zge, E.; Emir, A.S. Validation of two UHPLC-MS/MS methods for fast and reliable determination of quinolone residues in honey. *Food Addit. Contam. Part A* **2021**, *38*, 807–819.
27. Yamaguchi, T.; Okihashi, M.; Harada, K.; Uchida, K.; Konishi, Y.; Kajimura, K.; Hirata, K.; Yamamoto, Y. Rapid and easy multiresidue method for the analysis of antibiotics in meats by ultrahigh-performance liquid chromatography-tandem mass spectrometry. *J. Agric. Food Chem.* **2015**, *63*, 5133. [CrossRef] [PubMed]
28. Zhang, L.; Shi, L.; He, Q.; Li, Y. A rapid multiclass method for antibiotic residues in goat dairy products by UPLC-quadrupole/electrostatic field orbitrap high-resolution mass spectrometry. *J. Anal. Sci. Technol.* **2021**, *12*, 14. [CrossRef]
29. Zhou, Y.; Guan, J.; Gao, W.; Lv, S.; Ge, M. Quantification and Confirmation of Fifteen Carbamate Pesticide Residues by Multiple Reaction Monitoring and Enhanced Product Ion Scan Modes via LC-MS/MS QTRAP System. *Molecules* **2018**, *23*, 2496. [CrossRef] [PubMed]

30. Li, Y.-Y.; Wang, H.; Zhao, C.; Huang, Y.-Q.; Tang, X. Identification and Characterization of Kukoamine Metabolites by Multiple Ion Monitoring Triggered Enhanced Product Ion Scan Method with a Triple-Quadruple Linear Ion Trap Mass Spectrometer. *J. Agric. Food Chem.* **2015**, *63*, 10785–10790. [CrossRef]
31. Lode, H. Development of clinically important resistance to quinolones and other antibiotic groups. *J. Int. Med. Res.* **1990**, *18*, 27–36.
32. Roberts, M.C. *Tetracyclines: Mode of Action and Their Bacterial Mechanisms of Resistance, (Bacterial Resistance to Antibiotics—From Molecules to Man)*; Wiley Online Library: Hoboken, NJ, USA, 2019; pp. 101–124.
33. Yan, M.; Chen, X.; Huang, Y.; Nie, H.; Wang, J. Tetracyclines, sulfonamides and quinolones and their corresponding resistance genes in the Three Gorges Reservoir, China. *Sci. Total Environ.* **2018**, *631–632*, 840–848. [CrossRef]
34. Shao, B.; Jia, X.; Wu, Y.; Hu, J.; Tu, X.; Zhang, J. Multi-class confirmatory method for analyzing trace levels of tetracyline and quinolone antibiotics in pig tissues by ultra-performance liquid chromatography coupled with tandem mass spectrometry. *Rapid Commun. Mass Spectrom.* **2007**, *21*, 3487–3496. [CrossRef]
35. Ajibola, A.S.; Tisler, S.; Zwiener, C. Simultaneous determination of multiclass antibiotics in sewage sludge based on QuEChERS extraction and liquid chromatography-tandem mass spectrometry. *Anal. Methods* **2020**, *12*, 576–586. [CrossRef]
36. Chen, L.; Qi, W.Y.; Zhao, Z.Y.; Zhao, X.Y.; Zhou, C.Y. Rapid Determination of 19 Fluoroquinolones Antibiotic Residues in Soil by QuEChERS/Ultra Performance Liquid Chromatography-Tandem Mass Spectrometry. *J. Instrum. Anal.* **2019**, *38*, 194–200.
37. Frenich, A.G.; Romero-González, R.; Gómez-Pérez, M.; Vidal, J. Multi-mycotoxin analysis in eggs using a QuEChERS-based extraction procedure and ultra-high-pressure liquid chromatography coupled to triple quadrupole mass spectrometry. *J. Chromatogr. A* **2011**, *1218*, 4349–4356. [CrossRef] [PubMed]
38. Gao, Y.; Zhang, T.; Huang, S.; Lin, X.; Chen, M. Screening of 49 antibiotic residues in aquatic products using modified QuEChERS sample preparation and UPLC-QToFMS analysis. *PeerJ Anal. Chem.* **2021**, *3*, e8. [CrossRef]
39. Wilkowska, A.; Biziuk, M. Determination of pesticide residues in food matrices using the QuEChERS methodology. *Food Chem.* **2011**, *125*, 803–812. [CrossRef]
40. Zhang, C.; Deng, Y.; Zheng, J.; Zhang, Y.; Luo, A. The Application of the QuEChERS Methodology in the Determination of Antibiotics in Food: A review. *TrAC Trends Anal. Chem.* **2019**, *118*, 517–537. [CrossRef]
41. Andre, I.K.; Siuli, M. Response Surface Methodology. *Comput. Stat.* **2010**, *2*, 128–149.
42. Myers, R.H.; Montgomery, D.C.; Anderson-Cook, C.M. *Response Surface Methodology: Process and Product Optimization Using Designed Experiments*, 4th ed.; John Wiley & Sons: Hoboken, NJ, USA, 2016.
43. Bezerra, M.A.; Santelli, R.E.; Oliveira, E.P.; Villar, L.S.; Escaleira, L.A. Response surface methodology (RSM) as a tool for optimization in analytical chemistry. *Talanta* **2008**, *76*, 965–977. [CrossRef]
44. Ferreira, S.L.C.; Bruns, R.E.; Ferreira, H.S.; Matos, G.D.; David, J.M.; Brandao, G.C.; da Silva, E.G.P.; Portugal, L.A.; Reis, P.S.; Souza, A.S.; et al. Box-Behnken design An alternative for the optimization of analytical methods. *Anal. Chim. Acta* **2007**, *597*, 179–186. [CrossRef]
45. Li, Y.W.; Mo, C.H.; Zhao, N.; Zhang, R.J.; Yi, R.H. Determination of Sulfonamides Antibiotics in Water and Soil Using High Performance Liquid Chromatography. *Chin. J. Anal. Chem.* **2008**, *36*, 954–958.
46. Luo, H.T.; Huang, X.L.; Wu, H.Q.; Zhu, Z.X.; Lin, X.S. Simultaneous Determination of 33 Medicine Residues in Aquatic Products by Rapid Resolution Liquid Chromatography-Tandem Mass Spectrometry. *Chin. J. Anal. Chem.* **2012**, *40*, 273–279. [CrossRef]
47. Yue, Z.; Qiu, Y.; Liu, X.; Ji, C. Determination of Multi-Residues of Tetracyclines and Their Metabolites in Milk by High Performance Liquid Chromatography-Tandem Positive-ion Electrospray Ionization Mass Spectrometry. *Chin. J. Anal. Chem.* **2006**, *34*, 1255–1259. [CrossRef]
48. Sapadin, A.N.; Fleischmajer, R. Tetracyclines: Nonantibiotic properties and their clinical implications. *J. Am. Acad. Dermatol.* **2006**, *54*, 258–265. [CrossRef] [PubMed]
49. Gao, J.; Xu, H.; Li, Q.J.; Feng, X.H.; Li, S. Optimization of medium for one-step fermentation of inulin extract from Jerusalem artichoke tubers using Paenibacillus polymyxa ZJ-9 to produce R,R-2,3-butanediol. *Bioresour. Technol.* **2010**, *101*, 7076–7082. [CrossRef] [PubMed]
50. Liu, J.Z.; Weng, L.P.; Zhang, Q.L.; Xu, H.; Ji, L.N. Optimization of glucose oxidase production by Aspergillus niger in a benchtop bioreactor using response surface methodology. *World J. Microbiol. Biotechnol.* **2003**, *19*, 317–323. [CrossRef]
51. Amin, M.K.; Hassan, G.F.; Phyllis, R.B.; Burnaby, M. Mass spectral characterization of tetracyclines by electrospray ionization, H/D exchange, and multiple stage mass spectrometry. *J. Am. Soc. Mass Spectrom.* **2002**, *13*, 543–557.
52. Bin, M.; Yuming, S.; Chen, X. Analysis of fluroquinolones by electrosprayion trap mass spectrometry. *J. Chin. Mass Spectrom. Soc.* **2006**, *27*, 163–167.
53. Matraszek-Zuchowska, I.; Wozniak, B.; Posyniak, A. Comparison of the Multiple Reaction Monitoring and Enhanced Product Ion Scan Modes for Confirmation of Stilbenes in Bovine Urine Samples Using LC–MS/MS QTRAP®System. *Chromatographia* **2016**, *79*, 1003–1012. [CrossRef]
54. *GB/T 27404-2008*; Criterion on Quality Control of Laboratories. Chemical Testing of Food. National Certification and Accreditation Standardization Technical Committee: Beijing, China, 2008.

55. Treiber, F.M.; Beranek-Knauer, H. Antimicrobial Residues in Food from Animal Origin-a Review of the Literature Focusing on Products Collected in Stores and Markets Worldwide. *Antibiotics* **2021**, *10*, 534. [CrossRef]
56. *GB 31650-2019*; National Food Safety Standard Maximum Residue Limits for Veterinary Drugs in Foods. Ministry of Agriculture and Rural Affairs; National Health Commission; State Administration for Market Regulation: Beijing, China, 2019.

molecules MDPI

Article

Ultra-High-Performance Liquid Chromatography Quadrupole Time-of-Flight Mass Spectrometry for Simultaneous Pesticide Analysis and Method Validation in Sweet Pepper

Han Yeol Bang [1,†], Yong-Kyoung Kim [2,†], Hyoyoung Kim [2], Eun Joo Baek [2], Taewoong Na [2], Kyu Sang Sim [2] and Ho Jin Kim [2,*]

1 Gyeongnam Provincial Office, National Agricultural Products Quality Management Service, Busan 47537, Republic of Korea; bbangkkc@korea.kr
2 Experiment Research Institute, National Agricultural Products Quality Management Service, Gimcheon-si 39660, Republic of Korea; ykkim79@korea.kr (Y.-K.K.); hyo02@korea.kr (H.K.); qrgh1004@naver.com (E.J.B.); naratw@korea.kr (T.N.); sim9612@naver.com (K.S.S.)
* Correspondence: rex7878@korea.kr; Tel.: +82-54-429-7871
† These authors contributed equally to this work.

Abstract: Pesticides effectively reduce the population of various pests that harm crops and increase productivity, but leave residues that adversely affect health and the environment. Here, a simultaneous multicomponent analysis method based on ultra-high-performance liquid chromatography quadrupole time-of-flight mass spectrometry (UHPLC-QTOF-MS) pretreated by the QuEChERS method was developed to control the maximum residual levels. Among the 140 pesticides with high frequency of detection in agricultural products in Gyeongnam region in Korea for 5 years, 12 pesticides with high detection frequency in sweet pepper were selected. The analytical method is validated, linearities are $r^2 > 0.999$, limit of detection (LOD) ranges from 1.4 to 3.2 μg/kg, and limit of quantification (LOQ) ranges from 4.1 to 9.7 μg/kg, and the recovery rate was 81.7–99.7%. In addition, it was confirmed that a meaningful value of these parameters can be achieved by determining the measurement uncertainty. The results proved that parameters such as recovery rate and relative standard deviation of the analysis method were within international standards. Using the developed method, better and safer sweet peppers will be provided to consumers, and effective pesticide residue management will be possible by expanding to other agricultural products.

Keywords: measurement uncertainty; method validation; pesticide residue; sweet pepper; UHPLC-QTOF-MS

Citation: Bang, H.Y.; Kim, Y.-K.; Kim, H.; Baek, E.J.; Na, T.; Sim, K.S.; Kim, H.J. Ultra-High-Performance Liquid Chromatography Quadrupole Time-of-Flight Mass Spectrometry for Simultaneous Pesticide Analysis and Method Validation in Sweet Pepper. *Molecules* **2023**, *28*, 5589. https://doi.org/10.3390/molecules28145589

Academic Editors: Chao Kang and Ronald Beckett

Received: 14 June 2023
Revised: 13 July 2023
Accepted: 21 July 2023
Published: 22 July 2023

1. Introduction

Pesticides help maintain production by efficiently reducing the population of various pests that harm crops. However, their use also leads to the formation of pesticide residues on crops, which adversely affect health and the environment. Therefore, it is essential to regulate their usage, for which various standards such as CODEX and EU have been developed to manage their maximum residue limits [1]. Pesticides are generally spread in the environment through agricultural water or rainfall, and when highly volatile, disperse as aerosols [2], causing a variety of environmental problems. Therefore, continuous and intensive use of pesticides pollutes the soil and reduces the diversity of plants and animals, thereby threatening the stability of the entire ecosystem [3]. In addition, humans exposed to pesticides can develop various life-threatening diseases such as cancer and genetic disorders [4]. Especially, oral intake is more dangerous than exposure through the skin [5].

Fresh fruits and vegetables are rich in antioxidants such as vitamins and polyphenols [6]. These antioxidants reduce the amounts of free radicals present in the body and prevent damage to DNA and cells of the human body [7]. Sweet pepper contains large amounts of polyphenols, flavonoids, aglycones, and glycosides [8], and these phytochemicals can prevent cardiovascular disease, Alzheimer's disease, and Parkinson's disease [9]. Therefore, approximately 3000 ton of sweet pepper is produced and consumed annually worldwide. Sweet peppers are consumed not only for their taste but also for their protective action against various diseases [10].

However, the increase in the production and consumption of sweet pepper has also increased the use of pesticides. Pesticide residues within fruits and vegetables in high concentrations is a major route of pesticide exposure [11], and it is also possible that the pesticide contents in fruits and vegetables may increase or transform into more toxic metabolites during the manufacturing process [12]. Some pesticides remain in the sweet pepper in large amounts even after cooking [13], affecting the health of consumers. Hence, it is essential to develop appropriate preprocessing and analysis methods to determine the presence of pesticide residues in sweet pepper.

Some phytochemicals such as flavonoids and polyphenols interfere with the detection of target analytes through matrix effects [14]. To overcome these effects, suitable preprocessing methods such as solid-phase extraction (SPE) and liquid–liquid extraction (LLE) [15] and QuEChERS for positive matrix effects [16] have been previously employed. Multiple studies have analyzed pesticide residues in sweet pepper by using QuEChERS along with gas chromatography triple quadrupole mass spectrometry (GC–MS/MS) and liquid chromatography (LC) MS/MS [17–19]. Nevertheless, unlike the analysis using MS/MS, very few studies have investigated the analysis of pesticide residues in sweet pepper using quadrupole time-of-flight mass spectrometry (QTOF-MS).

QTOF-MS is known to reduce the deterioration of the peak shape that occurs when multiple compounds are screened. In addition, QTOF-MS has a relatively high resolving power that helps minimize the false-positive phenomenon that occurs when similar elements are analyzed [20]. While only a limited number of analytes can be simultaneously investigated through MS/MS, QTOF-MS has a relatively broad spectrum, high sensitivity, and allows for retrospective analysis. Therefore, QTOF-MS is gaining increasing attention as a highly useful tool [21,22]. It has already been used in various fields such as for the analysis of veterinary drug and pesticide residues in pig muscle [23] and phenolic compounds present in plums [24].

In this study, QuEChERS was used as a preprocessing method for pesticide residue analysis in sweet pepper, and UHPLC-QTOF-MS was used to obtain more reliable quantitative and qualitative results than previously developed methods using LC–MS/MS and GC–MS/MS. Method validation was performed using different parameters including measurement uncertainty, and the significance of the experiment was demonstrated by analyzing error factors that may occur during the experiment.

2. Results and Discussion

2.1. Simultaneous Multicomponent Analyses

From 2015 to 2020, 140 pesticides with a history of detection in the Gyeongnam region of Korea were selected through UHPLC-QTOF-MS analysis, and multiple reaction monitoring (MRM) for pesticides is shown in Table S1. The precursor ions of all analytes were successfully analyzed within the range of 163.05–746.48 m/z. Fragment ions of the five analytes with the highest sensitivity were selected. Among these five, the top two fragment ions with the highest sensitivity were selected as representative ions, which were then screened to determine a suitable retention time.

Thereafter, the 12 pesticide residues were selected and used to perform optimization (Table 1). In the case of acequinocyl, cyflumetofen, and procymidone, the experimental m/z is different from the calculated m/z, because of the presence of an NH_4 adduct that increases their stability. No significant difference was observed in the retention time from that reported in previous MRM settings. The most sensitive fragment of each material was selected and used for validation.

Table 1. Selected 12 pesticides and their analysis conditions.

Compound	Formula	Calculated m/z	Experimental m/z	Ionization Mode	Fragment Ion (m/z)	Mass Error (ppm)
Acequinocyl	$C_{24}H_{32}O_4$	384.2295	407.2638	$[M + NH_4]^+$	343.2288	0.3
Boscalid	$C_{18}H_{12}Cl_2N_2O$	342.0321	343.0399	$[M + H]^+$	307.0651	2.9
Cyflumetofen	$C_{24}H_{24}F_3NO_4$	447.1615	465.1995	$[M + NH_4]^+$	173.0222	0.9
Dinotefuran	$C_7H_{14}N_4O_3$	202.1060	203.1138	$[M + H]^+$	129.0911	−0.1
Flonicamid	$C_9H_6F_3N_3O$	229.0457	230.0535	$[M + H]^+$	203.0442	0.5
Fluopyram	$C_{16}H_{11}ClF_6N_2O$	396.0458	397.0536	$[M + H]^+$	173.0222	−1.8
Procymidone	$C_{13}H_{11}Cl_2NO_2$	283.0161	301.0505	$[M + NH_4]^+$	284.0272	3.7
Propamocarb	$C_9H_{20}N_2O_2$	188.1519	189.1597	$[M + H]^+$	102.0559	−1.3
Pyridaben	$C_{19}H_{25}ClN_2OS$	364.1370	365.1448	$[M + H]^+$	309.0840	0.7
Spirodiclofen	$C_{21}H_{24}Cl_2O_4$	410.1046	411.1124	$[M + H]^+$	313.0398	0.2
Spirotetramat	$C_{21}H_{27}NO_5$	373.1883	374.1962	$[M + H]^+$	330.2078	0.1
Spirotetramat-enol	$C_{18}H_{23}NO_3$	301.1672	302.1750	$[M + H]^+$	216.1031	−0.4

2.2. Method Validation

Method validation was performed following the guidelines of the European Commission [25] and EURACHEM guides [26]. The method was validated for selectivity, linearity, sensitivity, accuracy, and precision to confirm its effectiveness for the analysis of pesticide residues in sweet pepper (Table 2).

Table 2. Validation parameters of the developed UHPLC-QTOF-MS method.

Compound	r^2	Coefficient of Variation (%)			Recovery (%)			LOD [a] (μg/kg)	LOQ [b] (μg/kg)	ME (%) [c]	MU (%) [d]
		10 μg/kg	50 μg/kg	100 μg/kg	10 μg/kg	50 μg/kg	100 μg/kg				
Acequinocyl	0.99985	7.1	2.1	3.3	97.7 ± 0.7	97.8 ± 1.0	99.5 ± 3.3	3.2	9.7	−13	9.1
Boscalid	0.99992	19.1	4.5	2.7	89.2 ± 1.7	98.2 ± 2.2	92.8 ± 2.5	2.1	6.3	−7	12.1
Cyflumetofen	0.99962	5.4	1.8	1.4	97.0 ± 0.5	99.7 ± 0.9	90.1 ± 1.2	2.3	7.0	3	16.2
Dinotefuran	0.99951	16.3	7.7	5.9	92.3 ± 1.5	95.7 ± 3.7	87.7 ± 5.2	2.8	8.4	10	16.3
Flonicamid	0.99960	10.1	1.8	4.3	95.3 ± 1.0	98.9 ± 0.9	94.3 ± 4.0	2.2	6.6	−1	15.2
Fluopyram	0.99958	19.0	3.7	1.8	94.7 ± 1.8	97.2 ± 1.8	89.5 ± 1.6	2.4	7.3	1	14.5
Propamocarb	0.99983	17.7	4.4	3.5	95.2 ± 1.7	98.4 ± 2.2	88.7 ± 3.1	2.1	6.2	1	11.2
Procymidone	0.99970	15.7	5.5	6.4	93.7 ± 1.5	98.0 ± 2.7	97.1 ± 6.2	2.9	8.7	7	12.8
Pyridaben	0.99996	15.1	4.2	3.2	92.6 ± 1.4	98.1 ± 2.1	92.4 ± 3.0	2.3	7.0	−5	13.9
Spirodiclofen	0.99977	12.7	3.6	0.8	86.9 ± 1.1	96.1 ± 1.7	93.0 ± 0.8	2.5	7.6	−5	18.6
Spirotetramat	0.99992	19.0	4.6	2.5	88.9 ± 1.7	97.6 ± 2.3	93.5 ± 2.3	3.0	9.0	−9	13.5
Spirotetramat-enol	0.99993	19.1	6.2	2.0	87.5 ± 1.7	90.4 ± 2.8	81.7 ± 1.6	1.4	4.1	−12	11.6

[a] LOD: limit of detection, [b] LOQ: limit of quantification, [c] ME: matrix effect, [d] MU: measurement uncertainty.

Selectivity was determined based on the presence or absence of interfering peaks in the chromatography. As shown in Figure 1, the selected 12 pesticide residues were observed by conducting separate chromatography analyses within 15 min. The selectivity was found to be excellent and the separation was successful.

Linearity was evaluated based on a calibration curve using five different concentrations (5, 10, 25, 50, and 100 μg/L) of the mixture of each standard. The mixed solution was injected three times for evaluation and a formula was derived using the obtained values. The selected materials showed positive results ($r^2 > 0.999$), possibly owing to the high selectivity of QTOF-MS. Hence, it can be suggested that matrix-matched external calibration using a standard can be used for quantitative purposes.

Sensitivity was evaluated by determining the limits of detection (LOD) and limits of quantification (LOQ). The LOD and LOQ were calculated using the standard deviation of the value obtained from multiple replicates of a sample with the lowest concentration (10 μg/kg). Acequinocyl showed the highest LOD, while spirotetramat-enol showed the lowest LOD. The LOQ also exhibited the same trend. The LOD of the 12 pesticide residues ranged from 1.4 to 3.2 μg/kg, while their LOQ ranged from 4.1 to 9.7 μg/kg. Therefore, all 12 residual pesticides used in the analysis were found to have suitable sensitivity for analyzing sweet pepper.

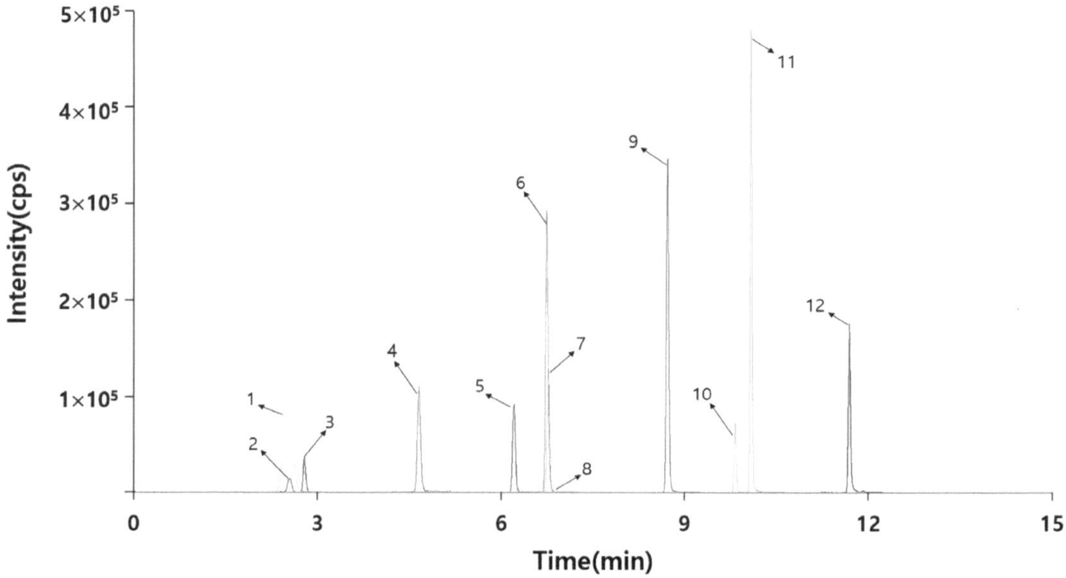

Figure 1. Total ion chromatogram (TIC) of 12 pesticide residues.

The precision of the experimental method was determined by determining their intraday precisions. The pesticide residue solution was measured three times a day, and the result was expressed as the percentage of the coefficient of variation (CV). The recovery rate for each residual pesticide was calculated by comparing the samples (10, 50, and 100 μg/kg) spiked with sweet pepper blank and standard mixture at each concentration. The following CV values were obtained for the pesticides at different concentrations: 5.4–19.1% for 10 μg/kg, 1.8–7.7% for 50 μg/kg, and 0.8–6.4% for 100 μg/kg. The recovery rate showed a slight difference depending on the concentration of each material, although all the recovery rates were between 80% and 110%. Hence, the proposed experimental method meets the criteria of the presented guideline [25].

Therefore, this method shows an appropriate level of LOD and LOQ, a CV of less than 20%, and a recovery rate of 70 to 120%, as specified in the EU guidelines [25], so that significant results can be obtained in the simultaneous multicomponent analysis of pesticide residues.

Other papers obtained analysis results only using a triple quadrupole, but this paper can more accurately identify the detected pesticide by quantifying it using a triple quadrupole and qualitatively confirming the molecular weight of the pesticide component to four decimal places using QTOF-MS.

2.3. *Measurement Uncertainty*

Measurement uncertainty was calculated according to the Guide to the Expression of Uncertainty in Measurement [27]. The measurement uncertainty is used as an indicator of the reliability of the analysis result by presenting a range estimated to be the actual value. In this experiment, the following uncertainty factors were considered during the analysis: sample weight, final volume, a stock standard used when preparing the calibration curve, and working standard. Various factors such as certification, temperature, and repeatability were also used to estimate uncertainty (Figure 2). First, sample weights and final volumes are common to all analyses. Balance, repeatability, and stability are generally used as uncertainty factors for sample weight, and 10 mL pipettes as uncertainty factors for the final volume (Table 3). The uncertainty in the calibration curve concentration arises owing to the following three factors: stock standard solution (100 mg/L), working standard solution (1 mg/L), and calibration. Standard material purity, balance, and volumetric flask are also considered uncertainty factors as stock standard solutions are prepared at 100 mg/L of the analyte. The working standard solution is a manufacturing process for diluting the stock solution to 1 mg/L, and the stock solution, pipette, and volumetric flask are considered the uncertainty factors. A calibration curve of 5–100 μg/L is prepared by appropriately diluting the working standard solution, and the result of the 10 μg/L addition test is used as an uncertainty factor (Table 4). The relative standard uncertainty is obtained by combining each standard uncertainty, and the relative combined standard uncertainty, combined standard uncertainty, and expanded uncertainty are sequentially obtained, and finally the measurement uncertainty is calculated. The result of the calculated measurement uncertainty was between 9.1% and 18.6% (Table 5). Spirodiclofen showed the highest measurement uncertainty of 18.6%, although the guidelines on measurement uncertainty suggested that a value of <44% was significant at a measurement concentration of 10 μg/kg [28]. Therefore, it can be concluded that all compounds meet the criteria for measurement uncertainty. By calculating the measurement uncertainty for 12 pesticides, errors that may occur during the experiment were confirmed and minimized.

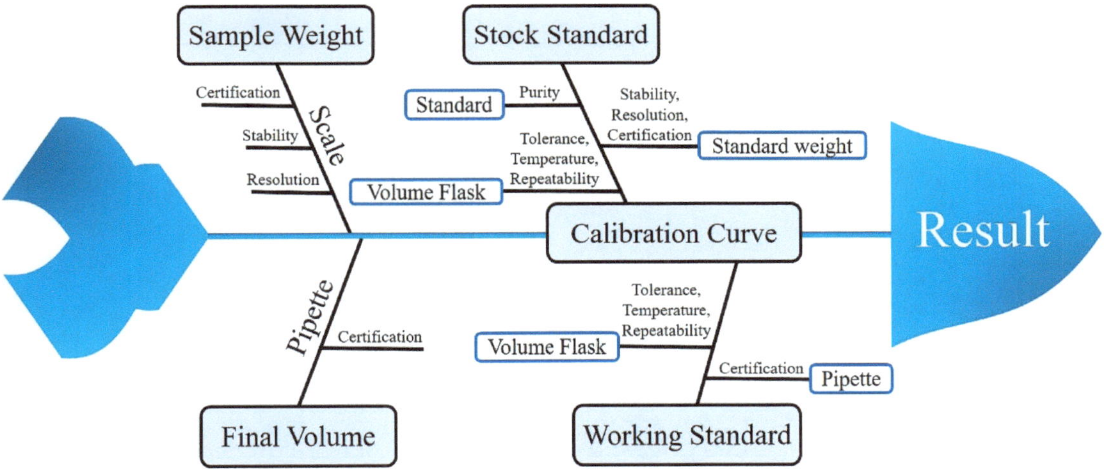

Figure 2. Measurement uncertainty diagram.

Table 3. Uncertainty of sample weight and final volume.

Parameter	Value (x_i)	Source		Type	Standard Uncertainty (u)	Combined Standard Uncertainty (u_c)	Relative Standard Uncertainty (u_r)
Sample weight	10.0335	Balance	Certification Readability Stability	B A A	0.000050 0.000029 0.000052	0.000078	0.000008
Final volume	10	Pipette	Certification	B	0.006500	0.006500	0.000650

Table 4. Uncertainty of calibration curve.

Source		Value (x_i)	Standard Uncertainty (u)	1st Combined Standard Uncertainty (u_c)	2nd Combined Standard Uncertainty (u_c)
Stock standard solution (100 mg/L)	Purity	0.999	0.000577	0.696323	
	Balance	0.01	0.000061		
	Volumetric flask	100	0.328927		
Working standard solution (1 mg/L)	Stock standard solution	100	0.696323	0.007717	
	Pipette	1	0.000500		
	Volumetric flask	100	0.328927		
Calibration curve concentration	Acequinocyl	9.77	0.433570	0.433570	0.445305
	Boscalid	8.92	0.530122	0.530122	0.538169
	Cyflumetofen	9.70	0.776454	0.776454	0.782973
	Dinotefuran	9.23	0.746118	0.746118	0.752261
	Flonicamid	9.53	0.710811	0.710811	0.717680
	Fluopyram	9.47	0.680736	0.680736	0.687816
	Propamocarb	9.52	0.519986	0.519986	0.529318
	Procymidone	9.37	0.591857	0.591857	0.599817
	Pyridaben	9.26	0.631509	0.631509	0.638802
	Spirodiclofen	8.69	0.799879	0.799879	0.804963
	Spirotetramat	8.89	0.590417	0.590417	0.597605
	Spirotetramat-enol	8.75	0.495735	0.495735	0.504009

Table 5. Measurement uncertainties of selected 12 pesticide.

Compounds	Uncertainty Factor	Standard Uncertainty (u)	Relative Standard Uncertainty (u_r)	Relative Combined Standard Uncertainty (u_{rc})	Combined Standard Uncertainty (u_c)	Extended Uncertainty (U)	Measurement Uncertainty (Confidence Level about 95%, k = 2)
Acequinocyl	Sample weight	0.000078 g	0.000008	0.045583	0.444 μg/L	0.888 μg/L	9.74 ± 0.89 μg/L (9.1%)
	Final volume	0.006500 mL	0.000650				
	Calibration curve	0.445305 μg/L	0.045579				
Boscalid	Sample weight	0.000078 g	0.000008	0.060336	0.536 μg/L	1.073 μg/L	8.89 ± 1.08 μg/L (12.1%)
	Final volume	0.006500 mL	0.000650				
	Calibration curve	0.538169 μg/L	0.060333				
Cyflumetofen	Sample weight	0.000078 g	0.000008	0.080721	0.780 μg/L	1.561 μg/L	9.67 ± 1.57 μg/L (16.2%)
	Final volume	0.006500 mL	0.000650				
	Calibration curve	0.782973 μg/L	0.080719				
Dinotefuran	Sample weight	0.000078 g	0.000008	0.081504	0.750 μg/L	1.500 μg/L	9.20 ± 1.50 μg/L (16.3%)
	Final volume	0.006500 mL	0.000650				
	Calibration curve	0.752261 μg/L	0.081502				
Flonicamid	Sample weight	0.000078 g	0.000008	0.075310	0.715 μg/L	1.431 μg/L	9.50 ± 1.44 μg/L (15.2%)
	Final volume	0.006500 mL	0.000650				
	Calibration curve	0.717680 μg/L	0.075307				
Fluopyram	Sample weight	0.000078 g	0.000008	0.072634	0.686 μg/L	1.371 μg/L	9.54 ± 1.38 μg/L (14.5%)
	Final volume	0.006500 mL	0.000650				
	Calibration curve	0.687816 μg/L	0.072631				
Propamocarb	Sample weight	0.000078 g	0.000008	0.055604	0.528 μg/L	1.055 μg/L	9.49 ± 1.06 μg/L (11.2%)
	Final volume	0.006500 mL	0.000650				
	Calibration curve	0.529318 μg/L	0.055601				
Procymidnoe	Sample weight	0.000078 g	0.000008	0.064018	0.598 μg/L	1.196 μg/L	9.34 ± 1.20 μg/L (12.8%)
	Final volume	0.006500 mL	0.000650				
	Calibration curve	0.599817 μg/L	0.064015				
Pyridaben	Sample weight	0.000078 g	0.000008	0.068988	0.637 μg/L	1.273 μg/L	9.23 ± 1.28 μg/L (13.9%)
	Final volume	0.006500 mL	0.000650				
	Calibration curve	0.638802 μg/L	0.068985				
Spirodiclofen	Sample weight	0.000078 g	0.000008	0.092633	0.802 μg/L	1.605 μg/L	8.66 ± 1.61 μg/L (18.6%)
	Final volume	0.006500 mL	0.000650				
	Calibration curve	0.804963 μg/L	0.092631				
Spirotetramat	Sample weight	0.000078 g	0.000008	0.067225	0.596 μg/L	1.191 μg/L	8.86 ± 1.20 μg/L (13.5%)
	Final volume	0.006500 mL	0.000650				
	Calibration curve	0.597605 μg/L	0.067222				
Spirotetramat-enol	Sample weight	0.000078 g	0.000008	0.057605	0.502 μg/L	1.005 μg/L	8.72 ± 1.01 μg/L (11.6%)
	Final volume	0.006500 mL	0.000650				
	Calibration curve	0.504009 μg/L	0.057601				

2.4. Application of the Developed Method to Sweet Peppers

A total of 276 sweet pepper samples were collected from 15 areas in the Gyeongnam region in Korea. The number of samples of each city are different; details of the sample number and collecting area are shown in Table 6. The results show that 12 pesticides were analyzed for all samples and 10 pesticides were detected in 234 samples. Two pesticides, spirodiclofen and spirotetramat-enol, were not detected in the analyzed sweet pepper, and procymidone was detected in only two samples from Jinju. The total number of detections was 101 for boscalid, 81 for flonicamid, and 60 for pyridaben and spirotetramat. Boscalid, flonicamid, pyridaben, and spirotetramat, which have high detection frequencies, were detected at concentrations of 0.011–1.316, 0.01–0.485, 0.01–0.964, and 0.016–1.626 mg/kg, respectively (Table 6 and Table S2). The concentration of total pesticides detected ranged from 0.01 to 1.626 mg/kg. The Haman samples showed 43.2 and 28.7% detection rates of flonicamid and boscalid, respectively (Table S2). The obtained analysis data were quantified using the quadrupole mode, and qualitatively confirmed using the QTOF mode to confirm the results. All 10 pesticides detected in sweet pepper are considered to be safely managed below the maximum residue limits (MRLs) in Korea, which are regulatory limits that set the level of pesticides allowed to remain in foods to protect human health.

Table 6. Concentration ranges of 12 pesticide residues analyzed in sweet peppers collected from 15 areas in Gyeongnam in Korea.

	Pesticide Detection Concentration Range (mg/kg)											
	Acequinocyl	Boscalid	Cyflumetofen	Dinotefuran	Flonicamid	Fluopyram	Procymidone	Propamocarb	Pyridaben	Spirodiclofen	Spirotetramat	Spirotetramat-Enol
Gangseo	-	-	-	0.012	-	-	-	-	-	-	-	-
Geoje	-	0.018–0.623	-	-	-	-	-	-	0.01	-	-	-
Geochang	0.047	-	-	-	0.056	-	-	-	-	-	-	-
Goseong	-	0.013–1.316	0.036–0.091	0.032–0.622	0.02–0.109	0.018–0.162	-	0.017–0.04	0.013–0.37	-	0.125–0.585	-
Gimhae	0.115–0.796	0.131–0.728	-	0.027–0.224	0.016–0.119	-	-	-	0.035–0.102	-	0.041–0.231	-
Miryang	-	0.016–0.3	0.501	0.019	0.036–0.119	-	-	-	0.048–0.166	-	0.016	-
Sancheong	-	0.154	-	0.093–0.278	-	-	-	-	0.052	-	0.848–1.626	-
Uiryeong	-	0.404	0.033	0.036–0.12	0.015–0.02	-	-	-	0.217–0.362	-	0.493	-
Jinju	0.015–0.035	0.011–0.885	0.27	0.012–1.335	0.017–0.485	0.024–0.306	0.02–0.041	0.012–0.028	0.029–0.118	-	0.032–1.041	-
Changnyeong	0.02–0.113	0.121–0.47	-	0.071–0.723	0.129	-	-	0.059	0.024–0.568	-	0.026–0.349	-
Changwon	0.018–0.393	0.048–0.755	0.16–0.309	0.013–1.24	0.012–0.056	0.087–0.227	-	0.01–0.02	0.013–0.964	-	0.017–1.236	-
Tongyeong	-	-	-	0.07–0.376	0.017	-	-	0.019–0.041	0.15–0.515	-	-	-
Hadong	0.163	-	-	0.01–0.187	0.012	-	-	0.078	0.32	-	-	-
Haman	0.021–0.072	0.03–0.733	0.208	0.035–0.513	0.01–0.167	0.153	-	0.019–0.04	0.01–0.198	-	0.021–0.471	-
Hapcheon	-	0.018	-	-	-	-	-	-	-	-	0.121–0.272	-
	0.015–0.796	0.011–1.316	0.033–0.501	0.02–0.041	0.01–0.485	0.018–0.306	0.02–0.041	0.01–0.078	0.01–0.964	-	0.016–1.626	-

3. Materials and Methods

3.1. Chemicals and Reagents

Water, acetonitrile, and methanol was purchased from Merck KGaA (Darmstadt, Germany) and used as the solvents in the overall experiment with steps such as extraction and dilution of the sample. Formic acid (98%) and ammonium acetate (99%), required for solvent composition, were purchased from Sigma-Aldrich (Steinheim, Germany). The pesticide residue standards were purchased from Agilent Technologies (Santa Clara, CA, USA). The QuEChERS extraction kit and 2 mL of QuEChERS dispersive SPE used for purification were obtained from Agilent (Boblingen, Germany). The sweet pepper was purchased from a market in Gyeongnam and kept refrigerated at −4 °C.

3.2. Instrumentation and Conditions

Analysis of pesticide residues using UHPLC-QTOF-MS was performed as follows. First, a 5200 NASCA2 HPLC (Osaka, Japan) with a Waters ACQUITY UPLC BEH C18 (2.1 mm × 100 mm, 2.7 μm) column was used. The composition of mobile phases and other conditions such as gradient compositions and ion mode are listed in Table 7. Mass spectrometry detection was performed using QTOF (AB Sciex X500R QTOF, Sciex, Framingham, MA, USA), and final data processing was performed with SCIEX OS software (version no. 1.7.0.36606).

Table 7. Analytical conditions of UHPLC-QTOF-MS.

Mobile Phase A	5 mM ammonium acetate & 0.1% formic acid in water			
Mobile Phase B	5 mM ammonium acetate & 0.1% formic acid in methanol			
Gradient	Time (min)	A (%)	B (%)	Flow (mL/min)
	Initial	100	0	0.1
	0.2	100	0	0.1
	0.3	100	0	0.3
	0.5	50	50	0.3
	2.5	45	55	0.3
	5.5	25	75	0.3
	7.5	15	85	0.3
	8.3	0	100	0.3
	12.0	0	100	0.3
	12.1	100	0	0.3
	14.8	100	0	0.3
	14.9	100	0	0.1
	15.0	100	0	0.1
Injection volume	10 μL			
Column temperature	40 °C			
Ionization mode	Electrospray ionization mode (positive mode)			
Source and gas parameters	Ion source gas 1–60 psi, curtain gas—30 psi, temperature—450 °C, ion source 2–40 psi, CAD gas—7			
QTOF, MS/MS	TOF start mass—100 Da, declustering potential—80 V, collision energy—10 V, TOF stop mass—1000 Da, DP spread—0 V, CE spread—0 V, accumulation time—0.25 s			

3.3. Sample Preparation

The sample was homogenized using a grinder (T 25 digital ULTRA-TURRAX®, IKA, Staufen, Germany). After weighing 10 g of the sample, 10 mL of acetonitrile was added to each weighed sample and shaken for 1 min. Thereafter, a QuEChERS extraction kit (magnesium sulfate: 98.5–101.5%; sodium chloride: ≥99.5%; sodium citrate: 99.9%; disodium citrate sesquihydrate: 99%) was added to the sample solution, followed by vigorous shaking for 1 min using a rotary mixer (DE/VIVA, Collomix GmbH, Gaimersheim, Germany). Subsequently, centrifugation was performed for 10 min at 4000 rpm using SORVALL LYNX 4000 (Thermo Scientific, Waltham, MA, USA). Then, 1 mL of the supernatant was put into the QuEChERS dispersive SPE kit (primary secondary amine, octadecyl silane end-capped, magnesium sulfate; 98.5–101.5%), mixed with Mixmate 5353 (Effendorf, Hamburg, Germany) for 1 min, and centrifuged again with Minispin plus 545 (Effendorf, Hamburg, Germany) at 10,000 rpm for 1 min. The liquid separated through this process was filtered with a 0.2 μm PTFE syringe filter (Whatman, Maidstone, UK) and was used as the final sample.

3.4. Standard Sample Preparation and Method Validation

Pesticide residues used as standards were prepared at a concentration between 1000 and 2000 mg/L, diluted with acetonitrile, and mixed to set the appropriate concentration (5, 10, 25, 50, and 100 μg/L). Next, the working standard was mixed with a blank extract to obtain a matrix-matched standard. Multiple simultaneous analysis conditions were established using the standard. Afterward, based on the monitoring results obtained under the set conditions according to Sections 2.2 and 2.3, the 12 most detected pesticide residues were selected as the main compounds for validating the method. Their calibration curves were prepared by matching them with those of the matrix working solutions. Working solutions were mixed with the sweet pepper extract to produce a matrix-matched sample, which was used as the final sample to determine the parameters (selectivity, precision, accuracy, sensitivity, and linearity) for method validation and to measure the uncertainty values of the experiment. The matrix effect (ME) was calculated by the following equation [29]:

$$\text{ME}(\%) = (\frac{\text{Slope of clibration curve in matrix}}{\text{Slope of clibration curve in solvent}} - 1) \times 100$$

4. Conclusions

Qualitative and quantitative results were obtained through a simultaneous analysis method using UHPLC-QTOF-MS for 12 residual pesticides found in sweet pepper. The analyte was quickly extracted through acetonitrile-based QuEChERS pretreatment, and the method was verified through various parameters such as selectivity, linearity, sensitivity, accuracy, and precision. As a result, all parameters conformed to international standards, proving the validity of the experimental method. In addition, the reliability of the measurement result was calculated as a quantitative indicator by calculating the measurement uncertainty, thereby proving that the experimental result was meaningful. When actual sweet peppers were analyzed using this verified method, 10 pesticides out of 12 were detected and all were detected below the MRLs in Korea.

Supplementary Materials: The following supporting information can be downloaded at: https://www.mdpi.com/article/10.3390/molecules28145589/s1, Figure S1. Molecular structures and QTOF fragments of 12 pesticides. (A) Acequinocyl, (B) Boscalid, (C) Cyflumetofen, (D) Dinotefuran, (E) Flonicamid, (F) Fluopyram, (G) Procymidone, (H) Propamocarb, (I) Pyridaben, (J) Spirodiclofen, (K) Spirotetramat, (L) Spirotetramat-enol. Table S1. Parameters for the analysis of 140 pesticides by UHPLC-QTOF. Table S2. Detection results of 276 samples in 15 regions of Gyeongnam for 12 pesticides (Black cells represent detection).

Author Contributions: Conceptualization, Y.-K.K. and H.J.K.; Methodology, H.Y.B., E.J.B., T.N. and K.S.S.; Formal analysis, H.K.; Investigation, H.Y.B., H.K., E.J.B., T.N. and K.S.S.; Resources, Y.-K.K., E.J.B., T.N. and K.S.S.; Data curation, H.K.; Writing—original draft, H.Y.B., H.K. and H.J.K.; Writing—review & editing, Y.-K.K.; Supervision, Y.-K.K. and H.J.K.; Project administration, H.J.K. All authors have read and agreed to the published version of the manuscript.

Funding: This research did not receive any specific grant from funding agencies in the public, commercial, or not-for-profit sectors.

Institutional Review Board Statement: Not applicable.

Informed Consent Statement: Not applicable.

Data Availability Statement: Not applicable.

Conflicts of Interest: The authors declare no conflict of interest.

Sample Availability: Not available.

References

1. Bhandari, G.; Zomer, P.; Atreya, K.; Mol, H.G.J.; Yang, X.; Geissen, V. Pesticide residues in Nepalese vegetables and potential health risks. *Environ. Res.* **2019**, *172*, 511–521. [CrossRef]
2. El-Nahhal, I.; El-Nahhal, Y. Pesticide residues in drinking water, their potential risk to human health and removal options. *J. Environ. Manag.* **2021**, *299*, 113611. [CrossRef]
3. Silva, V.; Mol, H.G.J.; Zomer, P.; Tienstra, M.; Ritsema, C.J.; Geissen, V. Pesticide Residues in European Agricultural Soils—A Hidden Reality Unfolded. *Sci. Total Environ.* **2019**, *653*, 1532–1545. [CrossRef]
4. Narenderan, S.T.; Meyyanathan, S.N.; Babu, B. Review of pesticide residue analysis in fruits and vegetables. Pre-treatment, extraction and detection techniques. *Food Res. Int.* **2020**, *133*, 109141. [CrossRef] [PubMed]
5. Kumari, D.; John, S. Health risk assessment of pesticide residues in fruits and vegetables from farms and markets of Western Indian Himalayan region. *Chemosphere* **2019**, *224*, 162–167. [CrossRef] [PubMed]
6. Al-juhaimi, F.; Ghafoor, K.; Özcan, M.M.; Jahurul, M.H.A.; Babiker, E.E.; Jinap, S.; Sahena, F.; Sharifudin, M.S.; Zaidul, I.S.M. Effect of various food processing and handling methods on preservation of natural antioxidants in fruits and vegetables. *J. Food Sci. Technol.* **2018**, *55*, 3872–3880. [CrossRef] [PubMed]
7. Rodrigues, C.A.; Nicácio, A.E.; Jardim, I.C.S.F.; Visentainer, J.V.; Maldaner, L. Determination of Phenolic Compounds in Red Sweet Pepper (*Capsicum annuum* L.) Using using aa Modified Modified QuEChERS QuEChERS Method Method and and

UHPLC-MS/MS UHPLC-MS/MS Analysis Analysis and and Its Relation to Antioxidant Activity. *J. Braz. Chem. Soc.* **2019**, *30*, 1229–1240. [CrossRef]

8. Kaur, R.; Kaur, K. Effect of Processing on Color, Rheology and Bioactive Compounds of Different Sweet Pepper Purees. *Plant Foods Hum. Nutr.* **2020**, *75*, 369–375. [CrossRef]
9. Kaur, R.; Kaur, K.; Ahluwalia, P. Effect of drying temperatures and storage on chemical and bioactive attributes of dried tomato and sweet pepper. *LWT* **2020**, *117*, 108604. [CrossRef]
10. da Costa Morais, E.H.; Collins, C.H.; Jardim, I.C.S.F. Pesticide determination in sweet peppers using QuEChERS and LC–MS/MS. *Food Chem.* **2018**, *249*, 77–83. [CrossRef]
11. Al-Nasir, F.M.; Jiries, A.G.; Al-Rabadi, G.J.; Alu'datt, M.H.; Tranchant, C.C.; Al-Dalain, S.A.; Alrabadi, N.; Madanat, O.Y.; Al-Dmour, R.S. Determination of pesticide residues in selected citrus fruits and vegetables cultivated in the Jordan Valley. *LWT* **2020**, *123*, 109005. [CrossRef]
12. Yigit, N.; Velioglu, Y.S. Effects of processing and storage on pesticide residues in foods. *Crit. Rev. Food Sci. Nutr.* **2020**, *60*, 3622–3641. [CrossRef] [PubMed]
13. Medina-Pastor, P.; Triacchini, G. The 2018 european union report on pesticide residues in food. *EFSA J.* **2020**, *18*, e06057. [CrossRef] [PubMed]
14. Sun, R.; Yang, W.; Li, Y.; Sun, C. Multi-residue analytical methods for pesticides in teas: A review. *Eur. Food Res. Technol.* **2021**, *247*, 1839–1858. [CrossRef]
15. Azam, S.M.R.; Ma, H.; Xu, B.; Devi, S.; Siddique, M.A.B.; Stanley, S.L.; Bhandari, B.; Zhu, J. Efficacy of ultrasound treatment in the and removal of pesticide residues from fresh vegetables: A review. *Trends Food Sci. Technol.* **2020**, *97*, 417–432. [CrossRef]
16. Kemmerich, M.; Rizzetti, T.M.; Martins, M.L.; Prestes, O.D.; Adaime, M.B.; Zanella, R. Optimization by Central Composite Design of a Modified QuEChERS Method for Extraction of Pesticide Multiresidue in Sweet Pepper and Analysis by Ultra-High-Performance Liquid Chromatography–Tandem Mass Spectrometry. *Food Anal. Methods* **2015**, *8*, 728–739. [CrossRef]
17. AlSaikhan, W.H.; Almatroodi, S.A.; Almatroudi, A.; Alsahli, M.A.; Rahmani, A.H. Pesticide residue measurement in commonly used vegetables using the QuEChERS method. *Pharmacogn. J.* **2021**, *13*, 142–149. [CrossRef]
18. Ferracane, A.; Zoccali, M.; Cacciola, F.; Salerno, T.M.G.; Tranchida, P.Q.; Mondello, L. Determination of multi-pesticide residues in vegetable products using a "reduced-scale" Quechers method and flow-modulated comprehensive two-dimensional gas chromatography-triple quadrupole mass spectrometry. *J. Chromatogr. A* **2021**, *1645*, 462126. [CrossRef]
19. Nguyen, V.-T.; Tran, L.-H.; Van, T.-K.; Le, D.-V. Determination of Chlorpyrifos Pesticide Residue in Bell Peppers Grown in Dalat (Vietnam) by GC-MS/MS Using QuEChERS Extraction. *J. Anal. Chem.* **2022**, *77*, 604–610. [CrossRef]
20. Saito-Shida, S.; Hamasaka, T.; Nemoto, S.; Akiyama, H. Multiresidue determination of pesticides in tea by liquid chromatography-high-resolution mass spectrometry: Comparison between Orbitrap and time-of-flight mass analyzers. *Food Chem.* **2018**, *256*, 140–148. [CrossRef]
21. Diallo, T.; Makni, Y.; Lerebours, A.; Thomas, H.; Guérin, T.; Parinet, J. Development and validation according to the SANTE guidelines of a QuEChERS-UHPLC-QTOF-MS method for the screening of 204 pesticides in bivalves. *Food Chem.* **2022**, *386*, 132871. [CrossRef] [PubMed]
22. Kottadiyil, D.; Mehta, T.; Thasale, R.; Sivaperumal, P. Determination and dietary risk assessment of 52 pesticide residues in vegetable and fruit samples by GC-MS/MS and UHPLC-QTOF/MS from Gujarat, India. *J. Food Compos. Anal.* **2023**, *115*, 104957. [CrossRef]
23. Pedersen, M.; Hakme, E.; Ninga, E.; Frandsen, H.L. Analysis of veterinary drug- and pesticide residues in pig muscle by LC-QTOF-MS. *Food Control* **2023**, *148*, 109656. [CrossRef]
24. Li, J.; Liu, H.; Mazhar, M.S.; Quddus, S.; Suleria, H.A.R. LC-ESI-QTOF-MS/MS profiling of phenolic compounds in Australian native plums and their potential antioxidant activities. *Food Biosci.* **2023**, *52*, 102331. [CrossRef]
25. European Commission. Guidance Document on Analytical Quality Control and Method Validation Procedures for Pesticide Residues and Analysis in Food and Feed. Available online: https://www.eurl-pesticides.eu/userfiles/file/EurlALL/SANTE_11312_2021.pdf (accessed on 1 January 2022).
26. EURACHEM. *Quantifying Uncertainty in Analytical Measurement*, 3rd ed.; EURACHEM: London, UK, 2012.
27. ISO. *Uncertainty of Measurement—Part 3: Guide to the Expression of Uncertainty in Measurement*; GUM: 1995; ISO: Geneva, Switzerland, 2008.
28. CODEX. *Guidelines on Measurement Uncertainty*; CAC/GL 54–2004; CODEX: Irvine, CA, USA, 2004.
29. Kim, H.; Joo Baek, E.; Kim, Y.-K.; Park, H.; Hye Hur, S.; Kim, J.-E.; Jin Kim, H. Development of a method for analysis and risk assessment of residual pesticides in ginseng using liquid and gas chromatography-tandem mass spectrometry. *Food Chem.* **2023**, *427*, 136675. [CrossRef] [PubMed]

separations

MDPI

Article

Validation of the Ultra-Performance Liquid Chromatography with Tandem Mass Spectrometry Method for Simultaneous Analysis of Eighteen Compounds in the Traditional Herbal Prescription, Sanjoin-Tang

Chang-Seob Seo * and Hyeun-Kyoo Shin

KM Science Research Division, Korea Institute of Oriental Medicine, Daejeon 34054, Republic of Korea; hkshin@kiom.re.kr

* Correspondence: csseo0914@kiom.re.kr; Tel.: +82-42-868-9361

Abstract: Sanjoin-tang (SJIT) is an ancient oriental medicine prescription listed in the *Jinguiyaolue* that is mainly used for the treatment of primary insomnia. This study was conducted to develop and validate an ultra-performance liquid chromatography with tandem mass spectrometry (UPLC–MS/MS) simultaneous analysis method for the quality control of SJIT using 18 target compounds. The 18 analytes were separated on an Acquity UPLC BEH C_{18} column maintained at 45 °C using a mobile phase composed of distilled water and acetonitrile. The MS system was used to simultaneously detect all analytes using the multiple reaction monitoring (MRM) method of Xevo TQ-XS coupled with an electrospray ionization source. The concentrations of the 18 analytes investigated in the SJIT samples ranged from below the limit of detection to 9.553 mg/g. In conclusion, the validated UPLC–MS/MS MRM analysis method can be used to obtain basic data to establish chemical-nonclinical linkage efficacy and for the clinical research and quality evaluation of SJIT.

Keywords: simultaneous quantitative analysis; traditional herbal prescription; Sanjoin-tang; UPLC–MS/MS

Citation: Seo, C.-S.; Shin, H.-K. Validation of the Ultra-Performance Liquid Chromatography with Tandem Mass Spectrometry Method for Simultaneous Analysis of Eighteen Compounds in the Traditional Herbal Prescription, Sanjoin-Tang. *Separations* **2023**, *10*, 411. https://doi.org/10.3390/separations10070411

Academic Editors: Chao Kang and Ronald Beckett

Received: 23 June 2023
Revised: 12 July 2023
Accepted: 14 July 2023
Published: 18 July 2023

1. Introduction

Insomnia is a common symptom in menopausal women and is one of the factors that determine quality of life [1]. Sanjoin-tang (SJIT, Suanzaoren-tang in Chinese, Sansoninto in Japanese) is one of the most frequently used prescriptions for treating insomnia in oriental medicine. SJIT was first recorded in *Jinguiyaolue* and consists of five traditional herbal medicines (*Ziziphus jujuba* Mill., *Cnidium officinale* Makino, *Anemarrhena asphodeloides* Bunge, *Poria cocos* Wolf, and *Glycyrrhiza uralensis* Fisch.) [1]. As a principal herb in SJIT, *Z. jujuba* has been reported to have insomnia, hypnotic, tranquilizing, and antianxiety effects [2–5]. In particular, jujubosides and spinosin, the main components of *Z. jujuba* seeds, have been reported to be effective for hypnosis and anxiety [2,6]. In addition, *G. uralensis* and *P. cocos*, which are other medicinal herbs that make up SJIT, have been reported to be effective for insomnia [7–10].

Various types of components such as flavonoids [11,12], terpenoids [13–15], alkaloids [13], miscellaneous [16,17], phenylpropanoids [17], xanthones [18], and steroids [19] were isolated from each medicinal herb of SJIT. Quantitative and qualitative analysis methods using high-performance liquid chromatography (HPLC), ultra-high-performance liquid chromatography, high-performance liquid chromatography coupled with quadrupole time-of-flight mass spectrometry, or ultra-performance liquid chromatography coupled with electrospray ionization (ESI) and quadrupole time-of-flight mass spectrometry have been developed for the efficient quality control of each herbal medicine based on the measurement of these isolated components [17,20–26]. As an analysis for the quality assurance of SJIT, Zhu et al. [27] reported a component profiling analysis using liquid chromatography coupled with quadrupole time-of-flight mass spectrometry and liquid chromatography–ion trap–mass spectrometry. Although the efficacy of SJIT and analysis methods for each component of herbal medicine have been reported, no simultaneous analysis method for the quality assurance of SJIT has been validated to date.

Ultra-performance liquid chromatography with tandem mass spectrometry (UPLC–MS/MS) has high specificity, sensitivity, reproducibility, resolution, and accuracy, and its use for the quantitative analysis of traditional herbal medicines and herbal formulas has been increasing compared to other analytical instruments [28,29].

In this study, a simultaneous analysis method was validated for 18 compounds in SJIT samples using UPLC–MS/MS multiple reaction monitoring (MRM). The compounds examined were neomangiferin (**1**), mangiferin (**2**), magnoflorine (**3**), spinosin (**4**), liquiritin apioside (**5**), liquiritin (**6**), ferulic acid (**7**), 6‴-feruloyl spinosin (**8**), isoliquiritin (**9**), ononin (**10**), liquiritigenin (**11**), jujuboside A (**12**), isoliquiritigenin (**13**), glycyrrhizin (**14**), jujuboside B (**15**), *Z*-ligustilide (**16**), dehydropachymic acid (**17**), and pachymic acid (**18**).

2. Materials and Methods

2.1. Chemicals and Reagents

The 18 standard compounds used for the evaluation of quality control of SJIT were purchased from professional high-purity phytochemical manufacturers: neomangiferin (CAS No. 64809-67-2, purity: 98.3%, catalog No. BP0993), mangiferin (CAS No. 4773-96-0, purity: 99.1%, catalog No. BP0922), spinosin (CAS No. 72063-39-9, purity: 99.3%, catalog No. BP1336), liquiritin (CAS No. 551-15-5, purity: 99.6%, catalog No. BP0874), 6‴-feruloyl spinosin (CAS No. 77690-92-7, purity: 99.7%, catalog No. BP1572), isoliquiritin (CAS No. 5041-81-6, purity: 98.6%, catalog No. BP0788), ononin (CAS No. 486-62-4, purity: 98.5%, catalog No. BP1031), liquiritigenin (CAS No. 578-86-9, purity: 99.8%, catalog No. BP0873), glycyrrhizin (CAS No. 1405-86-3, purity: 99.1%, catalog No. BP0682), and dehydropachymic acid (CAS No. 77012-31-8, purity: 98.8%, catalog No. BP1676) from Biopurify Phytochemicals (Chengdu, China); magnoflorin (CAS No. 2141-09-5, purity: 98.8%, catalog No. CFN98071) from ChemFaces (Wuhan, China); liquiritin apioside (CAS No. 74639-14-8, purity: 99.6%, catalog No. DR10690), jujuboside A (CAS No. 55466-04-1, purity: 99.1%, catalog No. D19052410), jujuboside B (CAS No. 565466-05-2, purity: 98.3%, catalog No. D102876), and pachymic acid (CAS No. 29070-92-6, purity: 99.7%, catalog No. DR11130) from Shanghai Sunny Biotech (Shanghai, China); ferulic acid (CAS No. 1135-24-6, purity: 98.0%, catalog No. 086-04282) from Fujifilm Wako Pure Chemical Co. (Osaka, Japan); and isoliquiritigenin (CAS No. 961-29-5, purity: 99.8%, catalog No. TB0235-0500) and Z-liqustilide (CAS No. 81944-09-4, purity: 99.6%, catalog No. TB0322-0100) from ChemNorm (Wuhan, China) (Figure S1). Solvents (acetonitrile, methanol, and distilled water) and reagent (formic acid) for analysis were LC–MS grade.

2.2. Plant Materials and Preparation of the SJIT Water Extract

The five raw herbal medicines (Table S1) that make up SJIT were purchased from Kwangmyungdang Pharmaceutical (Ulsan, Korea) and used in the study after morphological identification by Dr. Goya Choi (Korea Institute of Oriental Medicine (KIOM), Naju, Republic of Korea). Each raw herbal medicine (KE–90–1 to KE–90–5) and SJIT water extract (KE90) were deposited at the Korean Medicine Science Research Division, KIOM (Daejeon, Republic of Korea).

After the SJIT water extract (SJIT–1) was mixed well according to the amount shown in Table S2 (total 5000.0 g: *Z. jujuba* 2666.67 g, *C. officinale* 666.67 g, *A. asphodeloides* 666.67 g, *P. cocos* 666.67 g, and *G. uralensis* 333.32), 50 L of primary distilled water was added and the mixture was extracted at 100 °C for 2 h using an electric extractor, COSMOS-660 (Kyungseo E&P, Incheon, Korea). The extract solution was filtered through a sieve (53 µm mesh) and freeze-dried using an LP100R freeze dryer (IlShinBioBase, Yangju, Korea) to obtain the powdered sample. Finally, 840.2 g (16.8% yield) of powdered SJIT water extract was obtained. Other samples (SJT–2, SJT–3, and SJT–4) were purchased from different pharmaceutical companies.

2.3. Analytical Method for UPLC–MS/MS Analysis

The analytical method for the simultaneous analysis of the 18 markers from SJIT samples using the UPLC–MS/MS system, consisting of an Acquity UPLC I-Class system coupled with a Xevo TQ-XS MS, was conducted with the parameters summarized in Table S2. All markers were separated on a UPLC BEH C_{18} column (2.1 mm × 100 mm, 1.7 μm, Waters) with gradient elution of a 0.1% formic acid in distilled water–acetonitrile mobile phase. Temperatures of the column, sample, ion source, and ion desolvation were maintained at 45, 5, 150, and 500 °C, respectively. The injection volume was 0.2 μL and the flow rate was 0.3 mL/min. Details of the conditions for UPLC–MS/MS MRM analysis are presented in Table 1.

Table 1. UPLC–MS/MS MRM analytical conditions for simultaneous analysis of the 18 analytes in SJIT samples.

Analyte [1]	Ion Mode	Molecular Weight	Precursor Ion	Product Ion	Cone Voltage (V)	Collision Energy (eV)
1	−	584.5	583.3 $[M-H]^-$	331.0	45	35
2	−	422.3	421.1 $[M-H]^-$	301.0	30	20
3	+	342.4	342.4 $[M]^+$	297.2	30	20
4	+	608.5	609.5 $[M+H]^+$	327.2	40	25
5	−	550.5	549.3 $[M-H]^-$	255.0	45	30
6	−	418.4	417.4 $[M-H]^-$	255.2	30	15
7	+	194.2	195.0 $[M+H]^+$	177.0	15	10
8	+	784.7	785.5 $[M+H]^+$	327.1	35	25
9	+	418.4	419.3 $[M+H]^+$	257.0	35	15
10	+	430.4	431.3 $[M+H]^+$	269.0	25	15
11	+	256.3	257.2 $[M+H]^+$	137.0	35	25
12	+	1207.3	1225.1 $[M+H_2O]^+$	473.5	30	20
13	+	256.3	257.2 $[M+H]^+$	137.0	15	20
14	−	822.9	821.9 $[M-H]^-$	351.2	45	40
15	−	1045.2	1043.8 $[M-H]^-$	911.5	30	35
16	+	190.1	191.0 $[M+H]^+$	91.0	30	25
17	−	526.7	525.7 $[M-H]^-$	59.0	30	35
18	−	528.8	527.6 $[M-H]^-$	465.4	45	35

[1] Neomangiferin (**1**), mangiferin (**2**), magnoflorine (**3**), spinosin (**4**), liquiritin apioside (**5**), liquiritin (**6**), ferulic acid (**7**), 6‴-feruloyl spinosin (**8**), isoliquiritin (**9**), ononin (**10**), liquiritigenin (**11**), jujuboside A (**12**), isoliquiritigenin (**13**), glycyrrhizin (**14**), jujuboside B (**15**), Z-ligustilide (**16**), dehydropachymic acid (**17**), and pachymic acid (**18**).

2.4. Preparation of Standard Stock and Sample Solutions

Standard stock solutions of the 18 reference standards were prepared at a concentration of 1000 μg/mL using methanol and stored in a refrigerator (approximately 4 °C) until use. To prepare the sample solution of the SJIT samples, about 50 mg of the prepared SJIT water extract or commercially available SJIT granules was accurately taken and made to a concentration of 5 mg/mL using 70% methanol, followed by ultrasonic extraction (for 5.0 min) and vortexing (for 1.0 min). The extracted sample solution was filtered with a 0.2 μm hydrophobic filter (SSOLKOREA Co., Ltd., Daejeon, Korea) before UPLC–MS/MS analysis.

2.5. Validation of the UPLC–MS/MS MRM Analytical Method

The UPLC–MS/MS analytical method developed in this study was validated with respect to specificity, linearity, sensitivity, recovery, and precision evaluation [30]. That is, specificity was confirmed by verifying the presence or absence of an interference peak in the chromatogram of each component. The linearity was evaluated by the coefficient of determination (r^2) value in the regression equation of the calibration curve prepared at different concentrations. The sensitivity of the method was verified by limit of detection (LOD) and limit of quantitation (LOQ) values and calculated as signal-to-noise (S/N) ratios of 3 and 10, respectively. Recovery (%) was evaluated by the standard addition method, which was evaluated by adding three different levels (low, medium, and high) of standard solutions to the known sample. This parameter was calculated using the following equation: Recovery (%) = found amount/spiked amount × 100. Precision was demonstrated by evaluating relative standard deviation (RSD) values of intraday (within a day) and interday (three consecutive days) precisions and repeatability. This parameter was calculated using the following equation: RSD (%) = standard deviation/mean × 100.

3. Results and Discussion

3.1. Selection of Target Marker Components for Simultaneous Analysis Using UPLC–MS/MS in SJIT Samples

From the SJIT consisting of five medicinal herbs, 18 components were selected from each raw material for simultaneous analysis using the UPLC–MS/MS system: magnoflorine, spinosin, 6'''-feruloyl spinosin, jujuboside A, and jujuboside B from *Z. jujuba*; ferulic acid and Z-ligustilide from *C. officinale*; neomangiferin and mangiferin from *A. asphodeloides*; dehydropachymic acid and pachymic acid from *P. cocos*; and liquiritin apioside, liquiritin, isoliquiritin, ononin, liquiritigenin, isoliquiritigenin, and glycyrrhizin from *G. uralensis* [11–19].

3.2. Multiple Reaction Monitoring Conditions for UPLC–MS/MS Analysis of SJIT Samples

As mentioned in Section 2.3, MRM conditions for the simultaneous analysis of the 18 marker components in SJIT were explored using a UPLC–MS/MS system. An ESI attachment was used as the ion source. Eight compounds (**1**, **2**, **5**, **6**, **14**, **15**, **17**, and **18**) showed high intensity in the $[M–H]^-$ form, and eight compounds (**4**, **7–11**, **13**, and **16**) showed good intensity in the $[M+H]^+$ form. Magnoflorine (**3**) and jujuboside A (**12**) showed an $[M]^+$ peak and an adduct peak in the form of $[M+H_2O]^+$, respectively (Figure 1). In case of jujuboside A, the adduct peak ($[M+H_2O]^+$) observed at *m/z* 1225.1 was detected more strongly than $[M+H]^+$ during MS measurement (Figure S2).

To simultaneously quantify the 18 components in SJIT samples using the UPLC–MS/MS MRM analysis mode, parameters such as precursor ion (Q1) peak and product ion (Q3) peak, cone voltage, and collision energy for each component were set as shown in Table 1. The Q3 peaks of xanthone types, neomangiferin, and mangiferin were set to *m/z* 331.0 and 301.0 in the form of $[M–H–Glu–C_4H_8O_4]^-$ and $[M–H–C_4H_8O_4]^-$, respectively [31,32]. These components showed a common feature of the Q3 peak being formed due to the cleavage of the sugar ring. For magnoflorine and alkaloids, a Q3 peak was observed at *m/z* 297.2 ($[M–(CH_3)_2NH]^+$), corresponding to the loss of two methyl groups, and an amine group was lost from the ion generating the Q1 peak [33]. For spinosin, 6'''-feruloyl spinosin, and liquiritigenin, Q3 peaks were observed at *m/z* 327.2, 327.1, and 137.0, arising from the fragmentation of the Q1 peak to $[M+H–Glu–C_4H_8O_4]^+$, with the loss of a glucose and a $C_4H_8O_4$ molecule; $[M+H–Feryloyl–Glu–C_4H_8O_4]^+$, with the loss of a feruloyl functional group, one glucose, and a $C_4H_8O_4$ molecule; and $[M+H–C_8H_8O_8]^+$, with the loss of a $C_8H_8O_8$ molecule, respectively [33,34]. In both liquiritin and ononin, Q3 peaks were generated as $[M–H–Glu]^-$ and $[M+H–Glu]^+$ with one molecule of glucose removed [33,35]. In liquiritin apioside, the Q3 peak was detected in the form of an aglycone, in which apiose and glucose were removed from the Q1 peak [33]. The Q3 peaks of ferulic acid (phenylpropanoid-type) and Z-ligustilide (benzoquinone derivative) were set to *m/z* 177.0 ($[M+H–H_2O]^+$) and *m/z* 91 ($[M+H–H_2O–CO–C_4H_6]^+$), respectively [36]. In the MRM transition of isoliquiritin and isoliquiritigenin, which are chalcone-type compounds, Q3 peaks were observed at *m/z* 257.0 ($[M+H–Glu]^+$) and *m/z* 137.0 ($[M+H–C_8H_7O]^+$), respectively, where *m/z* 257.0 was produced by the release of one molecule of glucose from the Q1 peak, and *m/z* 137.0 was generated by the cleavage of the B ring in the molecular structure [37]. The Q3 peak of jujuboside A, triterpenoids, was observed at *m/z* 473.5 in the form of aglycone in which five sugars were cleaved from the Q1 peak [38], whereas the Q3 peak of glycyrrhizin was observed at *m/z* 351.2 in the form of two glucuronic acid conjugates, excluding the aglycone [39]. For jujuboside B, the Q3 peak was observed at m/z 911.5 ([M–H–Xyl]$^-$), where one molecule of xylose was removed from the Q1 peak [40]. In dehydropachymic acid and pachymic acid, the highest intensities were observed at m/z 59.0 and 465.4, respectively, as each Q1 peak was fragmented by cone voltage and collision energy, and these were set as Q3 peaks [24]. Under the optimized UPLC–MS/MS MRM analysis conditions, all components were completely eluted within 13.0 min; representative total ion chromatograms (TICs) are shown in Figure 1. The TIC of the other samples (SJIT–2 to SJIT–4) is shown in Figure S3. In addition, the extracted ion chromatograms of the 18 reference markers and samples (SJIT–1 ~ SJIT–4) are presented in Figure S4.

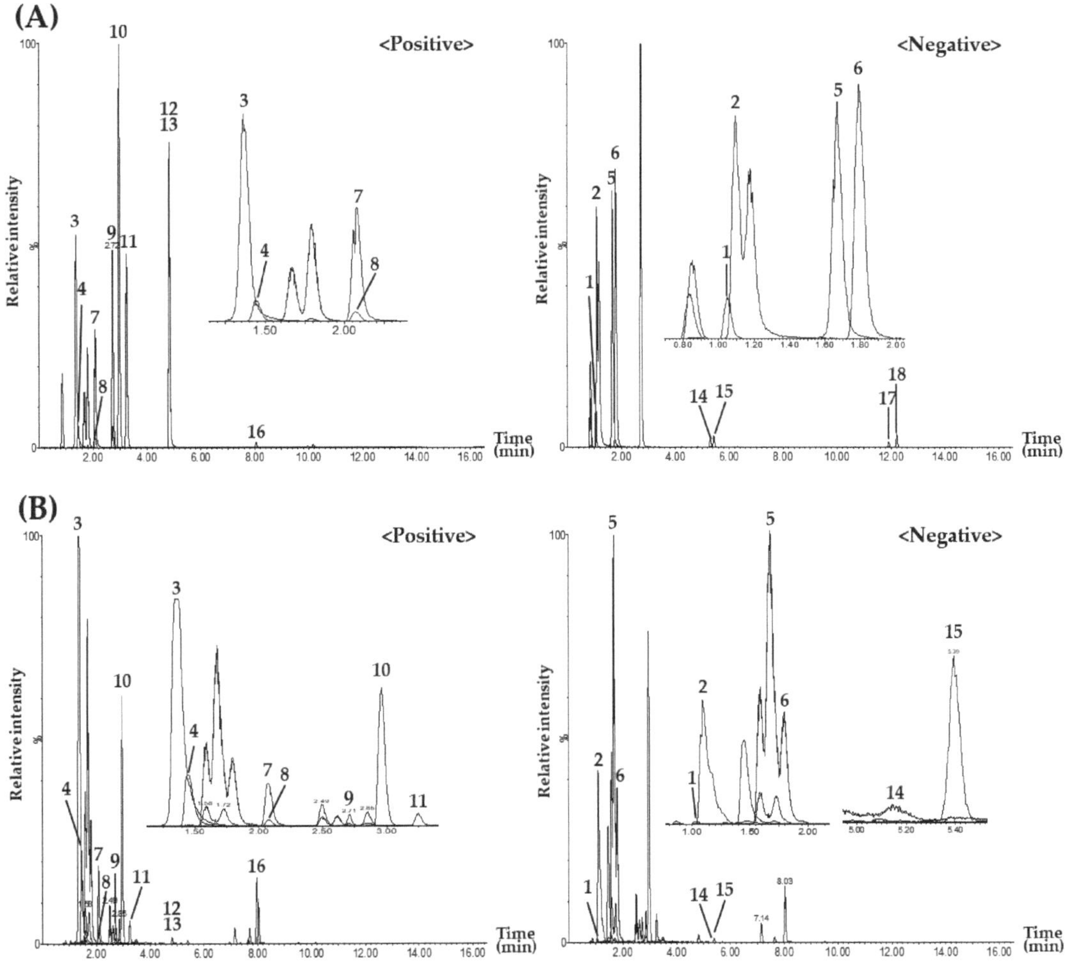

Figure 1. Representative total ion chromatograms of the standard mixtures (**A**) and SJIT–1 sample (**B**) by UPLC–MS/MS MRM analytical method mode. Neomangiferin (**1**), mangiferin (**2**), magnoflorine (**3**), spinosin (**4**), liquiritin apioside (**5**), liquiritin (**6**), ferulic acid (**7**), 6‴-feruloyl spinosin (**8**), isoliquiritin (**9**), ononin (**10**), liquiritigenin (**11**), jujuboside A (**12**), isoliquiritigenin (**13**), glycyrrhizin (**14**), jujuboside B (**15**), Z-ligustilide (**16**), dehydropachymic acid (**17**), and pachymic acid (**18**). The concentration of each compound in the standard mixtures (A) was 100.00 μg/L.

3.3. Validation of the Developed UPLC–MS/MS MRM Analytical Method

The developed UPLC–MS/MS MRM simultaneous analysis method was validated by assessing specificity, linearity, sensitivity (LOD and LOQ), recovery, and precision. Specificity results ensured the absence of any significant interference at the retention time of each component (Figure S4). The linearity was evaluated by the coefficient of determination (r^2). As shown in Table 2, all components showed acceptable linearity in the assay, with r^2 values $\geq$0.9968 in the regression equations tested at 10.00–500.00 and 250.00–5000.00 µg/L. In addition, LOD and LOQ values for sensitivity check were calculated as S/N ratios and were 0.002–11.825 and 0.005–35.474 µg/L, respectively (Table 2). Recovery tests for the 18 compounds in SJIT samples were performed to assess the accuracy of the developed analytical assay. Recovery (%) tested at three different concentration levels (low, medium, and high) was measured to be 85.26–114.21% with RSD values <10.0% for all marker components (Table 3). The tolerance of the recovery test was set within ±15% and the recovery of 85.26–114.21% proved that the newly developed UPLC–MS/MS method was acceptable. Precision results evaluated by RSD values are presented in Table 4. The intraday precision, evaluated by measuring five times within one day, was 0.69–9.94%, and the interday precision, evaluated by measuring on three consecutive days, was 2.11–9.24%. The repeatability of peak retention time and peak area was evaluated from the RSD values calculated by measuring the standard solution six times; the components had RSD values of 0.45–1.82% and 1.10–8.73%, respectively. The tolerance of precision (% RSD) was set to less than 15%, and the test results demonstrated that the developed UPLC–MS/MS method was suitable for less than 10%. Finally, based on all the validation results above, the developed UPLC–MS/MS MRM method proved to be suitable and precise as an analytical method for the simultaneous analysis of SJIT.

Table 2. Parameters for simultaneous quantification of the 18 analytes in SJIT samples using the UPLC–MS/MS MRM method.

Analyte [1]	Retention Time (min)	Linear Range (µg/L)	Regression Equation [2] $y = ax + b$	r^2	LOD (µg/L)	LOQ (µg/L)
1	1.06	10.00–500.00	$y = 296.00x + 3082.84$	0.9992	0.071	0.213
2	1.08	10.00–500.00	$y = 1550.26x + 24{,}896.50$	0.9980	0.058	0.174
3	1.35	10.00–500.00	$y = 23{,}534.20x + 432{,}415.00$	0.9996	0.011	0.032
4	1.45	10.00–500.00	$y = 2543.55x - 4216.33$	0.9968	0.017	0.050
5	1.65	10.00–500.00	$y = 2188.31x + 13{,}166.60$	0.9993	0.073	0.218
6	1.79	10.00–500.00	$y = 3298.34x + 23772.50$	0.9992	0.010	0.030
7	2.06	10.00–500.00	$y = 11{,}957.40x + 44{,}248.70$	0.9988	0.075	0.224
8	2.07	10.00–500.00	$y = 1041.87x + 3654.02$	0.9989	0.118	0.353
9	2.69	10.00–500.00	$y = 19{,}090.10x + 157{,}649.00$	0.9991	0.028	0.083
10	2.92	10.00–500.00	$y = 37{,}870.70x + 471{,}200.00$	0.9990	0.002	0.005
11	3.25	10.00–500.00	$y = 19{,}416.40x + 221{,}079.00$	0.9994	0.015	0.045
12	4.83	250.00–5000.00	$y = 0.30x - 49.79$	0.9978	11.825	35.474
13	4.83	10.00–500.00	$y = 28{,}495.50x + 368{,}896.00$	0.9992	0.008	0.024
14	5.26	10.00–500.00	$y = 71.87x + 65.45$	0.9985	0.975	2.925
15	5.42	10.00–500.00	$y = 78.53x + 378.80$	0.9989	2.148	6.444
16	8.04	10.00–500.00	$y = 609.03x + 5526.79$	0.9991	6.276	18.827
17	11.91	10.00–500.00	$y = 44.84x + 99.89$	0.9985	1.207	3.622
18	12.21	10.00–500.00	$y = 109.76x + 523.95$	0.9980	1.582	4.747

[1] Neomangiferin (**1**), mangiferin (**2**), magnoflorine (**3**), spinosin (**4**), liquiritin apioside (**5**), liquiritin (**6**), ferulic acid (**7**), 6‴-feruloyl spinosin (**8**), isoliquiritin (**9**), ononin (**10**), liquiritigenin (**11**), jujuboside A (**12**), isoliquiritigenin (**13**), glycyrrhizin (**14**), jujuboside B (**15**), *Z*-ligustilide (**16**), dehydropachymic acid (**17**), and pachymic acid (**18**). [2] *y*: peak area of each reference compound; *x*: concentration (µg/L) of each reference compound.

Table 3. Recovery (%) of the 18 target components using the UPLC–MS/MS MRM method.

Analyte [1]	Spiked Amount (μg/L)	Found Amount (μg/L)	Recovery (%)	SD [2]	RSD (%)
	40.00	37.15	92.86	8.87	9.55
1	80.00	81.21	101.52	4.82	4.75
	160.00	168.60	105.37	4.70	4.46
	120.00	119.47	99.56	4.22	4.24
2	240.00	250.58	104.41	8.73	8.36
	480.00	525.41	109.46	6.23	5.69
	40.00	38.28	95.71	6.98	7.30
3	80.00	82.12	102.65	6.59	6.42
	160.00	170.78	106.74	4.11	3.85
	40.00	39.64	99.10	9.13	9.21
4	80.00	89.88	112.35	7.36	6.55
	160.00	172.27	107.67	7.90	7.34
	80.00	80.70	100.88	1.17	1.16
5	160.00	170.10	106.31	1.85	1.74
	320.00	344.46	107.65	4.54	4.22
	80.00	78.89	98.61	3.32	3.36
6	160.00	169.20	105.75	2.88	2.72
	320.00	343.38	107.31	3.80	3.54
	10.00	9.78	97.81	6.26	6.40
7	20.00	20.73	103.63	7.33	7.08
	40.00	42.51	106.28	4.34	4.09
	8.00	8.16	102.06	6.69	6.55
8	16.00	17.12	106.99	0.94	0.88
	32.00	33.49	104.65	4.54	4.34
	80.00	82.71	103.39	2.56	2.48
9	160.00	168.82	105.51	1.83	1.73
	320.00	333.96	104.36	1.86	1.78
	4.00	3.94	98.60	5.32	5.39
10	8.00	8.18	102.19	3.49	3.42
	16.00	16.43	102.71	2.89	2.81
	80.00	78.72	98.39	3.91	3.98
11	160.00	162.03	101.27	3.91	3.86
	320.00	332.91	104.03	3.93	3.78
	600.00	526.21	87.70	8.04	9.16
12	1200.00	1200.96	100.08	9.37	9.36
	2400.00	2741.03	114.21	4.83	4.23
	20.00	20.28	101.42	1.56	1.54
13	40.00	40.46	101.16	1.62	1.60
	80.00	79.62	99.52	4.00	4.02
	200.00	191.87	95.94	3.62	3.78
14	400.00	399.28	99.82	4.39	4.39
	800.00	828.35	103.54	5.10	4.93
	100.00	109.39	109.39	6.00	5.48
15	200.00	211.04	105.52	6.13	5.81
	400.00	396.60	99.15	4.10	4.13
	80.00	84.39	105.49	2.99	2.84
16	160.00	171.35	107.10	3.93	3.67
	320.00	337.40	105.44	4.00	3.80
	20.00	17.05	85.26	8.02	9.41
17	40.00	41.81	104.54	6.59	6.30
	80.00	87.37	109.21	4.01	3.67
	20.00	18.54	92.72	8.45	9.11
18	40.00	41.77	104.43	1.55	1.49
	80.00	85.50	106.88	7.62	7.13

[1] Neomangiferin (**1**), mangiferin (**2**), magnoflorine (**3**), spinosin (**4**), liquiritin apioside (**5**), liquiritin (**6**), ferulic acid (**7**), 6‴-feruloyl spinosin (**8**), isoliquiritin (**9**), ononin (**10**), liquiritigenin (**11**), jujuboside A (**12**), isoliquiritigenin (**13**), glycyrrhizin (**14**), jujuboside B (**15**), *Z*-ligustilide (**16**), dehydropachymic acid (**17**), and pachymic acid (**18**). [2] Standard deviation.

Table 4. Precision test of the 18 target components using the UPLC–MS/MS MRM method.

Analyte [1]	Conc. (μg/L)	Intraday (n = 5)			Interday (n = 5)			Repeatability (n = 6)	
		Observed Conc. (μg/L)	Precision (RSD, %)	Accuracy (%)	Observed Conc. (μg/L)	Precision (RSD, %)	Accuracy (%)	RSD (%) of Retention Time	RSD (%) of Peak Area
1	40.00	39.57	3.67	98.93	39.20	7.08	97.88	1.67	3.94
	80.00	82.34	4.32	102.93	84.10	4.29	105.17		
	160.00	174.14	5.00	108.84	175.40	4.44	109.61		
2	120.00	122.09	5.22	101.74	119.90	6.09	99.92	1.82	8.05
	240.00	263.26	4.93	109.69	259.70	6.08	108.21		
	480.00	522.62	4.86	108.88	526.20	4.96	109.62		
3	40.00	37.43	7.34	93.58	38.60	7.13	96.49	1.21	4.17
	80.00	85.02	5.89	106.28	86.10	5.68	107.58		
	160.00	170.04	5.65	106.27	175.30	4.21	109.55		
4	40.00	36.00	8.20	90.01	39.00	7.54	97.55	1.15	8.73
	80.00	77.88	6.15	97.35	87.50	4.57	109.36		
	160.00	153.80	8.73	96.12	170.90	5.81	106.80		
5	80.00	79.53	3.95	99.41	79.40	3.19	99.26	1.31	4.24
	160.00	169.62	2.78	106.01	167.50	2.21	104.67		
	320.00	340.06	3.41	106.27	338.70	3.28	105.83		
6	80.00	78.11	3.19	97.64	79.57	3.91	99.46	1.14	6.01
	160.00	169.50	2.34	105.94	170.58	2.11	106.61		
	320.00	341.85	1.95	106.83	342.48	3.20	107.02		
7	10.00	9.53	5.30	95.33	9.80	5.33	98.34	1.43	4.75
	20.00	19.96	8.27	99.80	20.20	6.50	101.18		
	40.00	39.47	5.70	98.67	40.60	6.01	101.52		
8	8.00	7.74	3.98	96.76	8.10	4.69	100.96	1.15	7.87
	16.00	16.36	3.10	102.26	16.60	2.39	104.02		
	32.00	33.04	3.71	103.24	33.40	3.06	104.37		
9	80.00	77.81	3.24	97.27	73.90	4.70	92.42	1.09	3.85
	160.00	172.29	0.69	107.68	164.60	3.78	102.86		
	320.00	329.53	2.64	102.98	328.30	3.92	102.61		
10	4.00	4.05	3.53	101.33	4.00	4.52	99.07	1.13	4.45
	8.00	8.28	3.72	103.54	8.20	2.98	102.27		
	16.00	16.55	2.51	103.43	16.70	2.44	104.21		
11	80.00	81.95	0.86	102.44	81.00	2.45	101.25	1.11	5.74
	160.00	166.62	2.54	104.14	164.10	3.99	102.53		
	320.00	333.34	3.16	104.17	329.80	3.98	103.07		
12	600.00	618.67	9.71	103.11	540.90	9.24	90.15	0.83	7.52
	1200.00	1245.69	3.77	103.81	1232.90	7.54	102.74		
	2400.00	1958.18	7.55	81.59	2441.50	7.09	101.73		
13	20.00	19.76	3.33	98.80	19.90	2.71	99.45	0.92	1.10
	40.00	41.96	1.84	104.90	41.60	2.53	104.11		
	80.00	82.51	3.05	103.14	82.20	3.66	102.77		
14	200.00	208.51	2.10	104.25	199.40	2.49	99.72	0.72	4.48
	400.00	442.18	2.55	110.55	414.30	3.79	103.56		
	800.00	898.90	3.48	112.36	835.60	3.64	104.45		
15	100.00	103.36	9.94	103.36	74.60	7.08	100.76	0.73	6.57
	200.00	205.13	6.23	102.56	150.10	6.60	103.41		
	400.00	441.67	1.12	110.42	301.40	3.68	103.52		
16	80.00	84.93	4.79	106.17	82.40	3.66	102.97	0.82	3.54
	160.00	175.47	4.59	109.67	169.70	3.50	106.06		
	320.00	332.33	1.46	103.85	333.20	2.52	104.12		
17	20.00	18.50	8.46	92.51	18.00	8.96	90.09	0.46	4.85
	40.00	44.91	3.44	112.28	43.00	5.52	107.60		
	80.00	84.25	8.21	105.32	87.20	5.35	108.99		
18	20.00	21.47	3.57	107.34	20.30	5.72	101.47	0.45	8.15
	40.00	42.54	3.67	106.36	41.90	3.42	104.73		
	80.00	82.83	6.02	103.54	83.40	5.88	104.22		

[1] Neomangiferin (**1**), mangiferin (**2**), magnoflorine (**3**), spinosin (**4**), liquiritin apioside (**5**), liquiritin (**6**), ferulic acid (**7**), 6‴-feruloyl spinosin (**8**), isoliquiritin (**9**), ononin (**10**), liquiritigenin (**11**), jujuboside A (**12**), isoliquiritigenin (**13**), glycyrrhizin (**14**), jujuboside B (**15**), *Z*-ligustilide (**16**), dehydropachymic acid (**17**), and pachymic acid (**18**).

3.4. Simultaneous Quantitation of the Eighteen Targets in SJIT Samples by the UPLC–MS/MS MRM Method

The UPLC–MS/MS MRM analytical method developed and validated in this study was applied for the quantitative analysis of SJIT samples. As shown in Table 5, for samples

SJIT–1 to SJIT–4, the concentrations of the components ranged from below the LOD to 9.553 mg/g. Liquiritin apioside (**5**) and glycyrrhizin (**14**), markers of *G. uralensis*, and neomangiferin (**1**) and mangiferin (**2**), markers of *A. asphodeloides*, were detected in relatively large amounts compared with other components. Dihydropachymic acid (**17**), a marker of *P. cocos*, was detected only in SJIT–4 (0.030 mg/g). Previously, Li et al. and Miyaoka et al. performed simultaneous analysis and fingerprinting to evaluate the quality of SJIT using HPLC [41,42]. In their assay, active ingredients such as mangiferin, liquiritin, ferulic acid, and glycyrrhizin were detected. Among the components detected, glycyrrhizin was most abundant, which showed similar results to our analysis. In the simultaneous analysis of SJIT, it is considered that an accurate and sensitive UPLC–MS/MS analysis method with a short operation time compared to previous HPLC analysis results is considered to be effective.

Table 5. Quantification (mg/g) of the 18 components in SJIT–1 to SJIT–4 samples by the UPLC–MS/MS MRM analytical method.

Analyte [1]	SJIT–1 [2]		SJIT–2		SJIT–3		SJIT–4	
	Mean (mg/g)	RSD (%)	Mean (mg/g)	RSD (%)	Mean (mg/g)	RSD (%)	Mean (mg/g)	RSD (%)
1	2.265	4.687	0.043	5.906	0.648	7.163	0.406	4.768
2	9.553	2.823	0.087	3.694	0.510	0.486	3.751	5.533
3	2.885	0.539	0.096	7.088	0.100	1.462	1.128	5.727
4	1.481	1.204	0.218	4.676	0.053	4.677	0.368	2.905
5	3.368	3.073	0.260	1.309	0.246	0.470	0.960	2.167
6	0.911	1.188	0.062	6.332	0.032	2.652	0.055	8.212
7	0.429	1.702	0.015	6.049	0.037	4.512	0.226	6.107
8	0.412	1.693	0.076	6.829	0.023	3.091	0.216	3.549
9	0.016	4.267	0.007	5.049	0.005	6.585	0.061	9.730
10	0.325	3.114	0.048	5.819	0.027	0.911	0.052	5.128
11	0.035	3.444	0.002	6.671	0.001	0.886	0.173	0.323
12	0.162	4.013	ND	–	ND	–	0.140	4.147
13	0.006	5.085	0.002	3.399	ND	–	0.032	2.157
14	9.084	1.003	1.097	9.726	0.816	2.173	2.564	2.621
15	0.050	7.549	ND	–	0.007	7.890	0.077	2.755
16	2.962	4.361	ND	–	0.174	0.651	2.019	1.653
17	ND [3]	–	ND	–	ND	–	0.030	3.253
18	ND	–	ND	–	0.001	4.623	0.080	5.340

[1] Neomangiferin (**1**), mangiferin (**2**), magnoflorine (**3**), spinosin (**4**), liquiritin apioside (**5**), liquiritin (**6**), ferulic acid (**7**), 6‴-feruloyl spinosin (**8**), isoliquiritin (**9**), ononin (**10**), liquiritigenin (**11**), jujuboside A (**12**), isoliquiritigenin (**13**), glycyrrhizin (**14**), jujuboside B (**15**), Z-ligustilide (**16**), dehydropachymic acid (**17**), and pachymic acid (**18**). [2] SJT–1: SJT water extract manufactured by Korea Institute of Oriental Medicine (KIOM), SJT–2 and SJT–3: commercial granules produced by Korean pharmaceutical companies, SJT–4: commercially available granule produced by Japanese pharmaceutical company. [3] ND: not detected.

4. Conclusions

In the present study, we developed a simultaneous analysis method using UPLC–MS/MS to quantify 18 components for the efficient quality control of SJIT, which is a prescription medicine used for insomnia. The developed assay was validated with respect to linearity, LOD, LOQ, recovery, and precision. The established analytical method was then applied to the simultaneous analysis of real SJIT samples, and the content of the 18 target components was analyzed simultaneously. Among them, neomangiferin, mangiferin, liquiritin apioside, glycyrrhizin, and Z-ligustilide, which are the main components of *A. asphodeloides*, *G. uralensis*, and *C. officinale*, were found to be relatively abundant compared with the other components. The method is therefore concluded to be suitable for collecting basic data for future clinical and efficacy studies.

Supplementary Materials: The following supporting information can be downloaded at: https://www.mdpi.com/article/10.3390/separations10070411/s1, Figure S1: Chemical structures of the eighteen target components in SJIT; Figure S2: Precursor ion spectrum of jujuboside A; Figure S3: Total ion chromatograms of SJIT–2 to SJIT–4 samples by the UPLC–MS/MS MRM method. Neomangiferin (**1**), mangiferin (**2**), magnoflorine (**3**), spinosin (**4**), liquiritin apioside (**5**), liquiritin (**6**), ferulic acid (**7**), 6‴-feruloyl spinosin (**8**), isoliquiritin (**9**), ononin (**10**), liquiritigenin (**11**), jujuboside A (**12**), isoliquiritigenin (**13**), glycyrrhizin (**14**), jujuboside B (**15**), *Z*-ligustilide (**16**), dehydropachymic acid (**17**), and pachymic acid (**18**); Figure S4: Extracted ion chromatograms of standard compounds (A) and SJIT–1 to SJIT–4 samples (B–E) by UPLC–MS/MS method. Compounds **1–18** as in Figure S2; Table S1: Composition of SJIT; Table S2: LC–MS/MS MRM analysis conditions for quantification of markers in SJIT.

Author Contributions: Conceptualization, C.-S.S. and H.-K.S.; methodology, validation, formal analysis, investigation, data curation, writing—original draft preparation, writing—review and editing, C.-S.S.; project administration, funding acquisition, H.-K.S. All authors have read and agreed to the published version of the manuscript.

Funding: This research was funded by the Korea Institute of Oriental Medicine (KSN2022310 and KSN1823311).

Institutional Review Board Statement: Not applicable.

Informed Consent Statement: Not applicable.

Data Availability Statement: All the data in this study can be found in this paper.

Conflicts of Interest: The authors declare no conflict of interest.

References

1. Yeh, C.H.; Arnold, C.K.; Chen, Y.H.; Lai, J.N. Suan zao ren tang as an original treatment for sleep difficulty in climacteric women: A prospective clinical observation. *Evid.-Based Complement. Altern. Med.* **2011**, *2011*, 673813. [CrossRef] [PubMed]
2. Cao, J.X.; Zhang, Q.Y.; Cui, S.Y.; Cui, X.Y.; Zhang, J.; Zhang, Y.H.; Bai, Y.J.; Zhao, Y.Y. Hypnotic effect of jujubosides from *Semen Ziziphi Spinosae*. *J. Ethnopharmacol.* **2010**, *130*, 163–166. [CrossRef] [PubMed]
3. Wu, H.C.; Chen, Y.H.; Lai, J.N.; Hwang, J.S.; Wang, J.D. Improving sleep quality in climacteric women with insomnia: A randomized, head-to-head trial between *Jia-Wei-Shiau-Yau San* (*JWSYS*) and *Suan-Zao-Ren Tang* (*SZRT*). *Eur. J. Integr. Med.* **2011**, *3*, 143–151. [CrossRef]
4. Peng, W.H.; Hsieh, M.T.; Lee, Y.S.; Lin, Y.C.; Liao, J. Anxiolytic effect of seed of *Ziziphus jujuba* in mouse models of anxiety. *J. Ethnopharmacol.* **2000**, *72*, 435–441. [CrossRef] [PubMed]
5. Park, J.H.; Lee, H.J.; Koh, S.B.; Ban, J.Y.; Seong, Y.H. Protection of NDMA-induced neuronal cell damage by methanol extract of Zizyphi Spinosi Semen in cultured rat cerebellar granule cells. *J. Ethnopharmacol.* **2004**, *95*, 39–45. [CrossRef]
6. Kuang, X.; She, G.; Ma, T.; Cai, W.; Zhao, J.; Liu, B.; Xu, F. The pharmacology, pharmacokinetics, and toxicity of spinosin: A mini review. *Front. Pharmacol.* **2022**, *13*, 938395. [CrossRef]
7. Hoffmann, K.M.; Beltrán, L.; Ziemba, P.M.; Hatt, H.; Gisselmann, G. Potentiating effect of glabridin from *Glycyrrhiza glabra* on $GABA_A$ receptors. *Biochem. Biophys. Rep.* **2016**, *6*, 197–202.
8. Cho, S.M.; Shimizu, M.; Lee, C.J.; Han, D.S.; Jung, C.K.; Jo, J.H.; Kim, Y.M. Hypnotic effects and binding studies for $GABA_A$ and 5-HT_{2C} receptors of traditional medicinal plants used in Asia for insomnia. *J. Ethnopharmacol.* **2010**, *132*, 225–232. [CrossRef]
9. Shah, V.K.; Choi, J.J.; Han, J.Y.; Lee, M.K.; Hong, J.T.; Oh, K.W. Pachymic acid enhances pentobarbital-induced sleeping behaviors via $GABA_A$-ergic systems in mice. *Biomol. Ther.* **2014**, *22*, 314–320. [CrossRef]
10. Kim, H.; Park, I.; Park, K.; Park, S.; Kim, Y.I.; Park, B.G. The positive effects of *Poria cocos* extract on quality of sleep in insomnia rat models. *Int. J. Environ. Res. Public Health* **2022**, *19*, 6629. [CrossRef]
11. Lee, S.Y.; Lee, J.Y.; Kim, J.S.; Lee, J.H.; Kang, S.S. Flavonoids from the seeds of *Zizyphus jujuba* var. *spinosa*. *Korean J. Pharmacogn.* **2012**, *43*, 127–136.
12. Yang, Y.N.; Liu, Y.Y.; Feng, Z.M.; Jiang, J.S.; Zhang, P.C. Seven new flavonoid glycosides from the roots of *Glycyrrhiza uralensis* and their biological activities. *Carbohydr. Res.* **2019**, *485*, 107820. [CrossRef] [PubMed]
13. Li, L.M.; Liao, X.; Peng, S.L.; Ding, L.S. Chemical constituents from the seeds of *Ziziphus jujuba* var. *spinosa* (Bunge) Hu. *J. Intergr. Plant Biol.* **2005**, *47*, 494–498. [CrossRef]
14. Li, G.; Xu, M.L.; Lee, C.S.; Woo, M.H.; Chang, H.W.; Son, J.K. Cytotoxicity and DNA topoisomerases inhibitory activity of constituents form the sclerotium of *Poria cocos*. *Arch. Pharm. Res.* **2004**, *27*, 829–833. [CrossRef]
15. Liu, Y.Y.; Yang, Y.N.; Feng, Z.M.; Jiang, J.S.; Zhang, P.C. Eight new triterpenoid saponins with antioxicant activity from the roots of *Glycyrrhaiza uralensis* Fisch. *Fitoterapia* **2019**, *133*, 186–192. [CrossRef] [PubMed]

16. Bae, K.E.; Choi, Y.W.; Kim, S.T.; Kim, Y.K. Components of rhizome extract of *Cnidium officinale* Makino and their in vitro biological effects. *Molecules* **2011**, *16*, 8833–8847. [CrossRef] [PubMed]
17. Baek, M.E.; Seong, G.U.; Lee, Y.J.; Won, J.H. Quantitative analysis for the quality evaluation of active ingredients in Cnidium Rhizome. *Yakhak Hoeji* **2016**, *60*, 227–234. [CrossRef]
18. Xu, L.; Li, A.; Sun, A.; Liu, R. Preparative isolation of neomangiferin and mangiferin form Rhizoma Anemarrhenae by high-speed countercurrent chromatography using ionic liquids as a two-phase solvent system modifier. *J. Sep. Sci.* **2010**, *33*, 31–36. [CrossRef]
19. Yang, B.Y.; Zhang, J.; Liu, Y.; Kuang, H.X. Steroidal saponins from the Rhizomes of *Anemarrhena asphodeloides*. *Molecules* **2016**, *21*, 1075. [CrossRef]
20. Kim, W.I.; Zhao, B.T.; Zhang, H.Y.; Lee, J.H.; Son, J.K.; Woo, M.H. Quantitative and pattern recognition analyses of magnoflorine, spinosyn, 6‴-feruloyl spinosyn and jujuboside A by HPLC in Zizyphi Semen. *Arch. Pharm. Res.* **2014**, *37*, 1139–1147. [CrossRef]
21. Niu, C.; Zhang, J. Quantitative analysis and chromatographic fingerprinting of the semen zizyphi spinosae by ultra-high-performance liquid chromatography coupled with diode-array detector. *J. Sep. Sci.* **2011**, *34*, 2989–2996. [CrossRef] [PubMed]
22. Seo, C.S.; Ha, H.; Kim, Y.J.; Jung, J.Y. HPLC–PDA simultaneous determination and protective effect of *Anemarrhena asphodeloides* against acute renal failure. *Nat. Prod. Commun.* **2014**, *9*, 829–832. [CrossRef] [PubMed]
23. Peng, Y.; Zhao, L.; Lin, D.; Liu, Y.; Zhang, M.; Song, S.J. Determination of the chemical constituents of the different processed products of *Anemarrhena asphodeloides* Rhzomes by high-performance liquid chromatography quadrupole time-of-flight mass spectrometry. *Biomed. Chromatogr.* **2016**, *30*, 508–519. [CrossRef]
24. Wu, L.F.; Wang, K.F.; Mao, X.; Liang, W.Y.; Chen, W.J.; Li, S.; Qi, Q.; Cui, Y.P.; Zhang, L.Z. Screening and analysis of the potential bioactive components of *Poria cocos* (Schw.) Wolf by HPLC and HPLC–MS^n with the aid of chemometrics. *Molecules* **2016**, *21*, 227. [CrossRef] [PubMed]
25. Zhou, S.; Cao, J.; Qiu, F.; Kong, W.; Yang, S.; Yang, M. Simultaneous determination of five bioactive components in Radix Glycyrrhizae by pressurized liquid extraction combined with UPLC–PDA and UPLC/ESI–QTOF–MS confirmation. *Phytochem. Anal.* **2013**, *24*, 527–533. [CrossRef] [PubMed]
26. Wu, Y.P.; Meng, X.S.; Bao, Y.R.; Wang, S.; Kang, T.G. Simultaneous quantitative determination of nine active chemical compositions in traditional Chinese medicine *Glycyrrhiza* by RP–HPLC with full-time five-wavelength fusion method. *Am. J. Chin. Med.* **2013**, *41*, 211–219. [CrossRef] [PubMed]
27. Zhu, Y.; Chen, L.; Gong, J.; Wang, S. Identification of constituents in Suanzaoren tang by LC-Q-TOF-MS and LC-IT-MS. *China J. Chin. Mater. Medica* **2014**, *39*, 457–465.
28. Pang, B.; Zhu, Y.; Lu, L.; Gu, F.; Chen, H. The applications and features of liquid chromatography-mass spectrometry in the analysis of traditional Chinese medicine. *Evid.-Based Complement. Altern. Med.* **2016**, *2016*, 3837270. [CrossRef]
29. Wang, X.; Zhang, A.; Yan, G.; Han, Y.; Sun, H. UHPLC-MS for the analytical characterization of traditional Chinese medicines. *Trends Anal. Chem.* **2014**, *63*, 180–187. [CrossRef]
30. International Conference on Harmonisation (ICH). *Guidance for Industry, Q2B, Validation of Analytical Procedures: Methodology*; Food and Drug Administration: Rockville, MD, USA, November 1996.
31. Ma, C.; Wang, L.; Tang, Y.; Fan, M.; Xiao, H.; Huang, C. Identification of major xanthones and steroidal saponins in rat urine by liquid chromatography–atmospheric pressure chemical ionization mass spectrometry technology following oral administration of Rhizoma Anemarrhenae decoction. *Biomed. Chromatogr.* **2008**, *22*, 1066–1083. [CrossRef]
32. Zhang, Y.Y.; Wang, Q.; Qi, L.W.; Qin, X.Y.; Qin, M.J. Characterization and determination of the major constituents in Belamcandae Rhizoma by HPLC–DAD–ESI–MS^n. *J. Pharm. Biomed. Anal.* **2011**, *56*, 304–311. [CrossRef] [PubMed]
33. Seo, C.S.; Shin, H.K. Liquid chromatography tandem mass spectrometry for the simultaneous quantification of eleven phytochemical constituents in traditional Korean medicine, Sogunjung decoction. *Processes* **2021**, *9*, 153. [CrossRef]
34. Yan, Y.; Liu, J.; Zhang, M.; Zhang, Y.; Shi, B.; Qin, X.; Du, C. A strategy to explore the quality markers of Ziziphi Spinosae semen by combining metabolic in vivo study with network pharmacology. *Biomed. Chromatogr.* **2023**, *37*, e5530. [CrossRef] [PubMed]
35. Xiao, H.B.; Krucker, M.; Albert, K.; Liang, X.M. Determination and identification of isoflavnolids in Radix *astragali* by matrix solid-phase dispersion extraction and high-performance liquid chromatography with photodiode array and mass spectrometric detection. *J. Chromatogr. A* **2004**, *1032*, 117–124. [CrossRef] [PubMed]
36. Chen, L.; Qi, J.; Chang, Y.; Zhu, D.; Yu, B. Identification and determination of the major constituents in traditional Chinese medicinal formula Danggui-Shaoyao-San by HPLC–DAD–ESI–MS/MS. *J. Pharm. Biomed. Anal.* **2009**, *50*, 127–137. [CrossRef]
37. Li, Z.; Wen, R.; Du, Y.; Zhao, S.; Zhao, P.; Jiang, H.; Rong, R.; Lv, Q. Simultaneous quantification of fifteen compounds in rat plasma by LC-MS/MS and its application to a pharmacokinetic study of Chaihu-Guizhi decoction. *J. Chromatogr. B* **2019**, *1105*, 15–25. [CrossRef]
38. Zhang, Y.; Xie, J.; Zhang, Y.; Zhang, M. Degradation kinetics of jujuboside A by rat intestinal flora and identification of the metabolites by HPLC-MS/MS. *Int. J. Food Prop.* **2014**, *17*, 1841–1849. [CrossRef]
39. Meng, X.; Li, H.; Song, F.; Liu, C.; Liu, Z.; Liu, S. Studies on triterpenoids and flavones in *Glycyrrhiza uralensis* Fisch. By HPLC-ESI-MS^n and FT-ICR-MS^n. *Chin. J. Chem.* **2009**, *27*, 299–305. [CrossRef]
40. Liu, X.; Zhu, X.; Zhu, H.; Xie, L.; Ma, J.; Xu, Y.; Zhou, Q.; Wu, Z.; Cai, B. Simultaneous quantification of six bioactive components in decoction of Ziziphi spinosae semen using ultrahigh performance liquid chromatography coupled with triple-quadrupole mass spectrometry. *J. Anal. Methods Chem.* **2018**, *2018*, 8397818. [CrossRef]

41. Li, J.; Wang, H.; Liu, J.; Du, C.; Yan, Y. Quality evaluation of Suanzaoren decoction by integrating fingerprint and quantitative analysis of multi-component. *Chin. Tradit. Herb. Drugs* **2022**, *53*, 4698–4708.
42. Miyaoka, T.; Kawano, K.; Furuya, M.; Wake, R.; Hashioka, S.; Liaury, K.; Limoa, E.; Tsuchie, K.; Fukushima, M.; Araki, T.; et al. Efficay and safety of Sansoninto in insomnia with psychiatric disorder: An open-label study. *Altern. Integr. Med.* **2015**, *4*, 181. [CrossRef]

molecules

MDPI

Article

The Residue and Dietary Risk Assessment of Spirotetramat and Its Four Metabolites in Cabbage Using Ultra-High-Performance Liquid Chromatography–Tandem Mass Spectrometry

Junli Cao *, Jindong Li, Pengcheng Ren, Yanli Qi and Shu Qin

Shanxi Center for Testing of Functional Agro-Products, Longcheng Campus, Shanxi Agricultural University, No. 79, Longcheng Street, Taiyuan 030031, China
* Correspondence: caojunli@sxau.edu.cn; Tel.: +86-351-7965708; Fax: +86-351-7639301

Abstract: Spirotetramat is a potential tetronic acid pesticide for controlling various pests with piercing–sucking mouthparts. To clarify its dietary risk on cabbage, we established an ultra-high-performance liquid chromatography–tandem mass spectrometry (UHPLC–MS/MS) method and then investigated the residual levels of spirotetramat and its four metabolites in cabbage collected from field experiments under good agricultural practices (GAPs). The average recoveries of spirotetramat and its metabolites in cabbage were 74~110%, while the relative standard deviation (RSD) was 1~6%, and the limit of quantitation (LOQ) was 0.01 mg kg^{-1}. The terminal residue of spirotetramat was in the range of <0.05~0.33 mg kg^{-1}, the chronic dietary risk (RQ_c) was 17.56%, and the acute dietary risk (RQ_a) was 0.025~0.049%, which means an acceptable dietary intake risk. This study provides data to guide on the use of spirotetramat and to establish the maximum residue limits (MRLs) of spirotetramat on cabbage.

Keywords: spirotetramat; cabbage; dissipation; dietary intake risk

Citation: Cao, J.; Li, J.; Ren, P.; Qi, Y.; Qin, S. The Residue and Dietary Risk Assessment of Spirotetramat and Its Four Metabolites in Cabbage Using Ultra-High-Performance Liquid Chromatography–Tandem Mass Spectrometry. *Molecules* **2023**, *28*, 4763. https://doi.org/10.3390/molecules28124763

Academic Editors: Ronald Beckett and Chao Kang

Received: 8 May 2023
Revised: 3 June 2023
Accepted: 9 June 2023
Published: 14 June 2023

1. Introduction

Cabbage (*Brassica oleracea var. capitata* Linnaeus), a Brassica vegetable of the Cruciferae family, is rich in antioxidant chemicals such as vitamin C, vitamin E, flavonoids, and carotenoids, so it has the effect of reducing chronic diseases [1]. China is the largest cabbage producer in the world, with an annual yield of more than 33 million g hm^{-2} from 2000 to 2021 [2]. Spirotetramat (STM, Figure 1), *cis*-4-(ethoxycarbonyloxy)-8-methoxy-3-(2,5-xylyl)-1-azaspirodec-3-en-2-one, is a new type of tetronic acid insecticide and acaricide [3,4] which was developed by Bayer CropScience in 2008 to control aphids [5,6]. STM is the only insecticide that has the dual guiding property of moving up and down the crop through both the xylem and phloem, killing larvae by inhibiting the biosynthesis of insect fat, and can effectively control a variety of pests with piercing mouthparts and harmful mites [7]. As there is no cross-resistance to existing insecticides and little negative effect on beneficial arthropods, 140 STM products have been registered in China [8].

The European Union, Codex Alimentarius Commission (CAC), and other countries set maximum residue limits (MRLs) as thresholds for monitoring pesticide residues and ensuring food safety [9,10]. However, due to non-standardized detection methods, the existing MRLs for spirotetramat in China were all "temporary". Establishing a sensitive method for identifying and quantifying spirotetramat in agricultural products is significant. Previous reports focused on the analysis methods of STM in fruits (apples, grapes, oranges, strawberries, mangoes), vegetables (cucumbers, Chinese cabbage, spinach, pepper, onions), and cotton by liquid chromatography or liquid chromatography–mass spectrometry [11–15]. However, there were some challenges in the detection of STM's four metabolites, namely spirotetramat-enol (STM-enol), spirotetramat-enol-glucoside (STM-enol-glu), spirotetramat-keto-hydroxy (STM-keto), and spirotetramat-mono-hydroxyl (STM-mono).

Han et al. determined the presence of STM and STM-enol in apple and apple processed products based on ultra-high-performance liquid chromatography–tandem mass spectrometry (UPLC–MS/MS) [16]. Mohapatra et al. and Singh et al. used the QuEChERS (Quick, Easy, Cheap, Effective, Rugged, Safe) method based on high-performance liquid chromatography (HPLC) to quantify STM and STM-enol in mango whole fruit, peel, pulp, grape, okra, brinjal, green chili, red chili, and soil [12,17,18]. However, there have been very few reports on determining the other three metabolites of STM. Only two reports demonstrated the residue and risk assessment of STM and four metabolites in citrus and pineapple [19,20]. To our knowledge, the residue and dietary risk assessment of STM and its four metabolites on cabbage have yet to be reported.

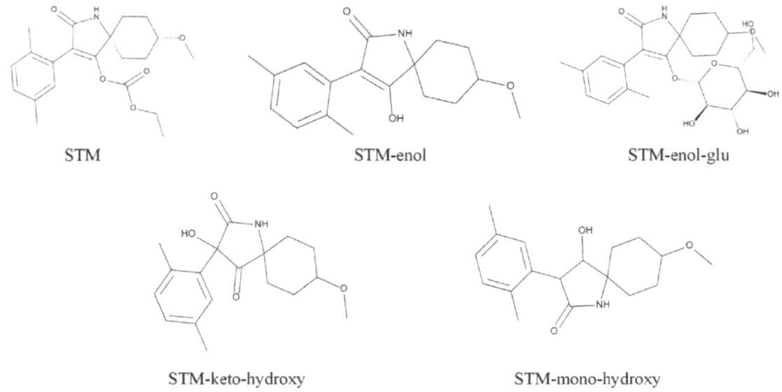

Figure 1. Structural formulas of spirotetramat and its metabolites.

We aim to (1) establish a simple and reliable ultra-high-performance liquid chromatography–tandem mass spectrometry (UHPLC–MS/MS) method for simultaneous determination of STM and its four metabolites in cabbage, (2) study the terminal residues of STM and its metabolites in cabbage, and (3) evaluate the acute and chronic dietary risks of supervised trials median residue (STMR) in cabbage based on field data. This work will provide primary data for guiding the rational use of STM and the risk to cabbage consumers.

2. Results and Discussion

2.1. Method Validation

Pesticide residue analysis includes pretreatment and instrumental analysis, among which sample pretreatment in complex matrices is the most critical step. In recent years, QuEChERS pretreatment technology has been widely used to extract STM from various matrices [15,21–23]. However, STM-enol and STM-mono-hydroxy are more polar than STM, which may cause extraction trouble. Some improvements should be made, such as pH adjustment and using formic acid to improve recovery [24]. In this study, the modified QuEChERS method was used for the pretreatment of STM and its metabolites, using 1% formic acid acetonitrile extraction, 4 g anhydrous magnesium sulfate, 1 g sodium chloride, 1 g sodium citrate, and 0.5 g disodium hydrogen citrate for salting out, and 20 mg primary secondary amine (PSA) along with 7.5 mg graphitized carbon black (GCB) for purifying. The multiple reaction monitoring (MRM) parameters are presented in Table 1, and the chromatograms showing the separation of STM and its metabolites are shown in Figure 2.

Table 1. MRM conditions of UHPLC–MS/MS for spirotetramat and its metabolites.

Compound	Retention Time (R_t, min)	Production (m z^{-1})	Declustering Potential (DP, V)	Collision Energy (CE, V)
STM	2.91	374.20 > 330.1 (quantitation)	66	47
		374.20 > 216.1 (confirmation)		21
STM-enol	2.65	302.30 > 270.2 (confirmation)	60	40
		302.30 > 216.0 (quantitation)		30
STM-enol-glu	2.10	464.40 > 302.2 (confirmation)	67	20
		464.40 > 216.0 (quantitation)		40
STM-keto-hydroxy	2.71	318.20 > 214.0 (quantitation)	40	20
		318.20 > 268.1 (confirmation)		20
STM-mono-hydroxy	2.49	304.30 > 254.1 (confirmation)	60	20
		304.30 > 211.1 (quantitation)		20

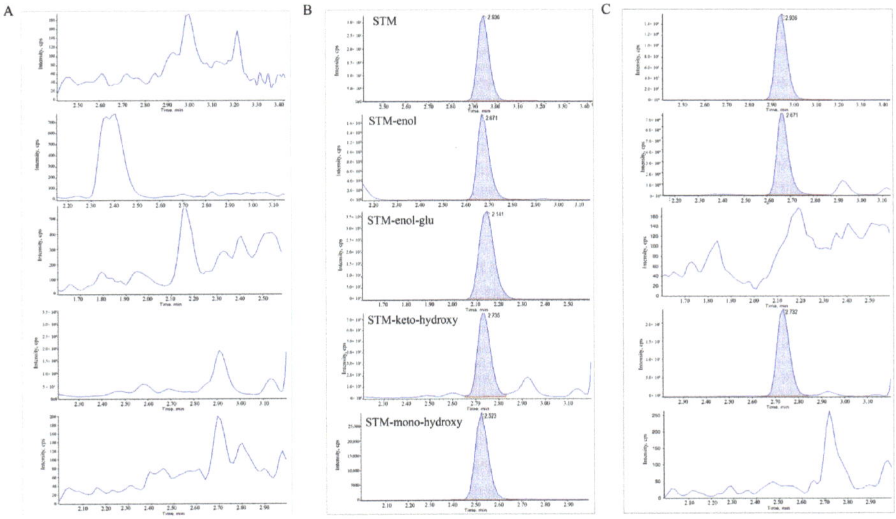

Figure 2. Representative UHPLC–MS/MS chromatograms: (**A**) chromatogram of blank cabbage sample, (**B**) chromatogram of spiked spirotetramat and its metabolite (0.01 mg kg^{-1}) spiked in blank cabbage, and (**C**) chromatogram of field cabbage sample.

The method of STM and its four metabolites in cabbage was verified by the linearity, correlation coefficient (R^2), matrix effect (ME), and LOQ. The matrix-matched standard curve was constructed with the standard solution concentrations of 0.005, 0.01, 0.02, 0.05, 0.1, and 0.2 mg L^{-1} as abscissa and the corresponding chromatographic peak area as ordinate. As shown in Table 2, the determination coefficients (R^2) of the standard curves of STM and its metabolites were greater than 0.99, indicating good linearity.

Table 2. The calibration curves, determination coefficient (R^2), and matrix effect of STM and its metabolites.

Compounds	Matrix	Calibration Curve	R^2	Matrix Effect (%)
STM	Acetonitrile	$y = 4.089 \times 10^6x + 355.0$	0.9998	-
	Cabbage	$y = 4.292 \times 10^6x + 389.9$	0.9998	5.0
STM-enol	Acetonitrile	$y = 9.598 \times 10^7x + 1.015 \times 10^5$	0.9979	-
	Cabbage	$y = 4.865 \times 10^7x - 4125$	0.9979	−29.2
STM-enol-glu	Acetonitrile	$y = 1.627 \times 10^7x + 3.406 \times 10^4$	0.9916	-
	Cabbage	$y = 1.023 \times 10^7x + -813.7$	0.9909	−49.3
STM-keto-hydroxy	Acetonitrile	$y = 2.822 \times 10^7x + 2.228 \times 10^4$	0.9989	-
	Cabbage	$y = 1.962 \times 10^7x + 3.421 \times 10^4$	0.9994	−37.1
STM-mono-hydroxy	Acetonitrile	$y = 1.322 \times 10^7x + 2.637 \times 10^4$	0.9953	-
	Cabbage	$y = 9.614 \times 10^6x + 1474$	0.9981	−29.9

As shown in Figure 3, the average recoveries of STM were 96% to 102% at three spiked levels of 0.01, 0.1, and 2.0 mg kg^{-1}, and the relative standard deviation (RSD, n = 5) was less than 2%. The average recoveries of STM-enol were between 83% and 90%, with RSD in the 2% to 3% range. The average recoveries of STM-enol-glu were between 79% and 84%, and the relative standard deviation was between 3% and 6%. The average recovery of STM-keto-hydroxy was 102~107%, and the RSD was 2~4%. The average recovery rate of STM-mono-hydroxyl was 95–105%, and the RSD was less than 2%. The average recovery of all compounds was in the range of 79% to 107%, and the RSD was less than 6%, which meets the requirements of "guideline on pesticide residue trials on crops (NY/T 788-2018)". The LOQs of the five compounds were all 0.01 mg kg^{-1}. Therefore, this method can be used for the residue analysis of STM and its metabolites in cabbage samples.

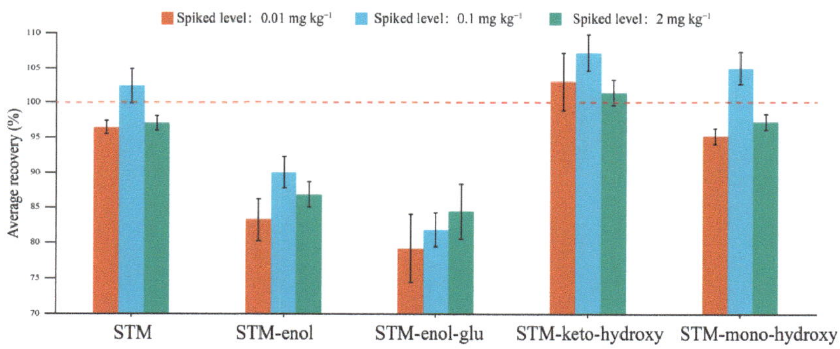

Figure 3. The average recovery and relative standard deviation of STM and its metabolites.

ME was caused by the co-ionization of the ESI source with other components in the matrix when analyzing the target compounds, which interferes with the quantitative accuracy of the analytes [25,26]. Except for STM (5%), the absolute ME values (Table 2) of STM-enol, STM-enol-glu, STM-keto-hydroxyl, and STM-mono-hydroxyl in cabbage were −29.2%, −49.3%, −37.1%, and −29.9%, respectively, all greater than 20%, indicating a prominent matrix weakening effect. Therefore, in this study, the matrix-matching standard curve was used for calibration as a compensation strategy for ME.

In conclusion, the modified QuEChERS pretreatment and UHPLC–MS/MS method was satisfactory for determining STM and its metabolites, so it can be used in field experiments.

2.2. The Terminal Residues

In 12 provinces, STM suspension was sprayed on open-field cabbage according to the recommended dosage (60 g ai hm^{-2}). The terminal residues of cabbage were collected at 7 d, 10 d, and 14 d after application, and the total residue of STM was calculated. As shown in Table 3, the terminal residues of STM in cabbage were between <0.01 and 0.108 mg kg^{-1}, those for STM-enol were in the range of <0.010 to 0.035 mg kg^{-1}, and those for the STM-keto-hydroxyl group were less than 0.14 mg kg^{-1}. The residues of STM-enol-glu and STM-mono-hydroxyl were all $\leq$0.01 mg kg^{-1} in actual cabbage samples. The total residue (risk assessment definition) of STM in cabbage was <0.050~0.33 mg kg^{-1}, which was lower than the maximum residue limit (MRL) of STM in cabbage as stipulated by the CAC (2 mg kg^{-1}) [27], European Union (7 mg kg^{-1}) [28], United States (2.5 mg kg^{-1}) [29], Japan (7 mg kg^{-1}) [30], and Australia (7 mg kg^{-1}) [31].

Table 3. Terminal residues of STM in cabbage.

Locations	Pre-Harvest Interval (Days)	Mean Residues (mg kg^{-1})					
		STM	STM-enol	STM-enol-glu	STM-keto-hydroxy	STM-mono-hydroxy	Total Residues
	7	0.039, 0.108	0.019, 0.035	<0.010, <0.010	0.081, 0.14	<0.010, <0.010	0.18, 0.33
Shanxi	10	0.043, 0.073	<0.010, <0.010	<0.010, <0.010	0.13, 0.11	<0.010, <0.010	0.22, 0.24
	14	0.023, 0.077	<0.010, <0.010	<0.010, <0.010	0.049, 0.051	<0.010, <0.010	0.11, 0.17
	7	<0.010, <0.010	<0.010, <0.010	<0.010, <0.010	<0.010, <0.010	<0.010, <0.010	<0.050, <0.050
Liaoning	10	<0.010, <0.010	<0.010, <0.010	<0.010, <0.010	<0.010, <0.010	<0.010, <0.010	<0.050, <0.050
	14	<0.010, <0.010	<0.010, <0.010	<0.010, <0.010	<0.010, <0.010	<0.010, <0.010	<0.050, <0.050
	7	<0.010, <0.010	<0.010, <0.010	<0.010, <0.010	0.011, <0.010	<0.010, <0.010	0.052, <0.050
Beijing	10	<0.010, <0.010	<0.010, <0.010	<0.010, <0.010	<0.010, <0.010	<0.010, <0.010	<0.050, <0.050
	14	<0.010, <0.010	<0.010, <0.010	<0.010, <0.010	<0.010, <0.010	<0.010, <0.010	<0.050, <0.050
	7	<0.010, <0.010	<0.010, <0.010	<0.010, <0.010	<0.010, <0.010	<0.010, <0.010	<0.050, <0.050
Shandong	10	<0.010, <0.010	<0.010, <0.010	<0.010, <0.010	<0.010, <0.010	<0.010, <0.010	<0.050, <0.050
	14	<0.010, <0.010	<0.010, <0.010	<0.010, <0.010	<0.010, <0.010	<0.010, <0.010	<0.050, <0.050
	7	<0.010, <0.010	<0.010, <0.010	<0.010, <0.010	<0.010, <0.010	<0.010, <0.010	<0.050, <0.050
Henan	10	<0.010, <0.010	<0.010, <0.010	<0.010, <0.010	<0.010, <0.010	<0.010, <0.010	<0.050, <0.050
	14	<0.010, <0.010	<0.010, <0.010	<0.010, <0.010	<0.010, <0.010	<0.010, <0.010	<0.050, <0.050
	7	<0.010, <0.010	<0.010, <0.010	<0.010, <0.010	0.033, 0.032	<0.010, <0.010	0.078, 0.078
Anhui	10	<0.010, <0.010	0.01, 0.02	<0.010, <0.010	0.032, 0.028	<0.010, <0.010	0.08, 0.087
	14	<0.010, <0.010	<0.010, <0.010	<0.010, <0.010	0.012, 0.014	<0.010, <0.010	0.055, 0.057
	7	<0.010, <0.010	<0.010, <0.010	<0.010, <0.010	<0.010, <0.010	<0.010, <0.010	<0.050, <0.050
Shanghai	10	<0.010, <0.010	<0.010, <0.010	<0.010, <0.010	<0.010, <0.010	<0.010, <0.010	<0.050, <0.050
	14	<0.010, <0.010	<0.010, <0.010	<0.010, <0.010	<0.010, <0.010	<0.010, <0.010	<0.050, <0.050
	7	0.018, <0.010	<0.010, <0.010	<0.010, <0.010	<0.010, <0.010	<0.010, <0.010	0.058, <0.050
Hunan	10	0.021, 0.039	<0.010, <0.010	<0.010, <0.010	<0.010, <0.010	<0.010, <0.010	0.061, 0.079
	14	<0.010, <0.010	<0.010, <0.010	<0.010, <0.010	<0.010, <0.010	<0.010, <0.010	<0.050, <0.050
	7	<0.010, <0.010	<0.010, <0.010	<0.010, <0.010	0.012, 0.016	<0.010, <0.010	0.054, 0.058
Guangxi	10	<0.010, <0.010	<0.010, <0.010	<0.010, <0.010	<0.010, <0.010	<0.010, <0.010	<0.050, <0.050
	14	<0.010, <0.010	<0.010, <0.010	<0.010, <0.010	<0.010, <0.010	<0.010, <0.010	<0.050, <0.050
	7	<0.010, <0.010	<0.010, <0.010	<0.010, <0.010	<0.010, <0.010	<0.010, <0.010	<0.050, <0.050
Guizhou	10	<0.010, <0.010	<0.010, <0.010	<0.010, <0.010	<0.010, <0.010	<0.010, <0.010	<0.050, 0.053
	14	<0.010, <0.010	<0.010, <0.010	<0.010, <0.010	<0.010, <0.010	<0.010, <0.010	<0.050, <0.050
	7	0.010, 0.013	<0.010, <0.010	<0.010, <0.010	0.028, 0.023	<0.010, <0.010	0.073, 0.069
Hainan	10	0.008, 0.012	<0.010, <0.010	<0.010, <0.010	0.028, 0.03	<0.010, <0.010	0.073, 0.077
	14	<0.010, <0.010	<0.010, <0.010	<0.010, <0.010	<0.010, <0.010	<0.010, <0.010	0.053, 0.065
	7	<0.010, <0.010	<0.010, <0.010	<0.010, <0.010	0.01, 0.014	<0.010, <0.010	0.052, 0.056
Guangdong	10	<0.010, <0.010	<0.010, <0.010	<0.010, <0.010	<0.010, <0.010	<0.010, <0.010	<0.050, <0.050
	14	<0.010, <0.010	<0.010, <0.010	<0.010, <0.010	<0.010, <0.010	<0.010, <0.010	<0.050, <0.050

2.3. Dietary Risk Assessment

The chronic dietary risk quotient (RQ_c) and acute dietary risk quotient (RQ_a) were used to assess the chronic and acute dietary risk of STM intake from cabbage, respectively. According to the China Pesticide Information Network, there are 136 products registered for cabbage, celery, tomato, eggplant, chili, cucumber, potato, citrus, apple, pear, peach, banana, watermelon, melon, and tea. The national estimated daily intake (NEDI) of STM was calculated based on the dietary group diet of different populations in China published by the Ministry of Health in 2002, combined with STMR. Since the STMR of crops other than cabbage could not be obtained, the MRL of each country was chosen instead of STMR. The MRLs established by China, the Commission, the United States, and Australia should be given priority. Based on risk maximization, the maximum value is selected for evaluation when there are multiple MRL values. The average weight of the general Chinese population was 63 kg. The ADI of STM was 0.05 mg kg^{-1} bw (GB2763-2021). STMR in cabbage and

MRLs in potato, celery, peach, and tea were used to calculate the NEDI of STM. As shown in Table 4, the RQ_c at 7, 10, and 14 days between harvesting periods was all 17.56%, much lower than 100%, indicating that the long-term dietary risk of STM would not threaten ordinary consumers.

Table 4. Chronic dietary intake risk assessment of STM based on Chinese dietary composition.

Crops	Food Classification	Fi (kg)	Residue (mg kg^{-1})	Sources	NEDI (mg)	ADI (mg)	Risk Quotient (RQ_c, %)
Potato	Tubers	0.0495	0.8	China, MRL	3.960×10^{-2}		1.26
Celery	Dark vegetables	0.0915	4	China, MRL	3.660×10^{-1}		11.62
Peach	Fruits	0.0457	3	China, MRL	1.371×10^{-1}		4.35
Tea	Salt	0.012	0.1	Australia, MRL	1.200×10^{-3}	0.05	0.04
			0.051	STMR1 (PHI = 7)	9.369×10^{-3}		0.30
Cabbage	Light vegetables	0.1837	0.050	STMR2 (PHI = 14)	9.185×10^{-3}		0.29
			0.050	STMR3 (PHI = 21)	9.185×10^{-3}		0.29
					5.533×10^{-1} (PHI = 7)		17.56 (PHI = 7)
	Total	0.3824			5.531×10^{-1} (PHI = 14)		17.56 (PHI = 14)
					5.531×10^{-1} (PHI = 21)		17.56 (PHI = 21)

The short-term dietary risk of STM after intake of cabbage was assessed (Table 5). According to the official data of the World Health Organization [32], the LP of cabbage in different age groups in China ranges from 0.0201 kg d^{-1} to 0.0515 kg d^{-1}. The high residue of STM in cabbage was 0.33 mg kg^{-1}. Therefore, in four different age groups, the national estimated short-term intake (NESTI) of STM was in the range of 2.46×10^{-4} to 5.39×10^{-4} mg (kg bw)$^{-1}$. The RQ_a was from 0.025% to 0.054%, much lower than 100%, indicating that the short-term dietary intake risk caused by STM in children and adults after eating cabbage was acceptable. Our experiment was significant in determining the residual status of STM, providing a scientific basis for reducing the dietary risk assessment and the supervision of agricultural authorities, and protecting people's consumption health.

Table 5. Acute dietary risk assessment of STM on cabbage for 4 representative ages.

Age	Weight (kg)	Food Consumption (kg d^{-1})	NESTI (mg (kg bw)$^{-1}$)	RQ_a (%)
2~10	12.3~22.9	0.0201~0.0343	4.94×10^{-4}~5.39×10^{-4}	0.049~0.054
11~17	34.0~46.9	0.0381~0.0440	3.10×10^{-4}~3.70×10^{-4}	0.031~0.037
18~59	52.1~64.9	0.0448~0.0515	2.62×10^{-4}~2.84×10^{-4}	0.026~0.028
≥60	51.0~61.5	0.0380~0.0472	2.53×10^{-4}~2.46×10^{-4}	0.025~0.025

3. Materials and Methods

3.1. Chemicals and Reagents

The certified standards, STM (purity 98.86%), STM-enol (purity 99.12%), STM-enol-glu (purity 95.7%), STM-keto-hydroxy (purity 94.24%), and STM-mono-hydroxyl (purity 99.48%) were provided by Dr. Ehrenstorfer (Augsburg, Germany). Analytical grade acetonitrile was from Tiandi Co., Ltd. (Ohio, USA). HPLC-grade acetonitrile and formic acid were purchased from Thermo Fisher Scientific Co., Ltd. (Shanghai, China). HPLC-grade methanol was purchased from Merck, Germany. HPLC-grade ammonium acetate came from Guangfu Institute of Fine Chemical Industry (Tianjin, China). Analytical grade formic acid was provided by Sinopharm Chemical Reagent Co., Ltd. (Shanghai, China). Analytical grade anhydrous magnesium sulfate, sodium chloride, disodium hydrogen citrate, and sodium citrate were purchased from Shimadzu Technology Trading Co., Ltd. (Shanghai, China). Disperse solid phase extraction purification tubes (20 mg PSA, 7.5 mg GCB, and 142.5 mg anhydrous $MgSO_4$, 2 mL) were provided by Aces Scientific. The polytetrafluoroethylene filter (0.22 μm) was purchased from GL Sciences Technology Trading Co., Ltd. (Shanghai, China).

The standard solution was prepared using acetonitrile to dissolve 10.1 mg, 10.1 mg, 10.4 mg, 10.6 mg, and 10.1 mg STM, STM-enol, STM-enol-glu, STM-keto-hydroxy, and STM-mono-hydroxyl in a 10 mL volumetric flask, respectively. These reserves were then stored in a refrigerator at −18 °C. The above standard solution was diluted with acetonitrile to obtain a mixed standard solution with a concentration of 100 g mL^{-1}. Then, six mixed standard solutions with a concentration of 0.005, 0.01, 0.02, 0.05, 0.1, and 0.2 μg mL^{-1} were serially diluted with acetonitrile. Matrix-matched standards were obtained by spiking an appropriate amount of standard to blank cabbage extract.

3.2. Field Trials and Sampling

According to "Guideline on Pesticide Residue Trials" (NY/T788-2018), STM's terminal residue experiment in cabbage was carried out in Jinzhong city in Shanxi province (E112°, N37°), Liaoyang city in Liaoning province (E123°, N41°), Changping District in Beijing (N40°, E116°), Qingdao city in Shandong province (N36°, E120°), Xinxiang city in Henan province (N35°, E113°), Suzhou city in Anhui province (N34°, E116°), Fengxian District in Shanghai (N30°, E121°), Liuyang city in Hunan province (N27°, E113), Nanning city in Guangxi province (N22°, E108°), Guiyang city in Guizhou province (N26°, E106°), Haikou city in Hainan province (N20°, E110°), and Foshan city in Guangdong province (N22°, E112°). These field experiments sites covered almost all the residual risks of cabbage planting areas, taking into account the effects of cabbage planting methods, varieties, soil types, cultivation methods, and climate on pesticide residues. The soil properties and climatic conditions of the field plots are presented in Table S1 (Supplementary Materials). Soil pH, cation exchange capacity, and organic matter were measured in accordance with the Agricultural Industry Standard of the People's Republic of China—NY/T1121 Part II, Part V, and Part VI. The mean temperature data and precipitation were continuously obtained by the field meteorological station during the experimental period. In the experiment, STM treatment and one control group were set up. Two replicates were set up for each treatment, and each treatment area was about 50 m^2. We kept a buffer zone between the treatment intervals to avoid cross-contamination. About 14 days before maturity, STM was sprayed on the cabbage according to the recommended dose (60 g ai hm^{-2}). The cabbage samples collected at 7 d, 10 d, and 14 d were used as terminal residual samples. After removing the wilted part, the cabbage sample was chopped with a knife, and two samples of no less than 200 g were taken by the quartering method, one for the experimental sample and one for the backup sample. All field samples were stored in a −20 °C freezer.

3.3. Sample Preparations

Cabbage samples were homogenized with a pulverizer. The weighed 10.0 g cabbage sample was put into a 50 mL PTFE centrifuge tube and 10 mL acetonitrile-formic acid (99:1, *v*:*v*) was added. The tube was vortexed for 10 min to extract the target compound. Then, 4 g of anhydrous magnesium sulfate, 1 g of sodium chloride, 1 g of sodium citrate, and 0.5 g of disodium hydrogen citrate were added. Then, the tubes were shaken for 5 min and centrifuged at 8000 rpm for 5 min.

Next, 1.5 mL of the supernatant was transferred to a 2 mL centrifuge tube containing 20 mg PSA, 7.5 mg GCB, and 142.5 mg anhydrous $MgSO_4$.The tube was vortexed for 3 min and then centrifuged at 5000 rpm for 2 min. Finally, a 0.22 μm organic membrane was used for filtration, to be determined by UHPLC–MS/MS.

3.4. UHPLC–MS/MS Analysis

A UHPLC–MS/MS system (Triple Quad 4500, AB Sciex) equipped with an electrospray ionization source was used to analyze STM and its metabolites in cabbage. The chromatographic column was a Kinetex® 2.6 μm EVO C18 100 μm chromatographic column (50 × 2.1 mm). The mobile phase comprised 4 mmol/L ammonium acetate aqueous solution with 0.1% formic acid (A) and methanol (B) at a flow rate of 0.3 m L/min. The gradient elution procedure was: 0~0.5 min constant, 90% A; 2.5~3.5 min, 5% A; and 3.6~5.1 min, 90% A. The column and sample room temperatures were 40 °C and 20 °C, respectively. The electrospray ionization source was scanned in positive ion mode. The ionization voltage was 5500 V, the collision gas was nitrogen, and the temperature of the heating module was

550 °C. The injection volume was 2 μL. The mass spectrometric parameters of STM and its metabolites are shown in Table 1.

3.5. Method Validation

The analytical methods of STM and its metabolites in cabbage samples were verified by the accuracy, precision, linearity, limit of quantitation (LOQ), and matrix effect (ME), according to SANTE/11312/2021 [33].

To evaluate the accuracy and precision of the analytical method, the standard of STM and its metabolites were spiked to blank cabbage samples at 0.01 mg kg^{-1}, 0.1 mg kg^{-1}, and 2.0 mg kg^{-1}, with five replicates per level. The recovery (%) and relative standard deviation (RSD, %) were calculated. The method had qualified accuracy and precision when the recovery was 70~120% and the RSD was less than 20%. The limit of quantitation (LOQ) was defined as the lowest spiked level.

The linearity was evaluated by analyzing the standard and matrix-matched standard solution curves in the concentration range of 0.005–0.2 mg L^{-1}. The matrix effect (ME) was calculated by comparing the slope of the matrix-matched calibration curve to the slope of the solvent calibration curve by the following formula:

$$ME\ (\%) = (S_m - S_s)/S_s \times 100\% \quad (1)$$

where S_m and S_s represent the slopes of the matrix-matched standard curve and the solvent standard curve, respectively. A positive ME value represents a matrix-enhancing effect, while a negative ME value shows a matrix-inhibiting effect. When the matrix effect is in the range of −20~20%, the matrix effect can be ignored; when the matrix effect is in the range of −50~−20% or 20~50%, it means a weak matrix effect; and when the matrix effect is lower than -50% or greater than 50%, it represents a strong matrix effect.

3.6. Definition of STM Residue

The residue definition for risk assessment of STM in plant foods was proposed as the "Sum of spirotetramat, STM-enol, STM-keto-hydroxy, STM-mono-hydroxyl, and STM-enol-glu, expressed as spirotetramat". In this study, the definition was the sum of STM and its four metabolites, which were expressed as spirotetramat. The sum of STM was calculated as follows:

$$C_{sum} = C_{STM} + 1.239 \times C_{STM\text{-}enol} + 0.806 \times C_{enol\text{-}glu} + 1.177 \times C_{keto\text{-}hydroxy} + 1.231 \times C_{mono\text{-}hydroxy} \quad (2)$$

where CSTM, C_{eno}, $C_{eno\text{-}glu}$, $C_{keto\text{-}hydroxy}$, and $C_{mono\text{-}hydroxy}$ were the residue concentrations of STM, STM-enol, STM-enol-glu, STM-keto-hydroxy, and STM-mono-hydroxy, respectively. The values 1.239, 0.806, 1.177, and 1.231 represent the ratio of the molecular weight of STM-enol, STM-enol-glu, STM-keto-hydroxy, and STM-mono-hydroxy to spirotetramat, respectively. When the residual concentration was less than the limit of quantitation (LOQ), 0.01 mg kg^{-1} was used for calculation.

3.7. Dietary Risk Assessment

In recent reports, the risk quotient method was used to assess the chronic dietary risk (RQ_c) and acute dietary risk (RQ_a) of pesticides. An RQ < 100% indicates that the dietary risk is acceptable to consumers, while an RQ > 100% indicates an unacceptable risk.

RQ_c was the ratio of the NEDI to ADI and was calculated as follows:

$$NEDI = Fi \times STMR/BW \quad (3)$$

$$RQ_c = NEDI/ADI \quad (4)$$

where NEDI is the national estimated daily intake, (mg kg^{-1} bw) d^{-1}; FI is the per capita daily intake of cabbage, kg d^{-1}; and STMR is the median residue of STM in cabbage obtained from field experiments, mg kg^{-1}. The field experiment showed that the STMR of STM was 0.051 (PHI = 7), 0.050 (PHI = 10), and 0.050 (PHI = 14) respectively. BW is the average body weight of Chinese adult, 63 kg. ADI represents the allowable daily intake, (mg kg^{-1} bw) d^{-1}. The ADI of STM is considered to be 0.05 mg kg^{-1} bw [34,35].

RQ_a was calculated as a percentage of NESTI to ARfD:

$$NESTI = Lp \times HR/bw \quad (5)$$

$$RQ_a = NESTI/ARfD \quad (6)$$

where NESTI is the national estimated short-term intake (mg kg^{-1} bw); LP is the highest consumption of cabbage per day, kg d^{-1}; HR is the highest terminal residue (0.33 mg kg^{-1}) of STM in cabbage obtained from field trials; BW is the average body weight of different age groups; and ARfD is the acute reference dose. The ARfD of STM was 1 mg kg^{-1} bw d^{-1} [35].

4. Conclusions

We verified a simple, sensitive, reliable quantitative method for determining STM and its four metabolites. The samples were extracted with acetonitrile-formic acid, purified by PSA and GCB, and determined qualitatively and quantitatively by UHPLC–MS/MS. The method's precision, accuracy, linearity, and LOQ all meet the requirements of the guidelines for pesticide residue analysis. In the supervised field experiment, the terminal residue range of STM was from <0.05 mg kg^{-1} to 0.033 mg kg^{-1}. The chronic dietary risk was 17.56%, and the acute dietary risk was 0.025~0.049%, all of which are less than 100%, indicating that the STM suspending agent is acceptable to the chronic dietary risk of cabbage and was sprayed according to the active ingredient at 60 g hm^{-2}.

Supplementary Materials: The following supporting information can be downloaded at: https://www.mdpi.com/article/10.3390/molecules28124763/s1, Table S1: Soil properties and climate conditions for field trials; Table S2: Terminal residues of STM in cabbage on 3d.

Author Contributions: Data analysis, writing—original draft preparation, writing—review and editing, visualization, funding acquisition, J.C.; methodology, software, resources, supervision, J.L.; methodology, validation, visualization, Y.Q.; methodology, validation, visualization, P.R.; project administration, resources, supervision, S.Q. All authors have read and agreed to the published version of the manuscript.

Funding: This research was funded by the Startup Foundation for Doctors of Shanxi Agricultural University (2021BQ127) and Shanxi Applied Basic Research Program Science–Youth Technology Research Fund (202103021223149).

Institutional Review Board Statement: Not applicable.

Informed Consent Statement: Not applicable.

Data Availability Statement: The data presented in this study are available from the authors upon request.

Acknowledgments: The field experiment was carried out with the help of Zhang Junwen, Su Hailing, and Zhu Li'ao. Our colleagues LV Ying and Wang Xia assisted in the indoor sample detection.

Conflicts of Interest: The authors declare no conflict of interest.

Sample Availability: Samples from our study are available from the authors upon request.

References

1. Franzke, A.; Lysak, M.A.; Al-Shehbaz, I.A.; Koch, M.A.; Mummenhoff, K. Cabbage family affairs: The evolutionary history of Brassicaceae. *Trends Plant Sci.* **2011**, *16*, 108–116. [CrossRef]
2. Food and Agriculture Organization of the United Nations. Available online: https://www.fao.org/faostat/en/#compare (accessed on 5 November 2022).
3. European Food Safety Authority. Conclusion on the peer review of the pesticide risk assessment of the active substance spirotetramat. *EFSA J.* **2013**, *11*, 3243. [CrossRef]

4. Vang, L.E.; Opperman, C.H.; Schwarz, M.R.; Davis, E.L. Spirotetramat causes an arrest of nematode juvenile development. *Nematology* **2016**, *18*, 121–131. [CrossRef]
5. Nauen, R.; Reckmann, U.; Thomzik, J.; Thielert, W. Biological profile of spirotetramat (Movento®)—A new two-way systemic (ambimobile) insecticide against sucking pest species. *Bayer Crop. J.* **2008**, *61*, 245–278.
6. Pesticides, A.; Authority, V.M. *Evaluation of the New Active SPIROTETRAMAT in the Product MOVENTO 240 SC INSECTICIDE*; APVMA: Kingston, Australia, 2009; pp. 9–10.
7. Salazar-López, N.-J.; Aldana-Madrid, M.-L.; Silveira-Gramont, M.-I.; Aguiar, J.-L. Spirotetramat—An alternative for the control of parasitic sucking insects and its fate in the environment. In *Insecticide Resistance*; IntechOpen: London, UK, 2016; pp. 41–55.
8. China Pesticide Information Network. Available online: http://www.chinapesticide.org.cn/zwb/dataCenter (accessed on 26 April 2022).
9. Ambrus, A.; Yang, Y.Z. Global harmonization of maximum residue limits for pesticides. *J. Agric. Food Chem.* **2016**, *64*, 30–35. [CrossRef]
10. Winter, C.K.; Jara, E.A. Pesticide food safety standards as companions to tolerances and maximum residue limits. *J. Integr. Agric.* **2015**, *14*, 2358–2364. [CrossRef]
11. Cui, S.; Li, Z.; Cheng, G.; Li, R.; Wang, Y.; Zhang, X.; Zhao, F. Determination of sulfoxaflor, pyrifluquinazon and spirotetramat residues in fruits and vegetables by UPLC-MS/MS. *Shipin Kexue/Food Sci.* **2018**, *39*, 302–308.
12. Mohapatra, S.; Deepa, M.; Lekha, S.; Nethravathi, B.; Radhika, B.; Gourishanker, S. Residue dynamics of spirotetramat and imidacloprid in/on mango and soil. *Bull. Environ. Contam. Toxicol.* **2012**, *89*, 862–867. [CrossRef] [PubMed]
13. Pandiselvi, S.; Sathiyanarayanan, S.; Ramesh, A. Determination of spirotetramat and imidacloprid residues in cotton seed, lint, oil and soil by HPLC UV method and their dissipation in cotton plant. *Pestic. Res. J.* **2010**, *22*, 168–173.
14. Kumar, B.; Kumaran, N.; Kuttalam, S. Determination of harvest time residues of spirotetramat on cotton using HPLC. *J. Plant Prot. Environ.* **2009**, *6*, 1–5.
15. Li, S.; Liu, X.; Dong, F.; Xu, J.; Xu, H.; Hu, M.; Zheng, Y. Chemometric-assisted QuEChERS extraction method for the residual analysis of thiacloprid, spirotetramat and spirotetramat's four metabolites in pepper: Application of their dissipation patterns. *Food Chem.* **2016**, *192*, 893–899. [CrossRef]
16. Han, Y.; Xu, J.; Dong, F.; Li, W.; Liu, X.; Li, Y.; Kong, Z.; Zhu, Y.; Liu, N.; Zheng, Y. The fate of spirotetramat and its metabolite spirotetramat-enol in apple samples during apple cider processing. *Food Control* **2013**, *34*, 283–290. [CrossRef]
17. Mohapatra, S.; Kumar, S.; Prakash, G. Residue evaluation of imidacloprid, spirotetramat, and spirotetramat-enol in/on grapes (Vitis vinifera L.) and soil. *Environ. Monit. Assess.* **2015**, *187*, 632. [CrossRef]
18. Singh, B.; Mandal, K.; Sahoo, S.K.; Bhardwaj, U.; Battu, R.S. Development and validation of an HPLC method for determination of spirotetramat and spirotetramat cis enol in various vegetables and soil. *J. AOAC Int.* **2013**, *96*, 670–675. [CrossRef]
19. Zhang, Q.; Chen, Y.; Wang, S.; Yu, Y.; Lu, P.; Hu, D.; Yang, Z. Dissipation, residues and risk assessment of spirotetramat and its four metabolites in citrus and soil under field conditions by LC-MS/MS. *Biomed. Chromatogr.* **2018**, *32*, e4153. [CrossRef]
20. Liang, Y.; Wu, W.; Cheng, X.; Hu, J. Residues, fate and risk assessment of spirotetramat and its four metabolites in pineapple under field conditions. *Int. J. Environ. Anal. Chem.* **2020**, *100*, 900–911. [CrossRef]
21. Faraji, M.; Noorbakhsh, R.; Shafieyan, H.; Ramezani, M. Determination of acetamiprid, imidacloprid, and spirotetramat and their relevant metabolites in pistachio using modified QuEChERS combined with liquid chromatography-tandem mass spectrometry. *Food Chem.* **2018**, *240*, 634–641. [CrossRef]
22. Lan, F.; Yao, J.; Zhou, X.; Liu, C.; Li, X.; Lu, Z.; Jiang, W.; Wang, Z. Residue of spirotetramat, diafenthiuron and their metabolites in fruits and vegetables by QuEChERS-ultra performance liquid chromatography-tandem mass spectrometry. *Chin. J. Pestic. Sci.* **2019**, *21*, 219–226.
23. Zhu, Y.; Liu, X.; Xu, J.; Dong, F.; Liang, X.; Li, M.; Duan, L.; Zheng, Y. Simultaneous determination of spirotetramat and its four metabolites in fruits and vegetables using a modified quick, easy, cheap, effective, rugged, and safe method and liquid chromatography/tandem mass spectrometry. *J. Chromatogr. A* **2013**, *1299*, 71–77. [CrossRef] [PubMed]
24. Food and Agriculture Organization. Spirotetramat. 2013. Available online: https://www.fao.org/fileadmin/templates/agphome/documents/Pests_Pesticides/JMPR/Report13/5.32_SPIROTETRAMAT__234_.pdf (accessed on 26 April 2022).
25. Liu, H.C.; Lin, D.L.; McCurdy, H.H. Matrix Effects in the Liquid Chromatography-Tandem Mass Spectrometry Method of Analysis. *Forensic Sci. Rev.* **2013**, *25*, 65–78. [PubMed]
26. Hayama, T. Matrix Effects in Mass Spectrometry Analysis. *Anal. Sci.* **2020**, *36*, 1151. [CrossRef] [PubMed]
27. Codex Online Databases. Available online: https://www.fao.org/fao-who-codexalimentarius/codex-texts/dbs/pestres/pesticide-detail/en/?p_id=234 (accessed on 26 April 2022).
28. EU Pesticide Database. Available online: https://ec.europa.eu/food/plant/pesticides/eu-pesticides-database/start/screen/mrls (accessed on 26 April 2022).
29. United States Pesticide MRLs. Available online: https://bcglobal.bryantchristie.com/db#/pesticides/results?q=eyJmaWx0ZXJzIjp7ImZVU1NlY3Rpb24xOFJlZ2lvbmFsIjowLCJmRmFjaWxpdHlVc2UiOjIsImZJbXBvcnRUb2xlcmFuY2UiOjAsInR5cGVTb3J0T3JkZXIiOlsxNSwyLDFdfSwiYm9keSI6W3siaXRlbVR5cGVJRCI6MTUsImxpc3QiOls0MzY2XX0seyJpdGVtVHlwZUlEIjoyLCJsaXN0IjpbNDU4XX0seyJpdGVtVHlwZUlEIjoxLCJsaXN0IjpbMjFdfV19&isSimpleViewProp=true (accessed on 26 April 2022).
30. The Japen Food Chemical Research Foundation. Available online: http://db.ffcr.or.jp/front/ (accessed on 26 April 2022).
31. Australian Goverment. Available online: https://www.legislation.gov.au/Details/F2023C00438 (accessed on 26 April 2022).

32. World Health Organization. Template for the Evaluation of Acute Exposure (IESTI). Available online: https://cdn.who.int/media/docs/default-source/food-safety/gems-food/guidance-iesti-2014.pdf?sfvrsn=9b24629a_2 (accessed on 26 April 2022).
33. European Commission. Guidance Document on Analytical Qualitycontrol and Method Validation Procedures for Pesticides Residues Andanalysis in Food and Feed. 2021. Available online: https://www.eurl-pesticides.eu/userfiles/file/EurlALL/SANTE_11312_2021.pdf (accessed on 26 April 2022).
34. *GB2763-2021*; National Food Safety Standard-Maximum Residue Limits for Pesticides in Food. National Health Commission of the People's Republic of China: Beijing, China, 2021.
35. Anastassiadou, M.; Bernasconi, G.; Brancato, A.; Cabrera, L.C.; Greco, L.; Jarrah, S.; Kazocina, A.; Leuschner, R.; Magrans, J.O.; Miron, I.; et al. Review of the existing maximum residue levels for spirotetramat according to Article 12 of Regulation (EC) No 396/2005. *EFSA J.* **2020**, *18*, e05960. [CrossRef]

foods

MDPI

Article

Changes in the VOC of Fruits at Different Refrigeration Stages of 'Ruixue' and the Participation of Carboxylesterase *MdCXE20* in the Catabolism of Volatile Esters

Dongmei Li, Jianhua Guo, Hai Ma, Linna Pei, Xiaojie Liu, Hui Wang, Rongxin Chen, Zhengyang Zhao and Hua Gao *

College of Horticulture, Northwest A & F University, Xianyang 712100, China; lidongmei@nwafu.edu.cn (D.L.); guojianhuahua@nwafu.edu.cn (J.G.); mahai@nwafu.edu.cn (H.M.); peilinna@nwafu.edu.cn (L.P.); echoliu@nwafu.edu.cn (X.L.); zhaozy@nwsuaf.edu.cn (Z.Z.)

* Correspondence: gaohua@nwafu.edu.cn; Tel.: +86-153-3234-7161

Abstract: Aroma is a crucial quality attribute of apple fruit, which significantly impacts its commercial value and consumer choice. Despite its importance the volatile aroma substances produced by the new variety 'Ruixue' after harvest remain unclear. In this study, we utilized headspace solid phase microextraction-gas chromatography-mass spectrometry (HS-SPME-GC-MS) to investigate the changes in volatile substances, fruit hardness, crispness, and related aroma synthase activity of commercially mature 'Ruixue' apples during cold storage. Our findings revealed a gradual decline in fruit firmness and brittleness of 'Ruixue' apples during cold storage, with hexyl acetate, hexyl caproate, and hexyl thiocyanate being the main hexyl esters detected. To gain a better understanding of the metabolic pathway of esters, we identified 42 MdCXE gene members that are associated with ester degradation. Through RT-qPCR analysis, we discovered that carboxylesterase *MdCXE20* exhibited higher expression levels compared to other MdCXE genes during cold storage. To confirm the role of *MdCXE20*, we conducted a transient injection of apple fruits and observed that overexpression of *MdCXE20* led to the degradation of esters such as hexyl hexanoate, butyl hexanoate, butyl 2-methylbutyrate, hexyl butyrate, and hexyl 2-methylbutyrate. The results of the study showed that the virus-induced gene silencing of *MdCXE20* found the opposite results. Additionally, the esters of OE-*MdCXE20* callus showed a lower content of ester VOC than the control callus, according to the homologous stable transformation of 'Wanglin' callus. Overall, these findings suggest that the *MdCXE20* gene plays a crucial role in the decrease of esters in 'Ruixue' apples, which ultimately affects their flavor.

Keywords: 'RuiXue'; apple; ester; *MdCXE20*

Citation: Li, D.; Guo, J.; Ma, H.; Pei, L.; Liu, X.; Wang, H.; Chen, R.; Zhao, Z.; Gao, H. Changes in the VOC of Fruits at Different Refrigeration Stages of 'Ruixue' and the Participation of Carboxylesterase *MdCXE20* in the Catabolism of Volatile Esters. *Foods* **2023**, *12*, 1977. https://doi.org/10.3390/foods12101977

Academic Editors: Chao Kang and Ronald Beckett

Received: 21 March 2023
Revised: 2 May 2023
Accepted: 8 May 2023
Published: 12 May 2023

1. Introduction

Apple (*Malus* × *domseica* Borkh.) is one of the four major fruits in the world. It is widely planted and loved by people because of its important edible value, economic value, social value and ecological benefits [1]. With the rapid development of economy, consumers are not only satisfied with external qualities such as appearance, color and mechanical damage, but also pay more attention to the internal quality of fruit nutrition and flavor. It is widely recognized that flavor is determined by a complex mixture of sugars, organic acids and volatile compounds (VOC). Among these, VOCs play a key role in determining fruit quality and consumer preference [2,3]. Presently, there are more than 2000 volatile aroma substances in plants, and more than 350 volatile aroma substances have been identified in mature apple fruits [3,4], including esters, alcohols, ketones, aldehydes and terpenes. The content of these volatiles is very low (ng/g), which is 1/1000 of the level of sugar acid (mg/g). Therefore, small changes in their components will greatly impact the flavor quality of apples. Among them, volatile esters are represent one of the important characteristics of apple fruit flavor [5,6]. For example, among the volatile compounds, methyl butyrate, ethyl butyrate, butyl butyrate, methyl caproate, ethyl caproate, butyl acetate, and hexyl acetate play an important role [7]. The synthesis of volatile substances in apple fruit is a dynamic process. During storage, the content and diversity of volatile substances in apples increased first and then decreased [8]. In order to prolong the fruit quality and preservation period and prevent decay after harvest, low temperature storage is usually used to inhibit the respiration, ethylene synthesis and pectin dissolution of apple fruit. However, it is found that the volatile substances of apple after cold storage have a certain impact, which seriously affects the commodity quality of apple [8]. Therefore, studying the aroma changes of chilled apples has important theoretical and practical significance for improving apple quality.

Esters are dominant in the aromatic substances of mature fruits such as apples, bananas, strawberries, peaches, and pears. As the fruit matures, esters increase significantly [9]. The specific components of ester aroma vary depending on the type of fruit. Apples are mainly composed of two types of esters, namely branched chain esters, like 2-methylhexyl butyrate and 2-methylbutyl acetate, and linear esters such as hexyl butyrate and hexyl acetate [10,11]; Ripe peach flavors are dominated by hexyl acetate, with linear esters such as vinyl ester E-2, being the most important [12]. Strawberry fruit is mainly composed of linear esters such as ethyl acetate and butyl acetate [13]. Ethyl acetate and butyl acetate are the main aromatic substances in Nanguo pears. Ester volatile substances have unique aroma, such as: butyl hexanoate has apple flavor, amyl acetate has banana, apple flavors; butyl butyrate has a fresh sweet fruit flavor [14].

Carboxylesterase (CXE, EC 3.1.1.1) is a metabolic enzyme found in a variety of organisms. It has the ability to catalyze the hydrolysis of carboxylic acid esters into corresponding alcohols and carboxylic acids. The function of this enzyme has been extensively studied in mammals, insects, and microorganisms [15,16]. Relatively speaking, there are relatively few studies on the CXE function of plants. The majority of these proteins are categorized under the α/β hydrolase folding superfamily. This particular superfamily shares a common core structure consisting of eight β chains that are interspersed with α-helixes and loops [17]. CXEs, in particular, have a catalytic triad that consists of serine (which is contained in the conserved common sequence GXSXG), an acidic amino acid, and histidine. This triad constitutes the active site of CXEs [17–19]. CXE enzymes exhibit a diverse range of substrate catalytic activity that is closely linked to various biochemical and physiological processes. These processes include the breakdown of exogenous substances and the conversion of volatiles, which play a crucial role in determining the aroma of mature fruits. Additionally, CXE enzymes are actively involved in plant secondary metabolism and plant defense responses [20,21]. Several studies have identified 20 CXE genes in the genome of *Arabidopsis thaliana*, which are found in various tissues [19]. Among these genes, *AtCXE8* overexpression in *Arabidopsis thaliana* has been shown to enhance plant resistance to gray mold invasion, while knockout of this gene renders plants susceptible to gray mold [22]. Additionally, tobacco *hsr203J* is involved in the detoxification of pathogen-derived compounds [23]. In a study conducted by [24], overexpression of *PepEST* in transgenic pepper fruits resulted in increased resistance to anthracnose. Fruit aroma and flavor are largely dependent on volatile esters, such as butyl acetate and hexyl acetate. *MdCXE1* in apple fruit was found to indirectly impact the hydrolysis of these esters, which are key components of mature fruit aroma [25]. Similarly, *SlCXE1* was identified as a regulator of tomato volatile acetate content [26]; functional analysis of 33 CXEs in peach fruit showed that 13 were expressed in fruit. The expression of six members (*PpCXE1*, *PpCXE2*, *PpCXE3*, *PpCXE6*, *PpCXE27* and *PpCXE32*) was related to ripening, and the transcription products of *PpCXE1*, *PpCXE2* and *PpCXE3* were the most abundant in mature fruits [27].

'Ruixue' is a significant apple variety in Shaanxi Province, and its quality is measured by its aroma. The aroma plays a crucial role in determining its economic prospects in the market. However, there is a lack of research on the aroma volatile substances of the new variety 'Ruixue' apple during cold storage. This study investigates the contribution of ester aromatic substances, including butyl acetate, hexyl acetate, and 2-methylbutyl acetate, to the aromatic quality of apple fruits. The hydrolysis of esters is facilitated by the carboxylesterase CXEs gene, which is known to play a crucial role in this process. However, there is a lack of experimental evidence regarding the function of this gene. To address this gap, the experimental material used in this study is 'Ruixue' apple fruit. This study investigated the changes in fruit hardness, brittleness, aroma key synthase activity, and volatile aroma substances in apples during cold storage. Additionally, the study screened the key gene *MdCXE20*, which is related to the hydrolysis of esters in apple fruits. This study provides a preliminary analysis of the changes in aroma and degradation of esters during cold storage. These findings serve as a basis for promoting a better understanding of these changes and their impact on molecular breeding in the future.

2. Materials and Methods

2.1. Plant Materials

The apple variety 'Ruixue' was collected in October 2020 at the Baishui Apple Experimental Station of Northwest A&F University (35°21′ N, 109°55′ E). A total of 150 fruits with the uniform size, uniform coloring, consistent maturity, no pests and diseases, and no mechanical damage were selected and bagged. After harvesting, they were placed in a room temperature (25 °C) environment for 12–24 h to dissipate the field heat. The next day, the plastic turnover box with single-layer paper and net belt filled with surrounding bedding newspapers (lined with fresh-keeping bags) was immediately transported back to the laboratory for processing. The fruits were stored at 1 ± 0.5 °C with relative humidity of 80–90% for 5 months, and sampled every other month for a total of 5 times. Each treatment had three biological replicates, and each biological replicate consisted of 10 fruits. After sampling, it was immediately frozen in liquid nitrogen and stored in a refrigerator at −80 °C for later use.

2.2. Determination of Fruit Physiological Characteristics

2.2.1. Firmness Measurement by Texture Analyzer

The texture analyzer (FTC, Washington, DC, USA) was utilized to measure the firmness and brittleness of the fruit. The measurement parameters included a 10 mm depth, a P/2 probe of 2 mm, and a measurement speed of 2 mm/s. From each treatment fruit, five fruits were randomly selected, and the skin was measured five times on the equatorial line of the central yin and yang surfaces of each fruit. The maximum peak force value measured each time was firmness, and finally the average value was taken as kg/cm^2. The ratio of the first peak force value to the movement distance was fruit brittleness, and the unit was Kg.sec.

2.2.2. Determination of Lipoxygenase (LOX)

The method of Chen [28] is improved. Extraction of LOX crude enzyme solution: 5.0 g of liquid nitrogen fully ground fruit sample tissue was added to 10 mL 100 mM phosphate buffer solution (pH 7.5, containing 2 mM DTT, 0.1% (*v*/*v*) Triton X-100, 1% (*w*/*v*) PVPP) under ice bath conditions, and fully shocked and mixed. The mixed homogenate was centrifuged at 25,000× *g* for 15 min at 4 °C after passing through 4 layers of gauze, and the supernatant was taken as crude enzyme solution.

The reaction substrate used in the experiment was 10 mM sodium linoleate. Configuration method: 70 mg of sodium linoleate, 70 μL Triton X-100, and 4 mL of anaerobic water were mixed (to avoid bubbles), and then titrated with 0.5 mol/L sodium hydroxide to clarify the solutio. The solution was diluted to 25 mL, packed into a 1.5 mL centrifuge tube, and stored at −20 °C.

Determination of lipoxygenase activity. In the 3 mL reaction system containing 25 μL of sodium linoleate mother liquor, 0.1 mM (pH 6.0) citric acid-phosphate buffer 2.775 mL, reaction temperature 30 °C (keep 10 min), 0.2 mL of crude enzyme solution was added and mixed. After 15 s of reaction, the lipoxygenase activity was measured at 234 nm. The data of at least 3 points were obtained every 30 s, and the changes of OD value within 1 min were recorded and repeated three times.

2.2.3. Determination of Hydroperoxide Lyase (HPL)

The method for extracting HPL crude enzyme solution was improved following Zhang [29]. 3.0 g of frozen fruit tissue was ground into powder in a mortar using liquid nitrogen. Then, 6 mL of pre-cooled extract at 4 °C containing 150 mM HEPES-KOH buffer (pH 8.0), 250 mM sorbitol, 10 mM EDTA, 10 mM $MgCl_2$, 1% *v*/*v* glycerol, and 4% PVPP was added. The homogenate was fully mixed and extracted, and then centrifuged at 25,000× *g* (4 °C) for 30 min. The supernatant was collected as the crude enzyme solution.

The reaction substrate was sodium hydroperoxylinoleate. The configuration method: 10 mL distilled water, 200 µL 10 mM sodium linoleate, 400 µL LOX enzyme solution (1 mg/10 mL boric acid buffer, pH 9.0), with substrate at 30 °C in a water bath 2 h.

The activity of HPL was determined through a reaction system of 3.5 mL. The system contained 2 mL of analytical buffer consisting of 150 mM HEPES-KOH pH 8.0, 250 mM sorbitol, 10 mM EDTA, and 10 mM $MgCl_2$. Additionally, the system contained 0.75 mL of substrate solution, 0.15 mL of 1.6 mM NADH, 0.1 mL of ADH enzyme solution (containing 1.5 mg/3 mL boric acid buffer, pH 8.6), and 0.5 mL of crude enzyme solution. The HPL activity was measured at 344 nm and 30 °C. The reaction was initiated 15 s after the addition of enzyme solution, and the change in OD value within 1 min was recorded and repeated three times.

2.2.4. The Activities of Keto Acid Decarboxylase (PDC) and Aldehyde Dehydrogenase (ADH) Were Determined

The method of Ke [30] is improved. Extraction of crude enzyme solution: 5 mL 100 mM MES buffer solution (containing 1% (*w*/*v*) PVPP, 2 mM DTT, pH 6.5) was added to the tissue of 3.0 g fruit samples fully ground in liquid nitrogen under ice bath conditions, and fully shaken and mixed. The mixed homogenate underwent centrifugation at 4 °C and 25,000× *g* for 15 min after passing through 4 layers of gauze. The resulting supernatant was collected as the crude enzyme solution.

Determination of PDC enzyme activity: 0.45 mL 100 mM pH 6.5 MES buffer, 0.1 mL 5 mM TPP, 0.1 mL 50 mM $MgCl_2$, 0.05 mL 1.6 mM NADH, 0.1 mL 50 mM pyruvate, 0.1 mL enzyme extract.

Determination of ADH activity: 0.8 mL 100 mM pH 6.5 MES buffer, 0.05 mL 1.6 mM NADH, 0.1 mL enzyme extract, 0.05 mL 80 mM acetaldehyde.

The absorbance change was measured at 340 nm for 2 min, starting 15 s after the enzyme solution was added, at a reaction temperature of 25 °C. The OD value change was recorded for each sample, which was repeated three times. The reaction solution without enzyme extract was used as a blank.

2.3. *VOC Extraction and GC-MS Analysis*

The determination was performed using Yang's modified SPME-GC-MS method [31]. To prepare the sample, fresh fruits were ground into powder using liquid nitrogen. 4.0 g of this powder was weighed and added to a 50 mL screw-cap vial containing 10 mL of saturated NaCl and 10 µL of 0.4 mg/mL 3-nonanone, which was used as an internal standard. The headspace vials were equilibrated for 10 min at 55 °C on a metal heating and stirring platform. A divinylbenzene/carboxy/polydimethylsiloxane (DVB/CAR/PDMS) coated SPME fiber with a thickness of 50/30 µm (Supelco, Bellefonte, PA, USA) was inserted into the headspace for 30 min under continuous heating and stirring (200 rpm) to adsorb VOC. The fiber was then inserted into the heating inlet of the chromatograph and desorbed at 250 °C for 2.5 min. VOCs were analyzed using a Thermo Trace GCU ultra gas chromatograph (Thermo Fisher Scientific, Waltham, MA, USA) equipped with an HP-INNO Wax 60 m × 0.25 mm × 0.25 mm capillary column.

GC conditions: Helium was used as the carrier gas and circulated in splitless mode at a flow rate of 1.0 mL/min. The initial oven temperature was 40 °C, held for 3 min, then ramped to 150 °C at a rate of 5 °C/min, then ramped to 220 °C at a rate of 10 °C/min and held for 5 min.

MS conditions: mass spectrometry ion source, transfer line temperature 240 °C. Mass spectrometry was performed in electron ionization mode at 70 eV with a scan range of 35–450 m/z.

The volatile aroma substances were analyzed by GC-MS. The mass spectrometry data were automatically retrieved and matched with the NIST 2014 standard spectrum. The data with a matching degree of more than 60% and a matching value of more than 800 were retained to determine the chemical composition.

Quantitative analysis: The relative content of volatile substances [32] was obtained by peak area normalization method. The calculation formula was as follows: the content of each component (μg/kg) = (component peak area/internal standard peak area) × internal standard concentration (g/L) × 10 μL/sample mass (g).

2.4. RNA Extraction, Gene Expression Analysis by RT-qPCR

According to previous studies [33], the modified CTAB method was used to isolate and purify RNA from 'Ruixue' fruits at different storage periods under different treatments. After RNA extraction, cDNA was synthesized using the EasyScript One-Step gDNA Removal and cDNA Synthesis Kit according to the manufacturer's instructions. The expression level of the candidate gene carboxylesterase, which is involved in the degradation of aroma esters, was analyzed using fluorescent quantitative real-time PCR (RT-qPCR) with cDNA as the template. Gene-specific primers were designed using Origin 2018 64 bit and NCBI. *MdActin* was selected as the reference gene for constitutive expression. All primer sequences are listed in Table 1.

Table 1. Lists of all the primer sequences used for RT-qPCR.

Name	Apple ID	Primers (F)	Primer (R)
MdCXE3	MD02G1275900	TGATGGCTCTGTCGACCGG	AAATATATTCGGACACGATGG
MdCXE5	MD02G1276600	TGGATTCGTTCGATCAGT	TAAAGACGGTGCTCTGGA
MdCXE6	MD03G1273300	CTCGTTTGATCACGTCGACAC	CTGCTCAAGATCCGAAATGCC
MdCXE9	MD05G1076400	AAGAACCAGCATTGTCCGT	AATGCAGTTTCAACCACGAA
MdCXE10	MD05G1078900	TGCCTACAATCTCCATACTCG	ATGAAACGCTCAATTGTACCA
MdCXE12	MD05G1191100	CAACCCGAAACCGGAGTCC	GTGGGGGAGGAAGCGCTTTC
MdCXE17	MD08G1226300	ATCATTCGAGTTCCCGACCCT	CTTTGGATTGGACCCCGGTT
MdCXE20	MD10G1068500	GAGGAGGTACTACCA GTGGTTGAAG	TCACACGGAAGCA ATGAAATCTTTG
MdCXE23	MD10G1091200	ATGGATTCAGCCTTGAGCAAC	ACATTCGACGCCTGTTTGTG
MdCXE25	MD10G1091900	ATGACAAACGAAGTAGCCCAT	CCTTGGATTGGACACCGATT
MdActin	MD01G1001600	GATATCTCCACTGAC GTAAGGGATG	AGGGTCAGCTTGCCG TAGGTGGCA

2.5. Bioinformatic Sequence Analysis

The members of the MdCXE gene family all contain a conserved core structure, which is composed of eight β chains interspersed with α-helices and rings [17], and contains a catalytic triad composed of serine (contained in the conserved common sequence GXSXG), acidic amino acids and histidine, constituting its active site. Using this sequence, we searched the GDR apple genome database (https://www.rosaceae.org/, accessed on 5 March 2022) using BLASTP, and used *Arabidopsis* CXEs with 10^{-5} as the cut-off value to screen apple MdCXE gene family members. A total of 42 members of the MdCXE family were retrieved, and CDD belonged to Abhydrolase-3, Pfam07859. The GXSXG conserved sequence of the MdCXE gene family was analyzed by MEME online software (http://meme-suite.org/tools/meme, accessed on 5 March 2022). In this study, we employed several bioinformatics tools to analyze amino acid sequences, predict domain and function using MEME 6.0 and InterPro, respectively. To determine the physicochemical properties of some genes, we used the pI/Mw tool in ExPASy Compute. We also utilized

Wolfpsort to predict subcellular location and calculated *p*I and Mw using the ExPASy Compute tool. Additional details can be found in Table S2. The phylogenetic tree was constructed using MEGA 6.0. The maximum likelihood tree was tested and analyzed 1000 times for accuracy.

2.6. Transient Injection of the MdCXE20 in Apple Fruit

The apple fruit was infected by injection, method reference by Wu [34]. The full-length CDS of *MdCXE20* gene was amplified by PCR using the cDNA of 'Ruixue' fruit as a template and the primers in Table 2. The BamHI and KpnI restriction sites were selected and the pCAMBIA2300-OE-*MdCXE20* overexpression vector was constructed by homologous recombination. The 400 bp fragment of *MdCXE20* was cloned into TRV2 as a virus-induced silencing vector, and KpnI and EcoRI were selected as restriction sites. The TRV1 + TRV2-*MdCXE20* silencing vector was constructed by homologous recombination. The recombinant vector confirmed by sequencing was transformed into EHA105 Agrobacterium competent. Positive single colonies were selected and activated by YEP liquid culture containing kanamycin sulfate (50 μg/mL) and rifampicin (50 μg/mL). Based on 28 °C culture to OD_{600} = 1.2~1.5, the bacteria were collected by centrifugation, and the equal volume of sterile MS culture medium (10 mM $MgCl_2$, 10 mM MES, 100 mM acetosyringone, pH 5.8) was added to re-suspend the bacteria. The empty pCAMBIA2300 and TRV1 + TRV2 vectors were used as controls, and the dark treatment was performed for 3 h. Choose the fruit with the same size and no mechanical damage, wipe the fruit surface with a wet cloth and wait for use. The apple fruit was divided into four parts by injecting the treated bacterial solution into the central axis epidermis of the fruit with a 1 mL syringe. Half of the fruits were injected with Agrobacterium containing pCAMBIA2300-OE-*MdCXE20*, and half were used for Agrobacterium infection containing pCAMBIA2300-OE, the silencing vector was the same as above. After labeling, it was placed in dark for 7 days, sampled, frozen in liquid nitrogen, and stored at −80 °C for VOC and gene analysis (15 fruits in each group).

Table 2. Construction of pCAMBIA2300 and TRV2 primers by *MdCXE20*.

Gene	Vector	Primers Sequences
MdCXE20	pCAMBIA2300-OE-Primers (F)	acgggggacgagctcggtaccATGACGACGTCGTTGGACTCC
	pCAMBIA2300-OE-Primer (R)	ggtgtcgactctagaggatccCACGGAAGCAATGAAATCTT
MdCXE20	TRV2-Primers (F)	gtgagtaaggttaccgaattcTTTTTCGGGGAGAGGAGC
	TRV2-Primer (R)	gagacgcgtgagctcggtaccCACGGAAGCAATGAAATCTT

2.7. Overexpression of 'Wanglin' Callus Transgenic

The Agrobacterium-mediated transformation method of 'Wanglin' callus is referenced [35]. The full-length CDS of *MdCXE20* gene was amplified by PCR using the cDNA of 'Ruixue' apple fruit as a template, combined with the primers in Table 2, and the CDS of *MdCXE20* gene were constructed by homologous recombination. The recombinant vector was introduced into Agrobacterium EHA105 through electroporation while the empty vector was utilized as a control. Positive single colonies were then chosen and grown in YEP liquid culture supplemented with kanamycin sulfate (50 μg/mL) and rifampicin (50 μg/mL). Based on the culture at 28 °C to OD_{600} = 0.6~0.7, the bacteria were collected by centrifugation, and the equal volume of sterile sterilized water containing 50 mg/L acetosyringone was added to re-suspend the bacteria, and placed in darkness for 3 h.

The callus of 'Wanglin', which grew for 20 days under sterile conditions, was infected with Agrobacterium containing pCAMBIA2300-OE-*MdCXE20* vector for 25 min. During the period of continuous shaking, the excess bacterial liquid was sucked with filter paper and placed on the subculture solid medium. After 3 days of dark culture, it was transferred to the screening medium containing 200 mg/L carbenicillin and 50 mg/L kanamycin sulfate, and cultured at 25 °C, 16 h/8 h in darkness. Observe their growth every 1 week until there is a small callus growth. Subculture medium components: 4.43 g/L MS, 7 g/L agar,

30 g/L sucrose, 0.5 mg/L 2,4-D, 1 mg/L 6-BA, pH 6.0. The differentiated callus was transferred to a new screening medium and subcultured every 2 weeks. DNA was extracted from the T8 generation of six lines for positive plant detection, and three lines were selected. The identified T8 transgenic 'Wanglin' callus was frozen in liquid nitrogen and stored at −80 °C for ester content and gene expression analysis. Three plants were selected from each line as three biological replicates, each containing five calluses. The callus *actin* gene was used as an internal reference gene to analyze the gene expression of *MdCXE20*.

2.8. Subcellular Localization Analysis

The analysis method is slightly modified based on [36]. The complete CDS sequence of *MdCXE20* was fused into the green fluorescent protein GFP to obtain the 35S-GFP-*MdCXE20* recombinant plasmid. The primer information is shown in Table 2. According to the method of described in Section 2.6, the fusion protein 35S-GFP-*MdCXE20* was transiently overexpressed in a *Nicotiana benthamiana* plant with MCherry nuclear localization marker by Agrobacterium infection. After 3 days of dark culture after injection, the laser confocal fluorescence microscope was used to observe and photograph.

2.9. Statistical Analysis and Data Processing

Excel 2010 software was used for data analysis, and IBM SPSS Statistics 26.0 software was used for one-way analysis of variance and significance analysis.

3. Results

3.1. Changes of Fruit Firmness and Brittleness of 'Ruixue' at Different Cold Storage Stages

Firmness is one of the important factors to evaluate apple quality [37]. As shown in Figure 1A, the firmness of apple fruit showed a decreasing trend with the increase of frozen days. The decrease of firmness of refrigerated fruit mainly occurred in the first 4 months of storage period, and the decrease was larger. The firmness of fruit decreased from 10.25 kg/cm^2 to 7.67 kg/cm^2 at 150 days of cold storage, which was 25.17% lower than that of harvest period. As shown in Figure 1B, the fruit brittleness showed a slow downward trend with the prolongation of storage time, indicating that refrigeration was beneficial to the fruit. It shows that cold storage can maintain fruit firmness quality attributes well.

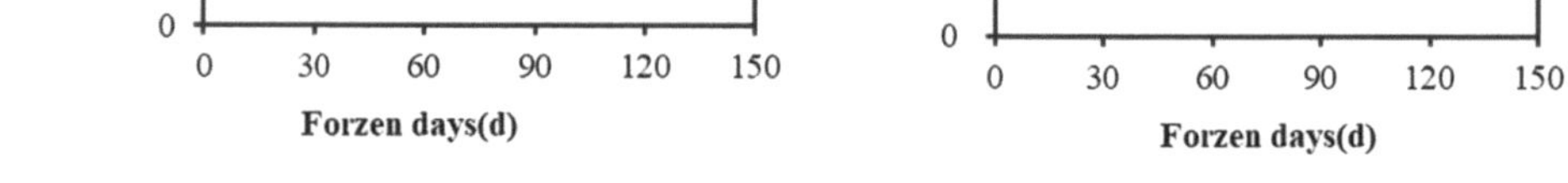

Figure 1. Changes of firmness and brittleness of apple fruit during storage. (**A**) Fruit firmness; (**B**) Fruit brittleness. Error bars show ± SE from three biological replicates. Different lowercase letters in columns denote significant differences between sampling dates for 'RuiXue' fruit by Duncan's multiple range test ($p < 0.05$).

3.2. Changes of Aroma Synthase Enzyme Activity in 'Ruixue' at Different Cold Storage Stages

C6 substances, such as 2-hexenal, are synthesized through the LOX-HPL pathway to produce hexyl esters. The content of 2-hexenal was the highest at 60 days of storage, the content of esters reached the maximum peak at 60 days of storage, and the ester substance 2-methyl hexyl butyrate also reached the maximum release amount, but the LOX and HPL activity reached the peak at 90 days (Figure 2A,B). It is speculated that this may be related to the content of linolenic acid and linoleic acid. It can be seen that the activity of PDC and ADH showed an overall upward trend in each storage stage during the whole storage period, and the rate of increase of enzyme activity was the fastest between 90 d and 120 d. The difference of PDC activity between 150 d and 1–30 d was the most obvious, and the difference of ADH activity between 120 d and harvest was the most obvious. (Figure 2C,D, $p < 0.05$).

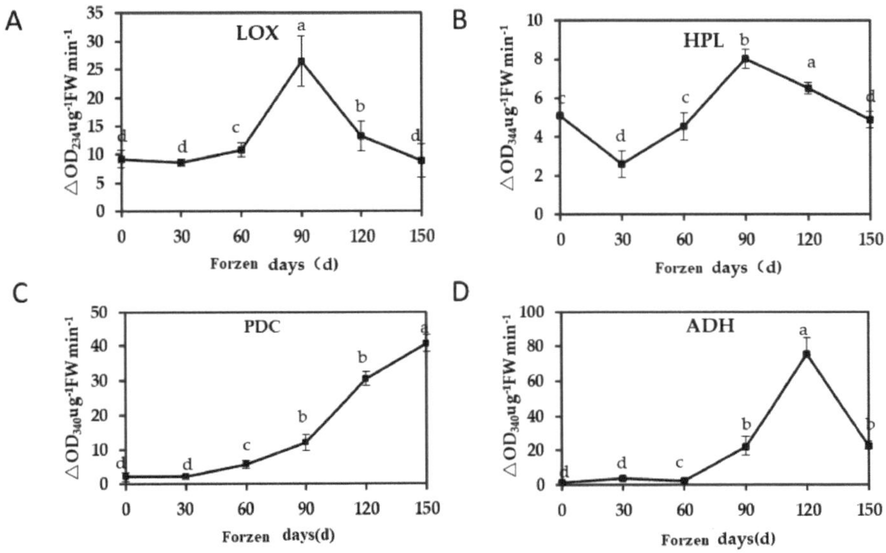

Figure 2. Changes of aroma synthase during apple cold storage. (**A**) Lipoxygenase (LOX); (**B**) Hydroperoxide lyase (HPL); (**C**) Keto acid decarboxylase (PDC); (**D**) Acetaldehyde dehydrogenase (ADH). Error bars show ± SE from three biological replicates. Different lowercase letters in columns denote significant differences between sampling dates for 'RuiXue' fruit by Duncan's multiple range test ($p < 0.05$).

3.3. VOC in Apple Fruits during Different Cold Storage Periods

In 'Ruixue' apple, the concentration of total aroma substances was a dynamic process during storage, which increased from 4839.45 μg/kg at harvest to 18,733.17 μg/kg at 60 days of cold storage (Figure 3A), and reached the maximum peak. A total of 51 volatile substances were detected during cold storage, and 36, 38, 48, 42, 43 and 40 aroma substances were detected respectively (Supplementary Table S1). With the prolongation of storage time, the types of aroma substances increased first and then decreased, among which the changes of esters were the most obvious. The types of esters during cold storage period were 19, 21, 24, 22, 21 and 20 respectively (Figure 3C). During the whole cold storage period, esters were the most abundant, followed by aldehydes, and were the least ketones, (Figure 2B) and contributed the most to the aroma of apple during cold storage, mainly hexyl esters and acetate esters. Some of the main compounds were hexyl 2-methylbutyrate, hexyl butyrate, hexyl hexanoate, butyl butyrate, 2-hexenal, hexanal, and 1-Hexanol (Table S1).

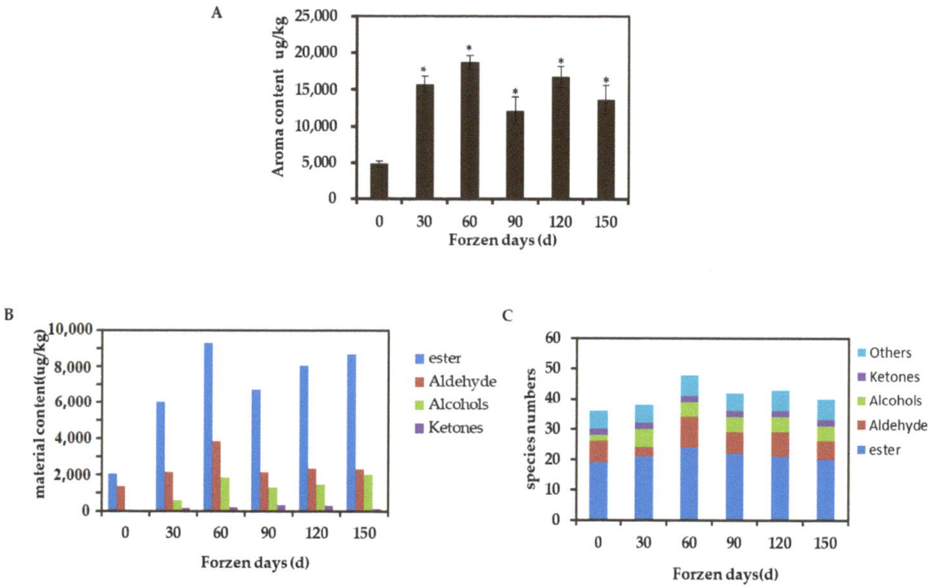

Figure 3. VOC in apple fruits during different cold storage periods. (**A**) Total volatile content, * indicates that the difference is statistically significant at the $p < 0.05$ level. Error bars were calculated from three biological experiments, and show the standard deviation of the mean; (**B**) The content of VOC in each component; (**C**) Number of VOC. Different colors represent different types of volatile substances.

3.4. *Bioinformatics and Phylogenetic Analysis of MdCXE Gene Family Proteins*

We identified 42 members of the apple CXE family from the GDR based on the published genome of 'Golden Delicious' apple, and found that all 42 MdCXE CDS contain complete predicted coding regions. Furthermore, the amino acid sequences of these proteins exhibit similar characteristics to those of the previously reported 20 AtCXE gene family members in *A. thaliana*. (*AtCXE1-20*, [19]), including the transmembrane region of the α/β hydrolase fold in the transmembrane region of the superfamily. As well as , oxygen anion pore and catalytic triad residues (Ser, acid, and His) predicted with conserved HGG sequence motifs, including the pentapeptide sequence around the nucleophilic serine (GXSXG) (Figure S1)

According to the location of the selected genes on the chromosome, we systematically named them as *MdCXE1-MdCXE42* (Table S3). We performed phylogenetic analysis on the predicted 42 apple CXE genes, 4 tomato CXE genes with known functions, 3 peach CXE genes and 20 *AtCXEs* genes previously reported (Figure 4). According to the classification of Arabidopsis [19], 42 members were divided into 7 clades (Group 1–7). In addition, as revealed by Blastp search, MdCXE has close similarities with carboxylesterases in other higher plants, ranging from 50% to 100%.

3.5. *MdCXE Gene Expression Analysis*

To identify additional genes associated with carboxylesterase hydrolysis, we conducted a blast alignment analysis using CXE genes from various plants and identified 10 genes with 100% homology to them. The genes involved in *MdCXE3, MdCXE5, MdCXE6, MdCXE9, MdCXE10, MdCXE12, MdCXE17, MdCXE20, MdCXE23*, and *MdCXE25*. During cold storage, the expression of these 10 candidate genes was analyzed by RT-qPCR using 'Ruixue' fruit as material, and a family member with high expression was selected as a candidate gene.

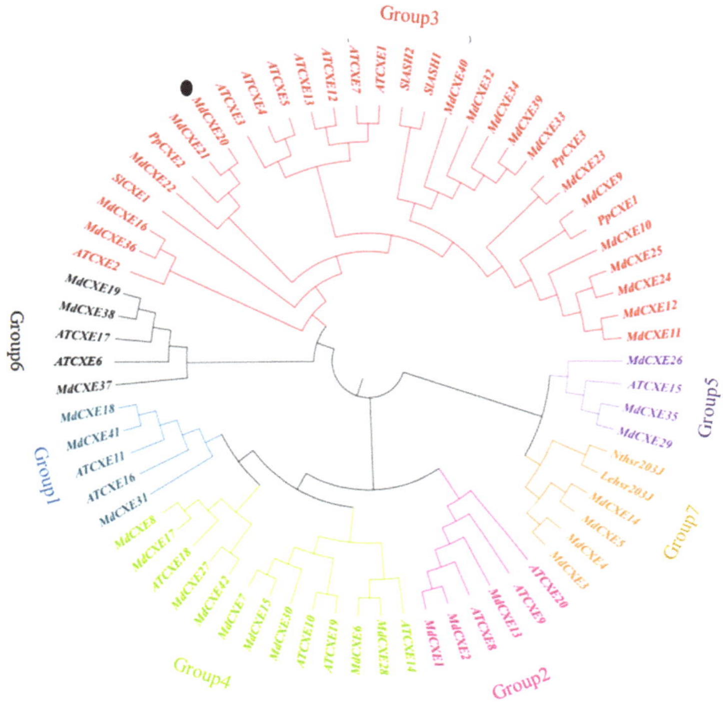

Figure 4. Phylogenetic analysis of CXE families. The full-length amino acid sequences were aligned to construct the phylogenetic tree. The accession members are *AtCXE1-20* from *Arabidopsis thaliana*, *SlCXE1*, *SlASH1-2* and *Lehsr203J* from tomato, *PpCXE1-3* from peach fruit and *Nthsr203J* from tobacco. Different colors represent different groups.

The results showed that *MdCXE10* was not expressed during refrigeration. However, *MdCXE20* had a significant high transcription level during fruit cold storage (Figure 5B); In addition, the phylogenetic tree (Figure 4) shows that *MdCXE20*, like other plant carboxylesterases, belongs to the Group3 group and is the largest branch. Previous studies have shown that the Group3 group contains the volatile ester *SlCXE1* involved in tomato and the volatile ester *PpCXE1* involved in peach fruit (Figure 4). Based on the above analysis, it is speculated that *MdCXE20* may be involved in the degradation of apple volatile esters.

Esters account for the highest proportion in 'Ruixue' apple varieties (Figure 3B). In order to further explore the degradation of volatile esters during storage, the contents of main characteristic esters, butyl butyrate, hexyl hexanoate, hexyl 2-methylbutyrate, and hexyl butyrate, C4 and C6 alcohols as precursors of ester synthesis and the expression of carboxylesterase *MdCXE20* gene involved in ester degradation were further analyzed. The results showed that the peak of hexyl caproate appeared at 120 d, and the peak of the corresponding substrate 1-Hexanol content was delayed. Similarly, the content of butanol peaked at 90 d, and the corresponding butyl ester peaked at 60 d. However, there was no significant correlation between hexyl 2-methylbutyrate, hexyl butyrate content and hexanol content. It was observed that the expression of *MdCXE20* gene in fruit decreased with the increase of esters, and it was speculated that the carboxylesterase *MdCXE20* may degrade some hexyl ester and butyl ester.

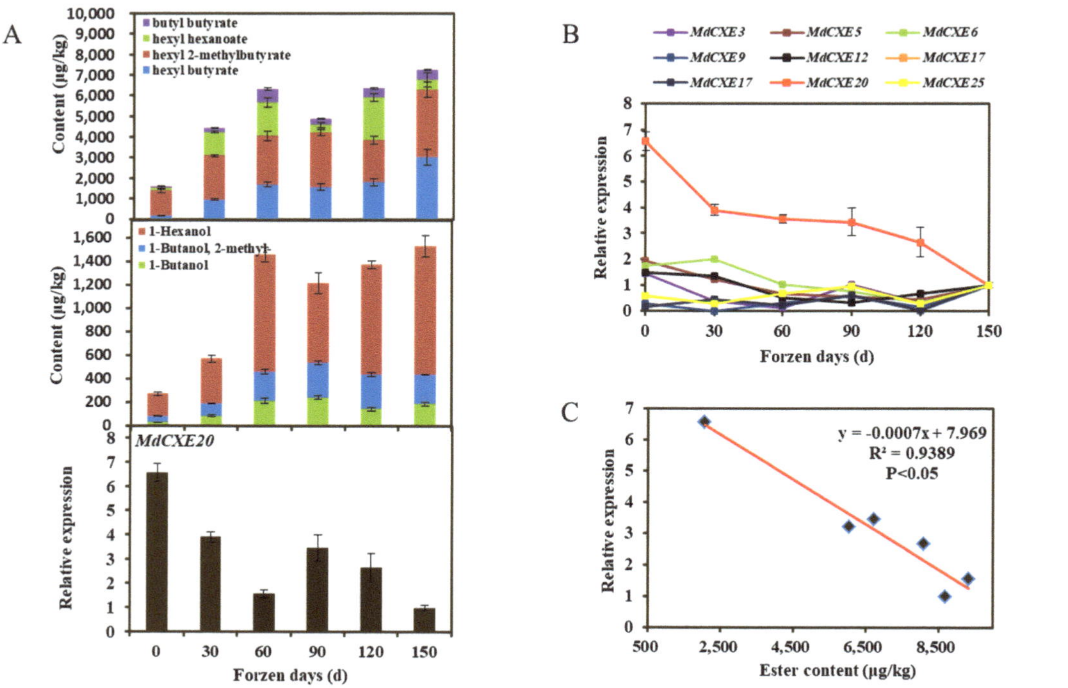

Figure 5. MdCXE gene expression analysis. (**A**) The content of volatile esters and alcohols and the expression of *MdCXE20* gene during apple cold storage. Error bars represent SE ($n = 3$); (**B**) Expression pattern of apple CXE gene during cold storage; (**C**) Transcript levels of CXEs during fruit cold storage which showed significant negative correlation to volatile esters ($p < 0.05$).

In order to explore the correlation between the candidate gene *MdCXE20* and the content of volatile esters, the correlation between the expression level and the content of volatile esters in 'Ruixue' fruit during cold storage was analyzed by RT-qPCR. Correlation analysis revealed that the transcription level of *MdCXE20* was negatively correlated with the content of esters (Figure 3C), and the correlation coefficient R^2 of volatile ester content of *MdCXE20* gene expression was 0.9389 ($p < 0.05$).

3.6. *Transient Overexpression and Silencing of MdCXE20 Changed the Content of Esters in Apple Fruit*

Homologous transient overexpression was used [35] and virus-induced gene silencing (VIGS) system to verify the function of *MdCXE20* in volatile substances in apple fruits. In this study, we created pGreen-OE-*MdCXE20* and TRV1 + TRV2-*MdCXE20* vectors, with pGreen-OE and TRV1 + TRV2 vectors as controls. These vectors were then injected into mature apple fruits for further analysis (Figure 6A). The expression of *MdCXE20* gene and the changes of VOC were detected 7 days after injection. As shown in Figure 6B, after overexpression of *MdCXE20* gene, the expression level increased significantly by about 2.5 times, the corresponding total volatile substances decreased by 2 times, and the esters decreased by about 1.5 times, while silencing *MdCXE20*, the gene expression decreased significantly. It decreased by about 0.8 times, and the esters increased by about 1.5 times. The contents of butyl butyrate, hexyl hexanoate, hexyl 2-methylbutyrate and hexyl butyrate in apple fruits of *MdCXE20* overexpression decreased significantly, and the opposite results were observed after silencing *MdCXE20* (Figure 6C). The above results indicated that *MdCXE20* gene was involved in the degradation of esters in apple fruit.

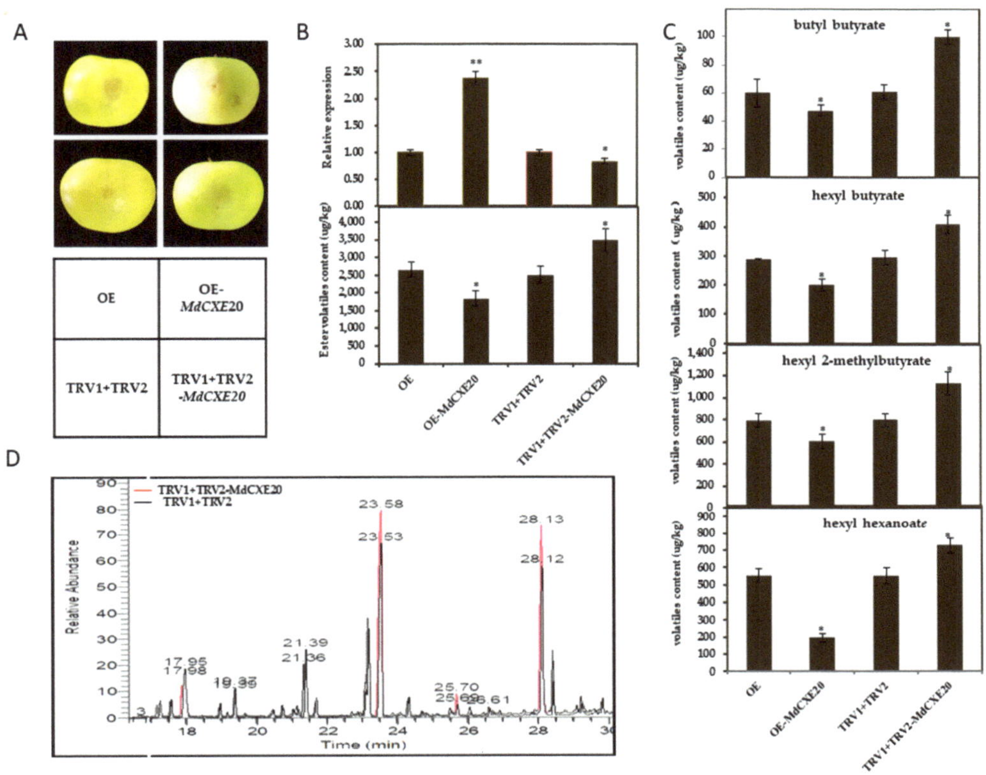

Figure 6. Instantaneous injection of *MdCXE20* 'Ruixue' fruit. (**A**) 'Ruixue' apple transient injection; (**B**) 'Ruixue' fruit *MdCXE20* gene expression and ester content change; (**C**) Instantaneous expression of *MdCXE20* 'Ruixue' apple 4 ester content; (**D**) Chiral GC-MS analysis of four esters in transiently silenced apple fruit, Red was virus-induced gene silencing, and black was blank control. Single and double asterisks (*), (**) represent significant differences (* t-test, $p < 0.05$; ** t-test, $p < 0.01$).

3.7. *MdCXE20* Transgenic Reduced the Content of Ester Volatile Substances in 'Wanglin' Callus

In order to further confirm the function of *MdCXE20* on ester volatiles, we performed homologous stable transformation experiments in 'Wanglin' callus. Agrobacterium containing pGreen-OE-*MdCXE20* construct was used to infect 'Wanglin' callus, and wild 'Wanglin' callus was used as control (Figure 7A). Three independent lines, *MdCXE20*-#OE1, *MdCXE20*-#OE2 and *MdCXE20*-#OE3, were obtained. Subsequently, the DNA detection and Western Blot detection of the three transgenic lines could detect the target band, while the control could not detect the target band (Figure 7C,D). The gene expression level in the three transgenic lines of overexpression *MdCXE20* was significantly higher than that in the wild-type 'Wanglin' callus (Figure 7B), the most significant was the *MdCXE20*-#OE3 line, the transcription level was about 18 times that of the control group, and SPEA-GC-MS analysis was performed. After overexpression of *MdCXE20*, the content of esters decreased by 78% (Figure 7E). In summary, these results indicate that overexpression of *MdCXE20* is involved in the degradation of ester volatiles in plants.

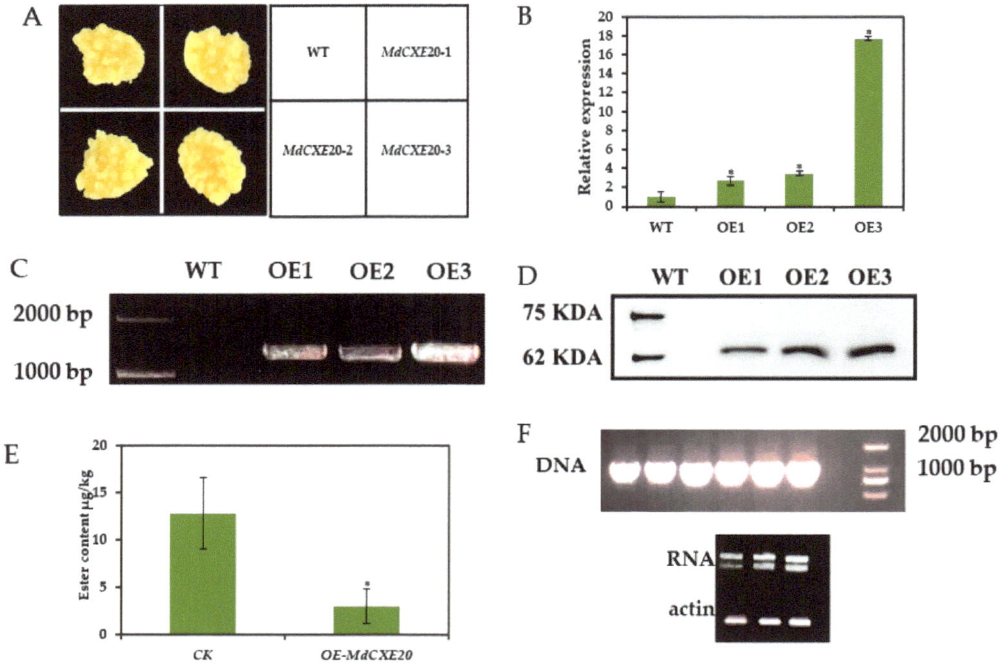

Figure 7. Changes in transcription level and ester content of *MdCXE20* transgenic 'Wang Lin' callus. (**A**) Schematic diagram of 'Wang Lin' callus; (**B**) Transcription level of *MdCXE20* transgenic 'Wang Lin' callus; (**C**,**D**,**F**) RNA, DNA, protein level identification of transgenic materials; (**E**) Change of ester content of *MdCXE20*-#OE3 transgenic 'Wang Lin' callus. WT: wild-type; *MdCXE20*-#OE1, *MdCXE20*-#OE2, *MdCXE20*-#OE3 a transgenic lines. Double asterisks (*) represent significant differences between wild-type and transgenic callus (t-test, $p < 0.01$).

3.8. Subcellular Localization Analysis

In order to determine the subcellular location of *MdCXE20* in apple, we verified the hypothesis that *MdCXE20* protein is located on the plasma membrane of plants by the expression of *MdCXE20* and GFP fusion protein. We transiently transformed Agrobacterium tumefaciens carrying pCAMBIA2300-GFP-*MdCXE20* and control construct pCAMBIA2300-GFP into tobacco leaf epidermal cells containing transgenic Mcherry. After injection, the leaf slices around the injection site were observed by laser scanning confocal microscopy after dark culture for 3 days. As shown in Figure 8, the control pCAMBIA2300-GFP constructs were distributed throughout the cell, while the GFP-*MdCXE20* protein fluorescence was targeted to the plasma membrane. These results indicate that *MdCXE20* is located in the cytoplasm where the hydrolysis of aromatic esters is likely to ocuur, thereby exerting its biological function.

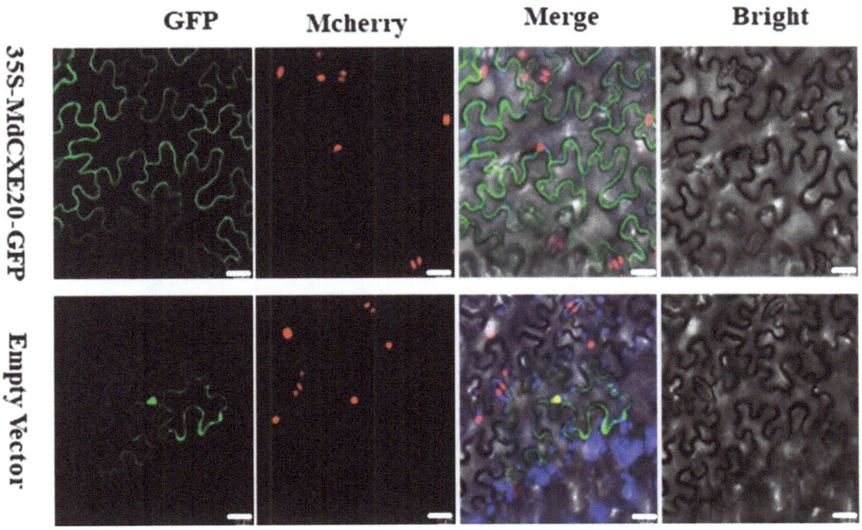

Figure 8. Subcellular localization of *MdCXE20*. GFP signals were observed in *Nicotiana benthamiana* plants. The pCAMBIA2300-GFP empty vector was used as a control. Scale bar = 10 μm.

4. Discussion

In China, apple is widely recognized as a key economic fruit tree. It serves as a major source of income for fruit farmers and has significantly contributed to the promotion of rural revitalization [38]. After harvesting, fruits are susceptible to rot and quality decline due to high water content and inadequate storage measures. This can result in significant economic losses. To address this issue and ensure year-round supply, refrigeration has become the most popular fruit storage method worldwide. Cold storage is a widely used and simple method for preserving fruits and vegetables. By lowering the temperature, the respiration rate and ethylene release rate of fruits are inhibited, which helps to maintain fruit firmness and achieve optimal preservation [39]. 'Ruixue' apple is a late-maturing variety that exhibits greater resistance to storage. Our study found that fruit firmness and crispness exhibited a significant decrease during storage, which is consistent with previous research.

Texture and volatile aroma compounds are both important factors in determining fruit quality and affecting final consumer acceptance of goods. These traits are crucial in quality, disease resistance breeding [40], and post-harvest processing and storage [41]. Metabolic pathways for volatile aroma substances are extensive and intricate, with numerous precursors and enzymes involved in the process. In plants, fatty acids and amino acids are typically the metabolic precursors. Key enzymes responsible for aroma metabolism include LOX, HPL, PDC, ADH, and CXE, among others. Studies have revealed a significant increase in hexyl esters and butyl esters in apple volatiles after treatment with hexanal and hexanoic acid [42]. The synthesis of aroma in melon fruit was significantly impacted by the cloning of *CmADH1* and *CmADH2*, as reported in [43]; Similarly, the production of VOC in 'Pink Lady' apples was stimulated by treatment with n-butanol during the early stages of development [44]; Exogenous fatty acid treatment of apple fruits resulted in a significant increase in hexyl esters and butyl esters, indicating that fatty acids are precursors of aroma synthesis. Additionally, the activity of LOX was found to be related to aroma synthesis, as it increased in response to the fatty acid treatment [42]. Although the activity of alcohol dehydrogenase involved in ethanol formation was not found to be related to ethylene regulation [45], a study showed that the content of total and monools in apples was induced by ethylene and inhibited by ethylene receptor inhibitor 1-MCP [10]. This suggests that volatile aroma compounds in apples are typically associated with substrate utilization and related enzyme activity [42]. This study examined the impact of aroma synthesis-related enzymes, including LOX, HPL, ADH, and PDC, on the quality of stored produce. Results indicated that the LOX-HPL enzyme activity remained consistent, with a significant increase observed before 90 days of cold storage. The production of volatile esters showed a gradual increase, peaking at 60 days, which was 30 days later than the peak in enzyme activity. It is speculated that it may be related to substrate content and availability of other compounds, and further research is needed.

The structure of volatile aroma substances in apples is complex and their quantity and distribution are influenced by various factors such as varieties, environment, cultivation, postharvest storage, and other conditions [46], particularly postharvest factors. Among the VOCs, only a few have a significant impact on fruit aroma, mainly esters, aldehydes, alcohols, and terpenes. Esters are considered the most important volatile substances in apple aroma [47,48]. During the cold storage of 'Ruixue' apple fruit, the content and types of volatile aroma substances underwent a dynamic change. As storage time increased, the aroma substances showed a trend of initially increasing and then decreasing, with the peak being reached at 60 days of storage. The esters constituted the largest proportion of the volatile substances, accounting for approximately 50% of the total. The study detected several main esters, including butyl butyrate, hexyl acetate, 2-methylbutyrate, hexyl hexanoate, hexyl thiocyanate, and isoamyl 2-methylbutyrate. The dominant esters were found to be hexyl esters, which is in line with previous research [4]. In the 'Ruixue' variety, compounds such as ethyl butyrate, butyl propionate, and propyl hexanoate were not detected until after 30 days of storage, indicating significant variations in the content and types of aromatic substances during different storage periods following cold storage.

Esters are significant volatile aroma compounds found in many mature fruits and are derived from amino acids and fatty acids. LOX-HPL catalyzes fatty acids to produce C6 and C9 aldehydes, which are then converted to alcohols through the dehydrogenation of ADH [42,49,50]. Finally, esters are produced through the action of AAT [51]. Tomato *SlAAT1* and peach *PpAAT1* are examples of enzymes related to the production of volatile esters during fruit ripening [52]. The synthesis of esters has been extensively studied in relation to the quality of apple fruit. However, there is a lack of research on the metabolic processes of esters during storage. This area presents an opportunity for further investigation and could provide valuable insights into the factors that influence the quality of stored apples.

CXEs, as a member of the α/β hydrolase superfamily, are known to catalyze various hydrolysis reactions and are typical ester bond hydrolases. However, there have been few studies on the hydrolysis of ester volatiles in fruits by carboxylesterase. The functions of some CXEs in plants, such as tomato, peach, and strawberry, have been identified [26]. The role of CXE protein in regulating the content of volatile esters in tomato has been widely studied, where *SlCXE1* has been found to hydrolyze volatile esters [26]; *PpCXE1* is known to be involved in the hydrolysis of volatile esters [25]. In this study, comparative analysis identified 42 members of the apple MdCXE gene family, and it was suggested that 10 of these genes may be responsible for the hydrolysis of volatile esters during the storage of 'Ruixue' fruit. The study confirmed that the gene *MdCXE20*, which exhibited the highest transcription level, plays a crucial role in the hydrolysis of volatile esters. Through homologous transient and stable transformation of 'Wanglin' callus, it was established that *MdCXE20* is involved in the decomposition and metabolism of volatile esters that are related to the flavor of apple fruit. Similar studies on tomato and peach fruits have revealed that acyltransferases AATs and CXEs work together to regulate the content of volatile esters in fruits [25,26,52]. During fruit storage, the level of CXE transcriptional metabolites significantly increased. Subcellular localization analysis revealed that *MdCXE20* and *MdAAT1* gene were both located in the cytoplasm [53]. The co-localization of these enzymes can create a cycle that regulates the balance of ester synthesis and hydrolysis in fruits through synergistic action [52]. The study revealed that the *MdCXE20* gene is capable of hydrolyzing esters. However, due to the current limitations in gene silencing or knockout transgenes in apple fruits for functional verification, it is difficult to determine whether other CXE gene members also play a role in regulating ester content.

Our research has identified *MdCXE20* as a crucial gene in the synthesis pathway of ester volatiles in apple fruit. The significant expression of *MdCXE20* in apple fruit has been found to cause a decline in hexyl ester content. This alteration in volatile substances has a direct impact on the aroma of the fruit, ultimately affecting its overall sensory profile. In summary, this study offers a comprehensive understanding of the formation and regulation of fruit aroma quality. It achieves this by exploring the regulation mechanism of ester aromatic substances in apple fruit. The findings provide an important theoretical basis for improving fruit aroma quality through molecular breeding and other methods.

5. Conclusions

During cold storage, 'Ruixue' apples are mainly composed of hexyl esters such as hexyl acetate, hexyl caproate, and hexyl thiocyanate which are the most important volatile substances after harvest. Carboxylesterases (CXEs) is a diverse and complex enzyme group that is capable of hydrolyzing esters. However, their natural substrates are not well understood. In this study, 42 MdCXE family members in apples were identified through phylogenetic analysis. The results of RT-qPCR revealed that *MdCXE20* had a high expression level in apple. The results from the transient overexpression of *MdCXE20* in apple fruit in *E. coli* and the stable overexpression of 'Wanglin' callus indicate that *MdCXE20* functions in the degradation of esters. These findings suggest that *MdCXE20* plays a crucial role in ester metabolism in mature apples, providing valuable insights into the potential role of CXE family members in apples and enhancing our comprehension of the molecular basis of fruit volatiles.

Supplementary Materials: The following supporting information can be downloaded at: https://www.mdpi.com/article/10.3390/foods12101977/s1, Table S1: Schedule Change of volatile substance content during storage period of 'Ruixue' apples (μg/kg); Table S2: Predicted properties of different carboxylesterases characterized in plants and of the intracellular locus of the proteins on websites; Table S3: Apple genome names corresponding to 42 MdCXE members; Figure S1: Multiple sequence alignment of CXE gene family in apple and *Arabidopsis thaliana*.

Author Contributions: Conceptualization, D.L. and H.G.; methodology, D.L. and J.G.; software, D.L.; validation, H.M., L.P., X.L., H.W. and R.C.; formal analysis, D.L.; investigation, H.M.; resources, L.P.; data curation, X.L., H.W. and R.C.; writing-original draft preparation, D.L.; writing-review and editing, Z.Z. and H.G.; funding acquisition, H.G. All authors have read and agreed to the published version of the manuscript.

Funding: The work was supported by the Shaanxi Provincial Key Research and Development (R&D) Program (2020zdzx03-06-02-02) and the Earmarked Fund for China Agriculture Research System (CARS-27).

Data Availability Statement: The datasets generated for this study are available on request to the corresponding author.

Acknowledgments: We would like to thank Jing Wei and Zhifeng Li for assisting with the data analysis, and thank all editors and reviewers for their comments on this manuscript.

Conflicts of Interest: The authors declare no conflict of interest.

References

1. Chen, Y.D.; Meng, X.X.; Zhao, R.X.; Dou, Y.T.; Xian, G.J.; Xu, Q. A preliminary exploration on the construction of a scientific data resource system for apple industry in China. *J. Fruit Sci.* **2016**, *33*, 719–726.
2. Kondo, S.; Setha, S.; Rudell, D.R.; Buchanan, D.A.; Mattheis, J.P. Aroma volatile biosynthesis in apples affected by 1-MCP and methyl jasmonate. *Postharvest Biol. Technol.* **2005**, *36*, 61–68. [CrossRef]
3. Echeverría, G.; Graell, J.; López, M.L.; Lara, I. Volatile production, quality and aroma-related enzyme activities during maturation of 'Fuji' apples. *Postharvest Biol. Technol.* **2004**, *31*, 217–227. [CrossRef]
4. Young, H.; Gilbert, J.M.; Murray, S.H.; Ball, R.D. Causal Effects of Aroma Compounds on Royal Gala Apple Flavours. *J. Sci. Food Agric.* **1996**, *71*, 329–336. [CrossRef]

5. Young, H.; Stec, M.; Paterson, V.J.; McMath, K.; Ball, R. Volatile Compounds Affecting Kiwifruit Flavor. *Am. Chem. Soc.* **1995**, *596*, 59–67.
6. Goodenough, P.W.; Entwistle, T.G. The hydrodynamic properties and kinetic constants with natural substrates of the esterase from Malus pumila fruit. *Eur. J. Biochem.* **1982**, *127*, 145–149. [CrossRef]
7. Ma, Y.Y.; Li, L.N.; Liu, Q.; Ding, Q.N.; Li, K.; Liu, P. Composition, Influencing Factors and Regulation Strategies of Aroma Substances in Apple. *North. Hortic.* **2023**, *520*, 119–127.
8. Liu, J.L. Analysis of Volatile Flavor Compounds and Their Genetic Characteristics in *Malus Domestica* 'Ruixue'. *Northwest AF Univ.* **2019**, *36*, 590–602.
9. Echeverría, G.; Graell, J.; López, M.L. Effect of Harvest Date and Storage Conditions on Quality and Aroma Production of 'Fuji' Apples. *Food Sci. Technol. Int.* **2002**, *8*, 351–360.
10. Yang, X.; Song, J.; Du, L.; Charles, F.; Campbell-Palmer, L.; Fillmore, S.; Paul, W.; Zhang, Z.Q. Ethylene and 1-MCP regulate major volatile biosynthetic pathways in apple fruit. *Food Chem.* **2016**, *194*, 325–336. [CrossRef]
11. Li, D.; Xu, Y.F.; Xu, G.M.; Gu, L.K.; Li, D.; Shu, H. Molecular cloning and expression of a gene encoding alcohol acyltransferase (*MdAAT2*) from apple (cv. Golden Delicious). *Phytochemistry* **2006**, *67*, 658–667. [CrossRef]
12. Zhong, R.; Lee, C.; Ye, Z. Global analysis of direct targets of secondary wall NAC master switches in *Arabidopsis*. *Mol. Plant* **2010**, *3*, 1087–1103. [CrossRef]
13. Cumplido-Laso, G.; Medina-Puche, L.; Moyano, E.; Hoffmann, T.; Sinz, Q.; Ring, L.; Studart-Wittkowski, C.; Caballero, J.L.; Schwab, W.; Munoz-Blaoco, J. The fruit ripening-related gene *FaAAT2* encodes an acyl transferase involved in strawberry aroma biogenesis. *J. Exp. Bot.* **2012**, *63*, 4275–4290. [CrossRef]
14. Williams, A.A.; Knee, M. The flavour of Cox's Orange Pippin apples and its variation with storage. *Ann. Appl. Biol.* **1977**, *87*, 127–131. [CrossRef]
15. Levisson, M.; Oost, J.V.D.; Kengen, S.W.M. Carboxylic ester hydrolases from hyperthermophiles. *Extremophiles* **2009**, *13*, 567–581. [CrossRef]
16. Wang, X.; Shi, J.; Zhu, H.J. Functional Study of Carboxylesterase 1 Protein Isoforms. *Proteomics* **2019**, *19*, 800288. [CrossRef]
17. Ollis, D.L.; Cheah, E.; Cygler, M.; Dijkstra, B.; Frolow, F.; Franken, S.M.; Harel, M.; Remington, S.J.; Silman, I.; Schrag, J. The alpha/beta hydrolase fold. *Protein Eng.* **1992**, *5*, 197–211. [CrossRef]
18. Ileperuma, N.R.; Marshall, S.D.; Squire, C.J.; Baker, H.M.; Oakeshott, J.G.; Russell, R.J.; Plummer, K.M.; Newcomb, R.D.; Baker, E.N. High-resolution crystal structure of plant carboxylesterase *AeCXE1*, from *Actinidia eriantha*, and its complex with a high-affinity inhibitor paraoxon. *Biochemistry* **2007**, *46*, 1851–1859. [CrossRef]
19. Marshall, S.D.G.; Putterill, J.J.; Plummer, K.M.; Richard, D.; Newcomb, R.D. The carboxylesterase gene family from *Arabidopsis thaliana*. *J. Mol. Evol.* **2003**, *57*, 487–500.
20. Schilmiller, A.L.; Gilgallon, K.; Ghosh, B.; Jones, A.D.; Last, R.L. Acylsugar Acylhydrolases: Carboxylesterase-Catalyzed Hydrolysis of Acylsugars in Tomato Trichomes. *Plant Physiol.* **2016**, *170*, 1331–1344. [CrossRef]
21. Nomura, T.; Murase, T.; Ogita, S.; Kato, Y. Molecular identification of tuliposide B-converting enzyme: A lactone-forming carboxylesterase from the pollen of tulip. *Plant J.* **2015**, *83*, 252–262. [CrossRef]
22. Lee, S.; Hwang, S.; Seo, Y.; Jeon, W. Molecular characterization of the *AtCXE8* gene, which promotes resistance to *Botrytis cinerea* infection. *Plant Biotechnol. Rep.* **2013**, *7*, 109–119. [CrossRef]
23. Pontier, D.; Godiard, L.; Marco, Y.; Roby, D. *hsr*203J, a tobacco gene whose activation is rapid, highly localized and specific for incompatible plant/pathogen interactions. *Plant J.* **1994**, *5*, 507–521. [CrossRef]
24. Ko, M.; Cho, J.H.; Seo, H.H.; Lee, H.H.; Kang, H.Y.; Nguyen, T.S.; Soh, H.C.; Kim, Y.S.; Kim, J.I. Constitutive expression of a fungus-inducible carboxylesterase improves disease resistance in transgenic pepper plants. *Planta* **2016**, *244*, 379–392. [CrossRef]
25. Souleyre, E.J.F.; Marshall, S.D.G.; Oakeshott, J.G.; Russell, R.J.; Plummer, K.M.; Newcomb, R.D. Show more Biochemical characterisation of MdCXE1, a carboxylesterase from apple that is expressed during fruit ripening. *Phytochemistry* **2011**, mboxemph72, 564–571. [CrossRef]
26. Goulet, C.; Mageroy, M.H.; Lam, N.B.; Floystad, A.; Tieman, D.M.; Klee, H.J. Role of an esterase in flavor volatile variation within the tomato clade. *Proc. Natl. Acad. Sci. USA* **2012**, *109*, 19009–19014. [CrossRef]
27. Cao, X.; Xie, K.; Wenyi, D.; Zhu, Y. Peach Carboxylesterase PpCXE1 Is Associated with Catabolism of Volatile Esters. *J. Agric. Food Chem.* **2019**, *67*, 5189–5196. [CrossRef]
28. Chen, K.S.; Xu, C.J.; Xu, W.P.; Wu, M.; Zhang, S.L. Establishment of a method for determining the activity of lipooxygenase in kiwi fruit and peach fruit. *J. Fruit Sci.* **2003**, *20*, 436–438.
29. Zhang, X.M. Physiological Studies on Aroma Volatile Formation in Peach Fruit. *Dep. Hortic. Zhejiang Univ.* **2005**. Available online: https://kns.cnki.net/kcms2/article/abstract?v=1UV6lUEUW658kDSswtdDD6HhzncyebpjLexctGdXWtVwW_7hfELsP1ddfk3Ny9ubzplLnUBQhLbOcjevA3xDQIcjlqRwPjPe9XW2kU1W0zrKH_YLWfayfyVa893Id2s9&uniplatform=NZKPT&language=CHS (accessed on 6 May 2023).
30. Ke, D.; Zhou, L.; Kader, A.A. Mode of Oxygen and Carbon Dioxide Action on Strawberry Ester Biosynthesis. *J. Am. Soc. Hortic. Sci.* **1994**, *119*, 971–975.
31. Yang, S.; Li, D.M.; Li, S.S.; Yang, H.J.; Zhao, Z.Y. GC-MS Metabolite and Transcriptome Analyses Reveal the Differences of Volatile Synthesis and Gene Expression Profiling between Two Apple Varieties. *Int. J. Mol. Sci.* **2022**, *23*, 2939. [CrossRef]
32. Meng, Z.P. Study on Genetic Characteristics of Volatile Compounds in Apple Fruit. *Northwest AF Univ.* **2022**. [CrossRef]

33. Gambino, G.; Perrone, I.; Gribaudo, I. A Rapid and effective method for RNA extraction from different tissues of grapevine and other woody plants. *Phytochem. Anal.* **2008**, *19*, 520–525. [CrossRef]
34. Wu, D.; Hao, J.F.; Zhang, L.Q.; Zhao, A.Q.; Ha, A. Analysis of Function of a-mannosidase Gene in Promoting Melon Fruit Ripening by Transient Expression. *Acta Hortic. Sin.* **2014**, *41*, 1601–1608.
35. Liu, X.; Li, D.M.; Li, Y.J.; Li, S.S.; Zhao, Z.Y. Brassinosteroids are involved in volatile compounds biosynthesis related to MdBZR1 in 'Ruixue' (Malus × domestica Borkh.) fruit. *Postharvest Biol. Technol.* **2022**, *189*, 111931. [CrossRef]
36. He, Y.; Liu, H.; Li, H.; Jin, M.; Wang, X.L.; Yin, X.R.; Zhu, Q.J.; Rao, J.P. Transcription factors DkBZR1/2 regulate cell wall degradation genes and ethylene biosynthesis genes during persimmon fruit ripening. *J. Exp. Bot.* **2021**, *72*, 6437–6446. [CrossRef]
37. Jia, C.S.; Sun, S.M.; Bao, A.M.; Wang, Z.H. Comparison of postharvest physiology and storage tolerance of four characteristic small apples. *Trans. Chin. Soc. Agric. Eng.* **2022**, *38*, 308–316.
38. Liu, Y.W.; Song, Z.B.; Chen, G.J.; Wang, B.J.; Yang, G.H.; Yuan, D.H. Chengde City develops cold apple industry to help rural revitalization. *Mod. Rural. Sci. Technol.* **2022**, *11*, 7–8.
39. Zhang, F.; Gong, X.M.; Ma, S.S.; Zhang, J.F.; Guan, J.F. Effect of 1-MCP storage and temperature on the quality and postharvest physiology of Qinyang Apple. *J. Northwest AF Univ. (Nat. Sci. Ed.)* **2009**, *37*, 115–119.
40. Fan, X.G. Cytological Studies on Texture Differ Ences in Fruit Development of 'Ruiyang', 'Ruixue' and Theie Parents. *Northwest AF Univ.* **2017**. Available online: https://kns.cnki.net/kcms2/article/abstract?v=1UV6lUEUW65E8nc7eaVi9OMyDQAEe86Wgwy9lccUR9HLIdat8wPLUhPbELp4qg_QlSJqbwYfm8uoStjf_afRqYyIF9_Do3xpH2CIN5c7TBq9A27_V3nb4SaupScZTm4gtrwxjBV2u3Q=&uniplatform=NZKPT&language=CHS (accessed on 6 May 2023).
41. Lewinsohn, E.; Schalechet, F.; Wilkinson, J.; Matsui, K.; Tadmor, Y.; Nam, K.H.; Amar, O.; Lastochkin, E.; Larkov, O.; Ravid, U.; et al. Enhanced Levels of the Aroma and Flavor Compound S-Linalool by Metabolic Engineering of the Terpenoid Pathway in Tomato Fruits. *Plant Physiol.* **2001**, *127*, 1256–1265. [CrossRef]
42. Dixon, J.; Hewett, E.W. Factors affecting apple aroma/flavour volatile concentration: A Review. *N. Z. J. Crop Hortic. Sci.* **2000**, *28*, 155–173. [CrossRef]
43. El-Sharkawy, I.; Manriquez, D.; Flores, F.B.; Regad, F.; Bouzayen, M.; Latche, A.; Pech, J.C. Functional characterization of a melon alcohol acyl-transferase gene family involved in the biosynthesis of ester volatiles. Identification of the crucial role of a threonine residue for enzyme activity. *Plant Mol. Biol.* **2005**, *59*, 345–362.
44. Tian, L.P.; Zhang, Q.; Li, R.; Ren, X.L. Effects of n-Butanol Treatment on Volatile Compounds of 'Pink Lady' Apple during Storage. *Sci. Technol. Food Ind.* **2022**, *43*, 337–345.
45. Defilippi, B.G.; Kader, A.A.; Dandekar, A.M. Apple aroma: Alcohol acyltransferase, a rate limiting step for ester biosynthesis, is regulated by ethylene. *Plant Sci.* **2005**, *168*, 1199–1210. [CrossRef]
46. Yan, D.; Shi, J.; Ren, X.L.; Tao, Y.S.; Ma, F.W.; Li, R.; Liu, X.R.; Liu, C.H. Insights into the aroma profiles and characteristic aroma of 'Honeycrisp' apple (Malus × domestica). *Food Chem.* **2020**, *327*, 127074. [CrossRef]
47. Rowan, D.D.; Allen, J.M.; Fielder, S.; Hunt, M.B. Biosynthesis of straight-chain ester volatiles in red delicious and granny smith apples using deuterium-labeled precursors. *J. Agric. Food Chem.* **1999**, *47*, 2553–2562. [CrossRef]
48. Fellman, J.K.; Rudell, D.R.; Mattinson, D.S.; Mattheis, J.P. Relationship of harvest maturity to flavor regeneration after CA storage of 'Delicious' apples. *Postharvest Biol. Technol.* **2003**, *27*, 39–51. [CrossRef]
49. Beekwilder, J.; Alvarez-Huerta, M.; Neef, E.; Verstappen, F.W.A.; Bouwmeester, H.J.; Aharoni, A. Functional Characterization of Enzymes Forming Volatile Esters from Strawberry and Banana. *Plant Physiol.* **2004**, *135*, 1865–1878. [CrossRef]
50. Aharoni, A.; Keizer, L.C.; Bouwmeester, H.J.; Sun, Z.; Alvarez-Huerta, M.; Verhoeven, H.A.; Blaas, J.; van Houwelingen, A.M.M.L.; De Vos, R.C.H.; Voet, H.V.D.; et al. Identification of the *SAAT* Gene Involved in Strawberry Flavor Biogenesis by Use of DNA Microarrays. *Plant Cell* **2000**, *12*, 647–661. [CrossRef]
51. Li, G.; Jia, H.J.; Li, J.H.; Wang, Q.; Zhang, M.J.; Teng, Y.W. Emission of volatile esters and transcription of ethylene- and aroma-related genes during ripening of 'Pingxiangli' pear fruit (*Pyrus ussuriensis* Maxim). *Sci. Hortic.* **2014**, *170*, 17–23. [CrossRef]
52. Goulet, C.; Kamiyoshihara, Y.; Lam, N.B.; Richard, T.; Taylor, M.G.; Tieman, D.M.; Klee, H.J. Divergence in the enzy-matic activities of a tomato and *Solanum pennellii* alcohol acyltransferase impacts fruit volatile ester composition. *Mol. Plant* **2015**, *8*, 153–162. [CrossRef]
53. Li, D.P. Expression Characterization of the *MdAAT2* gene and Its Regulation Mechanism of Volatile Ester Biosynthesis in Apple (cv. Golden Delicious). *Shandong Agric. Univ.* **2005**. Available online: https://kns.cnki.net/kcms/detail/detail.aspx?FileName=2006100793.nh&DbName=CDFD2006 (accessed on 6 May 2023).

foods

Communication

Simultaneous Analysis of Bergapten and Schinifoline in *Zanthoxylum schinifolium* Seeds Using HPLC and UPLC-MS/MS Systems

Chang-Seob Seo

KM Science Research Division, Korea Institute of Oriental Medicine, Daejeon 34054, Republic of Korea; csseo0914@kiom.re.kr; Tel.: +82-42-868-9361

Abstract: *Zanthoxylum schinifolium* Siebold et Zuccarini belongs to the Rutaceae family and has been widely used as a spice in East Asian countries such as Korea, China, and Japan. The present study focused on developing and validating a simultaneous analytical method for marker substances (bergapten and schinifoline) in *Z. schinifolium* seeds. This was achieved using high-performance liquid chromatography with a photo-diode array detector (HPLC-PDA) and ultra-performance liquid chromatography with tandem mass spectrometry (UPLC-MS/MS) systems. In the regression equation, all markers showed a coefficient of determination of ≥0.9990. Marker recovery was 96.90–105.16% (relative standard deviation (RSD) ≤ 2.23), and the intra- and interday precision was RSD < 3.00. Bergapten and schinifoline were detected in the seeds at 1.70–2.85 mg/g and 0.19–0.94 mg/g, respectively. This analytical method will improve quality control of *Z. schinifolium* seeds. Additionally, this assay will provide basic data and quality assurance for future biological activity experiments or clinical applications.

Keywords: simultaneous analysis; *Zanthoxylum schinifolium*; bergapten; schinifoline

Citation: Seo, C.-S. Simultaneous Analysis of Bergapten and Schinifoline in *Zanthoxylum schinifolium* Seeds Using HPLC and UPLC-MS/MS Systems. *Foods* **2023**, *12*, 1355. https://doi.org/10.3390/foods12071355

Academic Editors: Chao Kang and Ronald Beckett

Received: 24 February 2023
Revised: 20 March 2023
Accepted: 21 March 2023
Published: 23 March 2023

1. Introduction

Zanthoxylum schinifolium Siebold et Zuccarini is a perennial plant belonging to the Rutaceae family and has been used as a spice in East Asian countries, including Korea, China, and Japan [1]. The *Zanthoxylum* genus consists of more than 200 *Zanthoxylum* species, among which *Z. piperitum* De Candolle, *Z. schinifolium* Siebold et Zuccarini, and *Z. bungeanum* Maximowicz are recorded in the Korean Pharmacopoeia [2–4].

A wide variety of phytochemical components such as coumarins (bergapten and umbelliferon) [5,6]; alkaloids (skimmianine, schinifoline, and sanshools) [3,5–7]; essential oils (limonene and linalool) [2,6]; fatty acids (oleic acid and palmitic acid) [2,6]; and glycosides (betulalbuside A and osmanthuside H) [7] have been isolated from these species.

Anti-inflammatory [2,6], antimicrobial [2], anticancer [8], analgesic [2,9], hepatoprotective [2,10], and antiviral [2,11] effects have been reported in studies of the *Zanthoxylum* species.

In the present study, an analytical method was developed for the simultaneous analysis of coumarin (bergapten), and an alkaloid (schinifoline) isolated from these *Zanthoxylum* species in *Z. schinifolium* seeds. Bergapten, a coumarin derivative, and schinifoline, a 4-quinolinone derivative, were first isolated from the leaves and pericarps of *Z. schinifolium*, respectively [5,12,13]. Bergapten has been reported to be effective in various areas such as neuroprotection, organ protection, anticancer, anti-inflammatory, antibacterial, and antidiabetic effects [14]. In comparison, schinifoline has been reported to have a radiosensitizing effect on human non-small cell lung cancer A549 cells [8] and an antifungal effect on *Candida albicans* [15].

Several studies on the analysis of Zanthoxylum species have been reported [2,3,9]. Li et al. [2] and Wu and Wu [9] analyzed the essential oils of *Z. myriacanthum* and *Z. schini-*

folium using gas chromatography with mass spectrometry. In addition, Zhao et al. [3] quantitatively analyzed components of alkylamides, which are fragrance components, from *Z. schinifolium* oil using liquid chromatography with mass spectrometry. However, to our knowledge, analytical methods for bergapten and schinifoline in *Z. schinifolium* seeds have not yet been reported.

Therefore, in the present study, we endeavored to develop a simultaneous analysis method for bergapten and schinifoline in *Z. schinifolium* seeds using high-performance liquid chromatography with a photo-diode array detector (HPLC-PDA) and ultra-performance liquid chromatography with tandem mass spectrometry (UPLC-MS/MS).

2. Results

2.1. HPLC-PDA Analysis

2.1.1. Selection of HPLC-PDA Analysis Conditions for Simultaneous Quantification

For quantitative analysis of bergapten and schinifoline in *Z. schinifolium* seeds, we used a Waters XBridge reverse-phase C_{18} column (250 length × 4.6 mm ID, 5 μm particle size) (Milford, MA, USA) with isocratic elution by mobile phases of a 1.0% (v/v) distilled water solution of acetic acid and 1.0% (v/v) solution of acetic acid in acetonitrile in a 1:1 ratio. Both marker components were monitored simultaneously at 235 nm (schinifoline) and 310 nm (bergapten) using a PDA. The flow rate was 1.0 mL/min, the injection amount was 10 μL during HPLC analysis. During the analysis, the column was maintained at 40 °C (Table S1 (Supplementary Materials)). A representative chromatogram is shown in Figure 1. The two markers, bergapten and schinifoline, were eluted with a resolution of ≥1.50 within 10 min (5.63 min and 8.15 min, respectively).

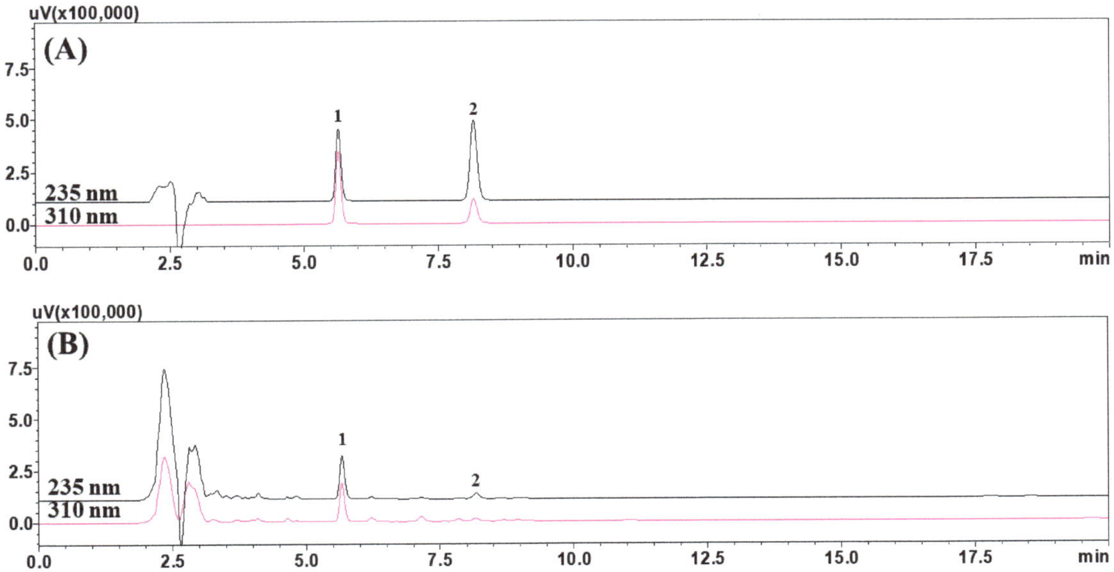

Figure 1. *Cont.*

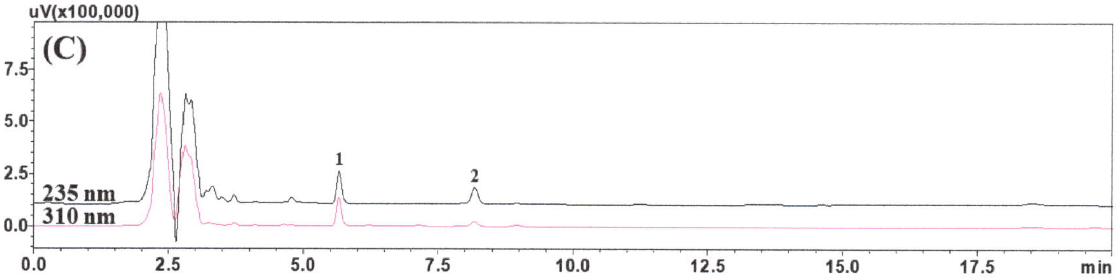

Figure 1. HPLC chromatograms of (**A**) a standard solution of the two marker substances, bergapten and schinifoline, (**B**) 70% ethanol extract of *Zanthoxylum schinifolium* seeds (2018ZSS), and (**C**) 70% ethanol extract of *Zanthoxylum schinifolium* seeds (2021ZSS). Bergapten (1) and schinifoline (2). The measured concentrations of the bergapten and the schinifoline in the standard solution (**A**) were each 50 μg/mL.

2.1.2. Validation of the HPLC-PDA Analytical Method

Table 1 shows the results of parameters used to verify the analytical method we developed. In the calibration curve regression equation for bergapten and schinifoline, the coefficients of determination (r^2) were both 0.9999 (Table 1 and Figure S1). Linearity of these two calibration curves was evaluated through the application of a Shapiro–Wilk test using the SigmaPlot 12.5 software (Systat Software, Inc., San Jose, CA, USA). As a result of the Shapiro–Wilk test, the p-value of both analytes was >0.05 at the 95% confidence level, indicating that the linear regression model had a normal distribution of the experimental data (Table 1). In addition, residuals of two markers show <2.0% (Figure S1). In the Shapiro–Wilk test for the residual distribution of the bergapten and schinifoline, the p values were 0.31 and 0.71, respectively, indicating that the data were normally distributed. The selectivity of each marker substance was confirmed by the UV spectrum of each compound (Figure S2) and the peak purity in the sample (Figure S3). The peak purity index values of bergapten and schinifoline were 1.0000, showing excellent selectivity, and impurities were not detected at each detection wavelength (235 nm for schinifoline and 310 nm for bergapten). The limit of detection (LOD) and the limit of quantitation (LOQ) of the two markers were 0.04–0.08 μg/mL and 0.12–0.24 μg/mL, respectively (Table 1). As shown in Table 2, it was validated by a standard addition method in which three different concentrations (low, medium, and high) were added. After adding the standard solution of each marker to the sample, ultrasonic extraction was performed for 30 min, followed by HPLC analysis. Both marker substances were measured at 96.02–100.96%, and the relative standard deviation (RSD) was calculated as ≤0.66 (Table 2). In intra- and interday precision tests for bergapten and schinifoline, RSD values were <0.40 (Table 3). In addition, the RSD value was <0.21% for the repeatability of retention time and peak area (Table S2). These results demonstrate that the precision of the developed assay is excellent.

Table 1. Linear range, regression equation, coefficients of determination (r^2), limit of detection (LOD), and limit of quantitation (LOQ) for simultaneous quantitation of two marker substances in Z. *schinifolium* seeds.

Marker	Linear Range (μg/mL)	Regression Equation [1]($y = ax + b$)	r^2	p-Value [2]	LOD (μg/mL) [3]	LOQ (μg/mL) [4]
Bergapten	1.56–100.00	$y = 42{,}108.21x + 24{,}433.44$	0.9999	0.30	0.08	0.24
Schinifoline	0.31–20.00	$y = 71{,}671.58x + 7668.61$	0.9999	0.28	0.04	0.12

[1] y: peak area of each marker substance, x: concentration of each marker substance. [2] p value of Shapiro–Wilk test (confidence level at 95%). [3] LOD: limit of detection. [4] LOQ: limit of quantitation.

Table 2. Results of recovery test of two marker substances in developed HPLC analytical method (n = 5).

Marker	Concentration (μg/mL)		Recovery (%)	SD [1]	RSD (%) [2]
	Spiked	Found			
Bergapten	5.00	4.94	98.73	0.27	0.27
	12.50	12.42	99.33	0.25	0.25
	25.00	24.93	99.74	0.07	0.07
Schinifoline	2.00	1.92	96.02	0.64	0.66
	4.00	4.04	100.96	0.35	0.35
	8.00	7.75	96.90	0.12	0.12

[1] SD: standard deviation. [2] RSD: relative standard deviation.

Table 3. The precision of the HPLC analytical method of the two markers.

Marker Substance	Conc. (μg/mL)	Intraday (n = 5)			Interday (n = 5)		
		Found Conc. (μg/mL)	Precision (RSD, %)	Accuracy (%)	Found Conc. (μg/mL)	Precision (RSD, %)	Accuracy (%)
Bergapten	25.00	25.42	0.05	101.69	25.43	0.22	101.71
	50.00	50.44	0.17	100.88	50.32	0.26	100.63
	100.00	100.04	0.09	100.04	100.17	0.39	100.17
Schinifoline	5.00	5.06	0.25	101.26	5.05	0.29	101.09
	10.00	10.03	0.27	100.26	10.00	0.29	100.02
	20.00	20.02	0.06	100.10	20.04	0.39	100.19

2.1.3. The Stability of the Markers

The stability of the two markers (bergapten and schinifoline) was measured at room temperature and under refrigeration (about 4 °C) for 10 days in standard and sample solutions, respectively. As shown in Table S3, the stability of the two markers was tested for 10 days (0, 1, 2, 3, 4, 7, and 10 days) at room temperature and under 4 °C refrigeration, respectively. As a result of comparing the bergapten and schinifoline for 10 days based on the first day (Day 0), 97.98–102.72% (RSD $\leq$ 1.55) and 98.30–103.99% (RSD $\leq$ 1.63) were measured, respectively. In the sample solution, the stability of the bergapten and schinifoline was measured as 97.58–100.09% (RSD < 1.00) and 99.08–100.52% (RSD < 1.00) compared to the initial data (Day 0) for 10 days at room temperature and under 4 °C refrigeration, respectively. We found that the two markers were stable for at least 10 days under either storage condition.

2.1.4. Simultaneous Quantitation of Two Markers in *Z. schinifolium* Seeds

The assay developed in the present study was efficiently applied to the quantitative analysis of two markers in *Z. schinifolium* seeds. In the optimized HPLC analytical method, bergapten and schinifoline were detected at 2.30–2.85 mg/g and 0.33–0.94 mg/g, respectively, in 70% ethanol extracts from samples collected in 2018 (2018ZSS) and 2021 (ZSS2021) (Table 4).

Table 4. Quantitation of bergapten and schinifoline in *Z. schinifolium* seeds using the optimized HPLC-PDA analytical method (n = 5).

Marker	2018ZSS			2021ZSS		
	Mean (mg/g)	SD $\times$ 10^{-1}	RSD (%)	Mean (mg/g)	SD $\times$ 10^{-1}	RSD (%)
Bergapten	2.85	0.13	0.44	2.30	0.12	0.54
Schinifoline	0.33	0.01	0.32	0.94	0.03	0.36

2.2. UPLC-MS/MS Quantitation

2.2.1. Setting Optimal UPLC-MS/MS Analytical Conditions for Quantitation

Quantitation of the two markers, bergapten and schinifoline, in *Z. schinifolium* seeds was performed using a UPLC-MS/MS (Waters, Milford, MA, USA) consisting of a Waters Acquity UPLC I-Class system and a tandem quadrupole MS detector system with an electrospray ionization source. Based on the analytical conditions presented in Table S4, bergapten and schinifoline were quantified by applying a multiple-reaction monitoring (MRM) method (Table 5). Bergapten and schinifoline were detected at 5.00 min and 6.64 min, respectively, in the form of $[M + H]^+$ in positive ion mode (Figure 2 and Figure S4). To quantify the two markers, the precursor ion (Q1) and product ion (Q3) were set as follows [16–19]: Q3 of bergapten was set at m/z 202.0 in the form of $[M + H-CH_3]^+$ in which the methyl group was removed from Q1, and Q3 of schinifoline was set at m/z 173.1 in the form of $[M + H-C_6H_{13}]^+$ in which the C_6H_{13} functional group was removed from Q1 (Table 5, and Figures S5 and S6).

Table 5. Parameters for UPLC-MS/MS MRM analysis of bergapten and schinifoline.

Marker	Ion Mode	Molecular Weight	MRM Condition	Cone Voltage (V)	Collision Energy (Ev)
Bergapten	Positive	216.04	216.9 → 202.0	30	20
Schinifoline	Positive	257.18	258.4 → 173.1	30	30

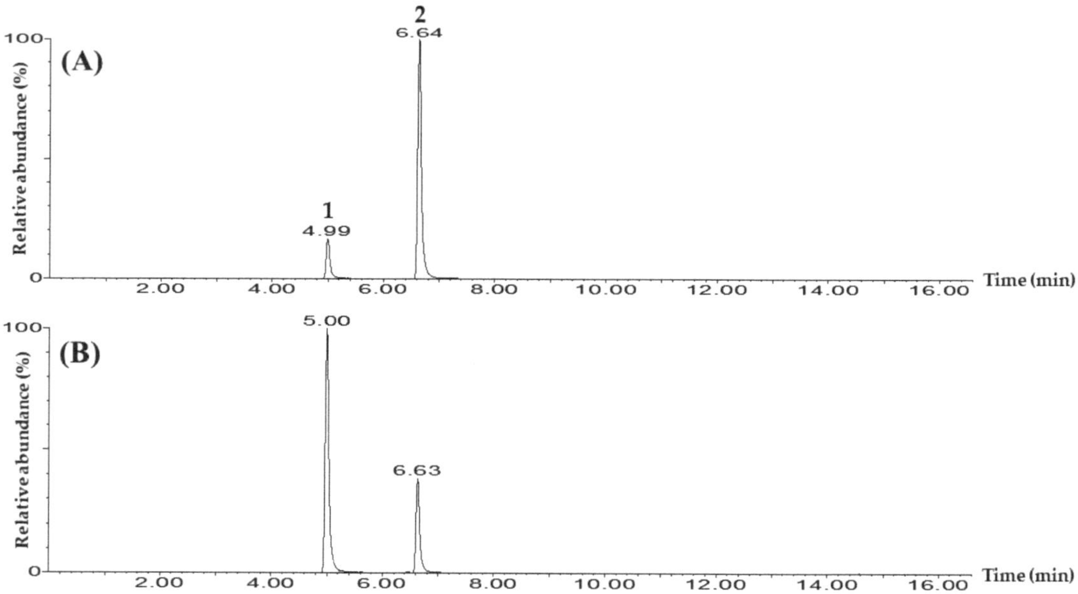

Figure 2. Total ion chromatography of (**A**) mixtures of the two markers, and (**B**) a 70% ethanolic extract of *Z. schinifolium* seeds measured by the UPLC-MS/MS MRM method in positive ion mode. Bergapten (1) and schinifoline (2). The measured concentrations of the bergapten and the schinifoline in the standard solution were both 0.5 μg/mL.

2.2.2. Validation of the UPLC-MS/MS MRM Analytical Method

For the quantitation of bergapten, the regression equation for the calibration curve prepared at the concentrations of 0.10–5.00 μg/mL was $y = 128{,}424.87x + 12{,}393.30$ ($r^2 = 0.9990$). In the case of schinifoline, the regression equation at the concentrations of 0.10–5.00 μg/mL was $y = 520{,}305.56x + 73{,}378.28$ ($r^2 = 0.9994$) (Figure S7). As a result of analyzing the calibration curves of bergapten and schinifoline with the Shapiro–Wilk test of SigmaPlot 12.5 software (Systat Software, Inc., San Jose, CA, USA), the *p*-values were 0.65 and 0.17, respectively. These values had a *p*-value greater than 0.05 at the 95% confidence level, indicating that the linear regression model had a normal distribution of the experimental data. The residual of two markers was <2.0% (Figure S7). In the Shapiro–Wilk test for the residual distribution of the bergapten and schinifoline, the *p*-values were 0.18 and 0.30, respectively, indicating that the data were normally distributed. LOD and LOQ of the two markers were 0.01–0.06 ($\times 10^{-2}$) μg/mL and 0.04–0.17 ($\times 10^{-2}$) μg/mL, respectively. The recovery in this developed assay was validated by a standard addition method. As shown in Table 6, standard solutions of three different concentrations were added. UPLC-MS/MS analysis was performed after consecutive pretreatment of 5 min of ultrasonic extraction and 1 min of vortexing. Recovery of bergapten and schinifoline was measured at 99.63–105.16% (RSD $\leq$ 2.23, Table 6). The RSD value showing intra- and interday precision was less than 3.00% (Table 7). Additionally, the RSD value of repeatability for retention time and peak area was less than 0.50% (Table 7).

Table 6. Results of recovery test of two marker substances in developed UPLC-MS/MS analytical method (n = 5).

Marker	Concentration (μg/mL)		Recovery (%)	SD	RSD (%)
	Spiked	Found			
Bergapten	0.40	0.40	99.63	2.23	2.23
	0.80	0.84	105.16	1.10	1.05
	1.60	1.66	104.02	1.39	1.34
Schinifoline	0.02	0.02	99.68	0.22	0.22
	0.04	0.41	102.02	0.10	0.10
	0.08	0.81	100.94	0.19	0.19

Table 7. The precision and repeatability of the UPLC-MS/MS analytical method of the two markers.

Marker	Conc. (μg/mL)	Intraday (n = 5)			Interday (n = 5)			Repeatability (n = 6)	
		Found Conc. (μg/mL)	Precision (RSD, %)	Accuracy (%)	Found Conc. (μg/mL)	Precision (RSD, %)	Accuracy (%)	RSD (%) of Retention Time	RSD (%) of Peak Area
Bergapten	0.40	0.37	2.83	93.11	0.39	2.82	98.48	0.11	0.41
	0.80	0.77	1.43	96.42	0.81	1.07	101.86		
	1.60	1.53	1.91	95.44	1.63	2.34	101.75		
Schinifoline	0.02	0.02	0.02	97.45	0.02	0.25	100.12	0.08	0.16
	0.04	0.04	0.07	97.14	0.04	0.10	100.87		
	0.08	0.08	0.02	94.89	0.08	0.21	100.39		

2.2.3. Quantitation of Bergapten and Schinifoline in *Z. schinifolium* Seeds

The developed UPLC-MS/MS MRM method was successfully applied to the simultaneous quantification of bergapten and schinifoline in *Z. schinifolium* seeds. Bergapten was detected at 1.70–2.85 mg/g, and schinifoline was 0.19–0.90 mg/g (Table 8).

Table 8. Quantitation of bergapten and schinifoline in *Z. schinifolium* seeds by UPLC-MS/MS MRM ($n = 5$).

Marker	2018ZSS			2021ZSS		
	Mean (mg/g)	SD	RSD (%)	Mean (mg/g)	SD	RSD (%)
Bergapten	2.85	0.15	5.30	1.70	0.06	3.45
Schinifoline	0.19	0.01	4.83	0.90	0.05	5.34

3. Discussion

In the present study, an analytical method for the simultaneous analysis of bergapten and schinifoline from *Z. schinifolium* seeds using HPLC and UPLC-MS/MS was developed and verified. Various constituents have been isolated and reported from *Zanthoxylum* species [2,3,5–7]. In particular, coumarins [5,6], alkaloids [3,5–7], and essential oils [2,6] have been reported.

Currently, HPLC is one of the most widely used analytical methods for natural product research in academia and industry, and it has the advantage of being relatively simple and easy to operate. However, UPLC-MS/MS is a more sensitive and accurate analytical method, and its use is increasing.

We conducted analyses to develop a quantitative method for the seeds of *Z. schinifolium* among several *Zanthoxylum* species using HPLC and UPLC-MS/MS systems. First, a simultaneous analytical method for bergapten and schinifoline was developed using an HPLC system. This method was validated through selectivity, linearity, recovery, and precision. Several types of reverse-phase C_{18} columns (SunFire, XBridge, and XTerr columns, 4.6 ID × 250 mm length, 5 μm particle size); acids (0.1% formic acid, 0.1% trifluoroacetic acid, 0.1% phosphoric acid, and 1.0% acetic acid); and column oven temperatures (30, 35, 40, and 45 °C) were tested for method development. When using a Waters XBridge reverse-phase C_{18} column maintained at 40 °C with isocratic elution by a distilled water-acetonitrile (both containing 1.0% acetic acid) mobile phase system, both markers were eluted within 10 min with a resolution of 12.15. Second, an analytical method for the two markers was developed using UPLC-MS/MS MRM with a Waters Acquity UPLC I-Class system and tandem quadrupole MS system. The assay developed by the two systems was validated through parameters such as selectivity, linearity, LOD, LOQ, recovery, and precision.

Based on the optimized HPLC and UPLC-MS/MS analytical methods described above, bergapten and schinifoline were successfully separated and quantitatively analyzed in two samples, 2018ZSS and 2021ZSS. In both assays, bergapten was more abundant than schinifoline. When comparing the content of both collection years, bergapten was more abundant in 2018ZSS than in 2021ZSS, while schinifoline was more abundant in 2021ZSS. The pattern was the same for both the HPLC and the UPLC-MS/MS analytical methods.

As shown in the results of this study, the UPLC-MS/MS method showed the advantages of shorter time, higher sensitivity, and separation with less solvent compared to the HPLC method. Nevertheless, the analysis method using HPLC, which is the most widely used and easy to operate to date, is considered a more appropriate analysis method for quality control of the two marker substances in *Z. schinifolium* seeds.

4. Materials and Methods

4.1. Chemicals and Reagents

The reference standard compounds for simultaneous analysis (Figure 3), bergapten (CAS No. 484-20-8, 98.0%, CFN98766) and schinifoline (CAS No. 80554-58-1, 99.1%, TBZ0836), were purchased from ChemFaces Biochemical Co. (Wuhan, China) and ChemNorm Biotech Co. (Wuhan, China), respectively. Solvents, methanol, acetonitrile, and distilled water were used at HPLC grade or LC–MS grade and were purchased from JT Baker (Phillipsburg, NJ, USA). Acetic acid (≥99.7%, A35-500) and formic acid (≥99.7%,

A117-50) were HPLC and LC-MS grades, respectively, and these were purchased from Fisher Scientific (Fair Lawn, NJ, USA).

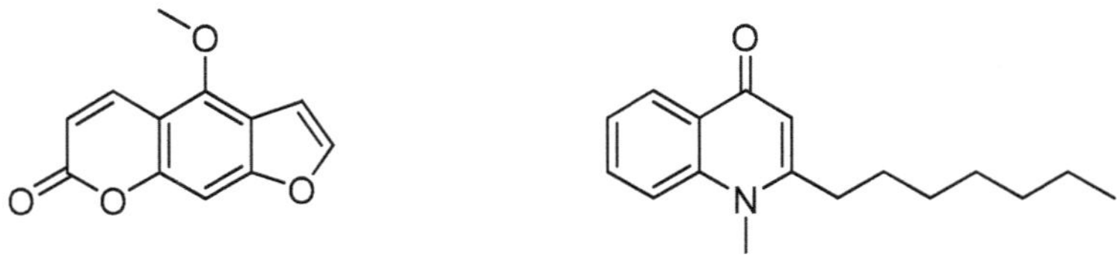

Bergapten Schinifoline

Figure 3. Chemical structures of the two markers selected of *Z. schinifolium* seeds.

4.2. Plant Materials and Preparation of the 70% Ethanolic Extract of Z. schinifolium Seeds

Dried *Z. schinifolium* seeds (sample numbers 2018ZSS and 2021ZSS) were purchased from the *Z. schinifolium* producer Woobosancho (Miryang, Republic of Korea) in 2018 and 2021, respectively. The scientific name of the sample was confirmed in The Plant List (www.theplantlist.org) [20] and morphologically identified by Dr Goya Choi, Korea Institute of Oriental Medicine (Naju, Republic of Korea). Extracts of *Z. schinifolium* seeds were prepared by KOC Biotech Co. (Daejeon, Republic of Korea). In brief, 2.0 kg of the dried *Z. schinifolium* seeds, which were removed from their seedcase, were extracted under reflux at 80 °C for 3 h using 20 L of 70% ethanol. After filtering the extract using a standard sieve (270 mesh), the organic solvent was removed using a rotary evaporator. Then, the residue was suspended in 1.0 L of distilled water and made into a powdered sample (about 3.7%) using a freeze-dryer (LP20, Daejeon, Republic of Korea).

4.3. Preparation of Standard Stock and Sample Solutions

Standard stock solutions for simultaneous analysis of bergapten and schinifoline in *Z. schinifolium* seeds were prepared in methanol at 1000 ppm and stored under refrigeration (approximately 4 °C) until use. Subsequently, each prepared solution was serially diluted before use. After accurately taking 100 mg of the 70% ethanol extract of the two markers prepared for quantitative analysis using HPLC from *Z. schinifolium* seeds, 10 mL of 70% methanol was added and extraction was conducted ultrasonically for 30 min. Then, the sample solution for UPLC-MS/MS analysis was separately prepared with 70% methanol at a concentration of 50 mg/10 mL, followed by ultrasonic extraction for 5 min and vortexing for 1 min. All the extracted solutions were used for quantitative analysis after 0.2 µm-membrane filtration (Pall Life Sciences, Ann Arbor, MI, USA).

4.4. HPLC-PDA Analytical Conditions to Quantify Bergapten and Schinifoline in Z. schinifolium Seeds

Simultaneous determination of two markers, a coumarin derivative and an alkaloid (bergapten and schinifoline), respectively, was performed using a Shimadzu Prominence LC-20A series system coupled to a PDA detector (Kyoto, Japan). This system was controlled by LabSolution software (version 5.54, SP3, Kyoto, Japan). Detailed analytical conditions such as column, column temperature, mobile phase, and elution conditions are presented in Table S1. Eluate was monitored at 235 nm and 310 nm considering the UV absorption maxima of the target components.

4.5. UPLC-MS/MS Analytical Conditions to Quantify Bergapten and Schinifoline in Z. schinifolium Seeds

The quantitation of bergapten and schinifoline in *Z. schinifolium* seeds was archived using a UPLC-MS/MS system consisting of an Acquity UPLC system and a tandem triple quadrupole MS system (Waters, Milford, MA, USA). In the UPLC-MS/MS analysis, an electrospray ionization source was used as an ion source, and MassLynx software (version 4.2, Milford, MA, USA) was used for data acquisition and processing. Detailed UPLC and

MS operating conditions for the quantitation are presented in Table S4, and UPLC-MS/MS MRM parameters are shown in Table 5.

4.6. Validation of the Developed Two Analytical Methods

The developed HPLC-PDA and UPLC-MS/MS analytical methods were validated by testing selectivity, linearity, LOD, LOQ, recovery, and precision.

First, its selectivity was determined by comparing the UV spectra of the markers in the standards and samples. In addition, the peak purity of the two components was evaluated in the sample solution. This was only verified using the HPLC-PDA method.

Second, the linearity of each marker was validated through the r^2 calculated from the regression equation of the calibration curve measured triplicate. In addition, LOD and LOQ were calculated using the following Equations (1) and (2).

$$\text{LOD}\ (\mu\text{g/mL}) = 3.3 \times \frac{\sigma}{S}\ \text{and} \tag{1}$$

$$\text{LOQ}\ (\mu\text{g/mL}) = 10 \times \frac{\sigma}{S} \tag{2}$$

where σ is the SD of the y-intercept in the regression equation for each marker, and S is the slope of the regression equation.

Third, the recovery of each marker was determined by the standard addition method. In brief, after accurately taking 100 mg of the sample, each of the two markers was added at three concentrations, prepared at 10 mg/mL using 70% methanol, and HPLC analysis was performed. The recovery (%) was calculated using the following Equation (3).

$$\text{Recovery}\ (\%) = \frac{\text{Found concentration}}{\text{Spiked concentration}} \times 100 \tag{3}$$

Finally, precision was demonstrated using RSD (%) values. Intra- and interday precision were measured on day 1 and for three consecutive days, and then the RSD values were calculated. For repeatability, RSD values were calculated after six repeated measurements using a standard solution in which the two markers were mixed. The RSD (%) value was calculated as in Equation (4).

$$\text{RSD}\ (\%) = \frac{\text{SD}}{\text{Mean}} \times 100 \tag{4}$$

4.7. Stability Test

The stability of bergapten (6.25 μg/mL) and schinifoline (1.25 μg/mL) was tested for 10 days (0, 1, 2, 3, 4, 7, and 10 days) at room temperature (23 ± 1 °C) and under refrigeration (approximately 4 °C) using standard solution. In addition, the stability of the two components was tested for 10 days using the sample solution prepared at a concentration of 10 mg/mL.

4.8. Statistical Analysis

All data used in this study were expressed as mean, SD, and RSD (%) using Microsoft Excel 2021 software (Microsoft, Redmond, WA, USA). Statistical analysis was performed using SigmaPlot software 12.5 (Systat Software, Inc., San Jose, CA, USA).

5. Conclusions

In the present study, bergapten and schinifoline were selected as marker substances for *Z. schinifolium* seeds, and an analytical method for the simultaneous analysis of these two markers from *Z. schinifolium* seeds was developed and validated for the first time, to our knowledge. This analytical method can be used as a basis for quality control of *Z. schinifolium* seeds and other herbal medicines. Furthermore, the method can be used to obtain basic data for biological activity research or clinical applications.

Supplementary Materials: The following supporting information can be downloaded at: https://www.mdpi.com/article/10.3390/foods12071355/s1, Figure S1: Calibration curves (A) and residual plots (B) for evaluating the linearity of the two markers in the HPLC-PDA method; Figure S2: Validation of selectivity through UV spectra comparison; Figure S3: Peak purity evaluation to demonstrate the selectivity of two components in a sample; Figure S4: Extracted ion chromatograms for each standard marker (A), 2018ZSS sample (B), and 2021ZSS sample (C) measured by LC-MS/MS MRM mode; Figure S5: Fragmentation of the two markers by the UPLC-MS/MS MRM method; Figure S6: Precursor ion (Q1) and product ion (Q3) peaks of bergapten (A) and schinifoline (B) by the UPLC-MS/MS MRM method; Figure S7: Calibration curves (A) and residual plots (B) of the two markers in the UPLC-MS/MS MRM method; Table S1: Chromatographic parameters for simultaneous quantitation of the two markers in *Z. schinifolium* seeds by HPLC-PDA; Table S2: Repeatability of retention time and peak area of the two markers (n = 6); Table S3: Stability (%) of the two markers measured at room temperature and under refrigeration using standard and sample solutions; Table S4: Parameters for simultaneous quantitation of the two markers in *Z. schinifolium* seeds by the UPLC-MS/MS MRM method.

Funding: This research was funded by the R&D Program for Forest Science Technology (Project Nos. 2016101C10-1919-AB02, NAN1614220, and D18260) and Korea Institute of Oriental Medicine (Project Nos. KSN2022310 and KSN1823311).

Data Availability Statement: All data supporting the present study can be found in this article.

Conflicts of Interest: The authors declare no conflict of interest.

References

1. Oh, M.; Chung, M.S. Effects of oils and essential oils from seeds of *Zanthoxylum schinifolium* against foodborne viral surrogates. *Evid. Based Complement. Alternat. Med.* **2014**, *2014*, 135797. [CrossRef] [PubMed]
2. Li, R.; Yang, J.; Shi, Y.; Zhao, M.; Ji, K.; Zhang, P.; Xu, Y.; Hu, H. Chemical composition, antimicrobial and anti-inflammatory activities of the essential oil from Maqian (*Zanthoxylum myriacanthum* var. *pubescens*) in Xishuanghanna, SW China. *J. Ethnopharmacol.* **2014**, *158*, 43–48. [CrossRef] [PubMed]
3. Zhao, Z.F.; Zhu, R.X.; Zhong, K.; He, Q.; Luo, A.M.; Gao, H. Characterization and comparison of the pungent components in commercial *Zanthoxylum bungeanum* oil and *Zanthoxylum schinifolium* oil. *J. Food Sci.* **2013**, *78*, C1516–C1522. [CrossRef] [PubMed]
4. Chun, I.K. *The Korean Pharmacopoeia*, 12th ed.; The Research Foundation for Pharmaceutical Quality: Seoul, Republic of Korea, 2019; p. 2301.
5. Wang, K.; Meng, X.H.; Chai, T.; Wang, C.B.; Sang, C.Y.; Wang, W.F.; Shang, X.Y.; Yang, J.L. Chemical constituents from the fruits of *Zanthoxylum bungeanum* and their chemotaxonomic significance. *Biochem. Syst. Ecol.* **2021**, *99*, 104356. [CrossRef]
6. Negi, J.S.; Bisht, V.K.; Bhandari, A.K.; Singh, P.; Sundriyal, R.C. Chemical constituents and biological activities of the genus *Zanthoxylum*: A review. *Afr. J. Pure Appl. Chem.* **2011**, *5*, 412–416.
7. Li, W.; Yang, S.Y.; Yan, X.T.; Sun, Y.N.; Song, S.B.; Kang, H.K.; Kim, Y.H. NF-kB inhibitory activities of glycosides and alkaloids from *Zanthoxylum schinifolium* stems. *Chem. Pharm. Bull.* **2014**, *62*, 196–202. [CrossRef] [PubMed]
8. Wang, C.F.; Fan, L.; Tian, M.; Qi, X.S.; Liu, J.X.; Feng, J.B.; Du, S.S.; Su, X.; Wang, Y.Y. Radiosensitizing effect of schinifoline from *Zanthoxylum schinifolium* Sieb et Zucc on human non-small cell lung cancer A549 cells: A preliminary in vitro investigation. *Molecules* **2014**, *19*, 20128–20138. [CrossRef] [PubMed]
9. Wu, Y.; Wu, H. Analgesia synergism of essential oil from pericarp of Zanthoxylum schinifolium and verapamil. *Evid. Based Complement. Alternat. Med.* **2014**, *2014*, 505876. [CrossRef] [PubMed]
10. Lee, S.J.; Lim, K.T. Glycoprotein of *Zanthoxylum piperitum* DC has a hepatoprotective effect via anti-oxidative character in vivo and in vitro. *Toxicol. Vitr.* **2008**, *22*, 376–385. [CrossRef] [PubMed]
11. Choi, H.J. Evaluation of antiviral activity of *Zanthoxylum* species against picornaviruses. *Osong Public Health Res. Perspect.* **2016**, *7*, 400–403. [CrossRef] [PubMed]
12. Fang, Z.; Jun, D.Y.; Kim, Y.H.; Min, B.S.; Kim, A.K.; Woo, M.H. Cytotoxic constituents from the leaves of *Zanthoxylum schinifolium*. *Bull. Korean Chem. Soc.* **2010**, *31*, 1081–1084. [CrossRef]
13. Liu, S.L.; Wei, L.X.; Wang, D.; Gao, C.Y. Studies on the chemical constituents from the peel of *Zanthoxylum schinifolium* Sieb et Zucc. *Acta Pharm. Sin.* **1991**, *26*, 836–840.
14. Liang, Y.; Xie, L.; Liu, K.; Cao, Y.; Dai, X.; Wang, X.; Lu, J.; Zhang, X.; Li, X. Bergapten: A review of its pharmacology, pharmacokinetics, and toxicity. *Phytother. Res.* **2021**, *35*, 6131–6147. [CrossRef] [PubMed]
15. Lu, L.; Li, Z.; Shan, C.; Ma, S.; Nie, W.; Wang, H.; Chen, G.; Li, S.; Shu, C. Whole transcriptome analysis of schinifoline treatment in *Caenorhabditis elegans* infected with *Candida albicans*. *Mol. Immunol.* **2021**, *135*, 312–319. [CrossRef] [PubMed]
16. Lehr, G.J.; Barry, T.L.; Franolic, J.D.; Petzinger, G.; Scheiner, P. LC determination of impurities in methoxsalen drug substance: Isolation and identification of isopimpinellin as a major impurity by atmospheric pressure chemical ionization LC/MS and NMR. *J. Pharm. Biomed. Anal.* **2003**, *33*, 627–637. [CrossRef] [PubMed]

17. Tang, D.Q.; Zheng, X.X.; Chen, X.; Yang, D.Z.; Du, Q. Quantitative and qualitative analysis of common peaks in chemical fingerprint of Yuanhu Zhitong tablet by HPLC-DAD–MS/MS. *J. Pharm. Anal.* **2014**, *4*, 96–106. [CrossRef] [PubMed]
18. Wang, X.X.; Zan, K.; Shi, S.P.; Zeng, K.W.; Jiang, Y.; Guan, Y.; Xiao, C.L.; Gao, H.Y.; Wu, L.J.; Tu, P.F. Quinolone alkaloids with antibacterial and cytotoxic activities from the fruits of *Evodia rutaecarpa*. *Fitoterapia* **2013**, *89*, 1–7. [CrossRef] [PubMed]
19. Su, C.H.; Cheng, Y.C.; Chang, Y.C.; Kung, T.H.; Chen, Y.L.; Lai, K.H.; Hsieh, H.L.; Chen, C.Y.; Hwang, T.L.; Yang, Y.L. Untargeted LC-MS/MS-based multi-informative molecular networking for targeting the antiproliferative ingredients in *Tetradium ruticarpum* fruit. *Molecules* **2022**, *27*, 4462. [CrossRef] [PubMed]
20. The Plant List. Available online: www.theplantlist.org (accessed on 3 January 2023).

foods MDPI

Article

Simultaneous Determination of Neonicotinoid and Carbamate Pesticides in Freeze-Dried Cabbage by Modified QuEChERS and Ultra-Performance Liquid Chromatography–Tandem Mass Spectrometry

Bingxin Yang [1,2], Sheng Wang [1], Wen Ma [3], Guanlin Li [4], Mengling Tu [1], Zhiyong Ma [2,*], Qinghe Zhang [1], Hongmei Li [1] and Xianjiang Li [1,*]

1 Key Laboratory of Chemical Metrology and Applications on Nutrition and Health for State Market Regulation, Division of Metrology in Chemistry, National Institute of Metrology, Beijing 100029, China
2 Beijing Advanced Innovation Center for Soft Matter Science and Engineering, State Key Laboratory of Organic-Inorganic Composites, College of Chemical Engineering, Beijing University of Chemical Technology, Beijing 10029, China
3 State Key Laboratory of Natural and Biomimetic Drugs, School of Pharmaceutical Sciences, Peking University, Beijing 100191, China
4 College of Plant Protection, Shandong Agricultural University, Tai'an 271018, China
* Correspondence: mazhy@mail.buct.edu.cn (Z.M.); lixianjiang@nim.ac.cn (X.L.)

Abstract: Dehydrated vegetables are popular in instant foods, but few reports have focused on their pesticide residues. This research developed and validated a modified QuEChERS method combined with ultra-performance liquid chromatography–tandem mass spectrometry to determine 19 kinds of neonicotinoid and carbamate pesticides in freeze-dried cabbage. Herein, acetonitrile/water (v/v = 2:1) was selected in the extraction step. Meanwhile, 4 g anhydrous magnesium sulfate and 1 g sodium chloride were applied to the partitioning step. Dispersive solid-phase extraction sorbents were selected, and liquid chromatography conditions were further optimized for dealing with the matrix effect. The limits of quantification ranged from 1.0 to 10.0 μg/kg. The validation results were acceptable, with average recoveries of 78.7–114.0% and relative standard deviations below 14.2%. The method recoveries were closely related to the volume proportion of water in the extractant. Finally, the developed method was applied to real freeze-dried cabbages and four pesticides (propamocarb, imidacloprid, acetamiprid, and thiacloprid) were detected in six samples.

Keywords: carbamate; freeze-dried cabbage; modified QuEChERS; neonicotinoid; ultra-performance liquid chromatography–tandem mass spectrometry

Citation: Yang, B.; Wang, S.; Ma, W.; Li, G.; Tu, M.; Ma, Z.; Zhang, Q.; Li, H.; Li, X. Simultaneous Determination of Neonicotinoid and Carbamate Pesticides in Freeze-Dried Cabbage by Modified QuEChERS and Ultra-Performance Liquid Chromatography–Tandem Mass Spectrometry. *Foods* **2023**, *12*, 699. https://doi.org/10.3390/foods12040699

Academic Editors: Chao Kang and Ronald Beckett

Received: 31 December 2022
Revised: 30 January 2023
Accepted: 1 February 2023
Published: 6 February 2023

1. Introduction

Fresh vegetables generally have a short shelf life and are not fit for long-term storage, which is a concern regarding long-term missions. Substantial studies have demonstrated that a drying process can restrain microbial growth and prolong storage time [1,2]. Among all drying processes, freeze-drying has been considered an efficient method to preserve nutrient compositions and antioxidant activities [3]. Recently, with promotion and improvement by the National Aeronautics and Space Administration, many freeze-dried (FD) ingredients have been introduced to the food market [4]. Cabbage is one of the most consumed vegetables in Asia and is very promising as a dehydrated vegetable due to its high nutrient value. Applying the freeze-drying process to cabbage has been dedicated to producing space shuttle goods, extreme-sport foodstuffs, and certified reference materials for multi-residue pesticide analysis [5,6].

In previous studies, occurrence of neonicotinoid (NEO) and carbamate (CBM) pesticides in fresh cabbage raised significant health concerns [7–9]. When NEOs bind to the α4β2 subtype of neuronal nicotinic acetylcholine receptors in humans, they may cause severe neurotoxicity, including tetralogy of Fallot, congenital anencephaly, autism spectrum disorder, memory loss, and finger tremor [10,11]. CBMs are reversible inhibitors of acetylcholine esterase enzymes, leading to abnormal function of nerve synapses and neuromuscular junctions, which will cause major risks to mammals. CBMs may cause impairments in the human immune system, and long-term exposure significantly increases risk of non-Hodgkin's lymphoma [12]. Freeze-drying is carried out at a low temperature; thus, pesticide residues may be retained in FD cabbage. Additionally, a pesticide would persist for quite a long time due to the low moisture content.

Because of those adverse effects on human health, many countries and organizations have set maximum residue limits (MRLs) for NEOs and CBMs in cabbage to guarantee consumer safety. For example, the European Union (EU) established MRLs for NEOs and CBMs ranging from 0.002 to 20.0 mg/kg [13]. In China, the MRLs ranged from 0.02 to 10.0 mg/kg [14], and the United States of America had MRLs below 21.0 mg/kg [15]. See attached Table S1 for more detailed information. Until now, MRLs have not yet been established for these pesticides in FD cabbage by the above-mentioned countries and organization.

So far as we know, there is no report about NEOs and CBMs determination methods in FD cabbage. Accordingly, a robust analytical method should be established to guarantee their safety. As one of the most common sample preparation methods for pesticides, QuEChERS (quick, easy, cheap, effective, rugged, and safe) can be highlighted owing to its inherent properties, including few steps, low organic solvent consumption, suitability for multi-pesticide residue analysis, and so forth [16]. Originally developed for analyzing pesticides in fresh fruits and vegetables, QuEChERS is also applied to dry commodities, with further development in reconstitution and clean-up procedures [17,18]. For multi-pesticides analysis, the target analytes cover a wide range of polarity [19]. Thus, they tend to exhibit partitioning differences in the aqueous and acetonitrile (ACN) phases. Furthermore, due to the discrepancy between fresh and FD cabbage in terms of water content, volatile compounds, and other matrix interference [3], it is necessary to precisely optimize the sample preparation method of NEOs and CBMs in FD cabbage in the extraction and clean-up procedure.

In the past two decades, several determination techniques have been published for detection of NEOs or CBMs in food of plant origin, such as gas chromatography–mass spectrometry (GC–MS) [20], high-performance liquid chromatography-diode-array detection (HPLC-DAD) [21], and ultra-performance liquid chromatography–tandem mass spectrometry (UPLC–MS/MS/MS) [22]. In view of its short separation time [23], increased peak capacity, and excellent sensitivity [24,25], UPLC–MS/MS/MS is very attractive to detect pesticide residues in FD cabbage.

The objective of the study was to develop and validate a high-throughput, sensitive, and robust method for determining 19 kinds of NEO and CBM pesticides in FD cabbage. A modified QuEChERS method combined with UPLC–MS/MS/MS was established for this purpose. For QuEChERS, the modifications were optimized with the goal of increasing recovery and removing matrix interference. Validation and applicability of both the extraction and determination methods were also developed and finally applied to real samples. To the best of our knowledge, this was the first method for simultaneous determination of NEOs and CBMs in FD cabbage.

2. Materials and Methods

2.1. Reagents and Materials

Standards of dinotefuran (99.47% purity), nitenpyram (98.84% purity), cycloxaprid (92.70% purity), thiamethoxam (99.65% purity), clothianidin (99.12% purity), imidacloprid (98.55% purity), acetamiprid (99.78% purity), imidaclothiz (97.46% purity), and thiacloprid (99.68% purity) were obtained from Dr. Ehrenstorfer (Augsburg, Germany). Standards of propamocarb (99.30% purity), oxamyl (99.40% purity), pirimicarb (99.50% purity), aldicarb

(97.00% purity), metolcarb (99.80% purity), propoxur (99.50% purity), carbofuran (99.70% purity), carbaryl (99.80% purity), isoprocarb (99.80% purity), and promecarb (99.80% purity) were obtained from Alta scientific Co., Ltd. (Tianjin, China). Chemical structures of the target NEO and CBM pesticides were exhibited in Figure S1.

ACN of HPLC grade was purchased from Merck KGaA (Darmstadt, Germany). For sample extraction, anhydrous sodium sulfate (Na_2SO_4) was purchased from FUCHEN Chemical Reagents (Tianjin, China); sodium chloride (NaCl) was purchased from Sinopharm Chemical Reagent Co., Ltd. (Shanghai, China). The QuEChERS CEN version extraction tube (containing 4 g of anhydrous magnesium sulfate ($MgSO_4$), 1 g NaCl, 1.0 g trisodium citrate dihydrate, and 0.5 g disodium hydrogen citrate sesquihydrate), AOAC version extraction tube (containing 6 g anhydrous $MgSO_4$ and 1.5 g sodium acetate), and original version extraction tube (containing 4 g anhydrous $MgSO_4$ and 1.0 g NaCl) were purchased from Waters Technologies (Milford, MA, USA). HPLC grade acetic acid was obtained from Innochem Company (Beijing, China). For sample purification, the QuEChERS dispersive kit (anhydrous $MgSO_4^+$ primary secondary amine (PSA)) was purchased from Agilent Technologies (Folsom, CA, USA); graphitized carbon black (GCB) was purchased from Agilent Technologies (Folsom, CA, USA). Carboxylated multi-walled carbon nanotubes (c-MWCNTs, 5–15 nm in outer diameter, 10–30 μm in length, 3.86% carboxyl content) and MWCNTs (5–15 nm in outer diameter, 0.5–2 μm in length) were purchased from Xianfeng Technology Co., Ltd. (Nanjing, China). For the mobile phase, LC–MS grade formic acid and ammonium formate were purchased from Honeywell (Shanghai, China). Purified water was purchased from Wahaha Group Co., Ltd. (Hebei, China). For the syringe filters, hydrophilic polytetrafluoroethylene (PTFE) with a pore size of 0.22 μm was manufactured by ANPEL Laboratory Technologies (Shanghai, China).

2.2. Pesticide-Free FD Cabbage Preparation

About 20 kg of organic-certified cabbages was purchased from Beicaiyuan Agriculture Technology Co., Ltd. (Beijing, China). After removing the roots and decomposed leaves, the samples were chopped into strips, followed by further homogenization with a food processor (VAKNOA QUALITAT, Germany) at room temperature for 5 min. Thereafter, the cabbage paste was collected and placed into a freeze dryer (VirTis Genesis/35 L Genesis Super XL-70, USA). The temperature and vacuum pressure in the drying chamber were set at −25 °C and 200 mTorr, respectively. The process took 48 h, with the stainless-steel plate temperature gradually raised to 25 °C. Finally, the pesticide-free FD cabbage samples were ground using a medicine pulverizer (Xulang, Guangzhou, China) and stored in a deep freezer at −80 °C until use.

2.3. Sample Fortification

Pesticide-free FD cabbage was collected for method validation. After homogenization, blank sample was fortified with the standard solution at the concentrations of 1.0, 10.0, 100.0, and 500.0 μg/kg. Then, the samples were allowed to equilibrate for over 30 min prior to further pretreatment.

2.4. Real Sample Analysis

Seven FD cabbage samples were obtained from the market in Jiangsu, and each sample was ground as before. The FD cabbage was prepared with a modified QuEChERS method as follows: first, the sample was taken out from the refrigerator and restored to room temperature; 1.0 g sample was weighed into a 50 mL centrifuge tube. Afterwards, 5 mL purified water and 10 mL ACN were added for liquid phase extraction, and the mixture was agitated with multi-tube vortex mixer (Lumiere., Beijing, China) for 3 min at 2500 rpm; for phase separation, 4 g of anhydrous $MgSO_4$ and 1 g NaCl were added to the tube and agitated with multi-tube vortex mixer for 1 min at 2500 rpm; the tube was then centrifuged for 5 min at 8000 rpm (3802 g) at 4 °C; for the dispersive solid-phase extraction (d-SPE) that followed, an aliquot of 2 mL of supernatant organic layer was collected into the purification tube, which contained 300 mg anhydrous $MgSO_4$, 50 mg PSA, and 20 mg GCB. The tube was agitated with multi-tube vortex mixer for 3 min at 2500 rpm and centrifuged for 5 min at 8000 rpm (3802 g) at 4 °C; finally, the upper layer (0.5 mL) was collected and mixed with 1 mL mobile phase A before filtration with PTFE membrane filter and transferred for UPLC–MS/MS/MS analysis.

2.5. Instrument and Apparatus

A UPLC–MS/MS/MS (Waters Technologies) was used for the instrumental analysis of the extract. ACQUITY UPLC® was used for LC separation and coupled to a triple quadrupole MS (TQ-S, Manchester, UK) equipped with orthogonal Z-spray electrospray ionization (ESI) interface. MassLynx 4.1 software with QuanLynx program was used for data acquisition and analysis. For chromatographic separation, an HSS T_3 column (1.8 μm particle size; 2.1 mm × 100 mm, Waters) was utilized, and the column temperature was maintained at 30 °C. The mobile phases were composed of a mixture of 0.1% formic acid, 5 mM ammonium formate in water (phase A), and ACN (phase B), and the flow rate was set at 0.3 mL/min. For the gradient elution, the initial eluent composition was 10% phase B, and phase B reached 50% at 3 min, progressively increasing to 80% at 5 min. Finally, it was lowered to 10% at 6 min and maintained for another 2 min for equilibration. The total time of a single run was 8.0 min, and 4.5 μL of the sample was injected for quantification.

The MS/MS was operated in positive ionization mode, and data were acquired in the multiple reaction monitoring (MRM) mode with two transitions per pesticide. The transition with the highest intensity was utilized for quantitation and the other was utilized for confirmation. Capillary voltage and source offset voltage were 3.0 kV and 55 V, respectively; desolvation and source temperatures were 500 and 150 °C, respectively; cone and desolvation N_2 gas flows were 150 and 1000 L/h, respectively. Table 1 listed the retention time (R_t), cone voltage (CV), precursor ion, product ion, and collision energy (CE) for the 19 kinds of pesticides in the UPLC–MS/MS/MS method.

Table 1. Parameters of MS/MS for NEO and CBM pesticides.

Analytes	R_t [(a)] (min)	CV [(b)] (V)	Precursor Ion (m/z)	Product Ion 1 (m/z)	CE [(c)]-1 (V)	Product Ion 2 (m/z)	CE-2 (V)
Propamocarb	1.71	35	189.0	102.0	17	144.0	10
Dinotefuran	1.85	30	203.0	129.0	10	113.3	10
Oxamyl	2.15	65	242.0	72.0	10	121.0	10
Nitenpyram	2.18	30	271.0	56.1	24	126.0	32
Cycloxaprid	2.56	30	323.0	125.8	36	150.9	22
Thiamethoxam	2.58	25	291.9	210.9	12	132.0	22
Clothianidin	2.89	25	250.1	168.9	12	132.0	14
Imidacloprid	3.04	35	255.9	175.0	20	209.0	20
Imidaclothiz	3.17	30	262.0	180.8	15	122.0	26
Acetamiprid	3.17	45	223.0	125.9	18	55.9	12
Pirimicarb	3.36	40	239.0	72.0	20	182.0	16
Thiacloprid	3.58	45	253.0	125.9	20	217.0	12
Aldicarb	3.69	35	213.0	89.0	16	116.0	12
Metolcarb	3.96	25	164.0	109.0	12	94.0	26
Propoxur	4.21	5	210.0	111.0	15	168.0	5
Carbofuran	4.27	30	222.0	123.0	22	165.0	12
Carbaryl	4.43	25	202.0	127.0	25	145.0	10
Isoprocarb	4.72	30	194.0	95.0	15	137.0	9
Promecarb	5.25	30	208.0	109.0	17	151.0	9

[(a)] retention time; [(b)] cone voltage; [(c)] collision energy.

2.6. Method Validation

Validation of the developed method was carried out according to the SANTE/11312/2021 guidance document for determination of pesticide residues [26]. The validation included selectivity, limit of quantification (LOQ), limit of detection (LOD), linearity, matrix effect

(ME), recovery, repeatability, and stability of the final extract. Blank extract from pesticide-free FD cabbage (prepared in 2.2) was used to evaluate method selectivity. Matrix-matched standard solutions were used for determining LOD, LOQ, and linearity. LOQ was determined as the concentration of signal-to-noise ratios larger than 10; LOD was defined as the lowest detectable concentration that was determined as the concentration of signal-to-noise ratios larger than 3. According to the prescription of SANTE/11312/2021, the linear range consisted of a minimum of five calibration points for each pesticide. MEs were calculated by flowing Equation (1): a positive value indicated matrix enhancement effects, while a negative value indicated the opposite. In addition, MEs were considered significant if they exceeded ±20%.

$$\text{MEs} = (\text{slope of matrix -matched standard/slope of solvent standard} - 1) \times 100\% \quad (1)$$

Recovery experiments were performed in five replicates at four spiked fortification levels of 1.0, 10.0, 100.0, and 500.0 μg/kg (taking the LOQ, 10 times LOQ, or the MRLs for each pesticide as reference values). Repeatability was termed as the relative standard deviation (RSD) calculated with samples prepared and analyzed in one day or over several days. The final extract stability was evaluated at 100 μg/kg after (1) storage at room temperature in the autosampler tray for 24 h; (2) storage in a refrigerator (4 °C for three days and seven days, respectively).

2.7. *Post-Column Infusion*

Post-column infusion system could directly reflect MEs occurring during a whole chromatographic run [27]. In our study, two sets of experiments were performed with pure ACN as a "reference" and the blank sample extract. Meanwhile, a pesticide infusion mixture (100 μg/kg) was permanently infused to the LC effluent after column with a constant flow (10 μL/min) via a T-piece in Intellistart^{TM} system. The intensity of each MRM transition was recorded for every 0.36 s; thus, 1383 data points were acquired during the total run time for each pesticide. By calculations according to Equation (2), the matrix profile was obtained from the two infusion profiles.

$$ME_i = [SI_i\ (\text{sample extract})/SI_i\ (\text{reference}) - 1] \times 100\% \quad (2)$$

Here, SI represented the signal intensity.

If no MEs occurred, the two infusion profiles were identical and MEs were close to zero. If matrix interference influenced the ionization efficiency of target analytes, the infusion profiles of the sample extract would differ from the reference.

3. Results and Discussion

3.1. *Optimization of the QuEChERS Method*

QuEChERS is a simple yet flexible method with many commercial reagents and sorbents for extraction and clean-up. To achieve accurate results, the QuEChERS parameters were optimized with the goal of increasing recovery, lowering RSD, and minimizing ME. The initial conditions used to perform the optimization were as follows: 1.0 g FD cabbage sample with a spiking level of 100 μg/kg, the extraction solvent volume was 20 mL (10 mL purified water and 10 mL ACN), and the extraction time was 3 min. The phase partitioning salts were from CEN version, and the d-SPE sorbents contained 300 mg anhydrous $MgSO_4$, 50 mg PSA, and 20 mg GCB.

3.1.1. Sample Extraction

For FD cabbage, addition of water was recommended, which swelled the pores and allowed the ACN access into sample tissue [28]. Figure 1A showed that an ACN/water mixture provided consistently higher extraction efficiency for all analytes, while extraction with ACN was only 4.2–14.6% of the former performance. Mastovska et al. found that use of ACN/water for single extraction showed comparable extraction efficiency with ACN extraction after water reconstitution [29]. Therefore, an ACN/water mixture was chosen for FD cabbage.

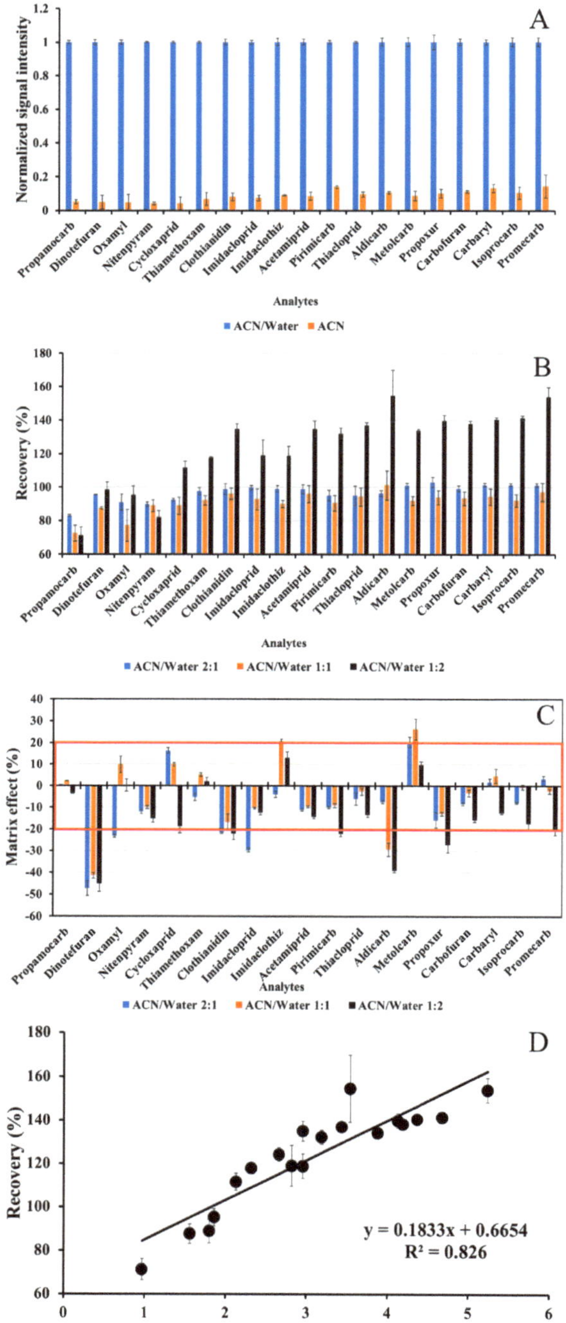

Figure 1. Effect of extraction parameters: (**A**) extraction efficiencies in different extraction solvents; (**B**) recoveries and (**C**) MEs in different volume proportions of ACN/water (the MEs in the red wireframe were negligible); (**D**) linear correlation between Rt and recoveries (v/v (ACN/water) = 1:2).

Moreover, recoveries of analytes may decrease and co-extract interference may increase with different water content in the extraction solvent [30,31]. Therefore, the volume of ACN was maintained at 10 mL while the volume proportion of ACN/water was further compared (*v/v* = 2:1, 1:1, 1:2, respectively). The experimental results shown in Figure 1B,C indicated that acceptable results were obtained in the range of 2:1–1:1 (*v/v*) ACN/water since there was no significant difference in recoveries and MEs. However, it was noteworthy that two considerable biases in recoveries were found for analytes when the proportion of ACN/water was 1:2 (*v/v*). In detail, negative bias for nitenpyram and propamocarb was mainly due to their hydrophilicity, while positive bias was found for relatively hydrophobic analytes (aldicarb, metolcarb, propoxur, carbofuran, carbaryl, isoprocarb, and promecarb). As a result, the recoveries of the target analytes in the ACN/water mixture (*v/v* = 1:2) exhibited a strong linear correlation with the R_t (R^2 = 0.826 in Figure 1D). The results indicated that an aqueous phase beyond the appropriate range would pry the "levers" of consistency in analyte recoveries. Finally, the ACN/water mixture (*v/v* = 2:1) was selected as the appropriate extraction solvent.

Inadequate extraction may lead to low recovery; then, it is imperative to optimize extraction time [32]. Thus, equilibrium extraction time was investigated by increasing the time from 1 min to 10 min. As shown in Figures S2 and S3, there was no significant difference in recoveries and MEs for most analytes, but the recoveries for four NEOs (nitenpyram, dinotefuran, cycloxaprid, and thiamethoxam) were moderately increased with the extraction time increasing from 1 min to 3 min. Therefore, 3 min was selected as the extraction time. The results indicated that extraction of target analytes in FD cabbage mainly depended on the volume proportion between the ACN and aqueous phases and extraction time was not a very critical factor.

3.1.2. Liquid–Liquid Phase Partitioning

After sample extraction, the ACN extract needed to be separated from the sample. This liquid–liquid phase partitioning was accomplished by adding partitioning salts. The partitioning salts not only control the polarity difference between the ACN and aqueous phases but also influence the pH of the sample environment, thus strongly influencing recoveries of target analytes [31]. Herein, the experiment was evaluated with two buffered partitioning salts (CEN version and AOAC version) and three unbuffered partitioning salts (original version, 4 g anhydrous Na_2SO_4+ 1 g NaCl, 3 g NaCl, respectively). As shown in Figure 2A, anhydrous $MgSO_4$ provided higher recoveries for propamocarb and nitenpyram, indicating that the recoveries of relative polar analytes primarily correlate with the water content in the ACN phase. In addition, the lower solubility of anhydrous Na_2SO_4 in water limited its water binding capacity. Thus, anhydrous $MgSO_4$ was selected as a component in partitioning salts to bind water from the sample.

Meanwhile, by evaluating MEs and matrix profiles, CEN version buffer salts had significant matrix interference on cycloxaprid and imidacloprid, as shown in Figure 2B, Figures S4 and S5. Judging from these results, the experiment demonstrated that there was no need to add buffer salts. Among them, the original version salts provided the best results in terms of lower MEs and consistently good recoveries.

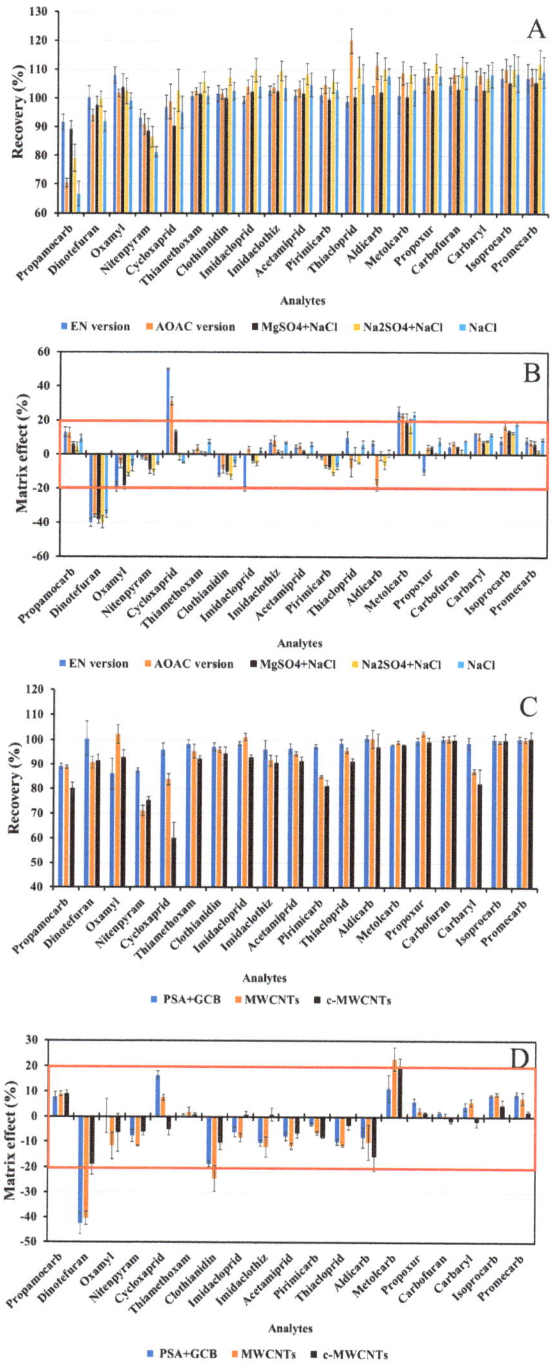

Figure 2. Effect of phase partitioning salts on (**A**) recoveries and (**B**) MEs; effect of d-SPE sorbents on (**C**) recoveries and (**D**) MEs (the MEs in the red wireframe were negligible).

3.1.3. Sample Clean-Up

In this study, the initiative was to keep the clean-up procedure as simple as possible, acquire consistently good recovery, and remove serious matrix interference simultaneously. Our previous study found that the combination of PSA and GCB has commendable clean-up efficiency for co-extracts in fresh cabbage [7]. MWCNTs and c-MWCNTs also demonstrated good clean-up performance for pesticides in cucumber, apple, and orange [33]. As shown in Figure 2C,D, when FD cabbage extracts were cleaned with c-MWCNTs, the MEs for dinotefuran, cycloxaprid, and clothianidin were decreased by 23.8%, 11.4%, and 8.2%, respectively. However, non-specific adsorption also occurred and the recovery for cycloxaprid fell to 60.2%. Compared with CBMs (except for carbaryl), pharmacophores (–NO_2 and –CN) in nine NEOs were electronegative, and carboxyl groups in the c-MWCNTs could form strong hydrogen bond interaction and increased the additional adsorption to NEOs. This was why PSA and GCB turned out to be generally applicable d-SPE sorbents for FD cabbage. Then, further optimization of the sample clean-up was not conducted; the study attempted to find better LC separation conditions to separate dinotefuran from co-eluting matrix interference in the following.

3.2. Optimization of UPLC–MS/MS/MS Conditions

In order to find better LC separation conditions to eliminate the ME of dinotefuran, the post-column infusion system was adopted. In our former work [7], the LC elution started with 15% phase B, reached 50% phase B within 3 min, 80% phase B in 5 min, and then dropped to the initial state within 1 min. Under this condition, most target analytes were well separated from each other. However, Figure 3A(a) showed that the signal of dinotefuran was significantly suppressed by the co-eluting matrix. After optimizing LC elution conditions, the proportion of phase B in the initial mobile phase decreased from 15% to 10%. Figure 3A(b,c) showed ME profiles of blank FD cabbage samples with optimized LC conditions. The results demonstrated that dinotefuran gradually separated from the matrix-suppressing zone as the proportion of phase B decreased. Therefore, the ME of dinotefuran in FD cabbage was eliminated with the optimized LC conditions and has no significant influence on other analytes (Figure 3B).

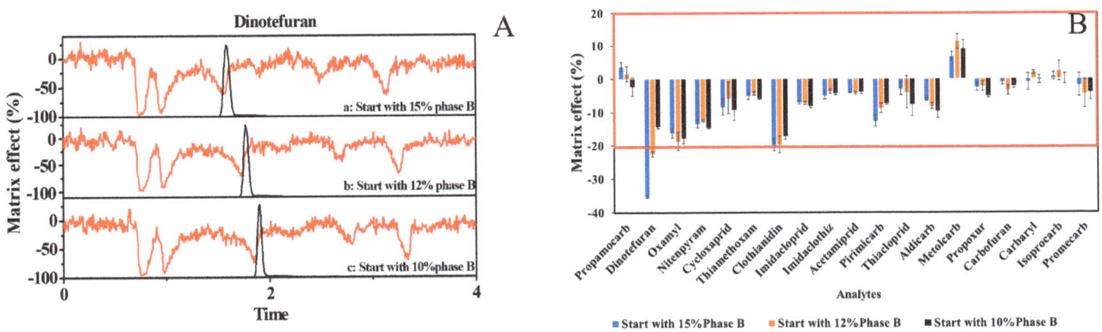

Figure 3. (**A**) Matrix profiles in different type LC methods from a post-column infusion of dinotefuran (a, b, c represented the gradient of the LC method); (**B**) MEs in different type LC method (the MEs in the red wireframe were negligible).

To improve the signal intensity of target analytes, mobile phase modifiers, such as ammonium acetate and formic acid in purified water, were discussed. Table S2 showed that adding ammonium formate and formic acid resulted in a 1.2- to 12.2-fold higher response than individually using formic acid for most analytes, except for oxamyl and aldicarb. The reasons why mobile phase modifiers influenced protonation of the target analytes may be ascribed to the following theory. Proton transfer was a fundamental chemical reaction occurring in both solution and the gas phase (Equations (3) and (4)) [34,35]. Adding a small mount of formic acid and ammonium formate can push reaction 3 and 4 and promote protonation of the analytes. Interestingly, oxamyl and aldicarb did not form a stable protonated molecule $[M + H]^+$ but formed a very stable sodiated ion $[M + Na]^+$ in formic-acid-containing ACN solution. Consequently, the observed suppression of Na adduct ions by ammonium formate was likely due to competition between Equations (4) and (5) in the gas phase [36]. Moreover, Figure S6 showed that adding extra ammonium formate produced a symmetric peak shape compared with adding only formic acid for nitenpyram. Eventually, purified water spiked with 0.1% formic acid and 5 mM ammonium formate was selected as phase A.

$$M\ (l) + H_3O^+\ (l) \rightarrow [M + H]^+\ (l) + H_2O\ (l) \tag{3}$$

$$M\ (g) + NH_4{}^+\ (g) \rightarrow [M + H]^+\ (g) + NH_3\ (g) \tag{4}$$

$$ACN{\bullet}Na^+\ (g) + M\ (g) \rightarrow [M + Na]^+\ (g) + ACN\ (g) \tag{5}$$

MS/MS conditions were optimized for unambiguous identification and accurate quantification of 19 kinds of NEO and CBM pesticides at trace levels. The MS parameters were optimized by directly injecting the standard solution. One precursor ion and the two most abundant product ions were selected for each compound as confirmation ions. Detailed parameters were shown in Table 1.

In preliminary experiments, a 1.5 µL injection volume was used to test the mobile phase composition, but a high proportion of organic solvent was apt to produce obvious solvent effect for propamocarb. The most common approach to prevent solvent effect is solvent replacement with the initial mobile phase. However, the concentration process led to instability of nitenpyram and quantitative errors. Therefore, directly diluting the final extract with phase A was attempted. Figure S7 demonstrated that solvent effects could be significantly lowered with two times dilution with phase A. Hence, two times dilution with phase A was selected and the injection volume was changed to 4.5 µL accordingly. QuEChERS is a simple yet flexible method, with many commercial reagents and sorbents for extraction and clean-up. To achieve accurate results, the QuEChERS parameters were optimized with the goal of increasing recovery, lowering RSD, and minimizing ME. The initial conditions used to perform the optimization were as follows: 1.0 g FD cabbage sample with a spiking level of 100 µg/kg, the extraction solvent volume was 20 mL (10 mL purified water and 10 mL ACN), and the extraction time was 3 min; the phase partitioning salts were from CEN version and d-SPE sorbents contained 300 mg anhydrous $MgSO_4$, 50 mg PSA, and 20 mg GCB.

3.3. Method Validation

3.3.1. LODs, LOQs, and Linearity

The LODs, LOQs, and linearity were experimentally determined by fortification of pesticide-free FD cabbage samples. Table 2 presented the LODs and LOQs for the NEO and CBM pesticides. Low LODs and LOQs were achieved in the range of 0.2 to 4.0 µg/kg and 1.0 to 10.0 µg/kg, respectively. The linearity of the method was evaluated by linear regression analysis of matrix-matched calibration curves. All calibration curves were constructed at the fortification levels of 1.0, 2.0, 5.0, 10.0, 50.0, 100.0, 200.0, 500.0, and 1000 µg/kg in pesticide-free FD cabbage samples. Table 2 showed that the method had good linearity, with satisfactory correlation coefficients ($R^2 > 0.9990$) for all target analytes.

3.3.2. Assay Selectivity

Pesticide-free FD cabbage samples were analyzed to evaluate the selectivity of the established method. No interfering endogenous peaks appeared at the R_t of the target analytes. The extracted MRM ion chromatograms of the FD sample with fortified NEOs and CBMs mixture at LOQ concentration levels were shown in Figure S8. The chromatogram demonstrated the selectivity and good chromatographic characteristics of the proposed method at a low fortification level.

Table 2. Linear range, coefficients (R^2), LODs, LOQs, MEs, average recoveries, RSD_r (n = 5), and RSD_R (n = 3) of NEO and CBM pesticides.

Analytes	Linear Range (μg/kg)	Regression Equation	R^2	LOD (μg/kg)	LOQ (μg/kg)	ME (%)	Fortification (μg/kg)	Recovery (%)	RSD_r (%)	RSD_R (%)
							1.0	92.2	6.9	1.8
Propamocarb	1.0–500.0	y = 11,641x − 21,927	0.9998	0.2	1.0	−3.4	10.0	88.7	4.5	7.0
							100.0	86.3	1.7	2.2
							10.0	97.9	5.7	3.7
Dinotefuran	10.0–500.0	y = 1026.8x − 258.23	0.9994	4.0	10.0	−14.2	100.0	97.7	1.5	2.1
							500.0	96.7	1.1	1.9
Oxamyl	10.0–500.0					−14.8	10.0	104.0	7.8	8.3
Oxamyl	10.0–500.0	y = 423.49x − 1011.8	0.9997	4.0	10.0	−14.8	100.0	98.6	3.5	2.6
							500.0	97.2	1.3	3.9
							10.0	83.7	6.8	11.2
Nitenpyram	5.0–500.0	y = 1710.8x − 3614.8	0.9997	2.0	5.0	−3.3	100.0	86.3	1.6	3.1
							500.0	85.4	1.4	3.6
							1.0	78.7	4.9	5.3
Cycloxaprid	1.0–100.0	y = 31,584x + 25,868	0.9993	0.2	1.0	16.1	10.0	81.6	3.9	5.9
							100.0	82.1	1.0	5.3
							1.0	99.4	7.1	4.8
Thiamethoxam	1.0–500.0	y = 3048.3x + 21,677	0.9998	0.2	1.0	2.8	10.0	98.2	3.3	1.9
							100.0	99.1	1.5	1.2
							10.0	93.8	7.3	6.3
Clothianidin	1.0–500.0	y = 787.1x + 1156.7	0.9998	4.0	10.0	−16.5	100.0	96.3	2.4	4.0
							500.0	97.8	2.9	2.7
							1.0	97.7	4.0	8.6
Imidacloprid	1.0–500.0	y = 1655.5x + 2906.1	0.9990	0.3	1.0	4.8	10.0	97.4	3.5	3.9
							100.0	97.9	0.9	2.2
							10.0	104.4	0.4	4.1
Imidaclothiz	10.0–500.0	y = 1554.4x −4607.2	0.9995	3.0	10.0	9.5	100.0	99.0	2.7	3.0
							500.0	98.7	0.9	2.4
							1.0	92.1	4.0	3.0
Acetamiprid	1.0–200.0	y = 15,870x + 17,493	1.0000	0.3	1.0	6.2	10.0	99.1	1.7	2.4
							100.0	99.6	0.8	0.9
	1.0–500.0		0.9998		1.0	−0.6	1.0	93.4	0.4	8.8
Pirimicarb	1.0–500.0	y = 16,692x + 30,784	0.9998	0.3	1.0	−0.6	10.0	98.7	1.2	0.7
							100.0	97.1	1.3	1.4
							1.0	99.8	1.7	1.0
ThiaclopridTHI	1.0–200.0	y = 22,576x + 21,938	1.00000	0.2	1.0	3.7	10.0	100.5	1.0	2.0
							100.0	99.3	1.4	1.3
							10.0	102.9	4.2	7.3
Aldicarb	10.0–500.0	y = 1200.4x − 3213.9	0.9995	3.0	10.0	−16.1	100.0	99.6	1.7	0.9
							500.0	100.6	0.6	0.9
							10.0	98.6	6.3	14.2
Metolcarb	10.0–1000.0	y = 242.56x − 308.51	0.9992	3.0	10.0	9.4	100.0	96.6	3.3	1.2
							500.0	97.9	3.8	2.3
							10.0	103.6	3.9	1.4
Propoxur	5.0–500.0	y = 1524.9x − 598.33	0.9996	2.0	5.0	−6.5	100.0	99.8	2.5	1.5
							500.0	99.8	0.9	1.8
							1.0	86.5	1.9	4.3
Carbofuran	1.0–200.0	y = 22,528x + 10,821	1.0000	0.3	1.0	3.8	10.0	102.9	2.7	2.8
							100.0	101.2	1.7	0.8
							10.0	105.7	4.4	6.3
Carbaryl	10.0–500.0	y = 1192.2x − 166.83	0.9997	3.0	10.0	−7.4	100.0	96.9	3.1	2.4
							500.0	98.2	1.5	3.7
							10.0	101.9	3.9	3.4
Isoprocarb	10.0–1000.0	y = 2748.7x + 6693.7	0.9996	3.0	10.0	−2.2	100.0	101.4	2.3	2.3
							500.0	101.0	2.5	2.1
							10.0	114.0	2.1	2.0
Promecarb	2.0–1000.0	y = 3294.2x + 13,084	0.9997	0.6	2.0	−9.4	100.0	108.5	2.4	7.1
							500.0	102.8	2.1	2.1

3.3.3. Matrix Effect

Some endogenous components in the FD cabbage might be extracted to the final extract, thereby interfering with ionization of target analytes. Then, this interference would be detrimental to maintenance of the UPLC–MS/MS/MS and affect quantification of the analytical method. According to the SANTE/11312/2021 guidelines, the MEs are considered significant if they exceed ±20%. The results were shown in Table 2. Thus, with values ranging from −16.5% to 16.1%, the MEs were not significant for all analytes. To further diminish MEs, the matrix-matched standard curves were applied to compensate the matrix interference and signal irreproducibility.

3.3.4. Recovery and Repeatability

For determination of recovery and repeatability, two levels of fortification must be considered: (i) a sample fortified with LOQs concentration for each analyte (n = 5); (ii) a sample fortified with 10 times LOQs or MRLs (set or proposed) level. Considering the difference in LOQs of target analytes, the method was validated in five replicates at four fortification concentrations: 1.0, 10.0, 100.0, and 500.0 μg/kg. Table 2 showed that the average recoveries ranged from 78.7% to 114.0%, with intraday RSD (RSD_r) values ≤7.8%. Interday RSD (RSD_R) values were slightly higher (≤14.2%) but lower than 15%, which was an acceptable level.

3.3.5. Stability of Final Extracts

A homogenized blank extract was separated into two aliquots to investigate the stability of the target analytes. Each aliquot was fortified at 100 μg/kg for each pesticide and prepared as in Section 2.4. One aliquot was analyzed at room temperature after one day, and the other was stored at 4 °C and analyzed three and seven days later. Figures S9 and S10 showed that no significant degradation (within the acceptable range of 70–120% [26]) was observed for all analytes in the final extracts for one-day room temperature storage and three-day 4 °C storage, but, evidently, degradation occurred for oxamyl, cycloxaprid, and aldicarb for seven-day 4 °C storage. In summary, the final extract should not be stored for more than three days.

3.4. Real Sample Analysis

In our research, seven batches of FD cabbage samples collected from the market in Jiangsu Province were used to validate the reliability of the developed method. The results were shown in Table 3. Three NEO and one CBM pesticides were detected in six FD cabbage samples. Among them, thiacloprid and propamocarb were the most frequently detected pesticides. In contrast, imidacloprid and acetamiprid were detected in one sample. In China, carbofuran and aldicarb are banned in fruits and vegetables (including cabbage) [37]. Table S3 summarized the publication of the analytical method and the detection frequency of tested pesticides in cabbage [7,8,38–40]. From the above results, the detection rate of NEOs was usually higher than CBMs. This may be ascribed to the heavy use of NEOs in agriculture.

Table 3. NEO and CBM pesticides in seven FD cabbage samples by the proposed method (X ± SD, n = 3).

Compounds	Sample 1 (μg/kg)	Sample 2 (μg/kg)	Sample 3 (μg/kg)	Sample 4 (μg/kg)	Sample 5 (μg/kg)	Sample 6 (μg/kg)	Sample 7 (μg/kg)
Propamocarb	<LOD	<LOD	<LOD	16.7 ± 1.1	15.5 ± 1.2	35.7 ± 2.1	<LOD
Dinotefuran	<LOD	<LOD	<LOD	<LOD	<LOD	<LOD	<LOD
Oxamyl	<LOD	<LOD	<LOD	<LOD	<LOD	<LOD	<LOD
Nitenpyram	<LOD	<LOD	<LOD	<LOD	<LOD	<LOD	<LOD
Cycloxaprid	<LOD	<LOD	<LOD	<LOD	<LOD	<LOD	<LOD
Thiamethoxam	<LOD	<LOD	<LOD	<LOD	<LOD	<LOD	<LOD
Clothianidin	<LOD	<LOD	<LOD	<LOD	<LOD	<LOD	<LOD
Imidacloprid	<LOD	<LOD	<LOD	<LOD	<LOD	<LOD	78.6 ± 6.0
Imidaclothiz	<LOD	<LOD	<LOD	<LOD	<LOD	<LOD	<LOD
Acetamiprid	<LOD	<LOD	<LOD	<LOD	<LOD	<LOD	4.0 ± 0.1
Pirimicarb	<LOD	<LOD	<LOD	<LOD	<LOD	<LOD	<LOD

Table 3. *Cont.*

Compounds	Sample 1 (μg/kg)	Sample 2 (μg/kg)	Sample 3 (μg/kg)	Sample 4 (μg/kg)	Sample 5 (μg/kg)	Sample 6 (μg/kg)	Sample 7 (μg/kg)
Thiacloprid	4.5 ± 0.2	5.7 ± 0.2	<LOD	17.1 ± 0.5	5.0 ± 0.3	<LOD	6.9 ± 0.5
Aldicarb	<LOD	<LOD	<LOD	<LOD	<LOD	<LOD	<LOD
Metolcarb	<LOD	<LOD	<LOD	<LOD	<LOD	<LOD	<LOD
Propoxur	<LOD	<LOD	<LOD	<LOD	<LOD	<LOD	<LOD
Carbofuran	<LOD	<LOD	<LOD	<LOD	<LOD	<LOD	<LOD
Carbaryl	<LOD	<LOD	<LOD	<LOD	<LOD	<LOD	<LOD
Isoprocarb	<LOD	<LOD	<LOD	<LOD	<LOD	<LOD	<LOD
Promecarb	<LOD	<LOD	<LOD	<LOD	<LOD	<LOD	<LOD

It should be remarked that MRLs for NEOs and CBMs have not been established in FD cabbage by the EU, China, and the United States. Therefore, the MRLs for fresh cabbage were taken as a reference. In such a case, the imidacloprid residue reached 78.6 μg/kg in a sample, which was above the recommended MRL in the EU (10.0 μg/kg in fresh cabbage). This result demonstrated the practicability of the method in real sample analysis, which showed great potential in daily food safety monitoring.

4. Conclusions

A robust and sensitive method was first established for simultaneous determination of pesticides in FD cabbage. The combination of modified QuEChERS and UPLC–MS/MS/MS provided high sample throughput to yield acceptable quantitative results for NEO and CBM pesticides. In this study, ACN/water (v/v = 2:1) was chosen as the extraction solvent, followed by anhydrous $MgSO_4$ and NaCl (4 g + 1 g) as partitioning salt. Moreover, post-column infusion was applied to reflect the matrix profiles of optimized LC conditions. Ultimately, the ME of dinotefuran was eliminated with an initial mobile phase of 10% ACN.

Furthermore, the reliability, robustness, and practicability of the method were verified according to the SANTE/11312/2021 document. Recoveries were achieved for all target analytes from 78.7% to 114.0%, with RSDs below 14.2% and MEs ranging from −16.5% to 16.1%. For the real samples, propamocarb, imidacloprid, acetamiprid, and thiacloprid were detected in six samples, which proved that FD cabbage also contains pesticide residues. This study can also be a practical reference for determining NEO and CBM pesticides in other dehydrated vegetables.

Supplementary Materials: The following supporting information can be downloaded at: https://www.mdpi.com/article/10.3390/foods12040699/s1, Figure S1: Chemical structures of the target NEO and CBM pesticides; Figure S2: Effect of extraction time on the recovery of NEO and CBM pesticides; Figure S3: Effect of extraction time on the MEs of NEO and CBM pesticides; Figure S4: Matrix profiles in different partitioning salts for cycloxaprid; Figure S5: Matrix profiles in different partitioning salts for imidacloprid; Figure S6: The influence of ammonium acetate on the peak shape of nitenpyram; Figure S7: The influence of the dilution with phase A on the peak shape of propamocarb; Figure S8: Extracted ion chromatogram of NEO and CBM pesticides at LOQs concentration; Figure S9: The stability of NEO and CBM pesticides in room temperature storage for one day; Figure S10: The stability of NEO and CBM pesticides in 4°C storage; Table S1: Current maximum residue limits (MRLs) for NEO and CBM pesticides in fresh cabbage (mg/kg); Table S2: Normalized signal intensity measured before and after use of 5 mM ammonium acetate. Table S3: Comparison of different methods for the analysis of NEO and CBM pesticides in fresh cabbage.

Author Contributions: Conceptualization, B.Y. and X.L.; methodology, B.Y.; validation, B.Y.; formal analysis, B.Y.; investigation, B.Y. and G.L.; data curation, B.Y.; writing—original draft preparation, B.Y.; writing—review and editing, S.W., W.M. and M.T.; supervision, Z.M. and Q.Z.; project administration, X.L.; funding acquisition, H.L. All authors have read and agreed to the published version of the manuscript.

Funding: The authors are grateful for financial support from the National Key R&D Program of China (2019YFC1604801).

Data Availability Statement: Not applicable.

Conflicts of Interest: The authors declare no conflict of interest.

References

1. Jin, X.; Oliviero, T.; van der Sman, R.G.M.; Verkerk, R.; Dekker, M.; van Boxtel, A.J.B. Impact of different drying trajectories on degradation of nutritional compounds in broccoli (*Brassica oleracea* var. italica). *LWT Food Sci. Technol.* **2014**, *59*, 189–195. [CrossRef]
2. Rajkumar, G.; Shanmugam, S.; Galvâo, M.D.S.; Dutra Sandes, R.D.; Leite Neta, M.T.S.; Narain, N.; Mujumdar, A.S. Comparative evaluation of physical properties and volatiles profile of cabbages subjected to hot air and freeze drying. *LWT Food Sci. Technol.* **2017**, *80*, 501–509. [CrossRef]
3. Xu, Y.; Xiao, Y.; Lagnika, C.; Li, D.; Liu, C.; Jiang, N.; Song, J.; Zhang, M. A comparative evaluation of nutritional properties, antioxidant capacity and physical characteristics of cabbage (*Brassica oleracea* var. Capitate var L.) subjected to different drying methods. *Food Chem.* **2020**, *309*, 124935. [CrossRef]
4. NASA Technology Transfer Program. Freeze-Dried Foods Nourish Adventurers and the Imagination. Available online: https://spinoff.nasa.gov/Spinoff2020/cg_2.html,2020 (accessed on 1 October 2022).
5. Bhatta, S.; Stevanovic Janezic, T.; Ratti, C. Freeze-Drying of Plant-Based Foods. *Foods* **2020**, *9*, 87. [CrossRef]
6. Kim, J.-H.; Choi, S.-G.; Kwon, Y.S.; Hong, S.-M.; Seo, J.-S. Development of cabbage reference material for multi-residue pesticide analysis. *Appl. Biol. Chem.* **2017**, *61*, 15–23. [CrossRef]
7. Yang, B.; Ma, W.; Wang, S.; Shi, L.; Li, X.; Ma, Z.; Zhang, Q.; Li, H. Determination of eight neonicotinoid insecticides in Chinese cabbage using a modified QuEChERS method combined with ultra performance liquid chromatography-tandem mass spectrometry. *Food Chem.* **2022**, *387*, 132935. [CrossRef]
8. Kecojević, I.; Đekić, S.; Lazović, M.; Mrkajić, D.; Baošić, R.; Lolić, A. Evaluation of LC-MS/MS methodology for determination of 179 multi-class pesticides in cabbage and rice by modified QuEChERS extraction. *Food Control* **2021**, *123*, 107693. [CrossRef]
9. Cesarino, I.; Moraes, F.C.; Lanza, M.R.V.; Machado, S.A.S. Electrochemical detection of carbamate pesticides in fruit and vegetables with a biosensor based on acetylcholinesterase immobilised on a composite of polyaniline–carbon nanotubes. *Food Chem.* **2012**, *135*, 873–879. [CrossRef] [PubMed]
10. Cimino, A.M.; Boyles, A.L.; Thayer, K.A.; Perry, M.J. Effects of Neonicotinoid Pesticide Exposure on Human Health: A Systematic Review. *Environ. Health Persp.* **2017**, *125*, 155–162. [CrossRef] [PubMed]
11. Marfo, J.T.; Fujioka, K.; Ikenaka, Y.; Nakayama, S.M.M.; Mizukawa, H.; Aoyama, Y.; Ishizuka, M.; Taira, K. Relationship between Urinary N-Desmethyl-Acetamiprid and Typical Symptoms including Neurological Findings: A Prevalence Case-Control Study. *PLoS ONE* **2015**, *10*, e0142172. [CrossRef]
12. Li, Q.; Kobayashi, M.; Kawada, T. Effects of carbamate pesticides on human natural killers and T lymphocytes. *Toxicol. Lett.* **2015**, *238*, S217–S218. [CrossRef]
13. European Food Safety Authority. Pesticide Residues and the MRLs That Apply for Such Residues in Food Products. 2022. Available online: http://ec.europa.eu/food/plant/pesticides/eu-pesticides-database/public/?event=homepage&language=EN (accessed on 1 October 2022).
14. NHC, MoA & SAMR of China. Maximum Residue Limits for Pesticides in Food. 2021. Available online: https://www.sdtdata.com/fx/fcv1/tsLibCard/183688.html (accessed on 3 March 2021).
15. The Electronic Code of Federal Regulations. Tolerances and Exemptions for Pesticide Chemical Residues in Food. 2022. Available online: https://www.ecfr.gov/cgi-bin/text-idx?c=ecfr&sid=1a0ecaf51aa3dba662c9cbf1b4336eaf&tpl=/ecfrbrowse/Title40/40cfr180_main_02.tpl (accessed on 1 October 2022).
16. Ma, L.; Wang, Y.; Li, H.; Peng, F.; Qiu, B.; Yang, Z. Development of QuEChERS-DLLME method for determination of neonicotinoid pesticide residues in grains by liquid chromatography-tandem mass spectrometry. *Food Chem.* **2020**, *331*, 127190. [CrossRef] [PubMed]
17. Végh, R.; Sörös, C.; Majercsik, N.; Sipos, L. Determination of Pesticides in Bee Pollen: Validation of a Multiresidue High-Performance Liquid Chromatography-Mass Spectrometry/Mass Spectrometry Method and Testing Pollen Samples of Selected Botanical Origin. J. Agric. *Food Chem.* **2022**, *70*, 1507–1515. [CrossRef]
18. Guan, K.; Huang, R.; Liu, H.; Huang, Y.; Chen, A.; Zhao, X.; Wang, S.; Zhang, L. Development of a Reliable ic-ELISA with a Robust Antimatrix Interference Capability Based on QuEChERS Technology for the Rapid Detection of Zearalenone in Edible and Medical Coix Seeds and Subsequent Risk Assessments. *Foods* **2022**, *11*, 2983. [CrossRef]
19. Octanol/Water Partition Coefficient. Available online: http://www.chemspider.com/ (accessed on 1 October 2022).
20. Dos Santos, E.O.; Gonzales, J.O.; Ores, J.C.; Marube, L.C.; Caldas, S.S.; Furlong, E.B.; Primel, E.G. Sand as a solid support in ultrasound-assisted MSPD: A simple, green and low-cost method for multiresidue pesticide determination in fruits and vegetables. *Food Chem.* **2019**, *297*, 124926. [CrossRef] [PubMed]

21. Farouk, M.; Hussein, L.A.E.A.; El Azab, N.F. Simultaneous determination of three neonicotinoid insecticide residues and their metabolite in cucumbers and soil by QuEChERS clean up and liquid chromatography with diode-array detection. *Anal. Methods* **2016**, *8*, 4563–4575. [CrossRef]
22. Ma, W.; Yang, B.; Li, J.; Li, X. Amino-functional metal-organic framework as a general applicable adsorbent for simultaneous enrichment of nine neonicotinoids. *Chem. Eng. J.* **2022**, *434*, 134629. [CrossRef]
23. Zuo, J.; Cai, R.; An, Y.; Tang, H. Simultaneous Quantification of Five Stereoisomeric Hexoses in Nine Biological Matrices Using Ultrahigh Performance Liquid Chromatography with Tandem Mass Spectrometry. *J. Anal. Test.* **2020**, *4*, 249–256. [CrossRef]
24. Li, X.; Ma, W.; Zhang, Q.; Li, H.; Liu, H. Determination of patulin in apple juice by amine-functionalized solid-phase extraction coupled with isotope dilution liquid chromatography tandem mass spectrometry. *J. Sci. Food Agric.* **2021**, *101*, 1767–1771. [CrossRef]
25. Ma, W.; Li, J.; Li, X.; Liu, H. Enrichment of diamide insecticides from environmental water samples using metal-organic frameworks as adsorbents for determination by liquid chromatography tandem mass spectrometry. *J. Hazard. Mater.* **2022**, *422*, 126839. [CrossRef]
26. European Commission. Guidance Document on Analytical Quality Control and Method Validation for Pesticide Residues Analysis in Food and Feed SANTE 11312/2021. 2021. Available online: https://ec.europa.eu/food/system/files/2022-02/pesticides_mrl_guidelines_wrkdoc_2021-11312.pdf (accessed on 24 February 2021).
27. Stahnke, H.; Reemtsma, T.; Alder, L. Cocompensation of Matrix Effects by Postcolumn Infusion of a Monitor Substance in Multiresidue Analysis with LC-MS/MS. *Anal. Chem.* **2009**, *81*, 2185–2192. [CrossRef] [PubMed]
28. Lehotay, S.J.; Sapozhnikova, Y.; Han, L.; Johnston, J.J. Analysis of Nitrosamines in Cooked Bacon by QuEChERS Sample Preparation and Gas Chromatography-Tandem Mass Spectrometry with Backflushing. *J. Agric. Food Chem.* **2015**, *63*, 10341–10351. [CrossRef]
29. Mastovska, K.; Dorweiler, K.J.; Lehotay, S.J.; Wegscheid, J.S.; Szpylka, K.A. Pesticide multiresidue analysis in cereal grains using modified QuEChERS method combined with automated direct sample introduction GC-TOFMS and UPLC-MS/MS techniques. *J. Agric. Food Chem.* **2010**, *58*, 5959–5972. [CrossRef]
30. Zhao, W.; Shi, Y. A porous boron nitride nanorods-based QuEChERS analysis method for detection of five neonicotinoid pesticide residues in goji berries. *J. Chromatogr. A* **2022**, *1670*, 462968. [CrossRef] [PubMed]
31. Anastassiades, M.; Lehotay, S.J.; Štajnbaher, D.; Schenck, F.J. Fast and Easy multiresidue method employing acetonitrile extraction partitioning and "dispersive solid-phase extraction" for the determination of pesticide residues in produce. *J. AOAC Int.* **2003**, *86*, 412–431. [CrossRef] [PubMed]
32. Islam, A.K.M.M.; Noh, H.H.; Ro, J.-H.; Kim, D.; Oh, M.-S.; Son, K.; Kwon, H. Optimization and validation of a method for the determination of acidic pesticides in cabbage and spinach by modifying QuEChERS procedure and liquid chromatography-tandem mass spectrometry. *J. Chromatogr. B* **2021**, *1173*, 122667. [CrossRef]
33. Wang, S.; Li, M.; Li, X.; Li, X.; Li, X.; Li, S.; Zhang, Q.; Li, H. A functionalized carbon nanotube nanohybrids-based QuEChERS method for detection of pesticide residues in vegetables and fruits. *J. Chromatogr. A* **2020**, *1631*, 461526. [CrossRef]
34. Thurman, E.M.; Ferrer, I.; Barceló, D. Choosing between Atmospheric Pressure Chemical Ionization and Electrospray Ionization Interfaces for the HPLC/MS Analysis of Pesticides. *Anal. Chem.* **2001**, *73*, 5441–5449. [CrossRef]
35. Ikonomou, M.G.; Blades, A.T.; Kebarle, P. Investigations of the Electrospray Interface for Liquid Chromatography/Mass Spectrometry. *Anal. Chem.* **1990**, *62*, 957–967. [CrossRef]
36. Yang, X.J.; Qu, Y.; Yuan, Q.; Wan, P.; Du, Z.; Chen, D.; Wong, C. Effect of ammonium on liquid- and gas-phase protonation and deprotonation in electrospray ionization mass spectrometry. *Analyst* **2013**, *138*, 659–665. [CrossRef]
37. Ministry of Agriculture and Rural Affairs of People's Repbulic of China. List of Prohibited Pesticides. Available online: http://www.zzys.moa.gov.cn/gzdt/201911/t20191129_6332604.htm (accessed on 19 January 2023).
38. Kunpatee, K.; Kalcher, K.; Chailapakul, O.; Chaiyo, S.; Samphao, A. A paper chromatographic- based electrochemical analytical device for the separation and simultaneous detection of carbofuran and carbaryl pesticides. Sensor. *Actuat. B Chem.* **2023**, *377*, 133116. [CrossRef]
39. Wang, M.; Wang, J.; Wang, K.; Zhang, L.; Cao, X.; Guo, C.; Wang, J.; Wu, B. Magnetic mesoporous material derived from MIL-88B modified by l-alanine as modified QuEChERS adsorbent for the determination of 6 pesticide residues in 4 vegetables by UPLC-MS/MS. *Food Chem.* **2022**, *384*, 132325. [CrossRef] [PubMed]
40. Zhang, Y.; Xu, J.; Dong, F.; Liu, X.; Li, X.; Li, Y.; Wu, X.; Liang, X.; Zheng, Y. Simultaneous determination of four neonicotinoid insecticides residues in cereals, vegetables and fruits using ultra-performance liquid chromatography/tandem mass spectrometry. *Anal. Methods* **2013**, *5*, 1449–1455. [CrossRef]

molecules

Article

Analysis of Chemical Constituents of Traditional Chinese Medicine Jianqu before and after Fermentation Based on LC-MS/MS

Yishuo Wang [1,2], Ruisheng Wang [1,2], Zhenling Zhang [1,2,*], Yitian Chen [1], Mengmei Sun [1], Jia Qiao [1] and Ziwei Du [1]

1 College of Pharmacy, Henan University of Chinese Medicine, Zhengzhou 450046, China
2 Henan Integrated Engineering Technology Research Center of Traditional Chinese Medicine Production, Zhengzhou 450046, China
* Correspondence: zhangzl6758@163.com

Abstract: Objective: To detect the chemical constituents in Jianqu samples under different fermentated states by using UPLC-QTOF-MS/MS technology, to conduct preliminary analyses, and to establish an HPLC method for the simultaneous determination of hesperidin and naringenin in Jianqu, and the variation of the two components during fermentation were compared. Methods: Waters ACQUITYTM UPLC HSST3 column (2.1 mm × 100 mm, 1.8 μm) was used; the mobile phase was 0.1% formic acid aqueous solution (A)-0.1% formic acid acetonitrile (B); The flow rate was 0.4 mL·min^{-1} with gradient elution; the column temperature was 45 °C; injection volume was 5 μL. The mass spectra of the samples were collected by negative ion mode under the electrospray ion source, and the data were screened and matched by UNIFI software. Hypersil gold C18 column (100 mm × 2.1 mm, 1.9 μm) was used; the mobile phase was acetonitrile (A)-0.1% acetic acid (B); the flow rate with gradient elution was 0.3 mL·min^{-1}; the column temperature was 30 °C; the injection volume was 2 μL. The content changes of hesperetin and naringenin in Jianqu at different fermentation time were detected. Results: A total of 54 compounds were identified, including flavonoids, amino acids, organic acids, terpenoids, coumarins, lignans, and other compounds. Under the selected HPLC conditions, the linear relationship between hesperidin and naringenin was discovered (r^2 = 0.9996). The content of hesperidin and naringenin changed significantly in the whole fermentation process. The highest concentration of content was observed at 36 h of fermentation and then decreased to varying degrees. Conclusion: This experiment can effectively identify various chemical components in Jianqu during different fermentation periods, and determine the content of the characteristic components, so as to provide a scientific basis for further study of Jianqu fermentation processing technology as well as a sound pharmacodynamic material basis.

Keywords: UPLC-QTOF-MS/MS technology; jianqu; fermentation; chromatographic peak identification; chemical composition

Citation: Wang, Y.; Wang, R.; Zhang, Z.; Chen, Y.; Sun, M.; Qiao, J.; Du, Z. Analysis of Chemical Constituents of Traditional Chinese Medicine Jianqu before and after Fermentation Based on LC-MS/MS. *Molecules* **2023**, *28*, 53. https://doi.org/10.3390/molecules28010053

Academic Editors: Chao Kang and Ronald Beckett

Received: 26 November 2022
Revised: 14 December 2022
Accepted: 15 December 2022
Published: 21 December 2022

1. Introduction

Fermentation processing of traditional Chinese medicine refers to the method of foaming and undressing the medicine after purification or treatment under certain temperature and humidity conditions through the catalytic decomposition of microorganisms and enzymes [1] (p. 416). The fermentation method is different from the traditional heating and processing methods such as frying, steaming, boiling, simmering, frying, refining, etc. It has a mild effect and mainly uses microorganisms and their secreted enzymes to change the chemical composition of medicines. Due to the variety of microorganisms and the different synergistic effects among them, the same drug fermented by different methods often has different therapeutic effects. Traditional Chinese medicine fermentation converts macromolecular substances that are difficult for the human body to absorb into small molecular components easier to absorb through microbial strains, which can improve the speed of drug absorption and promote the effectiveness of drugs. Moreover, a variety of secondary metabolites are generated during fermentation, which can enhance or change the efficacy. Drug components decompose after fermentation to a certain extent, thus bringing reduced toxicity and making it relatively safe to take in. The taste of Chinese medicine is usually tough for patients, while fermentation changes the original taste into something more acceptable, and the patient's compliance might be better. It can be seen that the fermentation processing of traditional Chinese medicine can provide new research ideas for the development of traditional Chinese medicine and has a very broad application prospect.

Jianqu, also known as Fanzhiqu and Baicaoqu, has a yellowish-brown surface, white mildew, fragrant smell, and slightly bitter taste. It has a long history of application. In addition to the effect of divine comedy, it also has the effect of deciphering the surface and reconciling it [2] (p. 385). It is produced all over the country, but Quanzhou, Fujian is the most famous, also known as Quanzhou Divine Comedy [3] (pp. 31–32). Jianqu, as a processed product fermented from dozens of fine medicinal powders, has been used by physicians in various dynasties intending to different therapeutic efficacy, and its prescriptions have changed accordingly. Jianqu is now fermented from Chinese medicines such as Artemisia annua, mint, Angelica dahurica, wheat bran, and flour. At present, there are more than 20 Jianqu products used in clinical applications. According to the literature, Jianqu can be used in the following four categories: (1) cough caused by cold. (2) cold, fever, dizziness, and vomiting caused by the plague. (3) chest stuffiness and fullness caused by qi stagnation, (4) eating and abdominal pain and indigestion caused by water and grain accumulation. At present, the research of Jianqu mainly focuses on the clinical prescription application and case analysis, and there are few reports on the chemical composition analysis of Jianqu, and the material basis of its efficacy is still unclear, which limits the clinical application of Jianqu. Therefore, clarifying the chemical composition of Jianqu before and after fermentation can provide a scientific basis for further improving the fermentation process of Jianqu and studying its pharmacodynamic material basis.

In this study, UPLC-QTOF-MS/MS technology was used to systematically analyze and study the chemical substances in Jianqu at different fermentation times, and HPLC technology was used to determine the content of the characteristic components.

2. Results

2.1. Linear Investigation Results

Taking the concentration X (mg/L) as the abscissa and the peak area Y as the ordinate, draw a standard curve to obtain the regression equations of naringenin and hesperetin. The results are shown in Table 1.

Table 1. Results of linear investigation of naringenin and hesperetin.

Element	Regression Equation	R^2	Linear Range ($\mu g \cdot mL^{-1}$)
Naringenin	$Y = 20.679X - 34.625$	0.9996	5.09–50.9
Hesperetin	$Y = 17.785X + 4.7475$	0.9996	5.02–50.2

2.2. Methodological Investigation

2.2.1. Precision Test

The test results showed that the RSDs of the peak areas of naringenin and hesperetin were 0.24% and 0.20%, respectively, indicating that the precision of the experimental instrument was good.

2.2.2. Stability Test

The test results showed that the RSDs of the peak areas of naringenin and hesperetin were 0.39% and 0.33%, respectively, indicating that the test solution was stable within 12 h.

2.2.3. Repeatability Test

The experimental results showed that the RSDs of the peak areas of naringenin and hesperetin were 1.95% and 1.82%, respectively, indicating that the experimental method was reproducible.

2.2.4. Sample Addition Recovery Test

The test results of the sample addition recovery showed that the average recoveries of naringenin and hesperetin were 99.19% and 99.88%, and the RSDs were 2.09% and 2.20%, respectively, indicating that the experimental method was accurate. The specific results are shown in Tables 2 and 3 below.

Table 2. Test results of naringenin sample recovery rate.

Sampling Volume/mg	Sample Content/mg	Addition/mg	Measured Amount/mg	Recovery Rate/%	The Average Recovery Rate/%	RSD/%
1.0003	0.2101	0.2001	0.4109	100.35	99.19	2.09
1.0002	0.2100	0.2001	0.4100	99.95		
1.0003	0.2101	0.2001	0.4084	99.10		
1.0009	0.2102	0.2001	0.4124	101.05		
1.0006	0.2101	0.2001	0.4006	95.20		
1.0003	0.2101	0.2001	0.4101	99.50		

Table 3. Hesperetin sample recovery test results.

Sampling Volume/mg	Sample Content/mg	Addition/mg	Measured Amount/mg	Recovery Rate/%	The Average Recovery Rate/%	RSD/%
1.0003	0.2292	0.2036	0.4396	103.05	99.88	2.20
1.0002	0.2291	0.2036	0.4279	97.64		
1.0003	0.2292	0.2036	0.4346	99.10		
1.0009	0.2293	0.2036	0.4376	100.88		
1.0006	0.2292	0.2036	0.4276	97.45		
1.0003	0.2292	0.2036	0.4351	101.13		

2.3. *UPLC-QTOF-MS/MS Analysis Results*

Comparing the basic shapes of the base peak ion chromatograms of the Jianqu samples with different fermentation times (see Figure 1), it can be intuitively found that the types of chemical components in each sample are similar. According to the response value, the chemical components in the Jianqu samples at different fermentation times were different. Data matching was conducted through UNIFI 1.8.1 software. The traditional Chinese medicine system pharmacology database and analysis platform (TCMSP) were used to search the ingredients of each single herb in Jianqu. The attribution of each component was preliminarily determined, and the component identification was carried out in combination with the reference substance and related literature information. A total of 54 compounds were identified, including 15 flavonoids, 3 amino acids, 14 organic acids, 5 terpenes, 4 coumarins, 2 lignans, and 11 other types. Among them, there are 20 species identified by the comparison of the reference substances. There are 28 unique compounds obtained through database screening. The other compounds were identified by database screening combined with literature reports. The UPLC-Q-TOF-MS/MS identification results of the compounds in Jianqu are shown in Table 4.

Jianqu mainly includes flavonoids and organic acids, Recent studies have shown that flavonoids have cough-relieving, expectorant, asthma-relieving and antibacterial activities, and at the same time have liver-protecting and detoxificating effects, and anti-free radical, antifungal, and antioxidant effects that may contribute to the treatment of hepatitis and liver cirrhosis. Organic acids have anti-inflammatory, antithrombotic, and antioxidant effects, inhibiting platelet aggregation, and inducing apoptosis of tumor cells, and the organic acids and esters in Jianqu may be the source of its wine aroma; At present, the role of these chemical components in the clinical application of Jianqu remains to be studied.

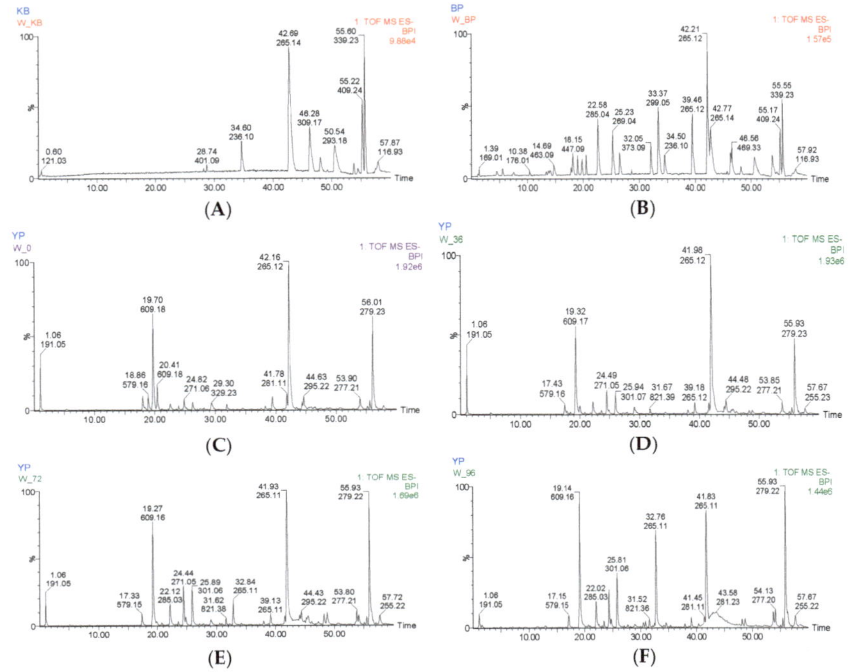

Figure 1. Base peak ion chromatograms of blank, standard, and Jianqu samples with different fermentation time in negative ion mode. (**A**) blank solution (**B**) reference solution (**C**) 0 h sample (**D**) 36 h sample (**E**) 72 h sample (**F**) 96 h sample.

Table 4. UPLC-Q-TOF-MS/MS identification of compounds in Jianqu.

Serial Number	tR/min	m/z[M-H]$^-$	Molecular Formula	Compound	Identification Method	References
1	1.06	191.05	$C_7H_{12}O_6$	quinic acid	database, literature	[4–6]
2	1.09	132.03	$C_4H_7NO_4$	Aspartic acid	database *	-
3	1.10	179.06	$C_6H_{12}O_6$	glucose	database *	-
4	1.13	149.01	$C_4H_6O_6$	tartaric acid	database *	-
5	1.17	180.07	$C_9H_{11}NO_3$	tyrosine	database *	-
6	1.32	169.01	$C_7H_6O_5$	Gallic acid	database *	-
7	1.32	237.04	$C_{10}H_8O_4$	coumarin	database *	-
8	2.66	203.08	$C_{11}H_{12}N_2O_2$	tryptophan	database *	-
9	4.42	175.06	$C_6H_{10}O_3$	3-Hydroxy-2,5-hexanedione	database *	-
10	5.19	179.03	$C_9H_8O_4$	caffeic acid	References, databases, literature	[6–8]
11	5.37	135.04	$C_8H_8O_2$	Not identified	database,	-
12	6.94	340.15	$C_{20}H_{23}NO_4$	Magnoflorine	References, databases, literature	[8]
13	7.42	785.25	$C_{35}H_{46}O_{20}$	Echinacea	database *	-
14	9.80	191.03	$C_{10}H_8O_4$	Scopolactone	References, databases, literature	[5,6,9]
15	12.52	417.12	$C_{21}H_{22}O_9$	liquiritin	References, databases, literature	[10,11]
16	13.02	463.09	$C_{21}H_{20}O_{12}$	isoquercitrin	References, databases, literature	[12]
17	13.80	609.15	$C_{27}H_{30}O_{16}$	Rutin	References, databases, literature	[4,8,12]

Table 4. *Cont.*

Serial Number	tR/min	*m/z*[M-H]$^-$	Molecular Formula	Compound	Identification Method	References
18	17.48	579.17	$C_{27}H_{32}O_{14}$	naringin rutin	database, literature	[13–15]
19	18.26	445.08	$C_{21}H_{18}O_{11}$	flavonoid glycosides	database *	-
20	18.48	579.16	$C_{27}H_{32}O_{14}$	Naringin	References, databases, literature	[4,13–15]
21	19.27	359.08	$C_{18}H_{16}O_8$	rosmarinic acid	reference, database	-
22	19.63	161.02	$C_9H_6O_3$	Umbelliferone	database *	-
23	19.85	609.18	$C_{28}H_{34}O_{15}$	Hesperidin	References, databases, literature	[4,13–15]
24	20.56	609.18	$C_{28}H_{34}O_{15}$	neohesperidin	References, databases, literature	[4,13–15]
25	22.30	301.03	$C_{15}H_{10}O_7$	Quercetin	References, databases, literature	[9,12,16,17]
26	22.50	285.03	C15H10O6	Luteolin	References, databases, literature	[17,18]
27	22.76	491.10	$C_{26}H_{20}O_{10}$	Salvianolic acid C	database *	-
28	24.82	271.06	$C_{15}H_{12}O_5$	naringenin	References, databases, literature	[4,14,15,19]
29	26.22	301.07	$C_{16}H_{14}O_6$	Hesperetin	References, databases, literature	[4,14,15,19]
30	26.34	315.05	$C_{16}H_{12}O_7$	Isorhamnetin	References, databases, literature	[18]
31	29.05	219.14	$C_{14}H_{20}O_2$	Thyme isobutyrate	database *	-
32	29.30	329.23	$C_{18}H_{34}O_5$	triangular acid	database, literature	[5,6]
33	29.80	315.05	$C_{16}H_{12}O_7$	flavonoids	database *	-
34	30.92	427.18	$C_{24}H_{28}O_7$	coumarin	database *	-
35	31.90	821.39	$C_{42}H_{62}O_{16}$	Glycyrrhizinate	References, databases, literature	[10,11]
36	31.92	373.10	$C_{19}H_{18}O_8$	cat's eye flavin	References, databases, literature	[12]
37	36.45	487.34	$C_{30}H_{48}O_5$	Terpenes	database *	-
38	38.14	313.33	$C_{18}H_{34}O_4$	Dibutyl Sebacate	database *	-
39	39.46	265.12	$C_{18}H_{18}O_2$	and honokiol	References, databases, literature	[8,20]
40	41.71	309.21	$C_{18}H_{30}O_4$	9,16-Dihydroxy-10,12,14-triene-octadecanoic acid	database *	-
41	41.78	281.11	$C_{18}H_{18}O_3$	Obovatol	database *	-
42	42.21	265.12	$C_{18}H_{18}O_2$	Magnolol	References, databases, literature	[8,20]
43	44.63	295.22	$C_{18}H_{32}O_3$	Coronaric acid	database *	-
44	45.52	293.21	$C_{18}H_{30}O_3$	(E,E)-9-Oxooctadeca-10,12-dienoic acid	database *	-
45	45.57	233.15	$C_{15}H_{22}O_2$	artemisinic acid	reference, database	-
46	48.78	295.22	$C_{18}H_{32}O_3$	Coronaric acid	database *	-
47	49.14	467.32	$C_{30}H_{44}O_4$	Triterpenoids	database *	-
48	51.89	439.25	$C_{27}H_{36}O_5$	(25R)-Spirosta-4-ene-3,6,12-trione	database *	-
49	53.90	277.21	$C_{18}H_{30}O_2$	linolenic acid	database, literature	[16,21]
50	55.89	307.26	$C_{20}H_{36}O_2$	Ethyl linoleate	database *	
51	56.01	279.23	$C_{18}H_{32}O_2$	Linoleic acid	database, literature	[16,21,22]
52	56.59	267.23	$C_{17}H_{32}O_2$	methyl palmitate	database *	-
53	57.05	489.36	$C_{30}H_{50}O_5$	Triterpenoids	database *	-
54	57.82	255.23	$C_{16}H_{32}O_2$	Palmitic acid	database, literature	[5,6,16,18]

Note: "Database ⁎" refers to the compound that has a unique corresponding in the database screening.

2.3.1. Flavonoids

The flavonoids in Jianqu mainly come from tangerine peel, citrus aurantium, licorice, and other medicinal materials. A total of 15 flavonoids and their glycosides were identified in this experiment. Among them, 13 components were accurately identified, namely, Liquiritin, Isoquercitrin, Rutin from Pomelo Peel, Naringin, Hesperidin, Rutin, Neohesperidin, Quercetin, Luteolin, Naringenin, Hesperidin, Isorhamnetin, and Ambrosin.

Compounds 23 and 24 are the main chemical components in Jianqu and are isomers of each other. Through the data screening of the UNIFI software, combined with its accurate relative molecular mass, it can be preliminarily inferred that the two are hesperidin and neohesperidin, according to the order in which they flow out of the chromatographic column [13] (pp. 58–63). Combined with reference substances and related literature reports [4,14,15] (pp. 40–45,136–140,899–909), compound 23 was identified as hesperidin, and compound 24 was identified as neohesperidin. The same method can accurately identify compound 18 as naringin and compound 20 as naringin. Free flavonoids can exist independently in plants, and free flavonoids and corresponding carbohydrates can also be generated after hydrolysis of flavonoid glycosides. The quasi-molecular ion peaks of compounds 28 and 29 are *m/z* 271.06[M-H]$^-$, *m/z* 301.07[M-H]$^-$, respectively. Through the data screening of the UNIFI software, it is speculated that its molecular formulas are $C_{15}H_{12}O_5$ and $C_{16}H_{14}O_6$, respectively, according to its exact relative molecular mass. Combined with the relevant literature reports [4,14,15] (pp. 40–45,136–140,899–909), it can be inferred that the two are naringenin and hesperetin, and it can be identified that compound 28 is naringenin and compound 29 is hesperetin, according to the sequence of the two eluted from the chromatographic column [19] (pp. 1751–1759).

2.3.2. Organic Acid Compounds

A total of 14 organic acid components were identified in this experiment, and 11 of them were identified as the follows: Quinic acid, gallic acid, caffeic acid, tartaric acid, Rosmarinic acid, Salvianolic acid C, Sparganic acid, (E,E)-9-Oxooctadeca-10,12-dienoic acid, linolenic acid, linoleic acid, and palmitic acid. The quasi-molecular ion peak of compound 52 is *m/z* 279.23[M-H]$^-$, filtered through the database, so it is speculated that their molecular formulas are $C_{18}H_{32}O_2$. Combined with accurate relative molecular mass and related literature reports [16,21,22] (pp. 20–34, 124–126, 2336–2342), it can be inferred that the compound is linoleic acid. Similarly, it can be inferred that compound 55 is palmitic acid. The quasi-molecular ion peak of compound 10 is *m/z* 179.03[M-H]$^-$. After database screening, combined with the accurate relative molecular mass and the comparison of the related literature [7,8,22] (pp. 81–84, 764–774, 2336–2342), it is inferred that it is caffeic acid.

2.4. Assay

2.4.1. HPLC Chromatogram

Take the test solution and the mixed reference solution 2, respectively, and inject them under the conditions of Section 4.3.1. The results are shown in Figure 2.

2.4.2. Sample Content Determination

Precisely weigh about 1 g of S1–S4 sample, prepare the test solution according to the method under Section 4.2.2, and inject and analyze according to the analysis conditions under Section 4.3.1. The contents of naringenin and hesperetin in each sample were calculated, and the results are shown in Table 5.

Table 5. Determination of naringenin and hesperetin in Jianqu samples with different fermentation times (mg/g).

Sample	Naringenin	Hesperetin
S1	0.1263	0.0652
S2	0.2821	0.2353
S3	0.2746	0.2018
S4	0.2100	0.2291

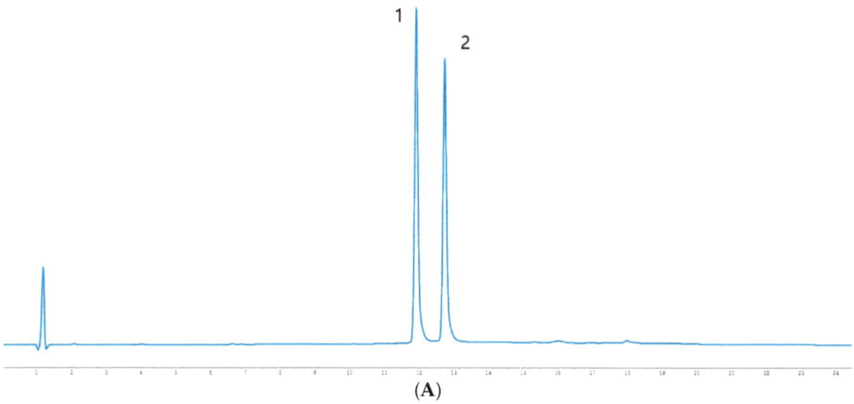

(**A**)

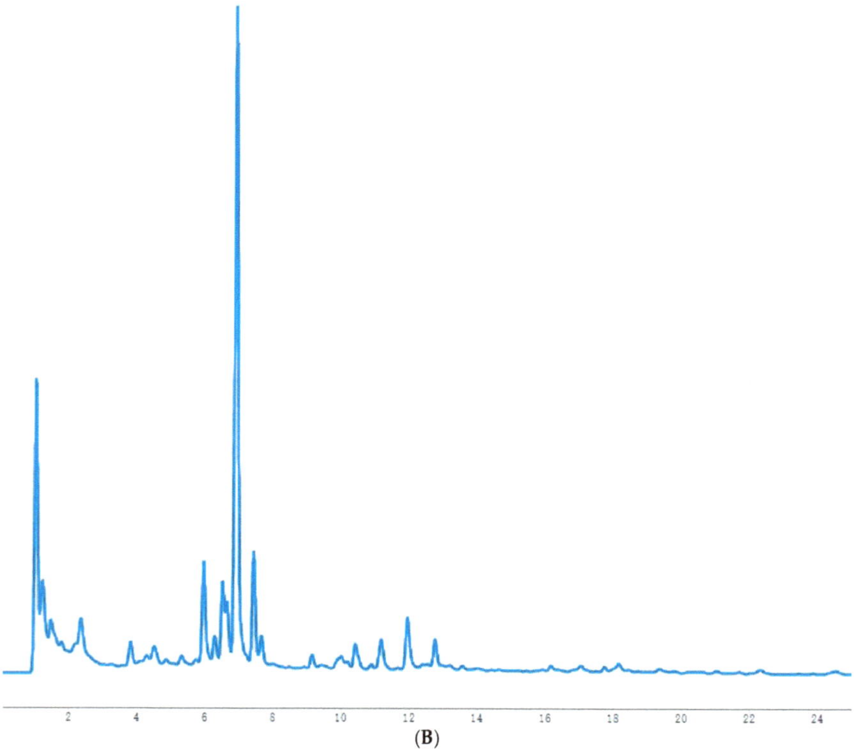

(**B**)

Figure 2. *Cont.*

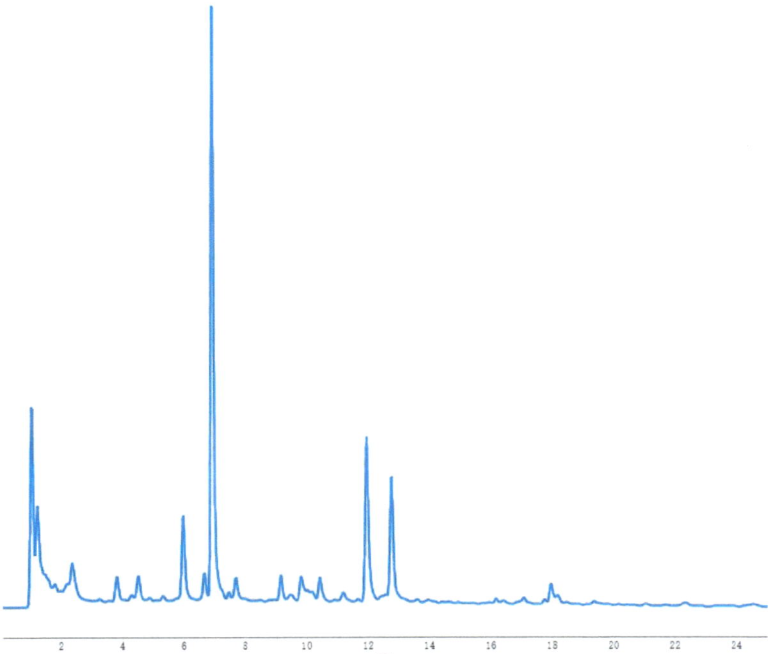

(C)

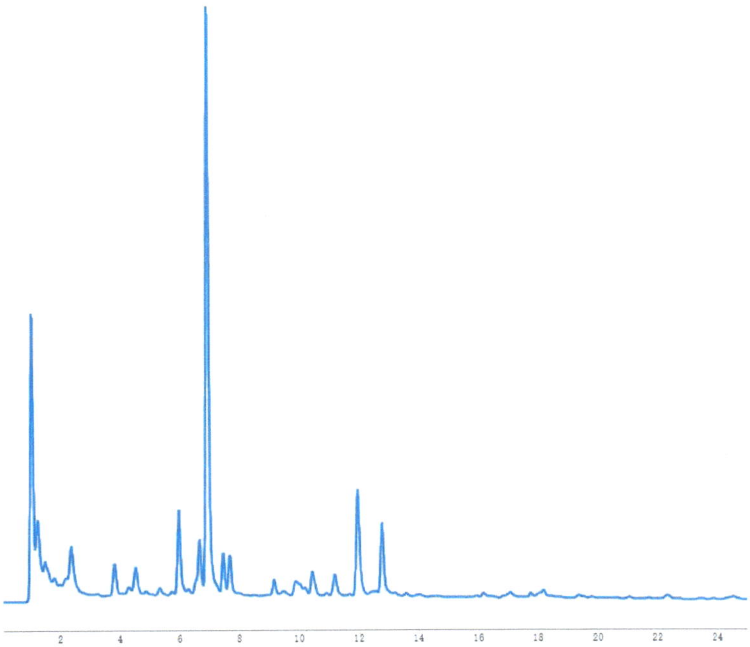

(D)

Figure 2. *Cont.*

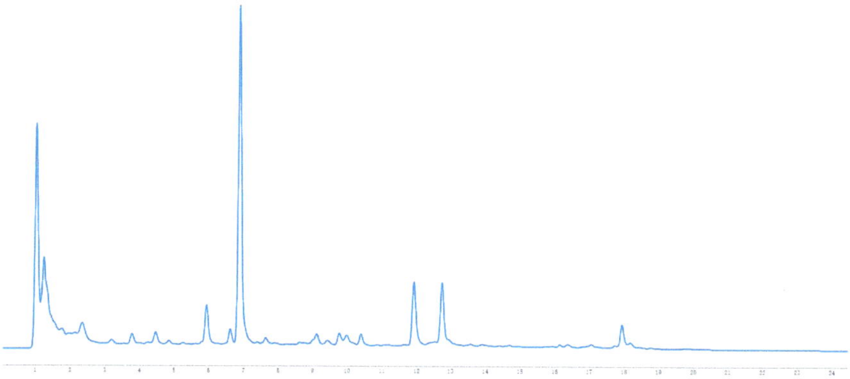

(E)

Figure 2. HPLC chromatograms of mixed reference substances and Jianqu samples with different fermentation times. (**A**). Mixed reference solution (**B**) 0 h sample (**C**) 36 h sample (**D**). 72 h sample (**E**) 96 h sample 1. naringenin 2. Hesperetin.

3. Discussion

In this experiment, positive ion mode and negative ion mode were used to analyze the extract of Jianqu with different fermentation times, and the two scanning modes were compared. It was found that under the negative ion mode, the response signal of the chemical components in the sample was stronger, the peak number was more, and the mass spectrum information was clearer. Therefore, the selection is carried out in the negative ion mode.

The HPLC chromatographic binding content determination results showed that the highest content of hesperidin and naringin reached 0.2353 mg/g and 0.2821 mg/g, respectively, at 36 h of fermentation. After 36 h, both are reduced to varying degrees and may be oxidized by microorganisms. It has been reported [23] (pp. 136–142) that the optimal fermentation time of Jianqu is 30–36 h, indicating that the samples used in this experiment are qualified, and it can also be speculated that hesperidin and naringin may be the material basis for the curative effect of Jianqu. It can be clearly seen from the mass spectrum that the response values of hesperidin and neohesperidin are high. Because both are derived from dried tangerine peel and fructus aurantii, hesperidin accounts for more than 50% of the total in the dried tangerine peel [24] (pp. 36–39,52). The response value of flavonoid glycosides, such as hesperidin, neophesperidin, and naringin, gradually decreases with the continuous fermentation, and it can be inferred that their content is decreasing. Combined with the response value changes and content determination results of hesperidin and naringin, it may be because they are used as raw materials in the fermentation process to synthesize other substances or are oxidized and degraded to form other substances. At present, only the content of two index components has been detected, and the optimal fermentation processing endpoint of Jianqu cannot be comprehensively judged according to the content of these two ingredients, and follow-up research is required.

After the fermentation of Jianqu, there is a slight aroma of wine. Relevant research on wine [25] (pp. 46–50) shows that the aroma of wine is mostly the smell of acid and ester compounds. Acid esters such as linoleic acid, palmitic acid, methyl palmioleate, and ethyl linoleate were detected before and after fermentation of Jianqu liquor, which may be the source of the aroma of Jianqu liquor. Linoleic acid and linolenic acid are both unsaturated fatty acids. With the fermentation process, the response of these substances to mass spectrometry first decreased and then increased. The main reason may be that some bacteria or fungi produced at the beginning of fermentation oxidized the unsaturated fatty acids. Related research shows that [26] (pp. 82–87) the content of oleic acid and linoleic acid in some plants was significantly negatively correlated. It is speculated that enzymes may be produced later in fermentation to convert oleic acid into linoleic acid and increase its content. The response values of some small molecular organic acids, such as quinic acid and caffeic acid, decrease continuously with the progress of fermentation, and some even cannot be detected in the late fermentation, which may be because other substances are synthesized as raw materials in the fermentation process or oxidized and degraded to produce other substances.

In addition, some amino acids, such as tyrosine, begin to appear at the later stage of fermentation, which may come from the decomposition of certain substances. Glucose has been decreasing since fermentation, because as fermentation progresses, the yeast and other microorganisms produced in the fermentation system continuously consume the reducing sugar. Triterpenoids were detected 72 h later, which may be due to the decomposition of enzymes that hydrolyze triterpenoid saponins during fermentation.

In this experiment, the chemical constituents of the different fermentation times of Jianqu were analyzed efficiently, accurately, and quickly. A total of 54 compounds were identified, including 15 flavonoids, 3 amino acids, 14 organic acids, 5 terpenes, 4 coumarins, 2 lignans, and 11 others. The main components were flavonoids and organic acids. The HPLC method was used to determine the content of characteristic components, and it was found that the contents of hesperidin and naringin were the highest after 36 h fermentation. It laid a foundation for further research on the pharmacodynamic material basis of Jianqu, the action mechanism of microorganisms in the fermentation process, and the fermentation process. After that, we will further study the role of the main components in Jianqu.

4. Materials and Methods

4.1. Materials

4.1.1. Experimental Instruments

Acquity UPLC-I-Class tandem Xevo-G2-SQ-TOF mass spectrometer, Masslynx4.1 mass spectrometer workstation (Waters Corporation, Milford, MS, USA), Agilent 1260 Infinity II liquid chromatograph (Agilent Technologies Co., Ltd., Santa Clara, CA, USA), UNIFY1.7 database (Waters, Milford, MS, USA); BT25S 1/100,000 balance, BSA224S-CW 1/10,000 balance (Sartoris Technology Instrument Co., Ltd., Gottingen Germany); TG16-WS high-speed desktop centrifuge (Hunan Xiangyi Laboratory Instrument Development Co., Ltd., Changsha, China); KQ-100DE Ultrasonic Cleaner (Kunshan Ultrasonic Instrument Co., Ltd., Kunshan, China).

4.1.2. Experimental Reagents

Chromatographic grade acetonitrile (batch number 19085176) and methanol (batch number 19085029) were produced by Thermofisher Company (Waltham, MA, USA), chromatography grade acetic acid (batch number 20180813) was produced by Tianjin Damao Chemical Reagent Factory (Tianjin, China), and ultrapure water was made by the laboratory. The products are shown in Table 6, all purchased from Chengdu Pusi-Biotechnology Co., Ltd., and the purity is ≥98%.

Table 6. Experimental reference substance and batch number.

Control	Batch Number
Scutellarin	PS010076
Artemisinin	PS010373
artemisinic acid	PS020402
Gallic acid	PS000688
Costunolide	PS012275
Isorhamnetin	MUST-16120110

Table 6. *Cont.*

Control	Batch Number
Luteolin	PS10320025
magnoflorine	PS012478
Quercetin	PS0605
isoquercitrin	PS010965
Pogostone	PS000401
Rutin	PS010207
Hesperidin	PS011588
neohesperidin	PS010413
Liquiritin	P29M7F12158
Glycyrrhizinate	150823
Naringin	PS012062
chrysosplenetin	PS012187
Magnolol	PS010353
Honokiol	PS011061
Atractylodin	PS011488
Scopolactone	PS011029
Nobiletin	PS012026
Kaempferol	MUST-16041502
Dehydroxylactone	PS010244
Tangeretin	PS012027
Beta-Cineol	PS010195
Oleanolic acid	PS0236-0025
Ursolic acid	MUST-15102905
rosmarinic acid	PS012101
caffeic acid	PS010522

4.1.3. Medicinal Materials

Wheat bran and flour were reserved in the laboratory, and the other medicinal materials used in the experiment are shown in Table 7, all purchased from Zhang Zhongjing Pharmacy(Bozhou, China).

4.2. Methods

4.2.1. Preparation of Jianqu Samples at Different Fermentation Times

The samples were prepared according to the Jianqu preparation method contained in the "Standard Set Prescription of Traditional Chinese Medicine Preparations of drugs Book XVII (Ministry of Health of the People's Republic of China)" [2] (p. 385). In addition to wheat bran and flour, we crushed the other twenty-one drugs such as Artemisia annua into fine powder; they were mixed with wheat bran and sieved, simultaneously made into a thin paste, then mixed well with the above medicinal powder while it was still hot. The appropriate mixing ratio means it could be kneaded into balls by hand easily and dispersed upon tossing. According to method, the mixture was shaped into cubes, put in a constant temperature and humidity fermentation box with a temperature of 30 °C and a humidity of 75%, and fermented for 0, 36, 72, and 96 h, respectively. Finally, they were taken out, dried at 60 °C, and pulverized which are denoted as S1–S4, respectively.

Table 7. Experimental medicinal materials and batch numbers.

Medicinal Herbs	Batch Number	Manufacturer
Spicy Polygonum	20190820	Bozhou Zhang Zhongjing Chinese Herbal Pieces Co., Ltd.
cocklebur	20200103	Bozhou Zhang Zhongjing Chinese Herbal Pieces Co., Ltd.
Artemisia annua	19601CP0330	Anguo Guangming Decoction Piece Processing Factory
Bitter almonds	191101	Bozhou Zhang Zhongjing Chinese Herbal Pieces Co., Ltd.
red bean	191001	Bozhou Zhang Zhongjing Chinese Herbal Pieces Co., Ltd.
malt	190520	Henan Lvhe Pharmaceutical Co., Ltd.
Hawthorn	190801	Shandong Luan Chinese Herbal Pieces Co., Ltd.
tangerine peel	190701	Shandong Luan Chinese Herbal Pieces Co., Ltd.
Patchouli	190601	Hebei Yihe Pharmaceutical Co., Ltd.
Atractylodes	20191224	Bozhou Zhang Zhongjing Chinese Herbal Pieces Co., Ltd.
Magnolia	20191231	Bozhou Zhang Zhongjing Chinese Herbal Pieces Co., Ltd.
Muxiang	191102	Bozhou Zhang Zhongjing Chinese Herbal Pieces Co., Ltd.
Angelica dahurica	20201109	Bozhou Zhang Zhongjing Chinese Herbal Pieces Co., Ltd.
betel nut	190101CP0655	Anguo Guangming Decoction Piece Processing Factory
Citrus aurantium	20200926	Bozhou Zhang Zhongjing Chinese Herbal Pieces Co., Ltd.
Basil leaves	20190801406	Hebei Xiuhe Pharmaceutical Co., Ltd.
Mint	190601CP0680	Anguo Guangming Decoction Piece Processing Factory
Valley sprouts	20200502	Bozhou Zhang Zhongjing Chinese Herbal Pieces Co., Ltd.
Cinnamon	190801	Bozhou Zhang Zhongjing Chinese Herbal Pieces Co., Ltd.
Cypress	20191107	Bozhou Zhang Zhongjing Chinese Herbal Pieces Co., Ltd.
Licorice	20200528	Bozhou Zhang Zhongjing Chinese Herbal Pieces Co., Ltd.

4.2.2. Preparation of the Test Solution

Samples 1 g of Jianqu were accurately weighed at different fermentation times, respectively, placed in an erlenmeyer flask, 25 mL of methanol were added, and then the total weight was weighed. The samples were sonicated for 30 min, cooled to room temperature, and methanol was added to make up for the lost weight, shaken well, and evaporated to dryness in a water bath at 70 °C. The methanol was added to the residue to dissolve and dilute to 10 mL in a volumetric flask. The solution was centrifuged at 16,149× g for 5 min., and the supernatant was separated and filtered through a 0.22 μm microporous membrane to obtain the test solution.

4.2.3. Preparation of Reference Solution

The appropriate amount of the reference substance under item Section 4.1.2 was precisely weighed, placed in a 10 mL volumetric flask, methanol was added to dissolve and reach up to the mark, and shaken up to obtain the mixed reference substance solution 1. Appropriate amounts of hesperetin and naringenin reference substances were precisely weighed, respectively, and methanol was added to prepare the mixed reference solution 2 containing 0.5020 mg/mL and 0.5090 mg/mL, respectively.

4.3. Analysis Conditions

4.3.1. HPLC Conditions

The column was Hypersil Gold C18 column (100 mm × 2.1 mm, 1.9 μm); the mobile phase was acetonitrile (A)-0.1% acetic acid (B); gradient elution is shown in Table 8; the flow rate was 0.3 mL·min^{-1}; the column temperature was 30 °C; injection volume was 2 μL.

Table 8. HPLC gradient elution method.

Time (min)	Acetonitrile (%)	0.1% Acetic Acid (%)
0→15	15→40	85→60
15→20	40→15	60→85
20→25	15	85

4.3.2. UPLC-QTOF-MS/MS Analysis Conditions

First Item: Chromatographic conditions

The Chromatographic column was Waters ACQUITYTM UPLC HSST3 (2.1 mm × 100 mm, 1.8 μm); the mobile phase was 0.1% formic acid aqueous solution (A)-0.1% formic acid acetonitrile (B); the gradient elution is shown in Table 9, and the flow rate was 0.4 mL·min^{-1}; the column temperature was 45 °C; injection volume was 5 μL.

Table 9. UPLC-QTOF-MS/MS gradient elution mode.

Time (min)	0.1% Formic Acid Aqueous Solution (%)	0.1% Acetonitrile Formate (%)
0→12	95→88	5→12
12→23	88→75	12→25
23→50	75→40	25→60
50→55	40→20	60→80
55→60	20→95	80→5

Second item: Mass spectrometry conditions

Data were collected in negative ion mode under electrospray ion source, ion source temperature: 120 °C, desolvation gas: N2, flow rate: 450 L·h^{-1}, temperature: 450 °C. Capillary voltage: 2000 V, cone hole voltage: 17 V, scanning range: *m/z* 100~1500. During low-energy scanning, the trap voltage was 6 eV and the transfer voltage was 4 eV. During high-energy scanning, the negative ion mode conditions were that the trap voltage was 40–60 eV, the transfer voltage was 15 eV, and the conditions of positive ion mode were that the trap voltage was 50–65 eV and the transfer voltage was 15 eV. Accurate mass LeucineenkEphalin was used as calibration solution.

4.4. Investigation of Linear Relationship

The mixed reference substance solution 2 under item Section 4.2.3 was accurately drawn in an appropriate amount, diluted by different times, and injected according to the analysis conditions under item Section 4.3.1. Taking the peak area as the ordinate (Y) and the concentration of the reference substance solution as the abscissa (X), we drew standard curves, respectively, to examine the linear relationship.

4.4.1. Precision Test

The powder of S4 sample was accurately weighed and an appropriate amount was prepared into the test solution according to the method under item Section 4.2.2, and the test solution was continuously injected for 6 times according to the analysis conditions under item Section 4.3.1. We calculated the RSD values of the peak areas of naringenin and hesperetin, respectively.

4.4.2. Stability Test

The powder of S4 sample was accurately weighed and an appropriate amount was prepared according to the method under Section 4.2.2 to prepare a solution of the test sample, which was injected at 0, 2, 4, 6, 8 and 12 h according to the analysis conditions under Section 4.3.1. We calculated the RSD values of the peak areas of naringenin and hesperetin, respectively.

4.4.3. Repeatability Test

The powder of S4 sample was precisely weighed in 6 parts, prepared into the test solution according to the method under Section 4.2.2, and injected according to the analysis conditions under Section 4.3.1. We calculated the RSD values of the peak areas of naringenin and hesperetin, respectively

4.4.4. Sample Addition Recovery Test

The powder of the Jianqu sample (S4) with known content was accurately weighed in 6 parts, each part was about 1 g, and an appropriate amount of mixed reference solution 2 was added, respectively, and prepared according to the method under Section 4.2.2. The test solution was injected according to the analysis conditions under Section 4.3.1. We calculated the mean recoveries and RSD values of naringenin and hesperetin, respectively.

4.5. Data Analysis and Processing

The database of UNIFI1.8.1 software was applied for data screening, and the intensity threshold of 3D peak detection parameters (30 counts for high energy, 200 counts for low energy) and the type of adduct ion peak (negative ion mode is $[M-H]^-$, [M+HCOO] were set$^-$; positive ion mode was $[M+H]^+$, $[M+Na]^+$). Compounds were subjected to univariate analysis using GraphPad Prism 5.01 software.

S1–S4 sample powders were prepared according to the method under Section 4.2.2 to create the test solution, and the solution were injected and analyzed according to the analysis conditions under Section 4.3.1. The peak areas of naringenin and hesperetin were measured, and the contents of each index component in the sample were calculated by regression equation.

Author Contributions: Conceptualization, Y.W. and Z.Z.; investigation and validation, R.W., Y.C., M.S., J.Q. and Z.D.; data curation, Y.W., Z.Z., R.W., Y.C., M.S., J.Q. and Z.D.; writing—original draft preparation, Y.W.; writing—review and editing, Z.Z.; project administration, Y.W. All authors have read and agreed to the published version of the manuscript.

Funding: This research was funded by Key scientific research projects of colleges and universities in Henan Province in 2022, Study on Optimization of fermentation process and improvement of quality standard of traditional Chinese medicine Jianqu, grant number 22B360004 and 2021 Henan Province Undergraduate University and Provincial College Students Innovation and Entrepreneurship Training Program Project, grant number S202110471027.

Institutional Review Board Statement: Not applicable.

Informed Consent Statement: Not applicable.

Data Availability Statement: All the data in the manuscript are available upon reasonable request.

Conflicts of Interest: The authors declare no conflict of interest.

Sample Availability: Jianqu samples of the compounds are available from the authors.

References

1. Gong, Q. *Traditional Chinese Medicine Processing: New Century*, 4th ed.; China Press of Traditional Chinese Medicine: Beijing, China, 2016; p. 416.
2. *Standard Set Prescription of Traditional Chinese Medicine Preparations of drugs Book XVII Ministry of Health of the People's Republic of China*; National Health Commission of the People's Republic of China: Beijing, China, 1998; p. 385.
3. Huang, J.; Jin, M. Chinese medicine Jianqu examination. *Asia-Pac. Tradit. Med.* **2007**, *12*, 31–32.
4. Gong, B.; Li, Q.; Hu, X. Advances in chemical constituents and pharmacological activities of *Fructus aurantii*. *South China For. Sci.* **2019**, *3*, 40–45.
5. Zhou, Y.; Gao, H.; Zhang, R. Chemical constituents from *Atractylodes japonica*. *Chin. Tradit. Patent Med.* **2020**, *10*, 2640–2643.
6. Zhou, J.; Tang, W.; Chen, J. 16Research on chemical constituents from *Artemisia annua*. *Pharm. Clin. Res.* **2020**, *5*, 321–328.
7. Lei, Y.; Li, W.; Cai, B. Research Progress of *Oleacea*. *Res. Pract. Chin. Med.* **2011**, *4*, 81–84.
8. Jing, W.; Du, J.; Wang, J. Review on Chemical Constituents of *Magnoliae officinalis* Cortex. *Mod. Chin. Med.* **2018**, *6*, 764–774.
9. Xiao, L.; Li, H.; Huang, Y. Study of the chemical composition of *Artemisia annua*. *China J. Chin. Mater. Med.* **2021**, *46*, 1160–1167.
10. Sun, C. Research progress on the chemical composition of licorice. *Sci. Technol. Inf.* **2020**, *2*, 64–65.
11. Zhao, Y.; Liu, S.; Zhang, C. Analysis on chemical constituents from *Glycyrrhizae redix* et Rhizoma by HPLC-Q-TOF-MS. *Chin. Tradit. Herb. Drugs* **2016**, *12*, 2061–2068.
12. Hu, S.; Liu, Q.; Liu, Y. The Study of *Polygonum hydropiper* L. 2014, pp. 15–24. Available online: http://kns.cnki.net-s.dres.hactcm.edu.cn/kcms/detail/detail.aspx?dbcode=CPFD&dbname=CPFDLAST2015&filename=ZHZY201407002002&uniplatform=NZKPT&v=CCGfSlyCWqsRH00hCCPtfQDCEqjvY4Meav7058IK--ujjIX0oMcr3fgdASh4vFp55P-7z19d1U8%3d (accessed on 25 November 2022).

13. Zhang, S.; Huang, X. Simultaneous determination of four flavonoid glycosides and limonin in pomelo by HPLC. *Food Mach.* **2021**, *3*, 58–63.
14. Mei, Z.; Zhang, R.; Zhao, Z. Research Progress in Chemical Components of *Citri reticulata* 'Chachi' and Its Quality Control Methods. *Chin. J. Inf. Tradit. Chin. Med.* **2020**, *7*, 136–140.
15. Zhang, K.; Xu, X.; Li, T. Chemome profiling of *Citri reticulatae* Pericarpium using UHPLC-IT-TOF-MS. *China J. Chin. Mater. Med.* **2020**, *4*, 899–909.
16. Hou, X.; Hao, E.; Qin, J. etc Chemical components and pharmacological action for *Cinnamomum cassia* and predictive analysis on Q-marker. *Chin. Tradit. Herb. Drugs* **2018**, *1*, 20–34.
17. Chen, X.; Liang, J.; Lin, J. Research Progress on Chemical Composition and Pharmacological Effects of Bohe (Menthae Haplocalycis Herba) and Predictive Analysis on Q—Marker. *Chin. Arch. Tradit. Chin. Med.* **2021**, *3*, 213–217.
18. Yang, Y.; Kong, W.; Sun, L. Research progress on chemical composition and pharmacological effects and clinical application of betel nut. *Mod. Tradit. Chin. Med. Mater. Med.-World Sci. Technol.* **2019**, *12*, 2583–2591.
19. Zhu, H.; Wang, K.; Shi, J. Simultaneous determination of ten components in Zhizi Houpo decoction by UFLC-MS/MS. *China J. Chin. Mater. Med.* **2020**, *10*, 1751–1759.
20. Zhang, W.; Yao, C.; Chen, X. Identification of chemical constituents of Xiaochengqi Decoction by UPLC-Q-TOF/Fast DDA combined with UNIFI software. *China J. Chin. Mater. Med.* **2022**, *47*, 2121–2133.
21. Li, K.; Shi, Q.; Zhu, H. Chemial Compositions in Bitter Almond. *J. Northwest For. Univ.* **2004**, *2*, 124–126.
22. Wu, F.; Wang, C.; Dong, Q. Rapid Analysis of Chemical Components of Zhitong Huazheng Capsules Based on UPLC-Q-TOF/MS Technology Combined with UNIFI Database. *China Pharm.* **2019**, *17*, 2336–2342.
23. Xu, M.; Liu, Y.; Jie, D. Optimal "Huohou" of fermentation process of Jianqu Leaven. *Chin. Tradit. Patent Med.* **2017**, *1*, 136–142.
24. Li, S.; Li, P.; Du, G. Progress in Composition and Determination Methods of Pericarpium *Citri reticulatae*. *China Fruit Veg.* **2019**, *10*, 36–39, 52.
25. Chen, G.; Li, S.; Ren, L. Comparative Analysis of Flavoring Components of Nongxiang Baijiu Produced in Gansu. *Liquor-Mak. Sci. Technol.* **2017**, *4*, 46–50.
26. Yao, L.; Sun, X.; Xian, S. Variation and correlation analysis of oleic acid, linoleic acid, linolenic acid and protein content in kale-type rapeseed grains. *J. Chin. Cereals Oils Assoc.* **2021**, *36*, 82–87.

MDPI AG
Grosspeteranlage 5
4052 Basel
Switzerland
Tel.: +41 61 683 77 34

MDPI Books Editorial Office
E-mail: books@mdpi.com
www.mdpi.com/books